S IN DIGITAL SIGNAL PROCESSING

RRUS and T. W. PARKS: *DFT/FFT AND CONVOLUTION ALGORITHMS:*
AND IMPLEMENTATION

TREICHLER, C. RICHARD JOHNSON, Jr., and MICHAEL G. LARIMORE:
AND DESIGN OF ADAPTIVE FILTERS

RKS and C. S. BURRUS: *DIGITAL FILTER DESIGN*

CHASSAING and DARRELL W. HORNING: *DIGITAL SIGNAL*
SING WITH THE TMS320C25

TOP

**Digital Signal Processing
with the TMS320C25**

C. S.
THE
JOH
THE
T. W
RUL
PRO

Digital Signal Processing with the TMS320C25

Rulph Chassaing
Darrell W. Horning

A WILEY-INTERSCIENCE PUBLICATION
JOHN WILEY & SONS
New York   Chichester   Brisbane   Toronto   Singapore

Library of Congress Cataloging in Publication Data:

Chassaing, Rulph.
    Digital signal processing with the TMS320C25 / Rulph Chassaing,
Darrell W. Horning.
        p.   cm.
    "A Wiley-Interscience publication."
    Bibliography: p.
    Includes index.
    ISBN 0-471-51066-1
    1. Signal processing—Digital techniques.   I. Horning, Darrell W.
II. Title.
TK5102.5.C473   1989
621.382'2–dc20   90-34031
                                                        CIP
Printed in the United States of America

10 9 8 7 6 5 4 3 2 1

To Our Families
Linda, Amber, Otis, Alice, Andrew, and Chloe

Contents

Preface

Digital signal processors, made possible through advances in integrated circuits, have added a new element to the environment of digital signal processing (DSP). With this new technology, a student can appreciate the concepts of digital signal processing through real-time implementation of experiments and projects.

This book was developed out of a digital signal processing course and a senior project course taught at Roger Williams College, as well as a digital signal processing laboratory course taught at the University of Bridgeport. The background assumed is an electrical engineering systems course and a knowledge of assembly language programming. Each chapter begins with a theoretical discussion, followed by representative examples. Thirty examples, many with program solution, are included throughout the book; and a variety of students' projects are described in Chapter 9.

This text is intended primarily for senior and first-year graduate students in electrical and computer engineering and as a tutorial for the practicing engineer.

In Chapter 1 we introduce the software development tools. These tools are demonstrated through short programming examples. Chapter 2 covers the architecture and the instruction set of the TMS320C25. Structures and special instructions that are useful in DSP are included. Chapter 3 focuses on input and output (I/O) methods. Two I/O alternatives are presented: The analog interface board (AIB) and the analog interface chip (AIC).

In Chapter 4 we introduce the Z-transform. A digital oscillator example is implemented and can be useful for later experiments and projects. Finite impulse response (FIR) filters are discussed in Chapter 5. Several window functions to improve the characteristics of FIR filters are demonstrated. In Chapter 6 we discuss infinite impulse response (IIR) filters illustrating different structures. The effect of quantization on IIR filters is examined. Two software design tools are covered in conjunction with FIR and IIR filters.

Chapter 7 includes both the decimation-in-time and the decimation-in-frequency fast Fourier transform (FFT). Special instructions for the implementation of the FFT are covered. In Chapter 8 we introduce an intuitive approach to adaptive filtering using the linear combiner structure and the least mean squared (LMS) algorithm. Laboratory examples demonstrate the usefulness of the adaptive approach. Chapter 9 covers a variety of projects, including multirate filtering, modulation techniques, and the FFT. This chapter can be used as a source of experiments, projects, and applications.

We feel that the principles of digital signal processing can best be mastered through interaction in a laboratory setting, with real-time algorithm implementations. This interaction can serve to enhance and enrich a student's understanding of DSP.

This book can be used in a variety of ways, such as:

1 For a senior or first-year graduate project course, using Chapters 1 to 7 (Chapter 3 partially) to provide general background, and selected materials from Chapters 8 and 9.
2 For a DSP lab course, covering Chapters 1 to 6 (Chapter 3 partially) and selected materials from Chapters 8 and 9. The beginning of the semester can be devoted to short experiments and miniprojects and the remainder of the semester used for a final project.

We would like to thank all our digital signal processing students who have made our project- and laboratory-oriented courses very rewarding; in particular, Peter Martin, for his work on adaptive filtering with the AIC, and Ken Zemlok for his work on the additional AIB channel. The suggestions made by Dr. Kun-Shan Lin of Texas Instruments and Dr. David P. Morgan of Sanders Assoc. are appreciated. The authors would like to acknowledge the National Science Foundation's equipment support through grants CSI-8851272, CSI-8650204, and USE-8851147 and the support of the Roger Williams College Research Foundation. A special thanks to Carol Reineke for typing the manuscript.

Bristol, Rhode Island   Rulph Chassaing
Bridgeport, Connecticut   Darrell W. Horning
January 1990

List of Examples and Exercises

1

A Digital Signal Processing Development System

CONCEPTS AND PROCEDURES

• *Use of the Software Development System (SWDS), commands and menus*
• *Creation, assembly, and execution of a TMS320C25 program*
• *Use of the debugging tools such as modify memory and single step*

In this chapter we introduce the use of development tools for real-time signal processing. Those tools include an SWDS, based on the second-generation TMS320C25 digital signal processor, an analog interface board (AIB) and an analog interface chip (AIC). In this chapter we show how to create a source file as well as an object file that can be downloaded into the SWDS and executed. A short example program (LOOP) illustrates how an input signal is brought through the A/D unit into the processor and then sent out to the D/A unit and output filter, where it is reconstructed.

1.1 INTRODUCTION

Signal processing can be split into two areas: nonreal-time signal processing and real-time signal processing. *Real-time processing* means that the processing must keep pace with some external event; *nonreal-time processing* has no such timing constraint. For signal processing, the external event to keep pace with is usually the analog input. This book and digital signal processing (DSP) processors such as the TMS320 family are concerned primarily with real-time signal processing.

The processing speed is often the paramount consideration in choosing design primitives for DSP applications. Figure 1.1 gives a general picture of the relative speeds of various primitives. In progressing from left to right on the diagram, the speed increases, as does the design difficulty. The degree of parallel processing also increases from left to right. On the far right is the fastest technology, fiber optic

1

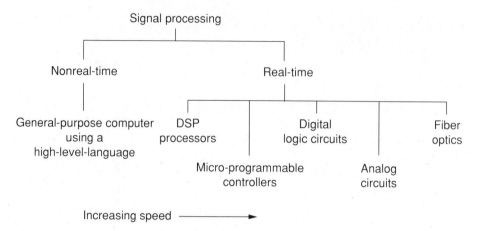

FIGURE 1.1. Different technologies for DSP applications

systems. These systems can perform analog multiplications, additions, frequency filtering, Fourier transforms, convolutions, and correlations in the range 1 MHz to 1 GHz [1]. Moving left on the diagram, we come to analog circuits. Analog electronic circuits are used routinely to implement such operations as filtering, modulation, and detection at microwave frequencies. Directly to the left of analog circuits are digital circuits. These design elements are used to build special-purpose signal processing systems such as FIR filters, or correlators, an example of which is the TDC1023J, made by TRW LSI Products, La Jolla, California, which does a correlation at 20 MHz [2]. The next entry to the left on the diagram is micropro-grammable controller/bit-slice elements. Designs built from these primitives are not as fast as digital logic, but they are done more easily and are more robust when it comes to modifications. The microprogramming approach generally leads to systems that are faster than DSP processors because more parallel operations are possible and the architecture can be tailored to match the specific calculation. Notice that DSP processors occupy the low end of the real-time spectrum, usually from 0 to 20 kHz.

Since DSP processors are just fast microprocessors with specialized instruction sets, they enjoy all the advantages of microprocessors: They are easy to use, flex-ible, and economical. We will deal almost exclusively with processor implemen-tation of DSP applications; therefore, we must study some of the features of the microprocessor as well as digital signal processing. Most of our work will involve designing a program to implement a DSP application. There are several software tools to aid us in this task. These tools are contained in what is called the Software Development System (SWDS).

1.2 SOFTWARE DEVELOPMENT SYSTEM

The Software Development System (SWDS) provides a user interface to the TMS320C25. In its environment, a TMS320C25 program can be developed, assem-

FIGURE 1.2. System with IBM PC and AIB

bled, run, tested, and saved. Debugging aids such as single stepping, break-points, and memory and processor register display/modification are supported in the SWDS. The SWDS supports both the second-generation TMS32020 and the C-MOS version TMS320C25 digital signal processors.·

The SWDS includes a board, containing the TMS320C25, which can be plugged into a large slot on an IBM (or compatible) personal computer. The board contains 24K words of partitionable program and data memory. Two 40-conductor cables attached to the SWDS connect, through an adapter, to the AIB for emulation, as shown in Figure 1.2. The SWDS can also be interfaced with the TLC32040 or TLC32041 analog interface chip (AIC) for emulation. Included in the SWDS package are the *TMS320C25 Software Development System User's Guide*, [3], two adapter boards to support the TMS32020 and the AIB, appropriate software such as an assembler/linker and a debugger, as well as a DSP software library. The SWDS can be installed following the procedure described in Appendix A or the TMS320C25 SWDS's *User's Guide*.

To perform the DSP experiments, the following equipment support is needed:

1. An IBM PC or compatible with a minimum of 384K. A co-processor is essential to run most commercial filter design packages.

2. Texas Instruments' Software Development System (SWDS), based on the second-generation C-MOS technology TMS320C25 digital signal processor, or an equivalent system from companies such as Hyperception, Atlanta Signal Processors, or Spectrum. The package includes:

   a. Two adapters to support Texas Instruments' AIB, containing A/D and D/A and the second-generation N-MOS technology TMS32020.

   b. Appropriate software such as a debug monitor, assembler and linker.

   c. Second-Generation user's guides [3–6].

   d. *Digital Signal Processing Applications with the TMS320 Family* [7].

3. Function generator.

4. Texas Instruments' analog interface board [8] or the module based on the TLC32040/TLC32041 analog interface chip described in Chapter 3 [3,9].

5. Two-channel oscilloscope.

6. Signal/spectrum analyzer (optional).

1.3 USING THE SWDS: A SIMPLE PROGRAM

The best way to learn the SWDS is to use it. Let us start by running some simple examples. Don't worry too much about what the programs are doing but concentrate on learning the features of the development tools. Upon completion of this section you should be able to enter a program directly into program memory, reverse assemble it, examine and change processor registers, examine and change data memory, and execute a program in the single-step mode.

Example 1.1   A Simple Addition Program: This example assumes a dual floppy system. Substitute the appropriate commands for a hard drive. Make sure that the DOS files AUTOEXEC.BAT and CONFIG.SYS have been modified as explained in Appendix A and DOS rebooted before proceeding. The symbols B> and CMD> are prompts and are not typed as part of the command which will be indicated in bold type. With the SWDS debug monitor installed and placed in drive B, proceed with the following steps (press <ENTER> after each command):

1. Run the SWDS by typing B>**SWDS**, which will put you at the SWDS monitor command level, resulting in the prompt CMD>.

2. Type CMD>**INIT**. The INITIALIZATION command initializes/partitions the 24K words of program and data memory. For example, assign 20K words for program memory and 4K words for data memory; then initialize the processor clock to INTERNAL. After each setting/command, press <ENTER> to save the initialization setting. You will not need to repeat this step when you access the SWDS again.

3. Type CMD>**DEBUG**, to evoke the debug monitor. Then proceed with the following steps to modify various windows.

4. Press the function key **F4**. The program state window is accessed. Initialize/set the program counter (PC) to >20. The ">" represents a hexadecimal (hex) value.

5. Press **F4** again, to access the data memory (DM) window. Note the MODIFY

DATA MEMORY command. Use the arrow key to modify/set DM location >60 to a data value of 1234.

6. Press **F5**. This toggles the screen from data memory (DM) to program memory (PM). Use the appropriate arrow key to enter the following hex codes in program memory starting at program memory address (PMA) >20.

PMA (in hex)	Content (in hex)
20	5588
21	C060
22	CA00
23	0080

7. Press **ESC** to exit from the MODIFY DATA MEMORY command.

8. Press **ESC** again to access the DEBUGGER. Your screen should now look like Figure 1.3. Use the **F4** key and arrows to center your program on the reverse assembler window. Note that the program has *already* been reverse assembled. Although it is not necessary to fully understand the instructions codes at this time, it is still worthwhile to examine the program. The first instruction, LARP 0, selects auxiliary register AR0; the next instruction, LARK 0,>60, points AR0 at data memory location >60, which contains the data value 1234 entered in step 5. The ZAC instruction zeros the accumulator, and the last instruction, ADD *,0, adds 1234 to the accumulator, previously set to zero.

```
DEBUGGER: Breakpoint, Count, Execute, Go, Halt, Load, Quit, Run, Show, sTep
================================== Program state ==================================
   PC = 20            AR0 = 0         AR4 = 0      S1 = 0         S5 = 0
  ACC = 0             AR1 = 0         AR5 = 0      S2 = 0         S6 = 0
    P = 0             AR2 = 0         AR6 = 0      S3 = 0         S7 = 0
    T = 0             AR3 = 0         AR7 = 0      S4 = 0        IMR = FFC0
 ARP    OV   OVM   INTM    DP   ARB   CNF   TC   SXM   C   HM   FSM   XF   FO   TXM   PM
  0      0    0      1     0     0     0    0     0    0    0     0    0    0     0    0

========= Status =========  ||=========== Hardware signal monitor ===========
 Processor halted           ||  INT0 = 1 INT1 = 1 INT2 = 1 BIO = 1 XF = 1 RESET = 1

========= Reverse assembler =========  ||========== Program memory ==========
 001C    NOP                            || 0010 =   FFFF    FFFF    FFFF    FFFF
 001D    NOP                            || 0014 =   FFFF    FFFF    FFFF    FFFF
 001E    NOP                            || 0018 =   5500    5500    5500    5500
 001F    NOP                            || 001C =   5500    5500    5500    5500
 0020    LARP   0                       || 0020 =   5588    C060    CA00    0080
 0021    LARK   0,60                     || 0024 =   5500    5500    5500    5500
 0022    ZAC                            || 0028 =   5500    5500    5500    5500
 0023    ADD    *,0                     || 002C =   5500    5500    5500    5500
 0024    NOP                            || 0030 =   5500    5500    5500    5500
 0025    NOP                            || 0034 =   5500    5500    5500    5500

F1 Help   F2 Toggle data/program memory   F3 Command card   F4 Modify windows   →
```

FIGURE 1.3. DEBUGGER screen for addition example

```
0001                      * ADDITION EXAMPLE
0002                      * ADD CONTENTS OF DATA MEM 60 (HEX) TO ACCUMULATOR
0003                      * ENTER DATA VALUE IN DMA 60
0004 0000                       AORG    0
0005 0000 FF80                  B       START   *INITIALIZE RESET VECTOR
     0001 0020
0006 0020          START        AORG    >20     *START PROG MEM AT 20 (HEX)
0007 0020 5588                  LARP    0        *LOAD AUX REG POINTER WITH 0,SELECT AR0
0008 0021 C060                  LARK    0,>60   *LOAD AR0 WITH 60 (HEX)
0009 0022 CA00                  ZAC              *INITIALIZE ACCUMULATOR TO 0
0010 0023 0080                  ADD     *        *ADD DATA AT DMA 60 TO ACC-INDIRECT ADD
0011                            END
NO ERRORS, NO WARNINGS
```

FIGURE 1.4. Program listing for addition example (ADD.LST)

9. Load the PC with >20, then press **T** to initiate single-stepping or press < SPACE BAR>. Repeated stepping can be invoked by pressing <SPACE BAR>. Step through the four program lines and observe the processor registers that change.

The complete program (listing file) is shown in Figure 1.4, with program code starting at location >20. An editor and an assembler were used to obtain this program. The second column lists the PM addresses (>20–23) and the third column shows the instruction codes. The mnemonics associated with the instruction codes are also shown. Figure 1.5 shows the source file that was created by the editor.

1.4 ANALOG INPUT/OUTPUT ALTERNATIVES

The two analog input/output alternatives offered by Texas Instruments (TI) will be used in this text: (1) The analog interface board (AIB) and (2) The analog interface chip (AIC). The AIB was originally designed to be connected to an evaluation module (EVM) based on the first-generation TMS32010 digital signal processor. The I/O structure of the TMS320C25, through the AIB adapter, can be converted to the TMS32010 I/O structure expected by the AIB. The AIB provides 12-bit analog-to-digital (A/D) and digital-to-analog (D/A) conversion as well as input and output filtering. To permit the AIB to be used with the faster TMS320C25-based SWDS, a jumper (P6) on the AIB adapter is placed in the WAIT position to extend all I/O operations. Make sure that this jumper is placed in the WAIT position for proper operation with the SWDS. Figure 1.6(a) shows the AIB connected to the AIB adapter, which in turn is connected to the SWDS through two cables. The

```
* ADDITIONAL EXAMPLE
* ADD CONTENTS OF DATA MEM 60 (HEX) TO ACCUMULATOR
* ENTER DATA VALUE IN DMA 60
             AORG    0
             B       START    *INITIALIZE RESET VECTOR
START        AORG    >20      *START PROGRAM MEM 20 (HEX)
             LARP    0        *LOAD AUX REG POINTER WITH 0, SELECT ARG
             LARK    0,>60    *LOAD ARG WITH 60 (HEX)
             ZAC              *INITIALIZE THE ACCUMULATOR TO 0
             ADD     *        *ADD DATA AT DMA 60 TO ACC-INDIRECT ADD
             END
```

FIGURE 1.5. Source program for addition example (ADD.ASM)

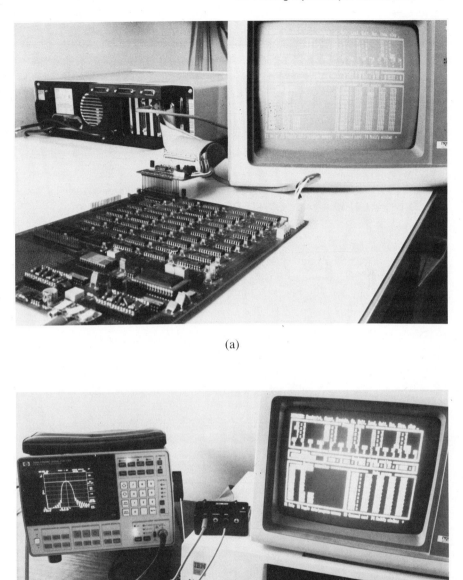

(a)

(b)

FIGURE 1.6. Analog interface connection to SWDS: (a) SWDS connected to AIB, (b) SWDS connected to AIC Module

SWDS package includes an adapter that connects the SWDS to an AIB available from Texas Instruments (part no. RTC/EVM 320C-06).

An inexpensive alternative to the AIB is the TI analog interface chip (part no. TLC32040/TLC32041), which allows I/O operations at full speed. The AIC includes *on-chip* 14-bit A/D and D/A, antialiasing and smoothing (input and output) filters, and two separate inputs for applications where more than one input is desirable. The AIC connects to the SWDS via the shorter of the two 40-conductor ribbon cables. Figure 1.6(b) shows a box containing the AIC connected to the cable. The AIC is discussed and used in Chapter 3.

1.5 TESTING THE ANALOG INTERFACE BOARD

Example 1.2. Loop Program. This example will give you an opportunity to observe a signal before and after it passes through the processor. In addition, another debugging technique, breakpoints, will be demonstrated in this example. The breakpoint option allows a programmer to halt the program at arbitrary points and observe the state of the machine.

Make the following connections in preparation for running the LOOP program:

1. With the power OFF, connect the two 40-conductor ribbon cables attached to the SWDS to connectors J1 and J2 on the AIB adapter (they can only go one way).

2. Jumper P6 in the AIB adapter board should be in the WAIT position to allow compatibility between the I/O structure of the second-generation TMS320C25 and the first-generation TMS32010 I/O structure expected by the AIB. Turn the power ON.

3. Connect a sinusoid source (amplitude less than 10V) to the analog input of the AIB (connector J2). Connect the output (connector J3) to an oscilloscope. Connect the input also to a second channel on the oscilloscope.

The program listed in Figure 1.7, which produces a delayed sinusoidal output of the same frequency as the input, will be analyzed later in more detail. Briefly, it initializes the AIB (done only once), waits for the converted data, then performs data read and write (IN/OUT) and repeats indefinitely. Proceed as follows to test this program:

1. Type B>**SWDS** to run the SWDS from drive B.

2. Type CMD>**INIT.** This is not necessary if already done in Example 1.1, unless a different partition of program and data memory spaces is desired. Entries used from before are saved into a file and reloaded each time the debug monitor is accessed.

3. Type CMD>**DEBUG** to select the debug monitor.

```
0001                    * LOOP PROGRAM
0002                    * INPUT INTO AIB IS OUTPUT - TEST FOR AIB
0003 0000                      AORG    0
0004 0000 FF80                 B       START   *INIT. RESET VECTOR
     0001 0022
0005 0020                      AORG    >20     *START AT PMA 20
0006                    * INITIALIZE AIB
0007 0020 03E7  RATE    DATA    999     *N=(10 MHZ/fs)-1 WITH fs=10 KHZ
0008 0021 00FA  MODE    DATA    >FA     *MODE FOR AIB
0009                    *
0010 0022 C800  START   LDPK    0       *INIT DATA PAGE POINTER TO 0
0011 0023 CA20          LACK    RATE    *GET SAMPLE RATE INTO ACC
0012 0024 5860          TBLR    >60     *TRANFER FROM PROG MEM TO DATA MEM 60
0013 0025 E160          OUT     >60,1   *OUTPUT RATE TO AIB PORT 1
0014 0026 CA21          LACK    MODE    *GET MODE INTO ACC
0015 0027 5860          TBLR    >60     *TRANSFER FROM PROG TO DATA MEM
0016 0028 E060          OUT     >60,0   *OUTPUT MODE TO AIB PORT 0
0017 0029 E360          OUT     >60,3   *DUMMY OUTPUT TO AIB PORT 3
0018 002A FA80  WAIT    BIOZ    ALOOP   *WAIT FOR END OF CONVERSION
     002B 002E
0019 002C FF80          B       WAIT    *LOOP BACK
     002D 002A
0020 002E 8260  ALOOP   IN      >60,2   *READ DATA FROM PORT 2
0021 002F E260          OUT     >60,2   *WRITE DATA OUT TO PORT 2
0022 0030 FF80          B       WAIT    *BRANCH TO CONTINUE
     0031 002A
0023                            END
NO ERRORS, NO WARNINGS
```

FIGURE 1.7. LOOP program listing (LOOP.LST)

4. Press **F4** to access the program state window. Set PC to 0. Press F4 again to access the data memory window.

5. Press **F5** after **F4** to toggle from data to program memory (if necessary). Then enter the following values in PM starting at PMA 0.

PMA (in hex)	Content (in hex)	
0	FF80	
1	0022	
20	03E7	
21	00FA	
.	.	
.		
.	.	as in the third column
		of Figure 1.7
30	FF80	
31	002A	

6. Press **ESC ESC**. Two escapes will bring you back to DEBUGGER.

7. Press **B** to select the breakpoint menu. **F2** allows you to toggle the breakpoint ON and OFF. Breakpoint addresses can be entered on the left of the menu. They will not be active until turned ON by pressing **F4**. Set a breakpoint at PMA >28 and turn it ON. Press **ESC** to exit the breakpoint menu.

8. Press **R** to run the program. Notice that the program does not break or halt at PMA >28. This is because the run (**R**) command ignores breakpoints.

9. Press **H** to halt the processor.

10. Press **E**. This command executes up to the breakpoint set above (location

>28 in PM). You can single step from here by first using the ESC key to get to the DEBUGGER command, then pressing **T**. The run (**R**) command again will continue running the program.

Vary the frequency of the function generator and observe the corresponding output waveform change. A sampling frequency of 10 kHz, initialized in PM location >20, can be checked by connecting the oscilloscope to pin 22 of the A/D on the AIB. Move the probe to the unfiltered output that is present on pin 1 of jumper E2. Observe the output.

In Chapter 3, the LOOP program is implemented with the analog interface chip (AIC) instead of the AIB.

1.6 PROGRAM DEVELOPMENT USING SWDS TOOLS

The SWDS provides an efficient way of developing a TMS320C25 executable program, which can be described in these steps:

1. Create a source program with an editor.

2. Assemble the program to create a listing and an object program.

3. Load the object program into program memory.

An editor and an assembler are the tools that are needed to create an executable program. If certain features of the assembler are used, a linker may be required. We will avoid this step for the moment. Almost any editor or word processor can be used to generate the source code. Several of them are available for the IBM and compatibles such as PCWRITE or the NORTON EDITOR. PCWRITE is an inexpensive, menu-driven, word processor which can be obtained from Quicksoft. The assembler and linker are supplied by TI with the SWDS.

Creating a Source File

Use an editor or word processor to type the program LOOP of Example 1.2, as shown in Figure 1.7. This will be the source program for the assembler, which understands certain keywords in addition to the TMS320C25 instruction set. Note that a "*" in column 1 means that the line is a COMMENT line. Comments can also be placed after an instruction. Labels such as RATE & MODE also start in column 1. The assembler expects the source code to be in a certain format. This format is demonstrated in the following line of code:

LABEL<TAB>INSTRUCTION OR DIRECTIVE<TAB>OPERAND<TAB>COMMENT

Note that a space can be used in lieu of <TAB>. This format is reflected in the lines of code of Figure 1.7. Use the TAB key for ease of aligning the labels, instructions and comments. An assembler directive is a message for the assembler; it is not a machine instruction, although it is aligned in the same way as an instruction. Notice some of the directives in the program:

AORG starts the program addresses at the address specified in the argument.

DATA assigns the data specified to a program memory address.

END marks the end of the program.

An additional useful directive is the EQU, which assigns a label to a number. When you have finished entering the program, save it as LOOP.ASM.

Creating an Object File

The assembler creates an object file. If data segments and external subroutines are not used, this is an executable program. To create an object file, type

```
B>XASM25
```

Assembler prompts will appear. Enter the name of the source file just created and select the defaults for the last two prompts by pressing <ENTER>:

> *Source file* [NUL.ASM]: **LOOP.ASM**<ENTER>
> *Listing file* [LOOP.LST]: <ENTER>
> *Object file* [LOOP.MPO]: <ENTER>

Version 3.0 or earlier does not support pathnames. This means that the file and the assembler must reside in the same directory. It does, however, support different devices. For example, the assembler can reside on B:, while the source file can reside on A:.

Loading the Program

An executable file has either a .MPO or a .LOD extension. The following steps show how to load the executable file into the SWDS, where it can be run by the TMS320C25 processor:

1. Run the SWDS.
2. Execute the debugger: CMD>**DEBUG**.
3. Press **L** to load. At the prompt, give the file pathname of the source. When loading is complete the software will prompt you for a return. The executable program is now loaded. All the commands of the DEBUGGER are available as before. Set PC = O with **F4**. Observe the program in the monitor window by single-stepping it (press **T** or <SPACE BAR>). Observe the auxiliary register AR0, the accumulator ACC, and data memory location >60 while

single-stepping the program. Execute the program at full speed to verify that it performs as in Example 1.2.

Editing and Assembling within the SWDS

All of the development tools can be run within the environment of the SWDS package. Although editing/creating a source file and assembling it from outside the SWDS is more suited to a student environment, where they may do the editing/assembling on a machine without an SWDS, it is generally more convenient to do all these operations within the environment of the SWDS.

Before performing these operations within the SWDS, the files must be placed in the proper directory or disk drive. For a dual floppy system:

Drive A: DOS
Drive B: SWDS debug monitor
           Editor—(for example, ED.EXE)
           Assembler—XASM25.EXE
           Linker—LINKER.EXE

For a hard drive system:

\SWDS   SWDS
\Default   Editor
           Assembler—XASM25.EXE
           Linker—LINKER.EXE

With a hard drive, the assembler, linker and source file must all reside in the same directory. Although the editor can be elsewhere, it seems to work best if it is also in the default directory. We are now ready to edit/assemble in the SWDS environment. The commands that follow demonstrate the procedure and assume a floppy disk system:

1. B>**SWDS.**
2. CMD>**SETUP.** This command will prompt you for a pathname for the editor, assembler and linker. For example:

   Editor pathname: B:**ED.EXE**
   Assembler pathname: B:**XASM25.EXE**
   Linker pathname: A:**LINKER.EXE** (Press ENTER since we will not use the linker now)

3. CMD>**EDIT**. This command will access the editor from the SWDS (don't type just ED) and prompt you for a file name. For example, you may

wish to edit/create the file A:LOOP.ASM as before from Figure 1.7. Upon completion of the editing session, save your file.

4. **CMD>ASM**. This command executes the assembler, XASM25, from the SWDS. From this point, the assembler execution proceeds exactly as described in the preceding section. Note for floppy disk users: You should leave the DOS in drive A while using this procedure. If your source file is in another disk, then temporarily replace your DOS with the disk containing your source file when prompted to do so. Alternatively, you can copy your source file into drive B.

5. **CMD>SHOW**. This command is used to display an ASCII file, usually a listing or source file—LOOP.LST or LOOP.ASM, for example.

6. **CMD>LOAD**. This command loads a file into the TMS320C25 program memory. It will prompt you for a file. The file must be an executable file such as A:LOOP.MPO. An object file can also be loaded by accessing first the debug monitor, then **L** (to load).

7. **CMD>XRA**. The reverse assembler is used to verify that loading was successful. Specify the program start and end address. You may wish to PRINT (from the REVERSE ASSEMBLER command) the listing file.

It is worthwhile to become familiar with the various commands available in the debug monitor. An extensive set of commands and utilities is available in the SWDS. The function key **F1** provides a HELP menu; for example, the utility CALC can be used for conversion of a number between hex and decimal.

What we have covered applies when using the assembler version 3.1 or earlier. In the next section we discuss the assembler version 5.04, which uses a common object file format (COFF) and requires linking before downloading into the SWDS. If you do not have assembler version 5.04, you may skip the following section.

1.7 ASSEMBLING AND DOWNLOADING USING COFF

The assembler, which we discussed in Section 1.6 (version 3.1), uses a TI-Tag format and creates a listing file (with extension LST) as well as an object file (with extension MPO) that can be downloaded into the SWDS. Texas Instruments has introduced an assembler and linker, version 5.04, which uses a common object file format (COFF). Included with this assembler version 5.04, from Texas Instruments, is a utility program, which converts a source file, with TI-Tag format, into a COFF version. This new version requires linking to create an executable file. The source file of the LOOP program (LOOP.ASM) of Example 1.2, among others, is used to illustrate an easy migration path between the two formats [4–6] , and is described in Appendix A.

REFERENCES

[1] H. F. Taylor, "Application of Guided-Wave Optics in Signal Processing and Sensing," *Proceedings of the IEEE* **75** (11), November 1987.

[2] W. D. Stanley, G .R. Dougherty, and R. Dougherty, *Digital Signal Processing;* Reston, Va., 1984.

[3] *TMS320C2x Software Development System User's Guide*, Texas Instruments Inc., Dallas, Tex., 1988.

[4] *TMS320C1x/TMS320C2x Assembly Language Tools User's Guide*, Texas Instruments Inc., Dallas, Tex., 1987.

[5] *Second-Generation TMS320 User's Guide*, Texas Instruments Inc., Dallas, Tex., 1987.

[6] *TMS320C1x/TMS320C2x Source Conversion Reference Guide*, Texas Instruments Inc., Dallas, Tex., 1988.

[7] K. S. Lin (editor), *Digital Signal Processing Applications with the TMS320 Family*, Vol. 1, Prentice-Hall, Englewood Cliffs, N.J., 1988.

[8] *TMS32010 Analog Interface Board User's Guide*, Texas Instruments Inc., Dallas, Tex., 1984.

[9] *TLC32040I, TLC32040C, TLC32041I, TLC32041C Analog Interface Circuits, Advanced Information*, Texas Instruments Inc., Dallas, Tex., September 1987.

2

The TMS320C25
Digital Signal Processor

CONCEPTS AND PROCEDURES

• *Architecture and instruction set of the TMS320C25*
• *Fixed-point arithmetic and computational basics*
• *TMS320C25 program structure and use of special instructions*

In this chapter we discuss the architecture, memory organization, addressing modes, and the instruction set of the TMS320C25. Fixed-point, binary, and two's-complement arithmetic are also covered, and examples provided. Special instructions, such as MACD in conjunction with RPTK, specially created to handle real-time signal processing, are illustrated. The chapter concludes with a detailed example program of a nonrecursive difference equation. This provides the necessary background to perform fast convolution, which will be useful in later chapters for implementing digital filters.

2.1 INTRODUCTION

The early 1980s marked the appearance of the first digital signal processing (DSP) microprocessors. The Intel 2920 appeared first, followed by the NEC μPD7720, and then the Texas Instruments TMS32010 in 1982. Two second-generation TMS320s followed: the TMS32020 (1985) and the TMS320C25 (1986). DSP microprocessors are characterized by fast multiply instructions, reduced instruction sets, and specialized instructions to make DSP algorithms execute fast and efficiently. They are considered general-purpose digital signal processors as opposed to a special-purpose digital signal processor such as an FFT digital signal processor or a finite impulse response (FIR) filter digital signal processor. Applications of these processors are limited to the frequencies from 0 to 20 kHz.

The TMS320C25 is a second-generation digital signal processor with an architecture based on the first-generation TMS32010. The TMS320C25 is a C-MOS version of the TMS32020 with a faster instruction cycle time and additional features. The performance of the TMS320C25 for DSP applications is a factor of two to three times that of the TMS32020, which in turn has a throughput two to three times that of the TMS32010. The TMS320C25 characteristics include an instruction cycle time of 100 ns with most instructions requiring only a single cycle. As a result, it is capable of executing 10 million instructions per second. With features such as a 16-bit by 16-bit integer multiply in one instruction cycle (100 ns), and a single-cycle instruction to multiply/accumulate (MAC) with a data move option, the TMS320C25 is well suited to implement many DSP applications, including digital filters, fast Fourier transform, and convolution [1–6]. In this chapter we consider the TMS320C25 architecture, memory organization, addressing modes, instruction set, and fixed-point arithmetic.

2.2 TMS320C25 ARCHITECTURE

The TMS320C25 is based on the Harvard architecture, which has separate program memory and data memory address spaces enabling overlap of instruction fetch and execution. Data transfer between program memory and data memory is possible only through special instructions. The TMS320C25 has 544 words (16 bits) of on-chip data RAM and 4K words of on-chip maskable program ROM. Of the 544 words of data RAM, a 256-word block can be addressed as a program or data memory. Both program and data memory have address spaces of 64K words. Additional on-chip features include a timer and a full-duplex serial port.

Figure 2.1 shows a programmer's view of the TMS320C25 processor. It has a file of eight 16-bit auxiliary registers (AR0–AR7), two status registers (ST0 and ST1), and three special-purpose registers: a 16-bit temporary register (TR), a 32-bit product register (PR), and a 32-bit accumulator (ACC). The status registers, ST0 and ST1, contain status and control bits for the processor. ST0 stores two pointers, the auxiliary register pointer (ARP) and the data memory page pointer (DP). The ARP selects which of the eight auxiliary registers will be used to reference data memory, and the DP selects which data memory page is active. The TR holds either an input to the multiplier or a shift code for the scaling shifter. The PR holds the product from the multiplier. The ACC holds the output of the ALU. The upper 16 bits (ACCH) or the lower 16 bits of the accumulator (ACCL) can be accessed. Included are three shifters to facilitate scaling, bit manipulation, and extended arithmetic: a shifter on the output of both the PR and the ACC, and one on the input of the ALU.

The TMS320 family of processors are single-address machines. For most of the two operand instructions, a register holds the implied operand. For example, the add operation

```
ADDL DA
```

takes the contents of a data memory location (DA) specified and the accumulator

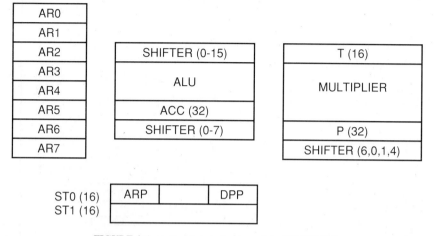

FIGURE 2.1. A Programmer's view of the TMS320C25

(ACC) as its two operands. The result of the add is placed into the ACC. Moves from data memory to data memory must pass through the accumulator as shown by the sequence of instructions

```
LAC    DA1    *LOAD CONTENT OF DA1 INTO ACC
SACL   DA2    *STORE LOWER ACC CONTENT ACC INTO DA2
```

where DA1 and DA2 are data memory locations.

The functional block diagram shown in Figure 2.2 presents a more detailed picture of the processor and shows the data flow from register to register. The top of the diagram contains the program memory units, while the bottom contains the data memory and data manipulation units. The separate data memory and program memory buses are clearly indicated with gateways between them. The upper part of the diagram shows the program counter and an 8 × 16-bit stack which will support eight nested subroutine or interrupt calls. Also shown in the lower part of the diagram are the three blocks (B0, B1, B2) of on-chip RAM; 288 words are always used as data memory, while 256 words (block B0) can be configured as either data or program memory.

The multiply operation demonstrates some of the allowed data paths of the processor. One input to the multiplier is the content of the TR and the other operand is either from the program or data memory. When one of the operands is a constant (immediate mode) the constant is embedded in the instruction and therefore comes from program memory as shown in the diagram. The 32-bit product is placed into the PR. An APAC instruction can then be used to add the content of the PR to the ACC or a PAC instruction can be used to simply load the PR into the ACC. The following program (on page 19) sequence multiplies the content of DA1 with the content of DA2 and puts the result in the accumulator:

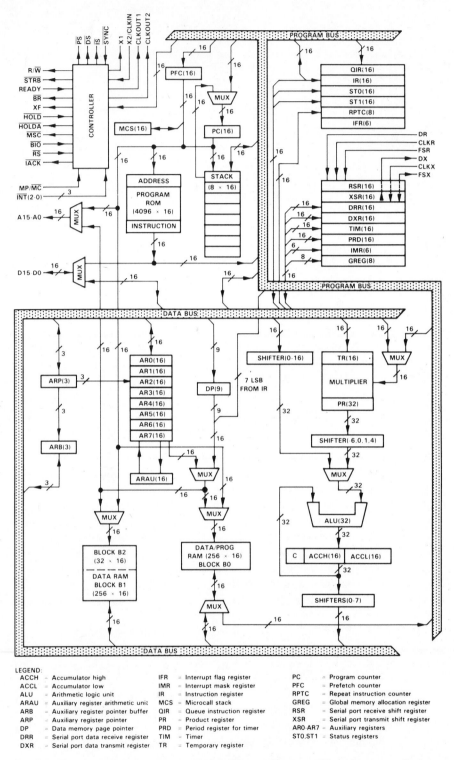

FIGURE 2.2. TMS320C25 functional block diagram. (Courtesy of Texas Instruments Inc.)

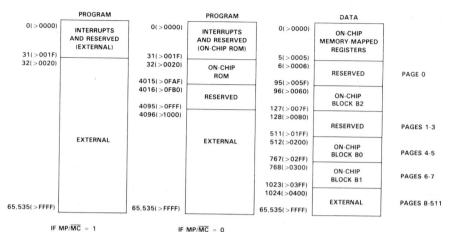

(a) MEMORY MAPS AFTER A CNFD INSTRUCTION

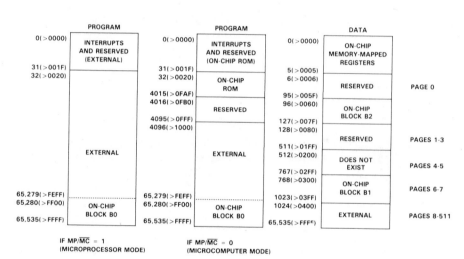

(b) MEMORY MAPS AFTER A CNFP INSTRUCTION

FIGURE 2.3. TMS320C25 memory maps. (Courtesy of Texas Instruments Inc.)

```
LT    DA1        *LOAD TR WITH CONTENT OF DA1
MPY   DA2        *MULTIPLY CONTENT OF TR BY CONTENT OF DA2
PAC              *MOVE RESULT TO ACC
```

2.3 MEMORY ORGANIZATION

The TMS320C25 provides separate address spaces for program and data memory as shown in Figure 2.3(*a*). Data memory is divided into 512 pages, with each page containing 128 words. Pages 0–7 are assigned to on-chip memory and the remaining 504 pages are designated external memory. Notice that some of the on-chip pages are reserved or partially reserved, thus not available. Several processor registers are memory-mapped to the lower six addresses of page 0, making these locations unavailable as data memory but allowing the programmer to address these registers in the same manner as any other data memory location.

Figure 2.3 demonstrates a useful feature of the processor. Block B0 can be mapped into program memory address space, thus allowing some instructions to address both program and data memory at the same time. Figure 2.3(a) shows a memory map after the instruction CNFD, configuring on-chip block B0 as data memory, and Figure 2.3(b) shows a memory map after the instruction CNFP, configuring B0 as program memory. After a CNFP instruction, the data that were at addresses on pages 4 and 5 are now at the last 256 locations in program memory. Instructions using this feature will be demonstrated later.

2.4 ADDRESSING MODES

Addressing modes are the different ways in which an instruction specifies its operand. The TMS320C25 supports three different addressing modes:

1. *Direct addressing.*   Data memory can be directly addressed with an instruction such as

       ADD DMA,2

where DMA is a data memory address relative to the beginning of the current page. The following diagram shows how the full address is derived from the relative page address and the data page pointer content:

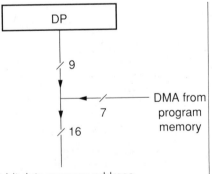

The content of DMA in the instruction above is shifted first left by 2, then added to the accumulator.

2. *Indirect addressing.*   The three forms of indirect addressing with the corresponding assembler notation are shown:

         *   indirect
        *+   indirect with autoincrement
        *−   indirect with autodecrement

One of the eight auxiliary registers contains the data memory address of the operand. The 3-bit code in the auxiliary register pointer (ARP) determines which of the eight ARs is selected. In the case of the autoincrement/autodecrement, the content of the AR is first used and then incremented/decremented by one. Data memory can be addressed indirectly with an instruction such as

ADD *+,2

where the current AR contains the address of the operand which is incremented after it is used to find the operand. The DM operand is left-shifted by 2 and then added to the accumulator.

Besides the use of indirect addressing with increment or decrement, a very useful mode of indirect addressing has been incorporated in the TMS320C25, specifically for efficient bit-reversal operations [1]. The bit-reversal process, encountered in implementing the fast Fourier transform (FFT), discussed in Chapter 7, results in a proper resequencing of data in an FFT program. This addressing mode is used with indexing by adding or subtracting the content of the auxiliary register AR0 to the content of another auxiliary register, such as AR1, with the carry bit propagated in the reverse direction. For example, the instruction ADD *BR0+ adds the content of AR0 to the current auxiliary register specified by *, with reverse carry propagation. Similarly, ADD *BRO− subtracts the content of AR0 from the current auxiliary register, again with reverse carry propagation.

3. *Immediate addressing.* In the immediate mode, the operand is contained within the opcode. Both short immediate (8-bit or 13-bit constants) and long immediate (16-bit constants) are available on the TMS320C25. The constant is a portion of the instruction. For example, in the instruction

ADDK >80

the constant embedded in the instruction, >80, is added to the content of the accumulator. This is a case of data, from program memory, being used as data in an instruction.

2.5 INSTRUCTION SET

The TMS320C25/TMS32020 instruction set is upwardly compatible with the first-generation digital signal processor TMS32010. The TMS320C25 contains a rich set of instructions, as shown in Table 2.1; at the bottom, the instructions which are not supported by the TMS32010 or TMS32020 are specified. Note that most instructions take only a single machine cycle time (100 ns).

Basic Instruction Types

1. Loading and storing instructions, such as the load auxiliary register (LAR)

TABLE 2.1 Instruction Set Summary (Courtesy of Texas Instruments Inc.)

ACCUMULATOR MEMORY REFERENCE INSTRUCTIONS				
Mnemonic and Description		Words	16-Bit Opcode	
			MSB	LSB
ABS	Absolute value of accumulator	1	1100 1110	0001 1011
ADD	Add to accumulator with shift	1	0000 SSSS	I DDD DDDD
ADDC‡	Add to accumulator with carry	1	0100 0011	I DDD DDDD
ADDH	Add to high accumulator	1	0100 1000	I DDD DDDD
ADDK†	Add to accumulator short immediate	1	1100 1100	KKKK KKKK
ADDS	Add to low accumulator with sign-extension suppressed	1	0100 1001	I DDD DDDD
ADDT†	Add to accumulator with shift specified by T register	1	0100 1010	I DDD DDDD
ADLK†	Add to accumulator long immediate with shift	2	1101 SSSS	0000 0010
AND	AND with accumulator	1	0100 1110	I DDD DDDD
ANDK†	AND immediate with accumulator with shift	2	1101 SSSS	0000 0100
CMPL†	Complement accumulator	1	1100 1110	0010 0111
LAC	Load accumulator with shift	1	0010 SSSS	I DDD DDDD
LACK	Load accumulator short immediate	1	1100 1010	KKKK KKKK
LACT†	Load accumulator with shift specified by T register	1	0100 0010	I DDD DDDD
LALK†	Load accumulator long immediate with shift	2	1101 SSSS	0000 0001
NEG†	Negate accumulator	1	1100 1110	0010 0011
NORM†	Normalize contents of accumulator	1	1100 1110	1010 0010
OR	OR with accumulator	1	0100 1101	I DDD DDDD
ORK†	OR immediate with accumulator with shift	2	1101 SSSS	0000 0101
ROL‡	Rotate accumulator left	1	1100 1110	0011 0100
ROR‡	Rotate accumulator right	1	1100 1110	0011 0101
SACH	Store high accumulator with shift	1	0110 1XXX	I DDD DDDD
SACL	Store low accumulator with shift	1	0110 0XXX	I DDD DDDD
SBLK†	Subtract from accumulator long immediate with shift	2	1101 SSSS	0000 0011
SFL†	Shift accumulator left	1	1100 1110	0001 1000
SFR†	Shift accumulator right	1	1100 1110	0001 1001
SUB	Subtract from accumulator with shift	1	0001 SSSS	I DDD DDDD
SUBB‡	Ssubtract from accumulator with borrow	1	0100 1111	I DDD DDDD
SUBC	Conditional subtract	1	0100 0111	I DDD DDDD
SUBH	Subtract from high accumulator	1	0100 0100	I DDD DDDD
SUBK‡	Ssubtract from accumulator short immediate	1	1100 1101	KKKK KKKK
SUBS	Subtract from low accumulator with sign extension suppressed	1	0100 0101	I DDD DDDD
SUBT†	Subtract from accumulator with shift specified by T register	1	0100 0110	I DDD DDDD
XOR	Exclusive-OR with accumulator	1	0100 1100	I DDD DDDD
XORK†	Exclusive-OR immediate with accumulator with shift	2	1101 SSSS	0000 0110
ZAC	Zero accumulator	1	1100 1010	0000 0000
ZALH	Zero low accumulator and load high accumulator	1	0100 0000	I DDD DDDD
ZALR‡	Zero low accumulator and load high accumulator with rounding	1	0111 1011	I DDD DDDD
ZALS	Zero accumulator and load low accumulator with sign extension suppressed	1	0100 0001	I DDD DDDD

†These instructions are specific to the TMS320C2x instruction set.
‡These instructions are specific to the TMS320C25 instruction set.

TABLE 2.1 (continued)

Mnemonic and Description		Words	16-Bit Opcode			
AUXILIARY REGISTERS AND DATA PAGE POINTER INSTRUCTIONS						
			MSB			LSB
ADRK‡	Add to auxiliary register short immediate	1	0111	1110	KKKK	KKKK
CMPR†	Compare auxiliary register with auxiliary register AR0	1	1100	1110	0101	00KK
LAR	Load auxiliary register	1	0011	0RRR	IDDD	DDDD
LARK	Load auxiliary register short immediate	1	1100	0RRR	KKKK	KKKK
LARP	Load auxiliary register pointer	1	0101	0101	1000	1RRR
LDP	Load data memory page pointer	1	0101	0010	IDDD	DDDD
LDPK	Load data memory page pointer immediate	1	1100	100K	KKKK	KKKK
LRLK†	Load auxiliary register long immediate	2	1101	0RRR	0000	0000
MAR	Modify auxiliary register	1	0101	0101	IDDD	DDDD
SAR	Store auxiliary register	1	0111	0RRR	IDDD	DDDD
SBRK‡	Subtract from auxiliary register short immediate	1	0111	1111	KKKK	KKKK
T REGISTER, P REGISTER, AND MULTIPLY INSTRUCTIONS						
Mnemonic and Description		Words	16-Bit Opcode			
			MSB			LSB
APAC	Add P register to accumulator	1	1100	1110	0001	0101
LPH†	Load high P register	1	0101	0011	IDDD	DDDD
LT	Load T register	1	0011	1100	IDDD	DDDD
LTA	Load T register and accumulate previous product	1	0011	1101	IDDD	DDDD
LTD	Load T register, accumulate previous product, and move data	1	0011	1111	IDDD	DDDD
LTP†	Load T register and store P register in accumulator	1	0011	1110	IDDD	DDDD
LTS†	Load T register and subtract previous product	1	0101	1011	IDDD	DDDD
MAC†	Multiply and accumulate	2	0101	1101	IDDD	DDDD
MACD†	Multiply and accumulate with data move	2	0101	1100	IDDD	DDDD
MPY	Multiply (with T register, store product in P register)	1	0011	1000	IDDD	DDDD
MPYA‡	Multiply and accumulate previous product	1	0011	1010	IDDD	DDDD
MPYK	Multiply immediate	1	101K	KKKK	KKKK	KKKK
MPYS‡	Multiply and subtract previous product	1	0011	1011	IDDD	DDDD
MPYU‡	Multiply unsigned	1	1100	1111	IDDD	DDDD
PAC	Load accumulator with P register	1	1100	1110	0001	0100
SPAC	Subtract P register from accumulator	1	1100	1110	0001	0110
SPH‡	Store high P register	1	0111	1101	IDDD	DDDD
SPL‡	Store low P register	1	0111	1100	IDDD	DDDD
SPM†	Set P register output shift mode	1	1100	1110	0000	10KK
SQRA†	Square and accumulate	1	0011	1001	IDDD	DDDD
SQRS†	Square and subtract previous product	1	0101	1010	IDDD	DDDD

†These instructions are specific to the TMS320C2x instruction set.
‡These instructions are specific to the TMS320C25 instruction set.

2. Math instructions to add, subtract, and multiply, such as the multiply and accumulate (MAC)

3. Logical instructions (AND/OR/XOR)

4. Input and output instructions (IN/OUT) to interface with the real world

5. Control instructions and branch instructions, such as branch if the accumulator is less or equal to zero (BLEZ)

Instruction Examples

Descriptions of several instructions useful in DSP follow:

MAC, MACD.

These instructions multiply a value from program memory by a value loaded into the TR from data memory, and accumulate. A data move option is included with

TABLE 2.1 (continued)

BRANCH/CALL INSTRUCTIONS

Mnemonic and Description		Words	16-Bit Opcode MSB LSB
B	Branch unconditionally	2	1111 1111 1DDD DDDD
BACC†	Branch to address specified by accumulator	1	1100 1110 0010 0101
BANZ	Branch on auxiliary register not zero	2	1111 1011 1DDD DDDD
BBNZ†	Branch if TC bit ≠ 0	2	1111 1001 1DDD DDDD
BBZ†	Branch if TC bit = 0	2	1111 1000 1DDD DDDD
BC‡	Branch on carry	2	0101 1110 1DDD DDDD
BGEZ	Branch if accumulator ≥ 0	2	1111 0100 1DDD DDDD
BGZ	Branch if accumulator > 0	2	1111 0001 1DDD DDDD
BIOZ	Branch on I/O status = 0	2	1111 1010 1DDD DDDD
BLEZ	Branch if accumulator ≤ 0	2	1111 0010 1DDD DDDD
BLZ	Branch if accumulator < 0	2	1111 0011 1DDD DDDD
BNC‡	Branch on no carry	2	0101 1111 1DDD DDDD
BNV†	Branch if no overflow	2	1111 0111 1DDD DDDD
BNZ	Branch if accumulator ≠ 0	2	1111 0101 1DDD DDDD
BV	Branch on overflow	2	1111 0000 1DDD DDDD
BZ	Branch if accumulator = 0	2	1111 0110 1DDD DDDD
CALA	Call subroutine indirect	1	1100 1110 0010 0100
CALL	Call subroutine	2	1111 1110 1DDD DDDD
RET	Return from subroutine	1	1100 1110 0010 0110
TRAP†	Software interrupt	1	1100 1110 0001 1110

I/O AND DATA MEMORY OPERATIONS

Mnemonic and Description		Words	16-Bit Opcode MSB LSB
BLKD†	Block move from data memory to data memory	2	1111 1101 I DDD DDDD
BLKP†	Block move from program memory to data memory	2	1111 1100 I DDD DDDD
DMOV	Data move in data memory	1	0101 0110 I DDD DDDD
FORT†	Format serial port registers	1	1100 1110 0000 111K
IN	Input data from port	1	1000 AAAA I DDD DDDD
OUT	Output data to port	1	1110 AAAA I DDD DDDD
RFSM‡	Reset serial port frame synchronization mode	1	1100 1110 0011 0110
RTXM†	Reset serial port transmit mode	1	1100 1110 0010 0000
RXF†	Reset external flag	1	1100 1110 0000 1100
SFSM‡	Set serial port frame synchronization mode	1	1100 1110 0011 0111
STXM†	Set serial port transmit mode	1	1100 1110 0010 0001
SXF†	Set external flag	1	1100 1110 0000 1101
TBLR	Table read	1	0101 1000 I DDD DDDD
TBLW	Table write	1	0101 1001 I DDD DDDD

†These instructions are specific to the TMS320C2x instruction set.
‡These instructions are specific to the TMS320C25 instruction set.

MACD to handle delays in such problems as convolution. The data being moved should be contained in the on-chip data memory blocks B0, B1, and B2, not in external memory; otherwise, the MACD instruction reduces to MAC.

LTA, LTD.

These instructions load TR with a data memory value, before a multiplication command. They also add the content of the product register (which would contain a previous product) into the accumulator. A data move option is included with LTD.

BLKP, BLKD.

BLKP performs a block move from program memory to data memory. BLKD performs a block move from data memory to data memory.

TABLE 2.1 (concluded)

CONTROL INSTRUCTIONS				
Mnemonic and Description		Words	16-Bit Opcode MSB LSB	
BITt	Test bit	1	1001 BBBB	I DDD DDDD
BITTt	Test bit specified by T register	1	0101 0111	I DDD DDDD
CNFDt	Configure block as data memory	1	1100 1110	0000 0100
CNFPt	Configure block as program memory	1	1100 1110	0000 0101
DINT	Disable interrupt	1	1100 1110	0000 0001
EINT	Enable interrupt	1	1100 1110	0000 0000
IDLEt	Idle until interrupt	1	1100 1110	0001 1111
LST	Load status register ST0	1	0101 0000	I DDD DDDD
LST1t	Load status register ST1	1	0101 0001	I DDD DDDD
NOP	No operation	1	0101 0101	0000 0000
POP	Pop top of stack to low accumulator	1	1100 1110	0001 1101
POPDt	Pop top of stack to data memory	1	0111 1010	I DDD DDDD
PSHDt	Push data memory value onto stack	1	0101 0100	I DDD DDDD
PUSH	Push low accumulator onto stack	1	1100 1110	0001 1100
RC‡	Reset carry bit	1	1100 1110	0011 0000
RHM‡	Reset hold mode	1	1100 1110	0011 1000
ROVM	Reset overflow mode	1	1100 1110	0000 0010
RPTt	Repeat instruction as specified by data memory value	1	0100 1011	I DDD DDDD
RPTKt	Repeat instruction as specified by immediate value	1	1100 1011	KKKK KKKK
RSXMt	Reset sign-extension mode	1	1100 1110	0000 0110
RTC‡	Reset test/control flag	1	1100 1110	0011 0010
SC‡	Set carry bit	1	1100 1110	0011 0001
SHM‡	Set hold mode	1	1100 1110	0011 1001
SOVM	Set overflow mode	1	1100 1110	0000 0011
SST	Store status register ST0	1	0111 1000	I DDD DDDD
SST1t	Store status register ST1	1	0111 1001	I DDD DDDD
SSXMt	Set sign-extension mode	1	1100 1110	0000 0111
STC‡	Set test/control flag	1	1100 1110	0011 0011

tThese instructions are specific to the TMS320C2x instruction set.
‡These instructions are specific to the TMS320C25 instruction set.

RPT, RPTK.

These repeat instructions can be used with instructions such as multiply and accumulate and table read/write. Multiple-cycle instructions become effectively single-cycle instructions by being pipelined when used in conjunction with the repeat feature.

The repeat instruction is very effective associated with instructions such as multiply/accumulate (MAC) and block move. For example, with the repeat instruction (RPT or RPTK), the instruction BLKP efficiently moves a block of data from program memory to data memory. An instruction can be repeated a maximum of 256 times, with the number set with the RPT (or RPTK) instruction and loaded in the 8-bit-wide repeat counter (RPTC). The following partial program illustrates the use of the MACD in conjunction with the RPTK instruction:

```
CNFP
RPTK   4
MACD   PMA, DMA
```

The CNFP instruction configures block B0 as on-chip program memory as shown in Figure 2.3(b). The RPTK 4 loads an immediate value of 4 into the repeat counter, which permits the instruction MACD to be executed five times (one time more than the value specified with RPTK). The MACD instruction is probably the most powerful TMS320C25 instruction. It has the following functions:

1. Loads into the prefetch counter (PFC) the program memory address PMA.
2. Adds the previous content of the PR to the accumulator.
3. Multiplies the content of data memory location specified by DMA by the content of program memory location specified by PMA.
4. Copies the data memory value into the location of the next-higher on-chip data memory address (DMA). This function cannot be performed with external data memory.
5. Increments PFC with each multiply/accumulate/data move.

The MACD instruction combines the LTD and multiply (MPY) instructions into one instruction. The LTD already combines the three instructions LT, APAC, and DMOV (LTD = LT + APAC + DMOV), where

LT loads TR with a value, anticipating a multiply instruction.

APAC adds the previous content of PR to the accumulator.

DMOV performs the on-chip data move into the next higher DMA.

The results of the RPTK and MACD pair of instructions could also have been accomplished with the LTD and MPYK (the immediate multiply mode) instructions. The data move is the operation used to implement the delay represented in a difference equation.

For a small number of multiplications (fewer than nine), the LTA/MPY combination of instructions is faster and takes less memory for fewer than four multiplications. However, the RPT/MAC combination each takes only one execution cycle time for each succeeding multiply and accumulate (after an initial overhead). For more than three multiplications, the RPT/MAC is efficient. With parallel addressing of both multiplicands, the program bus is fetching one multiplicand while the data bus is fetching the other multiplicand.

2.6 FIXED-POINT ARITHMETIC

Binary and Two's-Complement Representation

Since the TMS320C25 supports two's-complement arithmetic, a review of the two's-complement system is in order. To make the illustrations manageable, a 4-bit system will be used rather than a 16-bit word length. A 4-bit word can represent the unsigned numbers 0–15, as shown in Table 2.2. The 4-bit unsigned numbers represent a modulo (mod) 16 system. If one (1) is added to the largest number (15), the operation wraps around to give 0 as the answer. Finite bit systems have the same modulo properties as do number wheels on combination locks. Therefore, a number wheel graphically demonstrates the addition properties of a finite bit

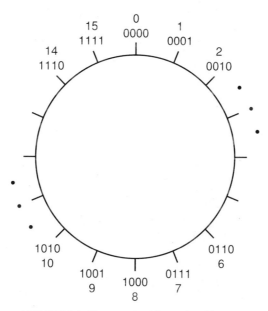

FIGURE 2.4. Number wheel for unsigned integers

system. Figure 2.4 shows a number wheel with the numbers 0–15 wrapped around the outside. For any two numbers x and y in the range, the operation amounts to the procedure: (1) find the first number x on the wheel; (2) step off y units in the clockwise direction, which brings you to the answer. For example, let us try the addition of two numbers such as $(5 + 7) \mod 16$, which yields 12. From the number wheel, locate 5, then step 7 units in the clockwise direction to arrive at the same answer, 12. As another example, try $(12 + 10) \mod 16 = 6$. First locate 12 on the number wheel, then step 10 units in the clockwise direction, which brings you past zero, to 6.

Negative numbers require a different interpretation of the numbers on the wheel. If we draw a line through 0 and 8 cutting the number wheel in half, the right half

TABLE 2.2   Unsigned Binary Number Representation

Binary	Decimal
1111	15
1110	14
.	.
.	.
.	.
0011	3
0010	2
0001	1
0000	0

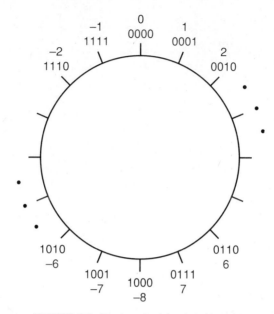

FIGURE 2.5. Number wheel for signed integers

will represent the positive numbers and the left half the negative numbers, as shown in Figure 2.5. This representation is the two's complement system. The negative numbers are the two's complement of the positive numbers, and vice versa. A two's-complement binary integer,

$$B = b_{n-1} \cdots b_1 b_0$$

is equivalent to the decimal integer

$$I(B) = -b_{n-1} \times 2^{n-1} + \cdots + b_1 \times 2^1 + b_0 \times 2^0 \qquad (2.1)$$

where the b's are binary digits. Notice that the sign bit has a negative weight, while all the others have positive weights. For example, consider the number -2,

$$1110 = -1 \times 2^3 + 1 \times 2^2 + 1 \times 2^1 + 0 \times 2^0 = -8 + 4 + 2 + 0 = -2$$

To apply the graphical technique to the operation $6 + (-2) \bmod 16 = 4$, locate the number 6 on the wheel, then step off (1110) or 14 units in the clockwise direction to arrive at the answer, 4. The binary addition of these same numbers,

```
  0 1 1 0
  1 1 1 0
1 0 1 0 0
C
```

shows a carry out of the MSB, which, in the case of finite register arithmetic, will be ignored. This carry corresponds to the wrap-around through zero on the number wheel.

The addition of these numbers results in correct answers, by ignoring the carry out of the MSB, provided that the answer is in the range of representable numbers -2^{n-1} to $(2^{n-1}) - 1$ in the case of an n-bit number and between -8 and 7 for the 4-bit number wheel example. When -7 is added to -8 (in the 4-bit system), we get an answer of $+1$ instead of the correct value of -15, which is out of range. When two numbers of like sign are added to produce an answer with opposite sign, overflow has occurred. The TMS320C25 has an overflow flag that indicates when overflow occurs on the accumulator. Subtraction with two's-complement numbers is equivalent to adding the two's complement of the number being subtracted to the other number.

Fractional Fixed-Point Representation

Rather than use the integer values just discussed, most DSP applications use a fractional fixed-point number which has values between $+0.99\ldots$ and -1. To obtain the fractional n-bit number, the radix point must be moved $n-1$ places to the left. This leaves one sign bit plus $n-1$ fractional bits. The expression

$$F(B) = -b_0 \times 2^0 + b_1 \times 2^{-1} + b_2 \times 2^{-2} + \cdots + b_{n-1} \times 2^{-(n-1)} \quad (2.2)$$

converts a binary fraction to a decimal fraction. Notice again that the sign bit has a weight of negative 1 and the weights of the other bits are positive powers

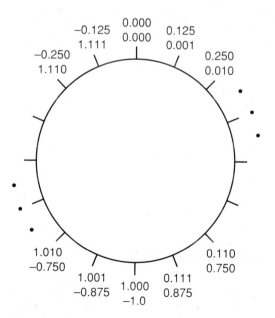

FIGURE 2.6. Number wheel for fixed-point representation.

of 1/2. The number wheel representation for the fractional two's-complement 4-bit numbers are shown in Figure 2.6. The fractional numbers are obtained from the two's-complement integer numbers of Figure 2.5 by scaling them by 2^3. Because the number of bits in a 4-bit system is small, the range is from -1 to 0.875. The TMS320 family processors use a 16-bit word; therefore, the signed integers range from -32768 (8000 hex) to $+32767$ (7FFF hex). To get the fractional range, scale those two signed integers by 2^{-15} or 32,768, which results in a range from -1 to 0.999969 (usually taken to be $+1$).

Multiplication

Multiplication of two signed numbers causes some interesting problems. If one multiplies two n-bit integer numbers, the common notion is that a $2n$-bit operand will result. Although this is true for unsigned numbers, it is not so for signed numbers. Recall that signed numbers need one sign bit with a weight of -2^{n-1}, followed by positive weights that are powers of 2. To find the number of bits needed for the result, multiply the two largest numbers together:

$$P = (-2^{n-1}) \times (-2^{n-1}) = 2^{2n-2} \qquad (2.3)$$

This number is a positive number representable in $(2n - 1)$ bits (the MSB of this resultant occupies the $(2n - 2)$ bit position counting from 0). Since this number is positive, its sign bit, which would show up as a negative number (a power of 2), does not appear. Unfortunately, we have stumbled upon the one exceptional case, which is usually treated as an overflow in fractional representation. Since the fractional representation requires that both operand and resultant occupy the range, $-1 \geq$ range $< +1$, the operation $(-1) \times (-1)$ produces an unrepresentable $+1$. Let us look at the next larger combination:

$$P = (-2^{n-1})(-2^{n-1} + 1) = 2^{2n-2} - 2^{n-1} \qquad (2.4)$$

Since the second number subtracts from the first, the product will occupy up to the $(2n - 3)$ bit position (counting from 0); thus it is representable in $(2n - 2)$ bits. With the exceptional case ruled out, this makes the bit position $(2n - 2)$ available for the sign bit of the resultant. Therefore, $(2n - 1)$ bits are needed to support an $(n \times n)$-bit signed multiplication. Most computers must support signed and unsigned multiplies and therefore provide $2n$ bits for the product.

To clarify equation (2.4), consider its 4-bit case. Substituting the 4-bit values into (2.4) gives

$$P = (-2^3)(-2^3 + 1) = 2^6 - 2^3$$

The number 2^6 occupies bit position 6. Since the second number is negative, the summation of the two is a number that will occupy only bit positions less than bit position 6:

$$2^6 - 2^3 = 64 - 8 = 56 = 00111000$$

Thus bit position 6 is available for the sign bit. The 8-bit equivalent would have two sign bits (bit positions 6 and 7).

Consider the multiplication of two fractional 4-bit numbers, with each number consisting of 3 fractional bits and 1 sign bit. Let the product be represented with an 8-bit number. The first number is equal to –0.5 and the second 0.75. The multiplication is shown as follows:

$$
\begin{array}{r}
-0.50 \;=\; 1.100 \\
\times 0.75 \;=\; \times 0.110 \\
\hline
11\overline{11}000 \\
\overline{1}11000 \\
\hline
\\
111.101000 \\
C
\end{array}
$$

$$= -2^1 + 2^0 + 2^{-1} + 2^{-3} = -0.375$$

The underlined bits of the multiplicand indicate sign extension. When a negative multiplicand is added to the partial product, it must be sign-extended to the left up to the limit of the resultant product, in order to give the proper larger bit version of the same number. To demonstrate that sign extension gives the correct expanded bit number, scan around the number wheel (Figure 2.5) in the counterclockwise direction from 0. Write the codes for 5-bit, 6-bit, 7-bit, . . . negative numbers. Notice that they would be correctly derived by sign-extending the existing 4-bit codes; therefore, sign extension gives the correct expanded bit number. The carry-out will be ignored; however, the numbers 111.101000 (9-bit word), 11.101000 (8-bit word), and 1.101000 (7-bit word) all represent the same number: –0.375. Thus the product of the preceding example could be represented by $(2n - 1)$ bits, or 7-bits for a 4-bit system.

The MPY instruction of the TMS320 family performs a multiply on the contents of the 16-bit-wide TR and the contents of a data memory location, also 16 bits wide, to produce a 32-bit result that is stored in the PR, which can be moved to the 32-bit accumulator under program control. With two's-complement operations, only $(2n - 1)$ bits are needed for the multiply operation. As a result, bit 30 is sign-extended to bit 31. The extended bits are frequently called sign bits.

Since all other operations on the processor require 16-bit operands, the 32-bit result must eventually be truncated or rounded to 16 bits. The most significant bits, along with the sign bit and its duplicate, are in the high end of the accumulator. The result in the high end of the accumulator is left-shifted to eliminate the extra sign bit and to give an additional bit of resolution, when moved to a 16-bit memory location. The following sequence of instructions performs the multiply and truncate operations just described:

```
*S1 IS THE ADDRESS OF THE MULTIPLICAND
*S2 IS THE ADDRESS OF THE MULTIPLIER
```

```
*R IS THE ADDRESS OF THE PRODUCT
        LT    S1       TR = CONTENT OF S1
        MPY   S2       (S1 X S2)→PR
        APAC           PR→ACC
        SACH R,1       SHIFT XTRA BIT, ACCH→R
```

A second multiply instruction, MPYK, multiplies the content of the T register (15-bit number with 1 sign bit) by a 13-bit constant (12-bit number with 4 sign bits), producing a 27-bit number with a sign bit and 4 extra sign bits. To get 15 bits of precision after using this instruction, a shift left by 4 is performed with the instruction

```
SACH   R,4
```

which stores the most significant 16 bits (15 bits and 1 sign bit) to data memory R, while removing the 4 extra sign bits.

2.7 REPRESENTATION OF NUMBERS GREATER THAN 1 AND OVERFLOW ON THE TMS320C25

Some situations arise that require the representation of coefficients greater than 1. These can be treated by scaling so that all coefficients are between -1 and $+1$, or they can be used directly provided that the end result lies in the permissible range. Coefficients greater than 1 can be implemented through shifts and adds. The following two examples illustrate how this is done.

Example 2.1 Multiplication by 10. Write a program segment to multiply the number in data memory location X by 10.

SOLUTION:

```
* MULTIPLY X BY 10
* FOR RESULT R TO BE VALID, |X| MUST BE LESS THAN 0.1
* X IS ADDRESS OF INPUT , AND R IS ADDRESS OF RESULT
        LAC   X,3       8X → ACC
        ADD   X,1       8X + 2X = 10X → ACC
        SACL R          STORE RESULT IN R
```

Example 2.2 Multiplication by -1.72416. Write a program segment to multiply the variable X by -1.72416.

SOLUTION:

```
* MULTIPLY X BY −1.72416
* RESULT R IS VALID FOR |X| <0.57999
A       EQU >60      DATA MEMORY LOCATION OF CONSTANT
AP      DATA −23729  PROGRAM MEMORY LOCATION OF CONSTANT
*                    −23729 IS THE −0.72416 FRACTIONAL
*                    PART OF CONSTANT SCALED BY 2**15
START LACK  AP       TRANSFER CONSTANT FROM PMA AP
      TBLR  A        INTO DMA >60
      . . .
      ZAC            ACC ← 0
      LT    X        TR ← X
      MPY   A        PR ← (X)(−0.72416)
      SUB   X,15     ACC ← −1.0 X
      APAC           ACC ← −1.0 X − .72416 X = −1.72416 X
      SACH  R,1      R ← ACCH
```

Overflow Management on the TMS320C25

Since the numbers in the fixed-point system lie between $+1$ and -1, the result for a multiplication will always be between $+1$ and -1 and overflow will not be a problem. However, with addition and subtraction, overflow is still a problem. To handle overflow, the TMS320C25 has certain features.

Accumulator Overflow Bit.
This bit is set when overflow occurs in the accumulator. It can be tested either with the instruction BV (branch on overflow) or by direct testing of the overflow bit in the status register STO.

Accumulator Saturation Overflow Mode.
This mode is set via the SOVM instruction, which allows overflow to behave similarly to an analog overflow. The overflow mode can be enabled with the instruction SOVM. On overflow, the accumulator content is replaced by >7FFFFFFF (the largest positive value) if the number overflowed is positive, or by >80000000 (the largest negative value) if the number overflowed is negative. This mode is disabled via the ROVM instruction.

Product and Accumulator Right Shifts.
The PR can be set to shift right automatically 6 bits when loaded. This mode is enabled with the SPM instruction, which sets the two bits (lower 2 bits in ST1) of the product register shift mode (PM). The accumulator can be shifted right one position using the SFR instruction.

Calculations with Overflows

Any chain of two's-complement additions, subtractions, or multiplications whose final result is in the representable range can be calculated correctly even if overflow has occurred in some of the intermediate results [7,8]. For example, consider the following calculation in the 4-bit number system whose range is -8 to +7:

$$+7 + 6 - 8 = +5$$

The chain of binary calculations is as follows:

```
0 1 1 1    +7
0 1 1 0    +6
1 1 0 1    (overflow)
1 0 0 0    -8
0 1 0 1    +5
```

Exercise 2.1   Calculation with Overflow. Write and test a TMS320C25 program to calculate

$$Y = (X1)(B1) + (X2)(B2) + (X3)(B3) + (X4)(B4)$$

where

$X1 = 0.80$	$B1 = 0.4$
$X2 = -0.95$	$B2 = 0.8$
$X3 = 0.90$	$B3 = 0.75$
$X4 = -1.0$	$B4 = 1.1$

Check the overflow bit at various points in the program. Run it with and without the saturation mode set.

2.8 FLOATING-POINT REPRESENTATION

Since the TMS320C25 performs operations in fixed-point arithmetic, in order to implement floating-point arithmetic, the operands must first be converted into fixed arithmetic, processed, and then converted back into floating point. The TMS320C25 handles floating-point arithmetic with instructions such as NORM, which normalizes a fixed-point number in the accumulator into a mantissa and an exponent. Additional instructions, such as ADDT/SUBT/LACT, which adds to/subtracts from and loads the accumulator with shift specified by the TR, are available. NORM can be used to convert a fixed-point number into a floating-point number, and LACT is used to convert it back into a fixed-point number.

2.9 DISCRETE EQUATION PROGRAMMING

Example 2.3   Example Using MACD and RPTK. Write a program to implement the following equation:

$$y(n) = h(0)x(n-3) + h(1)x(n-2) + h(2)x(n-1) + h(3)x(n)   (2.5)$$

where

$$h(0) = h(3) = 1 \text{ and } h(1) = h(2) = 2.$$

SOLUTION:

1. Load the program shown in Figure 2.7 into the SWDS.
2. Access the data memory window and set the data values starting at data memory address (DMA) >3FC. All values are in hex.

DMA	Inputs	Content At Time n
3FC	x(n)	4
3FD	x(n-1)	3
3FE	x(n-2)	2
3FF	x(n-3)	1

```
0001                    * DIFFERENCE EQUATION PROGRAM
0002                    * CONVOLUTION USING MACD,RPTK
0003                    * LOAD DATA VALUES (SAMPLES) STARTING AT DMA 3FC
0004 0000                      AORG    0
0005 0000 FF80                 B       START
     0001 0024
0006 0020          TABLE ⓐAORG   >20              *START PROGRAM AT PMA >20
0007 0020 0001     H0     DATA    >1               *COEFFICIENTS
0008 0021 0002     H1     DATA    >2
0009 0022 0002     H2     DATA    >2
0010 0023 0001     H3     DATA    >1
0011 0024 5588     START ⓑLARP   AR0              *SELECT AR0
0012 0025 D000           LRLK    AR0,>200         *POINT TO BLOCK B0
     0026 0200
0013 0027 CB03          ⓒRPTK   3                *EXECUTE BLOCK MOVE 4 TIMES
0014 0028 FCA0           BLKP    TABLE,*+         *P.M. TO D.M. B0 START >200
     0029 0020
0015 002A CE05          ⓓCNFP                     *CONFIGURE B0 AS P.M.
0016            *
0017 002B CA00          ⓔZAC                      *CLEAR ACCUMULATOR
0018 002C A000           MPYK    0                *CLEAR PRODUCT REGISTER
0019 002D D100           LRLK    AR1,>3FF         *POINT TO BOTTOM OF BLOCK B1
     002E 03FF
0020 002F 5589           LARP    AR1              *SELECT AR1 FOR INDIR ADDR
0021            * THE BEEF
0022 0030 CB03          ⓕRPTK   3                *EXECUTE MACD 4 TIMES
0023 0031 5C90           MACD    65280,*-         *MULT(PMA>FF00)(DMA>3FF),ETC
     0032 FF00
0024 0033 CE15          ⓖAPAC                     *ADD LAST PRODUCT TO ACCUM
0025                     END
NO ERRORS, NO WARNINGS
```

FIGURE 2.7. Difference equation program (CONVO.LST)

3. Set the program counter (PC) to 0 and single-step through the program. Observe the following:

a. The absolute origin of the program starts at PMA >20. H0, H1, H2, and H3 are specified at PMA >20–23.

b. Auxiliary register AR0 is selected and set to point at DMA >200, the top of block B0, in the first two instructions of the program.

c. With the repeat instruction RPTK set to 3, the block move instruction (BLKP) is executed four times to transfer the values of H0, H1, H2, and H3 from PMA >20–23 to DMA >200–203. AR0 contains the DMA, which was selected previously. Indirect addressing is specified by * and incremented using +.

d. B0 is then configured as PM with CNFP, mapping the 256 data memory addresses >200–2FF into program memory addresses >FF00–FFFF (locations 65280–65535), as shown in Figure 2.3(*b*).

e. Then both the accumulator and product register (PR) are initialized to zero. AR1 is then selected to point at the bottom of block B1, at DMA >3FE.

f. MACD is executed four times, multiplies H1 (the content of PMA >FF00) by $x(n - 3) = 1$ (at DMA >3FF), accumulates, then multiplies H2 (the content of PMA >FF01) by $x(n - 2) = 2$ (at DMA >3FE), and so on, to implement equation (2.5). After each multiplication, a data move is also performed; DMA >3FF now contains the value 2, previously at DMA > 3FE. This value had been "moved down" one higher data memory location. Similarly, DMA >3FE now contains the value 3, set previously for $x(n -1)$ at DMA >3FD. The data move effectively implements the delays associated with $x(n$ - k$)$.

g. The last instruction, APAC, adds the contents of the product register containing $(H3)x(n) = 4$ into the accumulator, effectively computing for a specified (time) n:

$$y = (1)(1) + (2)(2) + (2)(3) + (1)(4) = F$$

To calculate

$$y(n + 1) = h(0)x(n - 2) + h(1)x(n - 1) + h(2)x(n) + h(3)x(n + 1)$$

a new input sample $x(n + 1)$ would be stored at DMA >3FC, which now becomes the current input, for example, $x(n + 1) = 5$. The four data moves executed previously in conjunction with the MACD instruction in step f allow for the computation of the output at the next sample time $(n + 1)$, since $x(n - 2)$, $x(n - 1)$, and $x(n)$ had each been shifted down one higher data memory location. Set $x(n + 1) = 5$ at DMA >3FC, and set the PC to PMA >2B to reset both the accumulator and the product register to zero when single-stepping through the ZAC and MPYK instructions. The data values at time $(n + 1)$ are as follows:

DMA (Hex)	Data Values at Time $n + 1$
3FC	5
3FD	4
3FE	3
3FF	2

Single-step through the program and verify that

$$y(n + 1) = (1)(2) + (2)(3) + (2)(4) + (1)(5) = 15(\text{hex})$$

Consider this alternative:

1. Set DMA >3FC–3FF with data values 4, 3, 2, and 1, as before. However, change PMA >2E from >03FF to >03FC, which changes the instruction to LRLK AR1,>3FC. Access the program memory by using the SWDS window.
2. Change PMA >31 from >5C90 to >5CA0, which changes the instruction to MACD >FF00,*+.

The first change points AR1 to 3FC instead of 3FF. The second change increments AR1 instead of decrementing. Is the function of the program the same as before? No, since the value 4 in DMA >3FC after the first multiplication of H0 $x(n)$ = (1)(4) is moved down before the second multiplication. Hence $x(n - 1)$ is replaced by 4 in DMA >3FD, and the second multiplication results in H1 $x(n - 1) = 2 \times 4 = 8$. The third multiplication is H2 $x(n - 2) = 2 \times 4 = 8$ and the fourth multiplication is $1 \times 4 = 4$. The overall result in the accumulator is then

$$(1)(4) + (2)(4) + (2)(4) + (1)(4) = 18(\text{hex})$$

Because of the DMOV instruction included in MACD, it is necessary that the first multiplication start with the last data sample.

In a filter program, an IN XN,PA instruction would be used (in conjunction with the AIB, for example) in lieu of the comment specified by the * (after the instruction CNFP). The input sample at time n would be obtained from port address PA and placed in DMA XN. A branch instruction in lieu of the comment (before the END instruction) would specify to "GO BACK" and collect a new sample. The loop would continue, effectively calculating $y(n + 2)$, $y(n + 3)$, and so on.

The equation in (2.5) is a convolution equation that will be used again later for the implementation of digital filters.

Discrete Equation Programming Using COFF

The program listing shown in Figure 2.7 implements the discrete equation (2.5). This program was assembled using a TI-Tag format. In Chapter 1 and Appendix A, we discuss version 5.04 of the assembler/linker, which uses a different format—

the common object file format (COFF)—and how to convert between TI-Tag and COFF [1,4]. Figures A.6 to A.8 show the equivalent program using COFF.

REFERENCES

[1] *Second-Generation TMS320 User's Guide*, Texas Instruments, Inc., Dallas, Tex., 1987.

[2] *TMS320C1x/TMS320C2x Assembly Language Tools User's Guide*, Texas Instruments, Inc., Dallas, Tex., 1987.

[3] *TMS320C2x Software Development System User's Guide*, Texas Instruments, Inc., Dallas, Tex., 1988.

[4] *TMS320C1x/TMS320C2x Source Conversion Reference Guide*, Texas Instruments, Inc., Dallas, Tex., 1988.

[5] K. S. Lin (editor), *Digital Signal Processing Applications with the TMS320 Family*, Vol. 1, Prentice-Hall, Englewood Cliffs, N.J., 1988.

[6] *TMS32010 Analog Interface Board User's Guide*, Texas Instruments, Inc., Dallas, Tex., 1984.

[7] S. Waser and M. Flynn, *Introduction to Arithmetic for Digital Systems Designers*, Holt, Rinehart and Winston, New York, 1982.

[8] H. L. Garner, "Theory of Computer Addition and Overflows," *IEEE Transactions on Computers*, **C-27**(4), April 1978, pp. 297–301.

[9] V. C. Hamacher, Z. G. Vranesic, and S. G. Zaky, *Computer Organization*, McGraw-Hill, New York, 1978.

3

Input/Output

CONCEPTS AND PROCEDURES

• *Basic operations of analog-to-digital (A/D) and digital-to-analog (D/A) input/output (I/O)*
• *I/O programming with the AIB*
• *I/O programming with the AIC*

In this chapter we discuss input and output alternatives with both the analog interface board (AIB) and the analog interface chip (AIC). The AIB includes 12-bit A/D and D/A, antialiasing, and reconstruction filters. The use of the instruction BIOZ to test the availability of an input is demonstrated. The LOOP program, introduced in Chapter 1, is covered in detail in this chapter. Two other examples, one that generates a triangular waveform and one that generates pseudorandom noise, are discussed. The pseudorandom-noise-generator example provides background for the project on multirate filtering, included in Chapter 9.

The AIC includes 14-bit A/D and D/A, input and output filters, as well as a set of two inputs, all on one chip. The AIC provides an inexpensive alternative for analog interfacing and is useful in implementing adaptive filters, discussed in Chapter 8, which require two inputs. The LOOP example is again covered with the AIC and extended to a two-input example. Support circuitry and the communication codes for interfacing the AIC with the SWDS are described. The use of MACROS as a programming alternative with the AIC is also covered.

3.1 INTRODUCTION

As with any processor, the TMS320C25 must communicate or exchange information with the outside world. It does this through 16 parallel input/output (I/O) ports and one serial I/O port. The parallel ports are addressed via the IN and OUT

39

instructions. Examples of direct mode addressing with the input/output instructions follow:

```
IN SUM,PA6
```

```
OUT SUM,PA7
```

where SUM is the address of data memory and PA6 and PA7 indicate port numbers. The serial port has two registers, the data receive register (DRR) and the data transmit register (DXR), which are memory-mapped to data memory addresses 0 and 1, respectively. Both registers are addressed the same way as is any other data memory location.

Synchronization of I/O events is supported directly by an external signal called the BIO and with processor interrupts. The I/O of the serial port is interrupt driven. The interrupt vectors for the serial port are located at addresses >26 and >28 in program memory. Separate interrupts are generated upon completion of a receive or transmit operation. The parallel ports, on the other hand, can be synchronized with either the BIO signal or interrupts.

The I/O ports of the processor accept digital signals, but most DSP applications need at least one analog input and one analog output. Adaptive applications usually require an additional analog input. With this in mind, let us look at some of the building blocks that are required to convert an analog signal to a digital signal, and vice versa.

Input Signal Path

The operations that must be performed along the input signal path are filtering, sample-and-hold (S/H), and analog-to-digital (A/D) conversion, as shown in Figure 3.1. The first block on the left is the antialiasing filter, which suppresses frequencies above one-half the sample rate, or the folding (or Nyquist) frequency. Aliasing is illustrated in Figure 3.2, which shows the conversion of a sine wave into

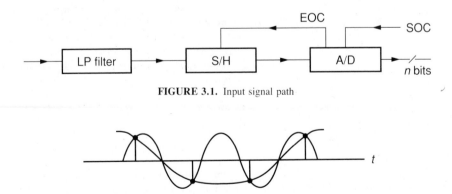

FIGURE 3.1. Input signal path

FIGURE 3.2. Aliasing of an undersampled sine wave

its sampled data representation but a nonunique conversion back to a continuous sine wave. This shows that a higher frequency, if present, can masquerade as a lower frequency if it is undersampled. The antialiasing filter will eliminate frequencies above one-half the sample rate.

The next block on the signal path is the sample-and-hold (S/H) function, described in Figure 3.3. When the pulse from the end-of-conversion (EOC) arrives, the S/H unit samples the voltage present at its input, then outputs this value and holds the output at this level until the next EOC pulse. This type of sampling is called zero-order sampling and is widely used in computer systems.

The final operation is the analog-to-digital (A/D) conversion, which takes the voltage level presented to it and converts it to a digital word. The range of input values is determined by the hardware used. In the case of the AIB, this range is ± 10 V. The voltage range is broken down into discrete voltage levels. The number of discrete levels is determined by the number of bits on the A/D device. The AIB, made by Texas Instruments, includes a 12-bit converter. The range of voltages is linearly mapped to the codes, with each voltage level being represented by one code.

Just how this is done varies from converter to converter. Two common codes are the offset binary and the two's complement. The two's complement scheme is illustrated in Figure 3.4 using eight levels. Since there are a small number of steps, the full scale voltage (V_F) in the positive direction is not closely approximated. In fact, the highest positive code will always be one step shy of full scale. To find the step size or quantization interval, q, divide the total range by the number of codes:

$$q = \frac{2V_F}{2^n} \tag{3.1}$$

where V_F is the full scale positive voltage swing and n is the number of bits.

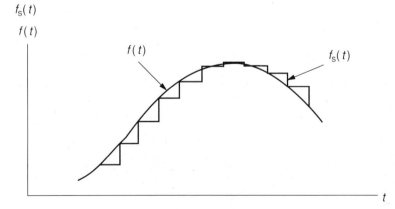

FIGURE 3.3. Sample-and-hold description

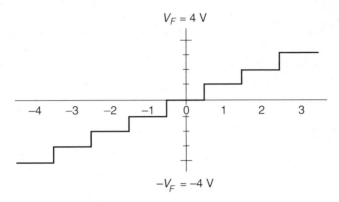

FIGURE 3.4. Voltage levels for 4 bits.

For the situation shown in Figure 3.4, the step size is conveniently chosen as 1 V. Resolution is defined as the step size divided by the range,

$$\text{resolution} = \frac{\text{step size}}{2V_F} = 2^{-n} \tag{3.2}$$

In the case of the AIB with a range of 20 V and a 12-bit word, the step size is

$$q = \frac{2(10)}{2^{12}} = 4.9 \text{ mV}$$

The most positive number is represented by

$$7\text{FFx} = 10 - q = 9.9958 \text{ V}$$

and the most negative number by

$$800\text{x} = -10 \text{ V}$$

On the AIB, the 12 most significant bits are loaded into the most significant 12 bits of data memory, discarding the 4 least significant bits.

Dividing the voltage range into discrete levels leads to an uncertainty in the voltage level within the quantization level, called the *quantization error*. Assuming a uniform distribution of error across the step leads to the following mean-square error:

$$E[e^2] = \frac{1}{q} \int_{-q/2}^{q/2} e^2 \, de = \frac{q^2}{(4)(3)} \tag{3.3}$$

which gives an rms value of

$$e_{rms} = (E[e^2])^{1/2} = \frac{q}{2\sqrt{3}} \tag{3.4}$$

With a sinusoidal signal of amplitude V_F and a quantization noise voltage of e_{rms}, the signal-to-noise ratio (S/N) becomes

$$S/N = 10\log_{10}\left[\frac{V_F/\sqrt{2}}{e_{rms}}\right]^2$$

Using (3.1) and (3.4),

$$S/N = 10\log_{10}\left[\frac{2^n q/(2\sqrt{2})}{q/(2\sqrt{3})}\right]^2$$

$$= 6.02n + 1.76 \text{ dB} \tag{3.5}$$

The first term of (3.5) is the signal power level and the second term is the quantization noise power level. For a 12-bit A/D converter ($n = 12$), these levels differ by approximately 70 dB. In other words, the quantization noise is down approximately 70 dB compared to the full-scale signal.

The path for the analog output signal is shown in Figure 3.5. Digital-to-analog (D/A) conversion undoes the operation of the A/D operation. The D/A block shown on the left takes the two's-complement number that represents a value inside the processor and converts it to a voltage level that is held as long as the digital word remains on the input of the D/A converter. The output of the D/A converter will try to track changes in the digital word and will change as fast as its internal time constants will allow. To avoid changes at random times, the input to the D/A usually has dual clocked buffers that synchronize to some event or are controlled by the software. The output of the D/A is a step-level signal like the output of the sample-and-hold unit shown in Figure 3.3. Each time a different digital word is loaded, the output of the D/A will step to a new voltage level. The voltage-level assignments are done in the same way as with the A/D operation except that the operation is reversed. After converting the signal sequence to a step-level signal, it is passed through a reconstruction filter which is a lowpass filter (LP) that smooths out the steps.

3.2 ANALOG INTERFACE BOARD

The analog interface board (AIB), introduced in Chapter 1, is a board with all the functional blocks just discussed integrated, to give one channel of input and

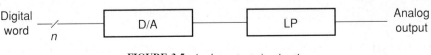

FIGURE 3.5. Analog output signal path

one channel of output. The AIB is connected to several of the parallel ports and communication is done using the IN and OUT instructions. The acquisition and reconstruction operations just described are done on the AIB and are a generic set that are present in some form in almost any analog I/O board.

The AIB communicates with the SWDS through four registers shown with their respective port addresses:

AIB control register   (PA0)
Sample rate register   (PA1)
Data input register   (PA2)
Data output register   (PA2)

The control register controls the various modes of operation of the A/D and D/A converters. See Appendix B or the *TMS32010 Analog Interface Board User's Guide* [1] for control register bit definitions and for more information in general on the AIB.

The frequency of the sample rate clock is controlled by the value written to port 1 (PA1). The binary output of the clock can be programmed to provide the start of conversion (SOC) signal (see Figure 3.1) for the A/D converter as well as other control signals. This sample rate is the sample rate of filters and other DSP structures. The sampling frequency is set by loading a positive constant, N, into the sample rate register. N is calculated according to the relationship

$$N = \frac{f_{clk}}{f_s} - 1 \qquad (3.6)$$

where f_{clk} is the system clock frequency and f_s is the sampling frequency. If it originates from the SWDS board, the system clock frequency may have one of two values,

$$f_{clk} = \begin{cases} 5 \text{ MHz} \\ 10 \text{ MHz} \quad \text{factory installed} \end{cases}$$

The AIB was not originally designed for the 10-MHz clock rate; consequently, when operating at this frequency, it drops words depending on the orientation of the connecting cable. Texas Instruments' solution at present is to provide a user-installed 5-MHz clock. Other solutions are discussed in Appendix B.

The clock frequency can be continuously programmable in the ranges

$$152.6 \text{ Hz to } 10 \text{ MHz} \qquad (f_s = 10\text{MHz})$$
$$76.3 \text{ Hz to } 5 \text{ MHz} \qquad (f_s = 5\text{MHz})$$

The sample rate constant must be loaded and the bit 0 (clock inhibit) of the control register cleared before the clock will run.

The mode of operation and the clock frequency are usually set once at the beginning of the program by writing to the registers using the following sequence of instructions:

```
OUT   CTL,PA0    TRANSPARENT MODE+CLK ENABLE
OUT   RATE,PA1   LOAD THE SAMPLE RATE
OUT   RATE,PA3   DUMMY WRITE TO DISABLE EXPAN PORT
```

The following is a typical set of constants in the data memory locations:

```
CTL   >A         TRANSPARENT AND AUTOMATIC MODES
RATE  999        10 KHz IF SYSTEM CLK IS 10 MHz
```

A read to port 2 will input data from the A/D converter, and a write to port 2 will output data to the D/A converter. Because of the differences in speed between the AIB and the processor, care must be taken to avoid overrunning the analog interface board, that is, doing multiple reads on the same sample or overwriting data before they are converted. To avoid this, the inputs and outputs must be synchronized so that they are done only when the data in the AIB are ready.

Synchronizing the I/O

As with most computer systems, the TMS320 family of processors is much faster than its I/O module, the AIB. The BIO signal or interrupts are used to synchronize reading and writing to the analog ports. The program segment in Figure 3.6 shows I/O synchronization using the BIO signal. Note that the end-of-conversion signal must be connected to the BIO (jumper pins 1 and 2 on E5 of the AIB). Both the input and the output are synchronized with the BIO signal, which is connected to the EOC. The BIOZ allows a programmer to test the BIO signal directly.

The interrupt synchronization program segment in Figure 3.7 accomplishes the same thing as the BIO version. The connection on jumper E5 must be switched to 2 and 3. Notice the following features in the program:

1. The interrupt vector for port 2 is located at program address 2, where the branch to the label ISR (interrupt service routine) is located.

```
*       I/O SYNCHRONIZATION
*       INITIALIZE THE AIB
WAIT    BIOZ                    *TEST THE BIO SIGNAL
        B                       *IF NOT ZERO LOOP
GO      IN      GO              *DO BOTH INPUT AND OUTPUT
        OUT     WAIT
        ...     X,PA2
        PROCESS Y,PA2
        ...
B       WAIT    DATA            *GET ANOTHER VALUE AND REPEAT
```

FIGURE 3.6. BIO Synchronization

```
            AORG 0
RESET   B     MAIN    *INITIALIZE THE RESET VECTOR
        B     ISR     *INITIALIZE THE INTERRUPT VECTOR
        ...
MAIN    ...
*             INITIALIZE THE AIB, FLAGS AND TABLES AS NEEDED
        ...
        EINT            *ENABLE INTERRUPT FOR THE FIRST TIME
LOOP    OTHER PROCESSING
        ...
WAIT    ZALS  FLAG
        BZ    WAIT    *LOOP UNTIL A SAMPLE ARRIVES
*             DISABLE INTERRUPTS AND PROCESS DATA
        ...
        ZAC             *RESET THE FLAG
        SACH  FLAG
        B     LOOP
* INTERRUPT SERVICE ROUTINE
ISR     IN    DATA,2  *DO I/O
        OUT   DATA,2
        LACK  1         *SET FLAG
        SACL  FLAG
        EINT            *ENABLE INTERRUPT
        RET             *RETURN
```

FIGURE 3.7. Interrupt synchronization

2. Interrupts are not enabled (with the EINT) until the AIB is set up.

3. The interrupt service routine (ISR) does both the input and output.

4. Interrupts must be reenabled after each one is received. (See the EINT instruction just before the RET instruction.)

5. A flag is used to give the main program the status of the I/O.

Example 3.1   Loop Program of Chapter 1 Revisited.   The LOOP program in Figure 1.7 is shown again in Figure 3.8. Follow the procedure in Chapter 1 to connect the SWDS to the analog interface board (AIB). Input a sinusoid to the analog port on the AIB and observe the resulting delayed output of the same input frequency.

Procedure. Edit/create the source program and save it as LOOP.ASM. Assemble the source file, and load the resulting object file LOOP.MPO into the SWDS. Access the debug monitor and observe the features of the LOOP program:

1. a. An absolute origin directive of 20 starts the program at address 20 (hex).

b. The values of both the RATE and MODE (specified as labels) are defined in program memory address (PMA) 20 and 21 (hex).

c. The RATE is calculated using

$$\text{RATE} = \frac{\text{clock frequency}}{\text{sampling frequency}} - 1$$

where the clock frequency of the TMS320C25 is 10 MHz and the desired sampling frequency is 10 kHz. The MODE $>$FA (or $>$A) initializes the AIB as described in the *Analog Interface Board User's Guide* and discussed in Section 3.2.

```
0001                        * LOOP PROGRAM
0002                        * INPUT INTO AIB IS OUTPUT - TEST FOR AIB
0003 0000                             AORG    0
0004 0000 FF80                        B       START   *INIT. RESET VECTOR
     0001 0022
0005 0020                             AORG    >20     *START AT PMA 20
0006                        * INITIALIZE AIB
0007 0020 03E7              RATE       DATA    999     *N=(10 MHZ/fs)-1 WITH fs=10 KHZ
0008 0021 00FA              MODE       DATA    >FA     *MODE FOR AIB
0009                        *
0010 0022 C800              START      LDPK    0       *INIT DATA PAGE POINTER TO 0
0011 0023 CA20                         LACK    RATE    *GET SAMPLE RATE INTO ACC
0012 0024 5860                         TBLR    >60     *TRANFER FROM PROG MEM TO DATA MEM 60
0013 0025 E160                         OUT     >60,1   *OUTPUT RATE TO AIB PORT 1
0014 0026 CA21                         LACK    MODE    *GET MODE INTO ACC
0015 0027 5860                         TBLR    >60     *TRANSFER FROM PROG TO DATA MEM
0016 0028 E060                         OUT     >60,0   *OUTPUT MODE TO AIB PORT 0
0017 0029 E360                         OUT     >60,3   *DUMMY OUTPUT TO AIB PORT 3
0018 002A FA80              WAIT       BIOZ    ALOOP   *WAIT FOR END OF CONVERSION
     002B 002E
0019 002C FF80                         B       WAIT    *LOOP BACK
     002D 002A
0020 002E 8260              ALOOP      IN      >60,2   *READ DATA FROM PORT 2
0021 002F E260                         OUT     >60,2   *WRITE DATA OUT TO PORT 2
0022 0030 FF80                         B       WAIT    *BRANCH TO CONTINUE
     0031 002A
0023                                   END
NO ERRORS, NO WARNINGS
```

FIGURE 3.8. LOOP program (LOOP.LST)

2. LDPK 0 selects page 0, where on-chip data memory (DM) block B2 is located.

3. TBLR DMA performs a table read instruction (TBLR), which transfers into DM (specified by DMA) the content of program memory (PM), with the PMA loaded previously into the accumulator. The RATE value is then sent to port 1 and the MODE value to port 0. The output instruction (OUT) can write the content of DM to one of the 16 possible ports supported by the TMS320C25. The MODE value sent to port 0, puts the AIB in the transparent mode. Port 1 is the timer counter port and port 2 is the data port.

4. BIOZ is to branch if the I/O flag equals zero. The BIO control pin on the TMS320C25 connected to the A/D end-of-conversion (EOC) pin is effectively polled to test its status. The TMS320C25 loops until a conversion takes place and the BIO flag goes to zero.

5. IN >60,2 reads a 16-bit value from the A/D (or a peripheral) on port address 2 and stores this value to DMA >60. The data are then written, with the OUT instruction, to port 2. The last two steps are repeated continuously with new input samples.

Example 3.2 Triangular Waveform Generation.   The program listed in Figure 3.9 generates a triangular waveform. Run this program and observe the output waveform shown in Figure 3.10. Notice the following features of the program:

1. The peak-to-peak amplitude depends on the step size and the number of cycle times (set in PMA >31) the step is added or subtracted from the output value.

2. The output frequency, 500 Hz, is proportional to the sampling frequency set at 10 kHz and inversely proportional to the number of cycle times. The initial

```
0001                    * TRIANGULAR WAVEFORM GENERATION
0002                    * NO EXTERNAL INPUT OR DATA INITIALIZATION REQUIRED
0003         0060  MODE    EQU    >60        MODE FOR AIB >A
0004         0061  RATE    EQU    >61        RATE = (10 MHZ/fS) -1 = 999
0005         0062  TRWAVE  EQU    >62        OUTPUT STORAGE FOR WAVE GENERATION
0006         0063  STEP    EQU    >63        STORAGE OF STEP INCREMENT VALUE
0007 0000               AORG   0
0008 0000 FF80         B      START
     0001 0025
0009 0020               AORG   >20        PROG MEM START AT PMA 20
0010 0020 000A  TABLE   DATA   >A         MODE VALUE FOR AIB
0011 0021 03E7         DATA   999        SAMPLE RATE = 10 KHZ
0012 0022 4000         DATA   16384      STARTING POINT OF WAVEFORM
0013 0023 F333         DATA   -3277      INITIAL STEP VALUE(WILL CHANGE SIGN)
0014 0024 5500         DATA   >5500      NOP-1st INSTRUCT SHOW REVERSED ASSEM
0015 0025 CA20  START   LACK   TABLE      ADDRESS OF DATA TABLE IN ACC
0016 0026 C003         LARK   AR0,03     AR0 =3 AS LOOP COUNTER
0017 0027 C160         LARK   AR1,MODE   AR1=DMA OF MODE
0018 0028 5589  LOP     LARP   AR1        SELECT AR1 FOR INDIRECT ADDRESSING
0019 0029 58A8         TBLR   *+,AR0     TRANSFER PMA 20-23
0020 002A CC01         ADDK   01         INTO DMA 60-63
0021 002B FB90         BANZ   LOP        IF AR0<>0 CONTINUE TRANSFER
     002C 0028
0022 002D E060         OUT    MODE,0     OUTPUT MODE = >A TO PORT 0
0023 002E E161         OUT    RATE,1     OUTPUT RATE = 999 TO PORT 1
0024 002F E361         OUT    RATE,3     DUMMY OUTPUT TO PORT 3
0025 0030 5588         LARP   AR0        SELECT AR0
0026 0031 C009  CYCLE   LARK   AR0,9      ADD/SUBTRACT 10 TIMES WITH LOOP
0027 0032 FA80  LOOP    BIOZ   LOOP1      BRANCH IF BIO LOW
     0033 0036
0028 0034 FF80         B      LOOP       IF NOT JUMP BACK AND WAIT
     0035 0032
0029 0036 E262  LOOP1   OUT    TRWAVE,2   OUTPUT CURRENT WAVE VALUE TO PORT 2
0030 0037 2062         LAC    TRWAVE     CURRENT WAVE VALUE IN ACC
0031 0038 0063         ADD    STEP       ADD STEP VALUE IN DMA 63
0032 0039 6062         SACL   TRWAVE     STORE NEXT VALUE OF WAVE
0033 003A FB90         BANZ   LOOP       IF AR0<>0 GO BACK WAIT FOR BIO LOW
     003B 0032
0034 003C 3C63         LT     STEP       CHANGE SIGN OF STEP SIZE
0035 003D BFFF         MPYK   -1         BY MULTIPLYING BY -1
0036 003E CE14         PAC               PRODUCT REGISTER RESULT INTO ACC
0037 003F 6063         SACL   STEP       RESTORE INTO STEP SIZE LOCATION
0038 0040 FF80         B      CYCLE      BACK FOR 10 LOOPS IN OTHER DIRECTION
     0041 0031
0039                    END
NO ERRORS, NO WARNINGS
```

FIGURE 3.9. Triangular waveform generation program (RAMP.LST)

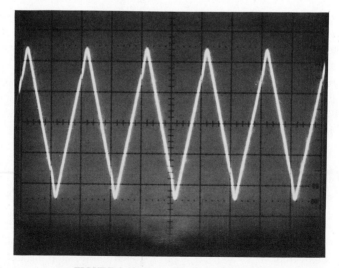

FIGURE 3.10. Triangular Waveform output

(starting) value of the waveform stored in DMA >62 can be used to "offset" the waveform.

3. Changing the number of cycle times at PMA >31 from 9 to 4 will double the output frequency of the waveform.

Example 3.3   Pseudorandom-Noise-Generator Using Interrupt.   The program in Figure 3.11 generates a pseudorandom-noise (PRN). The timing of each random noise sample is controlled using the TMS320C25 interrupt timer. Figure 3.12 shows the software implementation of a 32-bit-wide random noise sequence. Run this program and verify that the output spectrum from the AIB, taken before the D/A output filter, is as shown in Figure 3.13 (using a signal analyzer). The output rate of each noise sample is determined by the rate of the on-chip timer interrupt, set at PMA >22, which corresponds to a frequency of 16,393 Hz. Note the notch in Figure 3.13 at the sampling frequency.

Single-step through the program in Figure 3.11 and verify the following:

1. At PMA >26, the ACC is loaded with the content of PMA >20, containing the upper "seed" number, which is then transferred into DMA >63. The lower 16-bit seed number is transferred into DMA >64. DMA >61 and >62 are loaded with >1000, the chosen scaled positive number and F000 the chosen scaled negative number, respectively.

2. EINT enables the interrupt. When single-stepping, reset the PC to >3A in order to "jump over" the enable interrupt instruction (EINT), and START at PMA >3A.

3. Starting at PMA >3A, the ACC is loaded with the constant >7E52; then, masking with >8000, bit 15 (a zero on the first time through) is selected. Bit 14 is then added to bit 15, then bit 12 is added to that sum. At PMA >40, bit 1 is then added to the previous sum in the ACC, and again bit 15 is selected. The section at PMA >3A-4D implements the block diagram of Figure 3.12. The sequence above is performed in the lower 16 bits of the accumulator for convenience, although this is a 32-bit noise-generation implementation. The instructions at PMA >47 and >48 provide for a new 32-bit seed number which will be used in conjunction with the next random noise bit generated. The resulting bit 15 is stored at DMA >66. If the resulting random bit is a 1, it is scaled as >1000 and is output. If the random bit is zero, the output would be negative >F000. Thus the output is a two-level random waveform.

4. The instruction EINT at PMA >52 allows the processor to be interrupted again. Interrupts must be reenabled after each occurrence.

5. The output rate is determined by the number N, which is written to the on-chip timer register at location 3.

$$N = \frac{f_{clk}}{f_s} - 1$$

```
0001                    * PSEUDO-RANDOM NOISE GENERATOR
0002                    * 32-BIT MAXIMUM LENGTH SHIFT REGISTER
0003                    * USE OF INTERRUPT
0004                            IDT     'NOISE'
0005                    *       DATA MEMORY ALLOCATION/DEFINITION
0006        0060  STAT0    EQU     >60           * STATUS REG STORAGE FOR ISR
0007        0061  LH1000   EQU     >61           * NOISE GEN POSITIVE SCALER
0008        0062  LHF000   EQU     >62           * NOISE GEN NEGATIVE SCALER
0009        0063  RNUMH    EQU     >63           * RANDOM NUM UPPER 16-BITS
0010        0064  RNUML    EQU     >64           * RANDOM NUM LOWER 16-BITS
0011        0065  MASK15   EQU     >65           * BIT-15 MASK
0012        0066  TEMP     EQU     >66           * SCRATCH PAD MEMORY AREA
0013        007E  SWDS1    EQU     >7E           * RESERVED FOR SWDS USAGE
0014        007F  SWDS2    EQU     >7F           * RESERVED FOR SWDS USAGE
0015                    *       MEMORY MAPPED REGISTER LOCATIONS
0016        0002  TIMER    EQU     >0002         * INTERVAL TIMER REGISTER
0017        0003  PERIOD   EQU     >0003         * PERIOD REGISTER FOR TIMER
0018        0004  INTMSK   EQU     >0004         * INTERRUPT MASK REGISTER
0019                    *
0020 0000              AORG    0
0021 0000 FF80         B       INIT
     0001 0024
0022                    *
0023 0018              AORG    >18                *TINT TIMER INTERRUPT VECTOR
0024 0018 FF80         B       START              *
     0019 003A
0025                    *
0026 0020              AORG    >20                * P.M. STORAGE
0027 0020 7E52  SEEDH  DATA    >7E52              * RANDOM NUM SEED,UPPER 16-BITS
0028 0021 1603  SEEDL  DATA    >1603              * RANDOM NUM SEED,LOWER 16-BITS
0029 0022 0262  RATE   DATA    >0262              * OUTPUT DATA RATE (16393 Hz)
0030 0023 0008  IMASK  DATA    >0008              * ENABLE TIMER INTERRUPT ONLY
0031                    *
0032 0024 CE04  INIT   CNFD                       * CONFIGURE B0 AS DATA MEMORY
0033 0025 C800         LDPK    0                   * POINT TO DATA PAGE 0(forever)
0034 0026 D001         LALK    SEEDH               * POINT TO P.M. STORAGE
     0027 0020
0035 0028 5863         TBLR    RNUMH               * XFER SEEDH TO RNUMH IN D.M.
0036 0029 CC01         ADDK    1                   * INCREMENT READ ADDRESS IN ACC
0037 002A 5864         TBLR    RNUML               * XFER SEEDL TO RNUML IN D.M.
0038 002B CC01         ADDK    1                   * INCREMENT READ ADDRESS IN ACC
0039 002C 5803         TBLR    PERIOD              * XFER RATE TO PERIOD REGISTER
0040 002D CC01         ADDK    1                   * INCREMENT READ ADDRESS IN ACC
0041 002E 5804         TBLR    INTMSK              * XFER IMASK TO MASK REGISTER
0042 002F D001         LALK    >1000
     0030 1000
0043 0031 6061         SACL    LH1000              * INITIALIZE LH1000 TO >1000
0044 0032 CE23         NEG
0045 0033 6062         SACL    LHF000              * INITIALIZE LHF000 TO >F000
0046 0034 D001         LALK    >8000
     0035 8000
0047 0036 6065         SACL    MASK15              * INITIALIZE MASK15 TO >8000
0048 0037 CE00         EINT                        * ENABLE INTERRUPT
0049                    *
0050 0038 FF80  WAIT   B       WAIT                * WAIT HERE UNTIL AN INTERRUPT
     0039 0038
0052                    * Modulo 2 sum of the 32-bit noise word bits 31,30,28,and 17
0053                    * yields a random bit (1 or 0) shifted into the LSB of the
0054                    * noise word, and scaled and output to a D/A converter. Since
0055                    * only the upper 16-bits are involved in determining feedback,
0056                    * the summation is done in lower accumulator for convenience.
0057 003A 2063  START  LAC     RNUMH               * GET RANDOM NUM UPPER 16-BITS
0058 003B 4E65         AND     MASK15              * MASK BIT-15
0059 003C 0163         ADD     RNUMH,1            * SHIFT 1,ADD BIT-14 TO BIT-15
0060 003D 4E65         AND     MASK15              * MASK BIT-15
0061 003E 0363         ADD     RNUMH,3            * ADD BIT-12 TO THE SUMMATION
0062 003F 4E65         AND     MASK15              * MASK BIT-15
0063 0040 0E63         ADD     RNUMH,14           * ADD BIT-1 TO THE SUMMATION
0064 0041 4E65         AND     MASK15              * MASK BIT-15
0065 0042 6966         SACH    TEMP,1             * SHIFT R 15(LEFT 1,RIGHT 16)
0066 0043 4063         ZALH    RNUMH               * RESTORE RNUMH TO UPPER ACCUM
0067 0044 0064         ADD     RNUML               * RESTORE RNUML TO LOWER ACCUM
0068 0045 CE18         SFL                         * SHIFT 32-BIT ACCUM LEFT 1
0069 0046 0066         ADD     TEMP                * ADD FEEDBACK BIT
0070 0047 6863         SACH    RNUMH               * STORE UPPER 16-BIT NOISE WORD
```

FIGURE 3.11. Pseudorandom-noise-generator program (NOISE.LST)

```
0071 0048 6064          SACL    RNUML       * STORE LOWER 16-BIT NOISE WORD
0072 0049 2066          LAC     TEMP        * RESTORE RANDOM BIT INTO ACCUM
0073 004A F680          BZ      MINUS       *
     004B 004F
0074 004C 2061          LAC     LH1000      * SET OUTPUT POSITIVE IF BIT=1
0075 004D FF80          B       OUTPUT      *
     004E 0050
0076 004F 2062  MINUS   LAC     LHF000      * SET OUTPUT NEGATIVE IF BIT=0
0077 0050 6066  OUTPUT  SACL    TEMP        * STORE SCALED OUTPUT VALUE
0078 0051 E266          OUT     TEMP,2      * OUTPUT SCALED NOISE TO D/A
0079 0052 CE00          EINT                * READY FOR NEXT INTERRUPT
0080 0053 CE26          RET                 * RETURN TO WAIT
0081 0054
NO ERRORS, NO WARNINGS
```

FIGURE 3.11. (concluded)

where f_{clk} is the processor clock (10 MHz) and f_s is the desired sampling frequency. The sampling frequency is chosen as 16,393 Hz, which gives a value

$$N = \frac{10\text{MHz}}{16,393} = 610 => 262$$

A sampling frequency (output rate) of 32,786 can be obtained by replacing the RATE value >262 with >131. This value is written to the period register in DMA 3. Note that since the A/D is not used, no initialization of the AIB is necessary.

6. On interrupt, control goes to location PMA >18, where a branch is executed to jump to the START label of the service interrupt routine.

The PRN program in Figure 3.11 will be used in Chapter 5 to provide the input noise into a filter, in lieu of an external noise source.

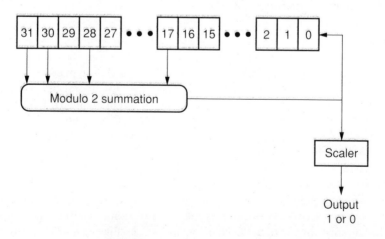

FIGURE 3.12. 32-bit Noise-generator

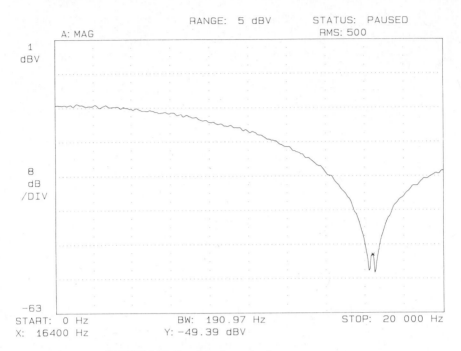

FIGURE 3.13. Pseudorandom-noise frequency response

3.3 ANALOG INTERFACE CHIP

This section is intended to provide the necessary support for adaptive filtering, discussed in Chapter 8. Although it is more difficult to use, the analog interface chip (AIC) provides an inexpensive option to the AIB. For applications requiring two sets of inputs, as in adaptive filtering, the AIC [2,3] is an effective alternative. The AIC functional block diagram is shown in Figure 3.14. The support circuitry shown in Figure 3.15 must be added to the AIC before it can be used with the TMS320C25. The AIC connects to the on-chip serial port of the TMS320C25 chip [2–6]. It can be interfaced to the SWDS board as shown in Figure 3.15. The AIC and support circuitry are enclosed in the box shown in Figure 1.6.

The AIC supports two analog input channels and one analog output channel with several different programmable sample rates (five rates ranging from 19.2 to 7.2 kHz). Both the A/D and D/A have 14 bits of resolution with 10-bit linearity over any 10-bit range. In addition, the chip contains an antialiasing and a reconstruction

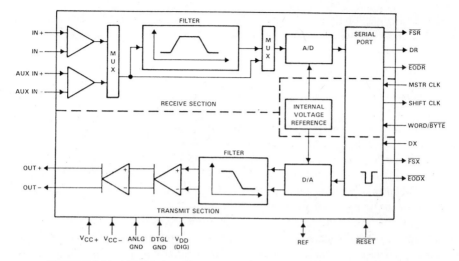

FIGURE 3.14. AIC functional block diagram. (Courtesy of Texas Instruments Inc.)

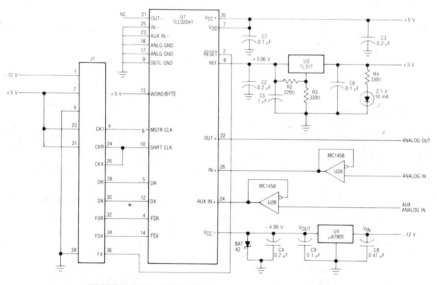

FIGURE 3.15. AIC/TMS320C25 interface circuit diagram

switched capacitor filter. The following list describes each filter's characteristics with a 288-kHz switched capacitor filter (SCF) frequency:

Input filter: bandpass, f_l = 300 Hz, f_h = 3.4 kHz

Output filter: lowpass, f_l = 0 kHz, f_h = 3.4 kHz

The input filter can be bypassed (under program control); however, the output filter is fixed in the circuit. The cutoff frequencies will be frequency-scaled accordingly if the SCF frequency is changed from 288 kHz. All data transmission takes place through two memory-mapped serial port registers, the data receive register (DRR) and the data transmit register (DXR), located at data memory addresses 0 and 1, respectively, as shown in the left side of Figure 3.16.

The serial port is interrupt driven. An interrupt occurs (provided that the interrupt was enabled) when a data word is received, and another interrupt occurs when the device is ready to transmit. Upon receipt of a serial port interrupt, the processor automatically jumps control to PMA >26 for the serial port receive interrupt (RINT) or PMA >28 for the serial port transmit interrupt (XINT), also shown in Figure 3.16. Both of these interrupts can be masked via bits in the interrupt mask register (IMR). Interrupts on the TMS320C25 have priorities 1 (highest) to 7 (lowest). The serial port interrupt priority is 6 for receive interrupt (RINT) and 7 for transmit interrupt (XINT). See Appendix C for the bit definitions of the interrupt mask register and the processor status register that pertain to the AIC.

Controlling the AIC

The AIC is controlled through the serial port transmit register DXR. Its 14 most significant bits are data bits, with the 2 least significant bits left to initiate communication functions. The AIC uses two types of communication: (1) primary and (2) secondary.

The primary communication commands pass the upper 14 bits directly to the D/A so that no data transmissions are missed during a primary communication. The

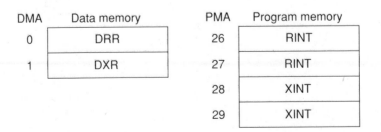

DMA	Data memory
0	DRR
1	DXR

PMA	Program memory
26	RINT
27	RINT
28	XINT
29	XINT

FIGURE 3.16. Registers associated with the serial port

four codes in the 2 least significant bits (LSBs) of the DXR allow the following functions:

$0 = 00$ normal transmission
$1 = 01$ increment/decrement the next clock period (with addition)
$2 = 10$ decrement/increment the next clock period (with subtraction)
$3 = 11$ initiate secondary communication

Normal transmission (0) and initiation of secondary communication (3) are the most frequently used.

The secondary communication mode is used to initialize and control the AIC. Transmission of code 3 in the 2 LSBs of the data word places the AIC in the secondary communication mode, which allows one secondary transmission before automatically switching back to the primary communication mode. The secondary communication word also uses the 2 LSBs to differentiate the control functions. These control functions are initiated by writing to several internal registers in the AIC. The four LSB codes write to the following AIC registers:

$0 = 00$ loads the A transmit and receive frequency divide registers (TA and RA)
$1 = 01$ loads the A' transmit and receive increment and decrement registers (TA'and RA')
$2 = 10$ loads the B transmit and receive frequency divide registers (TB and RB)
$3 = 11$ loads the control register

The letters A and B designate the location of control: A represents a filter control register, and B is associated with the A/D or D/A control. Each of the bits on the four secondary communication control words is defined in Appendix C.

The control register specifies the input port, the mode of operation, gain control, insertion of the bandpass input filter, and activation of the loopback test. The A registers control the frequency of the switched capacitor filters, and the B registers control the sample rate of the A/D and D/A converters. Below are described the location of the values in the control word and the calculation of the values.

TA/RA: A Transmit/Receive Frequency Divide Registers

$$\text{TA} \leftarrow [b13\text{-}b9] \quad \text{RA} \leftarrow [b6\text{-}b2] \quad [b1,b0] = [0,0]$$

These registers divide the master clock (10 MHz) down to an intermediate frequency, which is then divided by 2 to give the LP/BP switched capacitor filter frequency (SCFF). An example of an A-value calculation follows: Find the A-value that will give a 288-kHz switched capacitor frequency (SCFF) if the system clock is 10.368 MHz.

$$A\text{-value} = \frac{\text{system frequency}}{\text{SCFF} \times 2}$$

$$A\text{-value} = \frac{10.368 \text{ MHz}}{288 \text{ kHz} \times 2} = 18$$

TB/RB:B Transmit/Receive Frequency Divide Registers

$$\text{TB} \leftarrow [b14\text{–}b9] \qquad \text{RB} \leftarrow [b7\text{–}b2], \qquad [b1, b0] = [1, 0]$$

These registers divide the switched capacitor filter frequency down to give the A/D and D/A conversion rate for the transmit and receive channels respectively. An example of a B-value calculation follows: Find the B-value that will give an 8-kHz conversion rate, assuming a SCFF of 288 kHz.

$$B\text{-value} = \frac{\text{SCFF}}{\text{sample frequency}}$$

$$B\text{-value} = \frac{288 \text{ kHz}}{8 \text{ kHz}} = 36$$

Note that the system clock on the SWDS is exactly 10 MHz. If the values just calculated are used, they will result in slightly lower frequencies.

Several macros are available that make the AIC easier to use:

1. INIT SR,FSR,CTL,IM,PTAB   initializes the AIC.
2. INN X   outputs data to the AIC.
3. OUTT Y   inputs data from the AIC.
4. OUTS Y,CTL   outputs data to AIC, switches to specified input (IN or AUX IN).

Appendix C contains a complete listing of these macros as well as a brief description of each.

AIC Programming Examples

Example 3.4   AIC Loop Program. Since the AIC LOOP program uses some features that have not been discussed previously, a few words about producing an executable program are in order. This program uses the macros mentioned previously. For the TI assembler, macros called external to the assembly file must reside in the default directory and must be declared in the program by the directive

```
TASK AICL
PROGRAM 32
DATA >60
INCLUDE AICL.MPO
INCLUDE INITS.MPO
END
```

FIGURE 3.17. Linker control file for AIC LOOP program

```
MLIB 'C:'
```

Since the macro group calls subroutines in the INITS.MPO file, it must also be resident in the default directory. The use of external subroutines and data segments dictates the need for a linker to produce an executable file with a .LOD extension. The linker needs a control file to direct it (see Figure 3.17). Lines 2 and 3 in the control file tell the linker where to place the program and the data segments, respectively. Lines 4 and 5 tell the linker which files to merge or link.

The linker is invoked by typing

```
LINK <ENTER>
```

at the command prompt in the SWDS. The linker will prompt you for a control file name. The default assumes a .CTL extension.

Data segments provide a more convenient and foolproof way of specifying data memory locations. Data segments are created by an assembler directive as shown in the programming section:

```
        DSEG    *DATA SEGMENT DIRECTIVE
A       BSS 1   *RESERVE ONE DATA MEMORY LOCATION
                *AND ASSIGN THE LABEL A TO IT
B       BSS 5   *RESERVE 5 DATA MEMORY LOCATIONS
                *AND ASSIGN THE LABEL B TO THE
                *FIRST ONE
        DEND    *DATA SEGMENT END DIRECTIVE
```

The data segment is delimited by DSEG and DEND. This directive relieves the programmer of the task of assigning a data memory address to each label. Instead, this task is left to the linker.

A brief explanation of the AIC LOOP program in Figure 3.18 follows, with each description number corresponding to a section or line in the program.

1. A macro call to INIT initializes the transmit interrupt vector, the processor, and the AIC. INIT uses the following argument values:

   a. CSR:control register value

   CSR = >18 = 011000

   The CSR register is embedded in the DXR and receive the values shown:

```
*******************************************************************
*    =====================================================        *
*                  AIC    LOOP  TEST                               *
*    =====================================================        *
*                                                                 *
* AICL.ASM--INPUTS A VALUE FROM THE IN PORT AND PUTS IT            *
*           DIRECTLY OUT. IF YOU PUT A SINE WAVE ON THE INPUT,     *
*           YOU WILL SEE A SINE WAVE ON THE OUTPUT                 *
*           HAVE SOMETHING TO SEE                                  *
* CALLS--- [SUBROUTINES: INIT                                      *
*           INTERRUPT ROUTINES: XINIT SUPPLIED WITH INITS.MPO      *
*           MACROS: [INIT,INN,OUTT]--DEFAULT DIR                   *
*******************************************************************
            IDT     'MAIN'
            MLIB    'C:'                    LOCATION OF THE MACROS

**** SETUP PARAMETERS FOR AIC************
CSR      EQU       >0063                 SELECT: PRIMARY INPUT,SYNC MODE,
*                                        FS=+3 VOLTS TO -3 VOLTS
SRATE    EQU       >4892                 SAMPLE RATE = 7.7 kHZ
FLTCLK   EQU       >2448                 SWITCHED CAP CLK RATE= 278 kHZ


**** INPUT STORAGE**********
            DSEG
X           BSS 1                         INPUT STORAGE
            DEND

**** JUMP TO ENTRY POINT
            AORG 0
            B    START

            PSEG

**** SETUP PROCESSOR AND AIC
START    INIT CSR,SRATE,FLTCLK,TABLE

**** MAIN PROCESSING LOOP
AGAIN
            IDLE                          WAIT ONE SAMPLE TIME
            INN   X                       INPUT A VALUE
            OUTT  X                       OUTPUT IT DIRECTLY

            B        AGAIN                REPEAT
            PEND
            END
```

FIGURE 3.18. AIC LOOP program using MACROS (AICLOOPM.ASM)

DXR $\rightarrow$ 16-bit word $\rightarrow$ 0000 0000(CSR)11 = >63

Loopback = disabled   [b3] = [0]

Input = primary input [b4] = [0]

Mode = synchronous   [b5] = [1]

Full Scale = ± 3 V   [b7,b6] = [01]

b. FLTCLK:switched capacitor filter clock

TA = [b13–b9] = >12,   RA = ([b6–b2]× 2) = >12,   [b1,b0] = [00]

TA and RA are embedded in DXR as shown:
DXR $\rightarrow$ 16-bit word $\rightarrow$ 00(TA)00(RA)00 = >2448

$$SCFF_T = \frac{10 \text{ MHz}}{[b13\text{--}b9] \times 2} = \frac{10 \text{ MHz}}{18 \times 2} = 278 \text{ kHz}$$

$$SCFF_R = \frac{10 \text{ MHz}}{[b6\text{--}b2] \times 2} = \frac{10 \text{ MHz}}{18 \times 2} = 278 \text{ kHz}$$

c. SRATE:conversion sample rates

TB = [b14--b9] = >24   RB = [b7--b2] = >24   [b1,b0] = [10]

TB and RB are embedded in DXR as shown:

$$DXR \rightarrow 16\text{-bit word} \rightarrow 0(TB)0(RB)10 = >4892$$

$$SR_T = \frac{278 \text{ kHz}}{[b14\text{--}b9]} = \frac{278 \text{ kHz}}{36} = 7.72 \text{ kHz}$$

$$SR_R = \frac{278 \text{ kHz}}{[b7 - b2]} = \frac{278 \text{ kHz}}{36} = 7.72 \text{ kHz}$$

2. IDLE, the first instruction of the main processing loop, ensures that the AIC is ready to transmit and receive. It enables the interrupts and waits until an interrupt occurs before executing the next instruction. Since the AIC's transmit and receive are synchronized to the sample rate clock, the IDLE takes one sample time.

3. INN inputs data from the AIC and stores them in the data memory location X.

4. OUTT outputs the data just stored in X to the AIC. Only one idle is needed for both the INN and the OUTT since the AIC is programmed so that both transmit and receive are ready at the same time.

5. Repeat steps 3 and 4.

The procedure and requirements to run the LOOP program, shown in Figure 3.18, are discussed in Section C.2 of Appendix C. A LOOP program without macros is shown in Figure 3.19.

Example 3.5 AIC Two-Input Program. This example demonstrates the use of two-input ports with the AIC. Voltages of $+3$ and -3 are, respectively, connected to the IN port and the AUX IN port of the AIC with the oscilloscope monitoring the output port as shown in Figure 3.20. The program that drives this hardware (see Figure 3.21) produces a square-wave output. The setup and initialization are identical to the previous AIC LOOP example. The main processing loop performs the following steps:

1. The first IDLE synchronizes both the INN and OUTS operations which immediately follow.

```
0001                    * LOOP PROGRAM USING AIC
0002                    * TLC32041 AIC - TMS320C25 COMMUNICATION PROGRAM
0003          0060  INPUT    EQU    >60      STORAGE FOR INPUT SAMPLES
0004          0061  OUTPUT   EQU    >61      STORAGE FOR OUTPUT SAMPLES
0005 0000                    AORG   0
0006 0000 FF80             B      INIT     BRANCH TO INITIALIZE TMS320C25
     0001 002C
0007 001A                    AORG   >1A      RECEIVE INTERRUPT (RINT) VECTOR
0008 001A FF80             B      RINT     BRANCH TO RECEIVE INTERRUPT ROUTINE
     001B 0029
0009 001C                    AORG   >1C      TRANSMIT INTERRUPT (XMIT) VECTOR
0010 001C FF80             B      XINT     BRANCH TO XMIT INTERRUPT ROUTINE.
     001D 0026
0011 0020                    AORG   >20      START PROGRAM AT >20
0012                    *SECONDARY COMMUNICATION XMIT DATA TABLE FOR AIC REGISTERS
0013 0020 0003  TABLE    DATA   >03      XMIT DATA FOR SECONDARY COM (TWO LSB=1)
0014 0021 0067  CONTRL   DATA   >0067    CONTROL REG XXXXXXX01100111,IN FILTER
0015                    *                         LOOPBACK/AUX INPUTS OFF,SYNC,GAIN 1
0016 0022 0003           DATA   >03      GET READY FOR NEXT SECONDARY COMM
0017 0023 4892  RATE     DATA   >4892    TB=RB=24,RATE =277.8K/REG.CONTENTS=7.7K
0018 0024 0003           DATA   >03      GET READY FOR NEXT SECONDARY COMM
0019 0025 2448  SET      DATA   >2448    TA=RA=12,DIVIDE 10 MHZ CLOCK FOR BP/LP
0020                    * TRANSMIT INTERRUPT ROUTINE
0021 0026 2061  XINT     LAC    OUTPUT   LOAD ACC WITH OUTPUT SAMPLE.
0022 0027 6001           SACL   >01      STORE IT IN DXR ,DMA 1
0023 0028 CE26           RET             RETURN FOR MORE SECONDARY DATA.
0024                    * RECEIVE INTERRUPT ROUTINE
0025 0029 2000  RINT     LAC    >0       LOAD ACC WITH INPUT SAMPLE
0026 002A 6060           SACL   INPUT    STORE IN DMA 60
0027 002B CE26           RET             RETURN FOR MORE PRIMARY DATA.
0028                    * INITIALIZE PROCESSOR
0029 002C D001  INIT     LALK   >03E0    INIT ST1,FRAME SYNC PULSES AS INPUT
     002D 03E0
0030 002E 6068           SACL   >68      TXM=0,XF=0(LOW) CONN TO AIC(RESET)PIN 2
0031 002F 5168           LST1   >68      FSM=1 FRAME SYNC PULSES ARE NOT IGNORED
0032 0030 CA00           ZAC             SET ACC=0
0033 0031 6001           SACL   >01      CLEAR DXR (DATA XMIT REG) IN DMA 01
0034 0032 6061           SACL   OUTPUT   INIT OUTPUT TO 0
0035 0033 CA20           LACK   >20      LOAD IMR(INTER MASK REG)= 0010 0000 TO
0036 0034 6004           SACL   >04      MASK ALL INTER EXCEPT 'XINT', IN DMA 4
0037 0035 5588           LARP   0        SELECT AR0
0038 0036 C062           LARK   AR0,>62  TRANSFER XMIT DATA
0039 0037 CB05           RPTK   >05      FROM PMA 20-25
0040 0038 FCA0           BLKP   TABLE,*+ INTO DMA 62-67
     0039 0020
0041                    * INITIALIZE AIC REGISTERS   ( SECONDARY COMMUNICATION )
0042 003A C105           LARK   AR1,05   SELECT AR1 AS LOOP COUNTER
0043 003B C062           LARK   AR0,>62  AR0 POINTS AT 62
0044 003C CE0D           SXF             DISABLE AIC RESET WITH EXT FLAG XF=1
0045 003D CE1F  SCND     IDLE            ENABLE AND WAIT FOR XMIT INTER(INTM=0)
0046 003E 20A9           LAC    *+,0,AR1 TRANSFER DATA STARTING AT DMA 62
0047 003F 6061           SACL   OUTPUT    FOR OUTPUT TO THE AIC
0048 0040 FB98           BANZ   SCND,*-,AR0  BRANCH BACK IF AR1 NOT 0,DEC AR1
     0041 003D
0049 0042 CE1F           IDLE             WAIT FOR LAST XMIT DATA
0050                    *
0051 0043 CA10           LACK   >10      LOAD IMR TO MASK ALL INTER EXCEPT RINT
0052 0044 6004           SACL   >04      STORE INTO DMA 4
0053 0045 2061  OUT      LAC    OUTPUT   LOAD OUTPUT SAMPLE IN THE ACC
0054 0046 D004           ANDK   >FFFC    MASK 2 LEAST SIG BITS SINCE 14 BITS D/A
     0047 FFFC
0055 0048 6001           SACL   >01      STORE IN DATA XMIT REG. DXR IN DMA 1
0056 0049 CE1F           IDLE             ENABLE AND WAIT FOR A RECEIVE INTERRUPT
0057                    * MAIN PROGRAM
0058 004A 2060           LAC    INPUT    LOAD ACC WITH INPUT SAMPLE FROM THE AIC
0059 004B 6061           SACL   OUTPUT   STORE ACC WITH OUTPUT SAMPLE TO AIC
0060 004C FF80           B      OUT      BRANCH TO WAIT FOR DATA TRANSFER.
     004D 0045
0061                    END
NO ERRORS, NO WARNINGS
```

FIGURE 3.19. AIC LOOP program without MACROS (AICLOOP.LST)

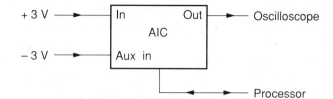

FIGURE 3.20. Configuration for the two-input test

```
************************************************************************
*                  AIC   TWO INPUT TEST PROGRAM                        *
*                                                                      *
* AICSQ--- MULTIPLEXES THE PRIMARY INPUT WITH THE AUXILARY INPUT,      *
*          THEN OUTPUTS THEM ALTERNATELY. IF CONSTANTS OF +3 AND       *
*          -3  ARE CONNECTED TO THE INPUTS, A 3 VOLT SQUARE WAVE       *
*          WILL RESULT                                                 *
* CALLS--- SUBROUTINES: INITS                                          *
*          INTERRUPT ROUTINES: XINIT--INITS.MPO                        *
*          MACROS: INIT,INN,OUTT,OUTS--DEFAULT DIR                     *
************************************************************************
         IDT     'MAIN'
         MLIB    'C:'            LOCATION OF MACROS
********SETUP CONSTANTS*************
CSR      EQU     >0063           SELECT PRIMARY INPUT,SYNC MODE
*                                FULL SCALE = +3 VOLTS TO -3 VOLTS
SRATE    EQU     >4892           SAMPLE RATE =7.7 KHZ
FLTCLK   EQU     >2448           SWITCHED CAPACITOR CLK RATE=278 KHZ
AUX      EQU     >73             SELECT AUX INPUT
PRI      EQU     >63             SELECT PRIMARY INPUT
********INPUT STORAGE *************
         DSEG
X        BSS 1                   PRIMARY INPUT
XA       BSS 1                   AUX INPUT
         DEND
********JUMP TO BEGINNING OF PROGRAM*************
         AORG 0
         B    START
         PSEG
**** INITIALIZE PROCESSOR AND AIC
START    INIT CSR,SRATE,FLTCLK,TABLE
*
********MAIN PROCESSING LOOP BEGINS*******************
*
AGAIN
**** INPUT TO PRIMARY PORT, OUTPUT X AND SWITCH TO AUX PORT
         IDLE                    WAIT UNTIL PORT IS READY
         INN     X               INPUT FROM PRIMARY PORT
         OUTS    X,AUX           OUTPUT AND SWITCH TO AUX INPUT
**** DELAY WHILE HOLDING THE OUTPUT AT X
         LALK    >7FFF           LOAD A LARGE CONSTANT INTO ACC
DLY1     SUBK    3               DEC BY 3
         BLZ     SKIP1           WAITED LONG ENOUGH?
         B       DLY1            NO, DEC SOME MORE
*
**** INPUT FROM AUX PORT, OUTPUT XA AND SWITCH TO PRIMARY INPUT
SKIP1    IDLE                    IS THE PORT AREADY
         INN     XA              INPUT FROM AUX PORT
         OUTS    XA,PRI          OUTPUT XA AND SWITCH TO PRI INPUT
**** DELAY WHILE HOLDING THE OUTPUT AT XA
         LALK    >7FFF           LOAD ACC WITH A LARGE CONSTANT
DLY2     SUBK    3               DEC BY 3
         BLZ     SKIP2           WAITED LONG ENOUGH?
         B       DLY2            NO, DEC SOME MORE
SKIP2    B       AGAIN           REPEAT AS NEEDED
         PEND
         END
```

FIGURE 3.21. AIC two-input program (AICTWO.ASM)

2. The INN macro inputs a value from the IN port.

3. The OUTS macro outputs the value just brought in. The OUTS macro then switches the input to the AUX port.

4. The output is held at this value for a time determined by the delay loop, thus producing one-half of the square wave. The delay loop used just decrements the accumulator to zero and then falls through to the next IDLE instruction.

5. The INN macro inputs a value from the AUX port.

6. The OUTS repeats the same function as before except that it switches the input back to the IN port.

7. Another delay is executed and then the sequence is repeated.

Exercise 3.1   AIC Two-Input Program without Delays. Replace the delays in the program shown in Figure 3.21 with IDLE instructions and run it. Observe the output waveform. Explain why you get this waveform. Remove the IDLE instructions one at a time. What happens? Explain.

REFERENCES

[1] *TMS32010 Analog Interface Board User's Guide*, Texas Instruments Inc., Dallas, Tex., 1984.

[2] *TLC32040I, TLC32040C, TLC32041I, TLC32041C Analog Interface Circuits, Advanced Information*, Texas Instruments Inc., Dallas, Tex., September 1987.

[3] *Second-Generation TMS320 User's Guide*, Texas Instruments Inc. , Dallas, Tex., 1987.

[4] *TMS320C1x/TMS320C2x Assembly Language Tools User's Guide*, Texas Instruments Inc., Dallas, Tex., 1987.

[5] *TMS320C2x Software Development System User's Guide*, Texas Instruments Inc., Dallas, Tex., 1988.

[6] *TMS320C1x/TMS320C2x Source Conversion Reference Guide*, Texas Instruments Inc., Dallas, Tex., 1988.

[7] R. Chassaing, "A Senior Project Course on Applications in Digital Signal Processing with the TMS320," *1988 ASEE Annual Conference Proceedings*, Vol. 1.

[8] K. S. Lin (editor), *Digital Signal Processing Applications with the TMS320 Family*, Vol. 1, Prentice-Hall, Englewood Cliffs, N.J., 1988.

[9] P. H. Garrett, *Analog I/O Design: Acquisition–Conversion–Recovery*, Reston, Va., 1981.

4

Introduction to the Z-Transform and Difference Equations

CONCEPTS AND PROCEDURES

* *Representation of discrete-time functions*
* *The z-domain representation of some common discrete functions, such as the unit impulse, the unit step, and the sinusoidal.*
* *Implementation of a difference equation with the TMS320C25*

In this chapter we introduce the Z-transform (ZT)[1–6], which is a generalization of the Fourier transform. While the Laplace transform is a tool used for the analysis of continuous-time signals, the ZT is used for discrete-time signals. In the last two decades, advances in technology have increased the importance of discrete-time systems, and hence the need for the analysis of such systems with the ZT. Mapping from the s-plane to the z-plane is discussed, and several examples are provided, including the ZT of a sinusoidal signal.

The inverse Z-transform (IZT) is also introduced in conjunction with difference equations. A difference equation, representing a digital oscillator, is presented. A program that implements the digital oscillator is discussed and demonstrates how sinusoidal waveforms of different frequencies can be generated. The oscillator program example provides a building block for future projects, such as modulation and measurement of pressure, discussed in Chapter 9.

4.1 INTRODUCTION

While an analog signal $x(t)$ can assume a continuous range of values, a discrete time signal $x(n)$ is represented by a sequence of numbers, defined only for integer values of n. Consider a continuous signal $x(t)$ ideally sampled and represented by

63

$$x_s(t) = \sum_{n=o}^{\infty} x(t)\delta(t - nT)$$

or

$$x_s(t) = x(t)\delta(t) + x(t)\delta(t - T) + x(t)\delta(t - 2T) + \cdots \qquad (4.1)$$

where $T = 1/f_s$ is the sampling interval and $\delta(t - nT)$ is the impulse function $\delta(t)$ delayed by nT and is nonzero only at $t = nT$. The Laplace transform of the sampled sequence is a sequence of exponentials,

$$X_s(s) = \int_0^{\infty} [x(t)\delta(t) + x(t)\delta(t - T) + x(t)\delta(t - 2T) + \cdots]e^{-st}dt \qquad (4.2)$$

Using $\int_0^{\infty} f(t)\delta(t - T)dt = f(T)$, equation (4.2) reduces to

$$X_s(s) = x(0) + x(T)e^{-sT} + x(2T)e^{-2sT} + \cdots = \sum_{n=0}^{\infty} x(nT)e^{-nsT} \qquad (4.3)$$

The term e^{-nsT} represents a pure time delay. If we let $z = e^{sT}$, then

$$X(z) = \sum_{n=0}^{\infty} x(nT)z^{-n} \qquad (4.4)$$

With the sampling interval T implied, $x(nT)$ can be represented by $x(n)$, and

$$X(z) = Z[x(n)] = \sum_{n-0}^{\infty} x(n)z^{-n} \qquad (4.5)$$

is the Z-transform (ZT) of $x(n)$. The ZT is a unique transformation; that is, there is a one-to-one correspondence between $x(n)$ and $X(z)$. We will now find the Z-transform of each of the following sequences.

Example 4.1 *Unit Sequence* $x(n) = 1, n \geq 0$. The Z-transform of $x(n)$ is

$$X(z) = \sum_{n=0}^{\infty} x(n)z^{-n} = \sum_{n=0}^{\infty} (1)z^{-n} = \sum_{n=0}^{\infty} (z^{-1})^n$$

Use the geometric series, $\sum_{n=0}^{\infty} u^n = 1/(1 - u)$ for $|u| < 1$, to reduce the expression for $X(z)$ to

$$X(z) = \frac{1}{1 - z^{-1}} = \frac{z}{z - 1} \qquad \text{for } |z^{-1}| < 1 \quad \text{or} \quad |z| > 1$$

Example 4.2 *Exponential Sequence* $x(n) = e^{-knT}$, $n \geq 0$. The ZT of $x(n)$ is

$$X(z) = \sum_{n=0}^{\infty} e^{-knT} z^{-n} = \sum_{n=0}^{\infty} (e^{-kT} z^{-1})^n = \frac{1}{1 - e^{-kT} z^{-1}}$$

$$= \frac{z}{z - e^{-kT}} \quad \text{for} \quad |e^{-kT} z^{-1}| < 1 \quad \text{or} \quad |z| > |e^{-kT}|$$

Example 4.3 *Sinusoidal Sequence* $x(n) = \sin n \omega T$. Using the identity $\sin u = (e^{ju} - e^{-ju})/(2j)$, the ZT of $x(n)$ can be written in terms of exponentials,

$$X(z) = \frac{1}{2j} \left[\sum_{n=0}^{\infty} (e^{+jn\omega T} - e^{-jn\omega T}) z^{-n} \right]$$

$$= \frac{1}{2j} \left[\sum_{n=0}^{\infty} (e^{j\omega T} z^{-1})^n - (e^{-j\omega T} z^{-1})^n \right]$$

Using the results of Example 4.2, with $-k = j\omega$ in the first summation and $k = j\omega$ in the second, yields

$$X(z) = \frac{1}{2j} \left(\frac{z}{z - e^{j\omega T}} - \frac{z}{z - e^{-j\omega T}} \right)$$

$$X(z) = \frac{1}{2j} \left(\frac{z^2 - ze^{-j\omega T} - z^2 + ze^{j\omega T}}{z^2 - z(e^{-j\omega T} + e^{j\omega T}) + 1} \right) = \frac{z \sin \omega T}{z^2 - 2z \cos \omega T + 1}$$

which is valid for $|z| > 1$. Substituting $C = \sin \omega T$, $A = 2 \cos \omega T$, and $B = -1$ produces a cleaner-looking expression:

$$X(z) = \frac{Cz}{z^2 - Az - B}$$

4.2 MAPPING FROM s-PLANE TO z-PLANE

Certain values of s have physical meaning and importance in engineering applications. For example, the $j\omega$ axis represents the sinusoidal frequency of ω radians, $j\omega = 0$ represents dc, and so on. What values of z have a similar meaning? To find out, we will consider mapping some points from the s-plane to the z-plane. The equation $z = e^{sT}$ maps regions from the s-plane to the z-plane. Since z can be expressed in terms of the complex variable $s = \sigma + j\omega$,

$$z = e^{\sigma T} e^{j\omega T} \tag{4.6}$$

with magnitude $|z| = e^{\sigma T}$ and phase $\theta = \omega T = 2\pi f / f_s$. Let us consider the mapping of three important regions in the s-plane to the z-plane.

Region 1: $\sigma = 0$

With $\sigma = 0$, equation (4.6) becomes

$$z = e^{0T} e^{j\omega T} = e^{j\omega T}$$

and $|z| = 1$. Thus the imaginary axis in the s-plane corresponds to a unit circle in the z-plane. A system with poles on the imaginary axis in the s-plane represents a sinusoidal oscillator with constant amplitude. Hence a digital oscillator would consist of poles located on the unit circle in the z-plane.

Later, we will program a difference equation to implement a sinusoid and will show that the poles associated with such difference equation lie on the unit circle in the z-plane.

Region 2: $\sigma < 0$

With $\sigma < 0$, $e^{\sigma T} < 1$ and $|z| < 1$. As σ goes from 0^- to $-\infty$, $|z|$ goes from 1^- to 0. The left-half s-plane corresponds to the z-plane region inside the unit circle, with the magnitude of z approaching zero as σ approaches $-\infty$. Systems with poles in the left-half s-plane represent a stable region, with responses that are either a decaying exponential (if real poles) or a decaying sinusoid (if complex poles). Hence poles inside the unit circle in the z-plane yield a stable system.

Region 3: $\sigma > 0$

If $\sigma > 0$ in (4-6), then $e^{\sigma T} > 1$ and $|z| > 1$. As σ goes from 0^+ to ∞, $|z|$ goes from 1^+ to ∞. The right plane in the s-domain corresponds to a region of $|z| > 1$ in the z-domain. Poles in the right-hand s-plane yield a response that is either an increasing exponential (if real poles) or a growing sinusoid (if complex poles). Hence the $|z| > 1$ region corresponds to an unstable region.

The angle of z corresponds to $\theta = \omega T = 2\pi f / f_s$ with $f < f_s / 2$. As f varies from 0 to $\pm f_s / 2$, θ varies from 0 to π. Notice that the z-value is periodic; the circle is completely traversed every $2\pi / T$ change in ω. Figure 4.1 shows the mapping from the s-domain into the z-domain.

4.3 SOLUTION OF DIFFERENCE EQUATIONS USING THE Z-TRANSFORM

A digital filter can be represented by a difference equation. We need to find the ZT of expressions such as $x(n - k)$, since equations containing $x(n - k)$ represent a kth order difference equation. The order of the delay term $x(n - k)$ corresponds to

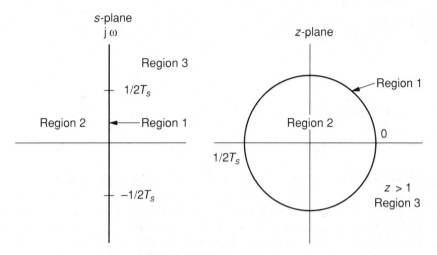

FIGURE 4.1. Mapping from $s \to z$

$d^k x(t)/dx^k$, the kth derivative of the analog signal $x(t)$. Consider the z-transform of $x(n)$.

$$Z[x(n)] = X(z) = \sum_{n=0}^{\infty} x(n)z^{-n} = x(0) + x(1)z^{-1} + x(2)z^{-2} + x(3)z^{-3} + \cdots$$
$$(4.7)$$

Next, consider the Z-transform of $x(n-1)$.

$$Z[x(n-1)] = \sum_{n=0}^{\infty} x(n-1)z^{-n} \qquad (4.8)$$

Considering only a causal system (i.e., $n \geq 0$),

$$Z[x(n-1)] = x(-1) + x(0)z^{-1} + x(1)z^{-2} + x(2)z^{-3} + x(3)z^{-4} + \cdots$$
$$= x(-1) + z^{-1}[x(0) + x(1)z^{-1} + x(2)z^{-2} + x(3)z^{-3} + \cdots]$$
$$= x(-1) + z^{-1}X(z)$$
$$(4.9)$$

$x(-1)$ represents the initial condition necessary on $x(n)$. Similarly,

$$Z[x(n-2)] = \sum_{n=0}^{\infty} x(n-2)z^{-n}$$
$$= x(-2) + x(-1)z^{-1} + x(0)z^{-2} + x(1)z^{-3} + \cdots$$
$$= x(-2) + x(-1)z^{-1} + z^{-2}[x(0) + x(1)z^{-1} + x(2)z^{-2} + \cdots]$$
$$= x(-2) + x(-1)z^{-1} + z^{-2}X(z)$$
$$(4.10)$$

In general,

$$Z[x(n-k)] = z^{-k} \sum_{m=1}^{k} x(-m)z^{m} + z^{-k}X(z)$$

If the initial conditions are zero, the equation above reduces to $Z[x(n-k)] = z^{-k}X(z)$.

4.4 DIGITAL OSCILLATOR FROM THE INVERSE Z-TRANSFORM

Since there is a one-to-one correspondence between $x(n)$ and $X(z)$, given $X(z)$, $x(n)$ can be uniquely recovered using $x(n) = \text{IZT} \{X(z)\}$. For example, with zero initial conditions, the IZT of $z^{-2}X(z)$ is $x(n-2)$, or

$$x(n-2) = \text{IZT}[z^{-2}X(z)]$$

From Example 4.3, the ZT of $x(n) = \sin n\,\omega T$ is

$$X(z) = \frac{Cz}{z^2 - Az - B} \tag{4.11}$$

where $C = \sin \omega T$, $A = 2\cos \omega T$, $B = -1$, $\omega = 2\pi f$, and $T = 1/f_s$. The transfer function $H(z)$ can be defined as

$$H(z) = \frac{Y(z)}{X(z)}$$

where $X(z)$ and $Y(z)$ are the input and output, respectively.
Let the Z-transform of the $\sin n\,\omega T$ be the transfer function $H(z)$,

$$H(z) = \frac{Y(z)}{X(z)} = \frac{Cz}{z^2 - Az - B} = \frac{Cz^{-1}}{1 - Az^{-1} - Bz^{-2}} \tag{4.12}$$

Note that the poles of $H(z)$ are the roots of $z^2 - Az - B = 0$, or

$$P_{1,2} = \frac{A \pm \sqrt{A^2 + 4B}}{2} = \frac{\cos \omega T \pm \sqrt{4\cos^2 \omega T - 4}}{2} = \cos \omega T \pm \sqrt{-\sin^2 \omega T}$$

and

$$P_{1,2} = \cos \omega T \pm j \sin \omega T$$

which represents a pair of complex poles with magnitude 1 and angle ωT. Poles with a magnitude of 1 are associated with the digital oscillator, with frequency determined by the coefficients A, B and C. From a design point of view, the desired frequency ω determines the values of A, B and C. From (4.12),

$$Y(z) - Az^{-1}Y(z) - Bz^{-2}Y(z) = Cz^{-1}X(z) \qquad (4.13)$$

Assuming all initial conditions equal to zero and performing the inverse Z-transform, we have

$$y(n) - Ay(n-1) - By(n-2) = Cx(n-1)$$

or

$$y(n) = Ay(n-1) + By(n-2) + Cx(n-1) \qquad (4.14)$$

which is a second-order difference equation whose unit impulse response is $\sin n\omega T$.

Applying the unit impulse $x(n-1) = 1$, for $n = 1$, to equation (4.14) results in the solution,

$$
\begin{aligned}
n &= 0 \quad y(0) = Ay(-1) + By(-2) + 0 = 0 \\
n &= 1 \quad y(1) = Ay(0) + By(-1) + C = C \\
n &= 2 \quad y(2) = Ay(1) + By(0) + 0 = Ay(1) \\
n &= 3 \quad y(3) = Ay(2) + By(1)
\end{aligned}
$$

$$n = k \quad y(k) = Ay(k-1) + By(k-2) \qquad (4.15)$$

For values of $n > 2$, $y(k)$ can be calculated in terms of $y(k-1)$ and $y(k-2)$. Equation (4.15) represents a recursive (with feedback) difference equation.

Example 4.4 Digital Oscillator. Design a variable oscillator with a sample frequency of 10 kHz. Test it at 1 kHz and 2 kHz. With $f_s = 10$ kHz and $f = 1$ kHz, the constants become,

$$A = 2\cos\omega T = 2\cos 36° \qquad B = -1 \qquad C = \sin\omega T = \sin 36°$$

Scaling the constants by 2^{15} and dividing them by 2 so that their values can be entered in the TMS320C25 produces the numbers

$$
\begin{aligned}
A/2 \times 2^{15} &= 26,510 => 678E \\
B/2 \times 2^{15} &= -16,384 => C000 \\
C/2 \times 2^{15} &= 9,630 => 259E
\end{aligned}
$$

```
0001                    * DIGITAL OSCILLATOR PROGRAM
0002                    * Y(n)=A Y(n-1) + B Y(n-2) + C X(n-1)
0003                    * Y0  =A Y1 + B Y2  FOR N>1 , X(n-1)=1 FOR n=1 (= 0 OTHERWISE)
0004                    * USES INDIRECT ADDRESSING ,  fs = 10 KHZ
0005                    * DATA MUST BE LOADED AT DMA 60-66 BEFORE RUNNING PROGRAM
0006      0060  MODE       EQU   >60       *TRANSPARENT MODE FOR AIB
0007      0061  RATE       EQU   >61       *RATE=(10 MHZ/fs)-1=999=03E7
0008      0062  Y0         EQU   >62       *INITIALLY Y0=0, OUTPUT DMA
0009      0063  Y1         EQU   >63       *Y(1)=C=SIN(2x3.14xFd/fs)
0010      0064  Y2         EQU   >64       *Y(2)=0
0011      0065  A          EQU   >65       *A=2 COS(WT)
0012      0066  B          EQU   >66       *B=-1
0013            *                          *A,B,C DIV BY 2 (MAX VALUE <1)
0014            *                          *A=678E,B=C000,C=259E  Fd=1KHZ
0015            *                          *A=278E,B=C000,C=3CDE  Fd=2KHZ
0016 0000                  AORG  >0
0017 0000 FF80             B     START
     0001 0020
0018 0020                  AORG  >20       *PM START AT >20
0019 0020 E060  START    ① OUT   MODE,0    *OUTPUT MODE = >A TO PORT 0
0020 0021 E161             OUT   RATE,1    *OUTPUT RATE =03E7 TO PORT 1
0021 0022 E361             OUT   RATE,3    *DUMMY OUTPUT TO PORT 3
0022 0023 FA80  WAIT     ② BIOZ  MAIN      *WAIT TIL BIO=0(EOC OF A/D=0)
     0024 0027
0023            *                          *BRANCH TO MAIN AFTER EOC
0024 0025 FF80             B     WAIT      *A/D PROVIDES TIMING
     0026 0023
0025           *MAIN SECTION
0026 0027 C064  MAIN     ③ LARK  0,Y2      *LOAD AUX REG AR0 WITH Y2
0027 0028 C166             LARK  1,B       *LOAD AUX REG AR1 WITH B
0028 0029 CA00             ZAC             *ZERO THE ACCUMULATOR
0029 002A 5588           ④ LARP  0         *SELECT AR0 FOR INDIR ADDR
0030 002B 3C99             LT    *-,1      *TR=Y2,DEC AR0,SELECT AR1
0031 002C 3898             MPY   *-,0      *PR=BxY2,DEC AR1,SELECT AR0
0032 002D 3F89             LTD   *,1       *TR=Y1,ACC=BxY2,MOVE"DOWN"
0033 002E 3888             MPY   *,0       *PR=AxY1
0034 002F CE15           ⑦ APAC            *ACC=BxY2 + AxY1
0035 0030 6962           ⑧ SACH  Y0,1      *SHIFT 1 DUE TO EXTRA BIT
0036           *                           *STORE UPPER 16 BITS
0037 0031 2062           ⑨ LAC   Y0        *SINCE A,B,C WAS DIVIDED BY 2
0038 0032 0062             ADD   Y0        *DOUBLE RESULT
0039 0033 6062             SACL  Y0        *STORE LOWER 16 BITS IN DMA 62
0040 0034 5662           ⑩ DMOV  Y0        *MOVE"DOWN"IN DM FOR NEXT N
0041 0035 E262             OUT   Y0,2      *OUTPUT (DM >62) TO PORT 2
0042 0036 FF80             B     WAIT      *BRANCH BACK FOR NEXT  N
     0037 0023
0043                       END
NO ERRORS, NO WARNINGS
```

FIGURE 4.2. Digital oscillator program using EQU assembler directive (OSCI.LST)

Since we halved the constants, care must be taken so that (4.15) remains correct.

Figure 4.2 shows the program that implements the digital oscillator above. To run this program, data memory locations >60 through >66 must be initialized as shown in Table 4.1. Use the data values for $f = 1$ kHz only.

Single-step through the program in Figure 4.2 and note the following, cross referenced by numbers in the program:

1. The first two instructions initialize the AIB for a 10-kHz sample rate using the transparency mode.

2. The wait loop sets the sample time for the difference equation. Shown below are the DMA, with the corresponding AR pointers.

DMA	Data Values	DMA	Data Values
>63	Y1	>65	A
>64	Y2 ← AR0	>66	B ← AR1

TABLE 4.1   Data Values for Digital Oscillator Program

		Data Values	
Label	DMA	$f = 1$ kHz	$f = 2$ kHz
Mode	>60	>FA	>FA
Rate	>61	>03E7	>03E7
Y0	>62	>0	>0
Y1	>63	>259E	>3CDE
Y2	>64	>0	>0
A	>65	>678E	>278E
B	>66	>C000	>C000

3. This instruction sequence points the auxiliary registers AR0 and AR1 to Y2 and B, respectively, zeros the accumulator, and selects AR0 for indirect addressing.

4. The LT *-,1 loads the T register with Y2, decrements AR0 to point to Y1, and selects AR1 for indirect addressing.

5. MPY *-,0 multiplies B (pointed to by AR1) by the content of the TR, then decrements AR1, pointing it at A, and finally selects AR0.

6. LTD *,1 loads the TR with Y1 (pointed to by AR0), decrements AR0, and selects AR1. This instruction also adds the last multiply to the ACC. A data move is performed on Y1 in anticipation for the calculation of the next value of n.

7. MPY *,0 multiplies A (pointed to by AR1) by Y1 (content of TR), decrements AR1, and selects AR0. Following this instruction is APAC, which adds the multiplication, (A)(Y1), to the accumulator.

8. SACH Y0, 1 stores the upper 16 bits of the accumulator in data memory location >62, which represents the output value Y0. A shift left by 1 eliminates the extra sign bit that results when multiplying two fractions.

9. This sequence of instructions doubles Y0 since the coefficients A and B were originally divided by 2, limiting them to a value of less than 1 in order to represent them.

10. These instructions perform the following: Y0 is MOVED DOWN to the next higher DMA in anticipation for the calculation of the next value of n. The output is then sent to the D/A through port 2. Next, branch back to WAIT and wait for the next sample time.

Run the program full speed and verify that a 1-kHz sinusoid is generated. Replace the values for Y0, Y1, Y2, A, and B using the 2-kHz values set in Table 4.1. Verify your results by generating a 2-kHz sinusoid.

4.5 DIGITAL OSCILLATOR USING DATA TRANSFER FROM PM TO DM

The digital oscillator program in the previous example uses the EQU assembler directive to assign values to corresponding symbols. Before the program is run, these values must be set in data memory. An alternative approach is to use the DATA Assembler Directive. The DATA directive can be used to place constant data values (such as coefficients in a filter program) in program memory. A BLKP/RPTK pair of instructions allows for a block transfer of data from PM to DM. The program in Figure 4.3 makes use of this alternative procedure. Run this program and observe an output sinusoid of 5 kHz.

```
0001                    * DIGITAL OSCILLATOR PROGRAM
0002                    * Y(n)=A Y(n-1) + B Y(n-2) + C X(n-1) ,fs= 20 KHZ
0003                    * YO  =A Y1 + B Y2  FOR N>1, X(n-1)=1 FOR n=1(= 0 OTHERWISE)
0004                    * NO EXTERNAL INPUT OR DATA INITIALIZATION REQUIRED
0005        0060  MODE     EQU    >60        *TRANSPARENT MODE FOR AIB
0006        0061  RATE     EQU    >61        *RATE=(10 MHZ/Fs)-1=499= >01F3
0007        0062  YO       EQU    >62        *INITIALLY YO=0, OUTPUT DMA
0008        0063  Y1       EQU    >63        *Y(1)=C=SIN(2x3.814xFd/fs)
0009        0064  Y2       EQU    >64        *Y(2)=0
0010        0065  A        EQU    >65        *A=2 COS(WT)
0011        0066  B        EQU    >66        *B=-1
0012              *                          *A=0 ,B=C000 , C=4000  Fd=5KHZ
0013 0000               AORG   >0
0014 0000 FF80          B      START
     0001 0027
0015 0020        TABLE  AORG   >20           *PM START AT >20
0016 0020 00FA   MD     DATA   >00FA         *MODE
0017 0021 01F3   CLK    DATA   >01F3         *RATE = 20 KHZ
0018 0022 0000   YYO    DATA   >0000         *YO=0  INITIALLY
0019 0023 4000   YY1    DATA   >4000         *Y1=(1/2)*2**15 ( MAX <1 )
0020 0024 0000   YY2    DATA   >0000         *Y2=0
0021 0025 0000   AA     DATA   >0000         *A =0
0022 0026 C000   BB     DATA   >C000         *B =(-1/2)*2**15 ( MAX < 1 )
0023 0027 5588   START  LARP   AR0           *SELECT AR0 FOR INDIRECT ADDR
0024 0028 D000          LRLK   AR0,>60       *AR0 POINT TO DMA 60
     0029 0060
0025 002A CB06          RPTK   >6            *REPEAT FOR SEVEN VALUES
0026 002B FCA0          BLKP   TABLE,*+      *TRANSFER PM 20-26 TO DM 60-66
     002C 0020
0027 002D E060          OUT    MODE,0        *OUTPUT MODE = >FA TO PORT 0
0028 002E E161          OUT    RATE,1        *OUTPUT RATE =03E7 TO PORT 1
0029 002F E361          OUT    RATE,3        *DUMMY OUTPUT TO PORT 3
0030 0030 FA80   WAIT   BIOZ   MAIN          *WAIT TIL BIO=0(EOC OF A/D LOW)
     0031 0034
0031             *                           *BRANCH TO MAIN AFTER EOC
0032 0032 FF80          B      WAIT          *A/D PROVIDES TIMING
     0033 0030
0033 0034 C064   MAIN   LARK   0,Y2          *AR0 POINT TO DMA 62
0034 0035 C166          LARK   1,B           *AR1 POINT TO DMA 66
0035 0036 CA00          ZAC                  *ZERO THE ACCUMULATOR
0036 0037 5588          LARP   0             *SELECT AR0 FOR INDIR ADDR
0037 0038 3C99          LT     *-,1          *TR=Y2,DEC AR0,SELECT AR1
0038 0039 3898          MPY    *,0           *PR=BxY2,DEC AR1,SELECT AR0
0039 003A 3F89          LTD    *,1           *TR=Y1 , ACC=BxY2 , MOVE"DOWN"
0040 003B 3888          MPY    *,0           *PR=AxY1 , SELECT AR0
0041 003C CE15          APAC                 *ACC = BxY2 + AxY1
0042 003D 6962          SACH   YO,1          *SHIFT 1 DUE TO EXTRA BIT
0043             *                           *STORE UPPER 16 BITS
0044 003E 2062          LAC    YO            *SINCE A,B,C WAS DIVIDED BY 2
0045 003F 0062          ADD    YO            *DOUBLE RESULT
0046 0040 6062          SACL   YO            *STORE LOWER 16 BITS IN DMA 62)
0047 0041 5662          DMOV   YO            *MOVE"DOWN"IN DM FOR NEXT N
0048 0042 E262          OUT    YO,2          *OUTPUT (DM >62) TO PORT 2
0049 0043 FF80          B      WAIT          *BRANCH BACK FOR NEXT  N
     0044 0030
0050                    END
NO ERRORS, NO WARNINGS
```

FIGURE 4.3. Digital oscillator program using DATA assembler directive (OSCIDATA.LST)

Digital Oscillator Example Using COFF

The common object file format (COFF) using assembler version 5.04 is discussed in Section 1.7 of Chapter 1 [8–10] and Appendix A. Figures A.9 to A.ll show the resulting COFF files of the digital oscillator example.

REFERENCES

[1] N. Ahmed and T. Natarajan, *Discrete-Time Signals and Systems*, Reston, Reston, Va., 1983.

[2] A. V. Oppenheim and R. Schafer, *Discrete-Time Signal Processing*, Prentice-Hall, Englewood Cliffs, N.J., 1989.

[3] D. J. DeFatta, J. G. Lucas, and W. S. Hodgkiss, *Digital Signal Processing: A System Approach*, Wiley, New York, 1988.

[4] L. R. Rabiner and B. Gold, *Theory and Application of Digital Signal Processing*, Prentice-Hall, Englewood Cliffs, N.J., 1975.

[5] J. G. Proakis and D. G. Manolakis, *Introduction to Digital Signal Processing*, Macmillan, New York, 1988.

[6] W. D. Stanley, G. R. Dougherty, and R. Dougherty, *Digital Signal Processing*, Reston, Reston, Va., 1984.

[7] *TMS320C1x/TMS320C2x Assembly Language Tools User's Guide*, Texas Instruments Inc., Dallas, Tex., 1987.

[8] *TMS320C1x/TMS320C2x Source Conversion Reference Guide*, Texas Instruments Inc., Dallas, Tex., 1988.

[9] *Second-Generation TMS320 User's Guide*, Texas Instruments Inc., Dallas, Tex., 1987.

[10] *TMS320C2x Software Development System User's Guide*, Texas Instruments Inc., Dallas, Tex., 1988.

[11] *TMS32010 Analog Interface Board User's Guide*, Texas Instruments Inc., Dallas, Tex., 1984.

5

Finite Impulse Response Filters

CONCEPTS AND PROCEDURES

- *Design and implementation of finite impulse response (FIR) filters*
- *Use of two filter design packages (one commercial and one homemade) for implementing FIR filters*

In this chapter, discrete-time and linear time-invariant systems are covered. From the discrete convolution equation,

$$y(n) = \sum_{k=0}^{N} h(k)x(n-k)$$

finite impulse response (FIR) filters are introduced [1–7]. The design of such filters is discussed using the Fourier series method.

An example of a lowpass FIR filter with 11 coefficients using both pairs of instructions, the MACD/RPTK and the LTD/MPY, is included. The effects of increasing the number of filter coefficients is demonstrated. The use and comparison of window functions (Hanning, Hamming, Blackman, Kaiser) to improve the characteristics of FIR filters [8–10] are covered with an example of a bandpass filter with 41 coefficients.

Two filter design packages are discussed.

1. *The Digital Filter Design Package (DFDP) available from Atlanta Signal Processors [11] can be used to design FIR filters and plot various responses. It also has the capability of generating TMS320C25 codes. Filter design packages are also available from other companies such as Hyperception.*

75

2. The Filter Design Package (FDP) is a homemade filter package which includes an FIR design package (FIRDP). Coefficients of an FIR filter can be calculated using the rectangular, Hanning, Hamming, Blackman, or Kaiser windows. The magnitude and phase of an FIR filter can also be calculated using the FDP. With a utility program included, the filter coefficients can be converted to a format for plotting.

5.1 INTRODUCTION

Digital filtering is one of the most useful operations in signal processing. Like the analog filter, the digital filter has frequency-selective characteristics. While analog filters operate on continuous signals and are typically built with amplifiers, resistors, capacitors, and inductors, digital filters operate on discrete samples or sequences and are realized with digital logic or special-purpose microprocessors such as the TMS320C25 digital signal processor.

The advantages of digital filters over analog filters include higher reliability, accuracy, and flexibility, with minimal sensitivity to temperature and aging. Digital filters can meet stringent magnitude and phase characteristics. Using a digital signal processor realization, filter parameters are easily modified to change characteristics such as bandwidth, resonant frequency, or filter type.

Digital filtering with a DSP microprocessor involves the use of an A/D converter to sample an analog signal and obtain a sequence of input samples, which are processed using a DSP microprocessor, then converted to a resulting analog signal using a D/A converter. The design of such digital filters consists of:

1. *Approximation:* Generation of a transfer function or its corresponding time function to satisfy desired specifications
2. *Realization:* Conversion of the transfer function into filter structures such as a coefficient table and delay variables
3. *Consideration of arithmetic errors:* Consideration of the effects of arithmetic approximation when finite-precision hardware such as the TMS320C25 is used (the accuracy with which the filter coefficients can be represented is limited by the finite word length of the processor)
4. *Implementation:* Build the actual filter using software and hardware

With VLSI technology, digital signal processors are now available to implement filters in real time, allowing a filter to operate on real-time signals. In a basic circuit course, we have seen how a differential equation can represent a filter, and how the Laplace transform enabled us to solve differential equations associated with continuous-time systems. Similarly, the Z-transform enables us to solve difference equations, associated with discrete-time systems, representing digital filters.

The systems we will be building are discrete, linear, and time invariant. Consequently, a convolution relationship exists which is used in the realization of the finite impulse response (FIR) filter. Since convolution is so closely related

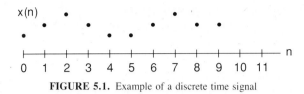

FIGURE 5.1. Example of a discrete time signal

to the FIR filter, it will be examined along with the design of FIR filters. There are two classes of digital filters that we will discuss: (1) finite impulse response (FIR) filters and, (2) infinite impulse response (IIR) filters. In this chapter we consider only FIR filters; in Chapter 6 we consider IIR filters.

5.2 DISCRETE SIGNALS

A discrete signal is a sequence of values that can be represented as

$$\{x(n)\} = \{x(1), x(2), x(3), \ldots\} \tag{5.1}$$

where n is the time index, with each sample being taken one sample time apart. The sequence $\{x(n)\}$ is shown graphically in Figure 5.1.

An important sequence is the unit sample or unit impulse sequence, which is defined as

$$\delta(n) = \begin{cases} 1 & n = 0 \\ 0 & \text{otherwise} \end{cases} \tag{5.2}$$

This function is an impulse located at the origin. A shifted unit impulse is written as

$$\delta(n - m) = \begin{cases} 1 & n = m \\ 0 & \text{otherwise} \end{cases} \tag{5.3}$$

This sample is shifted m units to the right or occurs m samples after the unit impulse at the origin. Examples of both are shown in Figure 5.2. A sequence $x(n)$ can be written in terms of the unit impulse:

$$x(n) = \sum_{m=-\infty}^{\infty} x(m)\delta(n - m) \tag{5.4}$$

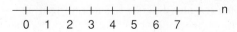

FIGURE 5.2. The unit impulse and the unit impulse shifted

5.3 LINEAR TIME-INVARIANT SYSTEMS

The digital systems we will be dealing with are linear and time invariant. Two important properties associated with these systems, superposition and shift invariance, will now be described.

Suppose that an input $x(n)$ yields a response $y(n)$, written as

$$x(n) \rightarrow y(n) \tag{5.5}$$

If we multiply the input by a constant, the output is multiplied by the same constant, as shown:

$$ax(n) \rightarrow ay(n) \tag{5.6}$$

Furthermore, if

$$ax_1(n) \rightarrow ay_1(n) \tag{5.7}$$

and

$$bx_2(n) \rightarrow by_2(n)$$

then

$$ax_1(n) + bx_2(n) \rightarrow ay_1(n) + by_2(n) \tag{5.8}$$

This last statement describes the *superposition property*, which states that the total response to a system is equal to the sum of the individual responses to each input.

The second property, *shift invariance*, is described as follows: If

$$x(n) \rightarrow y(n)$$

then

$$x(n - m) \rightarrow y(n - m) \tag{5.9}$$

Delaying the input by m samples delays the output by m samples.

5.4 CONVOLUTION

Using the two properties just described, superposition and shift invariance, the convolution summation can readily be obtained. Suppose that we introduce a unit impulse to the input of a system; the response $h(n)$, which is called the *unit impulse response*, results:

$$\delta(n) \rightarrow h(n) \tag{5.10}$$

If this input is delayed, the output will be delayed similarly:

$$\delta(n - m) \rightarrow h(n - m) \tag{5.11}$$

If the unit impulse is multiplied by a constant, the response will be multiplied by a constant, to yield

$$x(m)\delta(n - m) \rightarrow x(m)h(n - m) \tag{5.12}$$

A general input signal can be written as a summation of weighted unit impulse responses as was shown in equation (5.4) and repeated here for convenience:

$$x(n) = \sum_{m=-\infty}^{\infty} x(m)\delta(n - m) \tag{5.13}$$

By superposition we can add the individual response of each weighted unit impulse to get the collective response:

$$y(n) = \sum_{m=-\infty}^{\infty} x(m)h(n - m) \tag{5.14}$$

If the system is causal, the response $h(n)$ cannot begin before the unit impulse is applied. Therefore, m cannot exceed n in equation (5.14). The convolution summation can be adjusted to reflect this:

$$y(n) = \sum_{m=-\infty}^{n} x(m)h(n - m) \tag{5.15}$$

With a change in variable $k = n - m$, the summation becomes

$$y(n) = \sum_{k=0}^{\infty} h(k)x(n - k) \tag{5.16}$$

In the design of FIR filters we use only a finite number of terms to represent $h(n)$, and therefore the convolution can be written as

$$y(n) = \sum_{k=0}^{N} h(k)x(n - k) = \sum_{m=0}^{N} h(m)x(n - m) \tag{5.17}$$

If $x(n) = \delta(0)$, the unit impulse, in equation (5.17), the response is

$$y(n) = \left. \sum_{m=0}^{N} h(m)\delta(0) \right]_{m=n} = h(n) \qquad (5.18)$$

The response to the unit impulse is the impulse response, $h(n)$, which is what is expected.

5.5 FREQUENCY RESPONSE

Since filter response specifications are usually given in terms of frequency response, we will find an expression for the frequency response of a system. Let the input of the system be a complex sinusoidal,

$$x(n) = X \exp(j\omega n) \qquad (5.19)$$

The response to this input is

$$y(n) = \sum_{m=0}^{\infty} h(m)X \exp[j\omega(n - m)] \qquad (5.20)$$

Since the sum is with respect to m, $\exp(j\omega n)$ can be taken out of the summation, giving

$$y(n) = \left[\sum_{m=0}^{\infty} h(m) \exp(-j\omega m) \right] X \exp(j\omega n) \qquad (5.21)$$

Thus the response to a sinusoidal is also a sinusoidal with its amplitude modified by the term inside the brackets. This term is called the *frequency response function*,

$$H(j\omega) = \frac{Y(j\omega)}{X(j\omega)} = \left[\sum_{m=0}^{\infty} h(m) \exp(-j\omega m) \right] \qquad (5.22)$$

Equation (5.22) is an interesting relationship, the time samples of the impulse response function end up being the magnitude of the discrete frequency components of $H(j\omega)$. This relationship will be used later in the design of FIR filters.
    Similarly, if $x(n) = Xz^n$, using (5.17),

$$y(n) = \sum_{m=0}^{\infty} h(m)Xz^{(n-m)} \qquad (5.23)$$

and

$$y(n) = \left[\sum_{m=0}^{\infty} h(m)z^{-m} \right] X z^{n} \qquad (5.24)$$

As in equation (5.21), the response is of the same form as the input but modified by the term inside the brackets. This term is the Z-transform of $h(n)$ and for a finite number of terms is given by

$$H(z) = \sum_{m=0}^{N} h(m)z^{-m} = h(0) + h(1)z^{-1} + \cdots + h(N)z^{-N} \qquad (5.25)$$

5.6 FINITE IMPULSE RESPONSE FILTERS

Equation (5.25) implies that

$$Y(z) = H(z)X(z) = h(0)X(z) + h(1)z^{-1}X(z) + \cdots + h(N)z^{-N}X(z) \quad (5.26)$$

Equations (5.25) and (5.26) can be represented by Figure 5.3, with the z^{-1} representing the delays. The transfer function $H(z)$ can be represented as

$$H(z) = \frac{h(0)z^N + h(1)z^{N-1} + h(2)z^{N-2} + \cdots + h(N)}{z^N}$$

with poles only at the origin. As a result, finite impulse response (FIR) filters are inherently stable. Such filters can be implemented without feedback or the knowledge of past outputs. A second important feature of an FIR filter is that it can be designed to guarantee linear phase. The time-delay function associated with a linear phase filter is a constant. Linear phase with FIR filters is discussed thoroughly in Refs. 1 and 5. With guaranteed linear phase, FIR filters are very

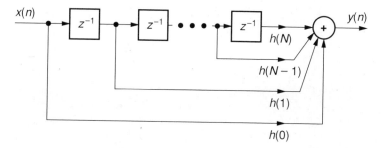

FIGURE 5.3. FIR structure showing delays

useful in applications where phase distortion is critical. The linear phase charac-
teristic causes all input sinusoidal components to be delayed by the same amount.

Consider an ideal sample delay difference equation $y(n) = x(n - k)$. Taking
the Fourier transform yields

$$Y(j\omega) = e^{-j\omega kT} X(j\omega)$$

The transfer function

$$H(j\omega) = \frac{Y(j\omega)}{X(j\omega)} = e^{-j\omega kT}$$

has a constant magnitude of 1 and a linear phase $\theta = -\omega kT$. The group delay function, defined as the derivative of the phase with respect to ω, is $-kT$, a constant.
FIR filters are also referred to as nonrecursive, moving average, transversal, and
tapped delay filters.

5.7 FIR IMPLEMENTATION USING FOURIER SERIES

A basic method of designing an FIR filter is to approximate the desired response
with a finite Fourier series. The desired response can be expanded in a Fourier
series,

$$H_d(\gamma) = \sum_{n=-\infty}^{\infty} C_n e^{jn\pi\gamma} \qquad |n| < \infty \qquad (5.27)$$

where γ is the normalized frequency variable, $\gamma = f/f_N$, with $|\gamma| < 1$; $f_N = f_s/2$
is the Nyquist frequency, f_s is the sampling frequency, $\omega T = 2\pi f/f_s = \pi\gamma$, and
$H_d(\gamma)$ is the desired transfer function. The C_n will be picked such that the transfer
function $H(z)$ approximates $H_d(\gamma)$ with least-mean-square fit. The C_n's are given
by

$$C_n = \frac{1}{2} \int_{-1}^{1} H_d(\gamma) e^{-jn\pi\gamma} d\gamma \qquad (5.28)$$

Assuming that $H_d(\gamma)$ is an even function for $|\gamma| < 1$,

$$C_n = \int_0^1 H_d(\gamma) \cos n\pi\gamma \, d\gamma \qquad n \geq 0 \qquad (5.29)$$

and $C_{-n} = C_n$.

The desired transfer function $H_d(\gamma)$ requires an infinite number of coefficients C_n. Realizable filters require a finite number of coefficients; hence we truncate the infinite series in (5.27) to obtain an approximate transfer function,

$$H_a(\gamma) = \sum_{n=-Q}^{Q} C_n e^{jn\pi\gamma} \tag{5.30}$$

with $|\gamma| < 1$ and Q is positive and finite. Letting $z = e^{j\pi\gamma}$, we have

$$H_a(\gamma) = \sum_{n=-Q}^{Q} C_n z^n \tag{5.31}$$

The impulse response of this approximated transfer function is given by the sequence $C_{-Q}, C_{-Q+1}, \ldots, C_0, \ldots, C_{Q-1}, C_Q$. However, the positive powers of z in (5.31), for $n > 0$, imply a noncausal filter, which produces an output that starts before the input is applied. To find the response at time n, the response at time $n + 1$ would be needed. Introducing a delay of Q samples into equation (5.31), gives us

$$H(z) = Z^{-Q} H_a(z) = z^{-Q} \sum_{n=-Q}^{Q} C_n z^n = \sum_{n=-Q}^{Q} C_n z^{n-Q} \tag{5.32}$$

A change in variable, $i = -(n - Q)$, gives

$$H(z) = \sum_{i=2Q}^{0} C_{Q-i} z^{-i} = \sum_{i=0}^{2Q} C_{Q-i} z^{-i} \qquad 0 \le i \le 2Q \tag{5.33}$$

By letting $h_i = C_{Q-i}$ and $N = 2Q$, $H(z)$ becomes

$$H(z) = \sum_{i=0}^{N} h_i z^{-i} \qquad 0 \le i \le N \tag{5.34}$$

where $h_0 = C_Q$, $h_1 = C_{Q-1}$, $h_2 = C_{Q-2}, \ldots$, $h_Q = C_0$, $h_{Q+1} = C_1, \ldots, h_{2Q-1} = C_{-Q+1}$, $h_{2Q} = C_{-Q}$.

Note that there are $2Q + 1 = N + 1$ impulse response coefficients h_i, $0 \le i \le N$, and the impulse response is symmetric about h_Q since $h_i = C_{Q-i}$ and $C_n = C_{-n}$. For example, if $Q = 5$, the filter will have $(2Q + 1)$ or 11 coefficients $h_0, h_1, \ldots, h_{10}$ which are symmetric about h_5 and are related as follows:

$$h_0 = h_{10} = C_5$$
$$h_1 = h_9 = C_4$$
$$h_2 = h_8 = C_3$$
$$h_3 = h_7 = C_2$$
$$h_4 = h_6 = C_1$$
$$h_5 = C_0$$

The convolution equation,

$$y(n) = \sum_{k=0}^{N} h(k)x(n-k)$$

determines the response of an FIR filter represented by $N + 1$ terms. Later in this chapter, we show how filter characteristics such as roll-off frequency and ripple can be improved through the use of so-called window functions. Before we do so, let us consider the design and implementation of an FIR filter using the Fourier series method.

Example 5.1   FIR Lowpass Filter. Design and implement an FIR lowpass filter of order 11 with cutoff frequency $f_c = 1$ kHz and sampling frequency $f_s = 10$ kHz.

SOLUTION:

From (5.29),

$$C_n = \int_0^{\gamma} H_d(\gamma) \cos n\pi\gamma d\gamma = \left. \frac{\sin n\pi\gamma}{n\pi} \right|_0^{\gamma} = \frac{\sin 0.2n\pi}{n\pi}$$

using $\gamma = f_c/f_N = 2f_c/f_s = 0.2$. Evaluation of the coefficients C_n above yields the following values and assignments with appropriate scaling:

$$C_0 = 0.2$$
$$C_n = \frac{\sin 0.2n\pi}{n\pi} \qquad n = 1, \ldots, N$$
$$C_0 = 0.2 \times 2^{15} = \; >199D$$
$$C_1 = 0.1592 \times 2^{15} = \; >17F6$$
$$\vdots$$
$$C_5 = 0$$

and

$$h_0 = C_5 = 0$$

$$h_1 = C_4 = >05FE$$

$$h_2 = C_3 = >0CED$$

$$h_3 = C_2 = >1363$$

$$h_4 = C_1 = >17F6$$

$$h_5 = C_0 = >199D$$

$$h_6 = C_{-1} = C_1 = >17F6$$

$$h_7 = C_{-2} = C_2 = >1363$$

$$.$$

$$h_{10} = C_{-5} = C_5 = 0$$

where $h_i = C_{Q-i}$ and $C_{-n} = C_n$.

These coefficients are shown in the table labeled TABLE in the program in Figure 5.4. This program implements a lowpass filter with a cutoff frequency of 1 kHz. It uses 11 coefficients scaled by 2^{15} and f_s is set to 10 kHz. This program can also be useful for implementing different filter characteristics by editing the TABLE section to set different coefficient values.

Testing the filter—a walk-through: Connect a variable sine-wave generator to obtain a sinusoidal input, in a fashion similar to the LOOP program in Chapter 2. Using an oscilloscope, observe the cutoff frequency at approximately 1 kHz.

Observe the following program characteristics in Figure 5.4. Note that single-stepping is the best way to do this.

1. A TABLE is set starting as PMA >20, where the 11 coefficients are defined, as well as the mode and the sampling rate for the AIB.
2. Since page 6 is the top of block B1 starting at DMA >300, the four symbols MODE, CLOCK, YN, and XN are defined in DMA $>300-303$, respectively. For example, XN EQU >3 represents >300 offset by 3. Observe the values of MODE and CLOCK being transferred from PM into DMA $>300-301$.
3. At PMA >36 (second column in Figure 5.4), AR0 is selected to point at the top of block B0 in DMA >200. At PMA >38, the instruction BLKP, which transfers a block of data from PM into DM is performed 11 times in conjunction with RPTK $>A$. When single-stepping, note that both the RPTK and BLKP are executed, transferring at once all the coefficients starting at DMA >200. The 11 coefficients H0, H1, . . . , H10 have been transferred from PMA starting at >20 into DMA $>200-20A$.
4. At PMA $>3B$ is the CNFP instruction, which configures B0 as program memory from $>FF00$ to $>FFFF$. When single-stepping through this instruc-

```
0001                    * LOWPASS (FIR) FILTER WITH 11 COEFFICIENTS USING MACD
0002                    * CUT-OFF FREQUENCY = 1 KHZ , fs = 10 KHZ
0003                    * y(n)=h(0)*x(n)+h(1)*x(n-1)+h(2)*x(n-2)+...+h(10)*x(n)
0004                    * y(n)=h(0)*x(n-10)+h(1)*x(n-9)+...+h(10)*x(n)
0005 0000                        AORG     0
0006 0000 FF80                   B        START
     0001 002D
0007 0020          TABLE  AORG    >20                    *P.M. START AT >20
0008                    **  COEFFICIENTS TABLE    **
0009 0020 0000     H0     DATA    >0                     *FIRST COEFFICIENT    H0=H10
0010 0021 05FE     H1     DATA    >05FE                  *SECOND COEFFICIENT   H1=H9
0011 0022 0CED     H2     DATA    >0CED                  *THIRD COEFFICIENT    H2=H8
0012 0023 1363     H3     DATA    >1363                  *      .
0013 0024 17F6     H4     DATA    >17F6                  *      .
0014 0025 199D     H5     DATA    >199D                  *      .
0015 0026 17F6     H6     DATA    >17F6                  *
0016 0027 1363     H7     DATA    >1363                  *
0017 0028 0CED     H8     DATA    >0CED                  *
0018 0029 05FE     H9     DATA    >05FE                  *
0019 002A 0000     H10    DATA    >0                     *H10=H0
0020                    **  SET UP PARAMETERS    **
0021 002B 000A     MD     DATA    >000A                  *MODE FOR AIB
0022 002C 03E7     RATE   DATA    >03E7                  *SAMPLING FREQUENCY fs=10 KHZ
0023      0000     MODE   EQU     >0                     *MODE FOR AIB(NO OFFSET)
0024      0001     CLOCK  EQU     >1                     *fs IN DMA+1 (OFFSET BY 1)
0025      0002     YN     EQU     >2                     *OUTPUT IN DMA+2(OFF 2)
0026      0003     XN     EQU     >3                     *NEW SAMPLE IN DMA+3 (OFF 3)
0027                    **  INITIALIZE AIB       **
0028 002D C806     START  LDPK    6                      *SELECT PAGE 6,TOP OF B1 >300
0029 002E CA2B            LACK    MD                     *LOAD ACC WITH (MD)
0030 002F 5800            TBLR    MODE                   *TRANSFER INTO DMA >300
0031 0030 E000            OUT     MODE,0                 *OUT VALUE MODE TO PORT 0
0032 0031 CA2C            LACK    RATE                   *LOAD ACC WITH (RATE)
0033 0032 5801            TBLR    CLOCK                  *TRANSFER INTO DMA >301
0034 0033 E101            OUT     CLOCK,1                *OUT VALUE CLOCK TO PORT 1
0035 0034 E301            OUT     CLOCK,3                *DUMMY OUTPUT TO PORT 3
0036                    ** LOAD FILTER COEFFICIENTS **
0037 0035 5588            LARP    AR0                    *SELECT AR0 FOR INDIR ADDRESS
0038 0036 D000            LRLK    AR0,>200               *AR0 POINT TO TOP OF BLOCK B0
     0037 0200
0039 0038 CB0A            RPTK    >A                     *REPEAT FOR 11 COEFFICIENTS
0040 0039 FCA0            BLKP    TABLE,*+               *XFER PM 20-2A INTO DM 200-20A
     003A 0020
0041 003B CE05            CNFP                           *CONFIGURE BLOCK B0 AS PM
0042                    ** MAIN PROCESSING LOOP **
0043 003C FA80     WAIT   BIOZ    MAIN                   *NEW SAMPLE WHEN BIO PIN LOW
     003D 0040
0044 003E FF80            B       WAIT                   *CONTINUE UNTIL EOC IS DONE
     003F 003C
0045                    **  INPUT    **
0046 0040 8203     MAIN   IN      XN,2                   *INPUT NEW SAMPLE IN DMA 303
0047 0041 D100            LRLK    AR1,>30D               *AR1=>X(n-10)   LAST SAMPLE
     0042 030D
0048 0043 5589            LARP    AR1                    *SELECT AR1 FOR INDIR ADDR
0049 0044 A000            MPYK    0                      *SET PRODUCT REGISTER TO 0
0050 0045 CA00            ZAC                            *SET ACC TO 0
0051                    ** THE BEEF **
0052 0046 CB0A            RPTK    >A                     *DO 11 MULTIPLY
0053 0047 5C90            MACD    >FF00,*-               *H0*x(n-10) + H1*x(n-9) +...
     0048 FF00
0054              *                                      *FIRST PMA FF00,LAST DMA 30D
0055 0049 CE15            APAC                           *LAST ACCUMULATE
0056 004A 6902            SACH    YN,1                   *SHIFT LEFT 1 DUE TO XTRA BIT
0057              *                                      *STORE UPPER 16 BITS
0058                    **  OUTPUT    **
0059 004B E202            OUT     YN,2                   *OUT RESULT Y(n) FROM DMA >302
0060 004C FF80            B       WAIT                   *BRANCH FOR NEXT SAMPLE
     004D 003C
0061                    END
NO ERRORS, NO WARNINGS
```

FIGURE 5.4. Lowpass FIR program using the MACD implementation (LP11RMAC.LST)

tion, observe that *all* the coefficients, which were in data block B0 starting at >200, have at once "disappeared," having been transferred into PMA >FF00–FFFF. It is essential to perform this step in order to use the powerful MACD instruction, since one of the operands must reside in PM.

5. The processor waits at the label WAIT until the A/D conversion finishes; then it takes a new sample. This sample is placed in DMA >303 since page 6 was selected (instruction at PMA >2D). The top of page 6 starts at >300 and is offset by 3 due to XN EQU 3. XN occupies the top of the delay table, which exists in data memory locations >303–30D.

6. At PMA >41, AR1 is set to point to >30D, where the "last" sample $x(n-10)$ is stored. The exact initial values of the samples are not essential, provided that you are not concerned with transients. The new input sample is stored at >303 and will gradually be moved down.

7. MACD in conjunction with RPTK is "the beef." All 11 multiplications are performed with appropriate data move of the samples in anticipation of the output calculation for the next value of n. The first multiplication is H0 $x(n-10)$, with H0 in PMA >FF00 and $x(n-10)$ in DMA >30D, specified indirectly by AR1. After each multiplication, the PMA is incremented to select the next operand (coefficient) and the DMA is decremented to select the appropriate sample. Also after each multiplication, the sample is moved down in the next-higher DMA and the previous content of the product register (PR) is added to the accumulator.

8. APAC adds the content of the PR into the accumulator and is needed since the last multiplication only places the result in the PR.

9. The accumulator is shifted left by one eliminating the extra sign bit and providing a full 16 bits of resolution. The output result is then sent to port 2 of the AIB. The program continues to calculate the output $y(n)$ for the next value of n, branching back to select a new input sample and repeat the procedure in steps 5 through 9.

The output of this filter (from the AIB) is shown in Figure 5.5. Noise from a signal analyzer or noise generator can be used as an input to the AIB. The transition region is rather poor due to the small number of coefficients used. The sharpness of this lowpass filter can be considerably improved using a higher-order filter. Figure 5.5 also shows the output of the filter using 41 coefficients. Window functions other than the currently used "rectangular" can also improve many of the characteristics of an FIR filter.

Example 5.2   Lowpass Filter Using LTD/MPY. Figure 5.6 shows a program that implements the 11 coefficients of the lowpass FIR filter in Figure 5.5, using LTD/MPY instructions. Such instructions can be quite useful, especially if one wishes to single-step through each multiplication. With the LTD/MPY pair of instructions, both (coefficient and sample) operands may reside in data memory. Using in-line coding takes less execution time than does setting up a loop with

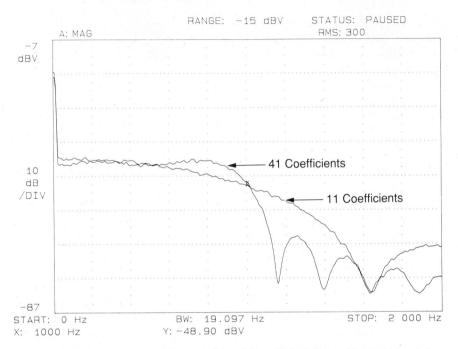

FIGURE 5.5. Frequency responses of lowpass (11- and 41-coefficient) filters with rectangular window

a branching instruction, because each branch instruction takes two machine cycle times. Since the repeat instruction RPTK carries an initial overhead of four machine cycle times, it is very worthwhile to consider an alternative to the MACD/RPTK instruction pair, even though more program memory space is used with this procedure. Run this program and observe the same cutoff frequency of 1 kHz. Halt the TMS320C25 and reset the program counter (PC) to zero. You may wish to reinitialize your data memory space. Select the data memory window. Use the fill instruction to "FILL" your data memory space with FFFF. This will allow you to see the initialization process when single-stepping.

Single-step through the program and observe/verify the actions:

1. Data page zero is selected (after the START label). The accumulator is loaded with the PMA >2B containing the value of MODE, and is then read and transferred directly into DMA >60 using the TBLR instruction. Similarly, the RATE associated with the sampling frequency value is transferred into DMA >61. This procedure eliminates the need to manually load the MODE and RATE.

2. After block B0 is configured as data memory, the auxiliary register AR0 is set to point at CH0, the top of B0. The transfer of coefficients from PM into DM takes place with the RPTK/BLKP. Note that H0 is used with BLKP instead of the label TABLE, to designate the first PMA.

```
0001                    * LOWPASS(FIR) FILTER WITH 11 COEFFICIENTS USING LTD/MPY
0002                    * CUT-OFF FREQUENCY = 1 KHZ , fs = 10 KHZ
0003                    * y(n)=h(0)*x(n)+h(1)*x(n-1)+...+h(10)*x(n-10)
0004                    * y(n)=h(0)*x(n-10)+h(1)*x(n-9)+...h(10)*x(n)
0005 0000                    AORG    0
0006 0000 FF80               B       START
     0001 002D
0007 0020          TABLE     AORG    >20                 *P.M. START AT >20
0008                    ** COEFFICIENTS TABLE  **
0009 0020 0000     H0        DATA    >0                  *FIRST COEFFICIENT   H0=H10
0010 0021 05FE     H1        DATA    >05FE               *SECOND COEFFICIENT  H1=H9
0011 0022 0CED     H2        DATA    >0CED               *THIRD COEFFICIENT   H2=H8
0012 0023 1363     H3        DATA    >1363               *   .
0013 0024 17F6     H4        DATA    >17F6               *   .
0014 0025 199D     H5        DATA    >199D               *   .
0015 0026 17F6     H6        DATA    >17F6               *
0016 0027 1363     H7        DATA    >1363               *
0017 0028 0CED     H8        DATA    >0CED               *
0018 0029 05FE     H9        DATA    >05FE               *
0019 002A 0000     H10       DATA    >0                  *H10=H0
0020                    *INITIALIZE AIB
0021 002B 000A     MODE      DATA    >000A               *MODE FOR AIB
0022 002C 03E7     RATE      DATA    >03E7               *SAMPLING FREQUENCY fs=10 KHZ
0023      0200     CH0       EQU     >200                *TOP OF B0 FOR COEFFICIENTS
0024      000B     YN        EQU     >B                  *OUTPUT IN DMA 20B(OFFSET >B)
0025      000C     XN        EQU     >C                  *NEW INPUT SAMPLE IN DMA 20C
0026               *                                     * >200 OFFSET BY >C
0027 002D          START
0028 002D C800               LDPK    0                   *SELECT PAGE 0
0029 002E CA2B               LACK    MODE                *LOAD ACC WITH (MD)
0030 002F 5860               TBLR    >60                 *TRANSFER INTO DM >60
0031 0030 E060               OUT     >60,0               *OUT VALUE MODE TO PORT 0
0032 0031 CA2C               LACK    RATE                *LOAD ACC WITH (RATE)
0033 0032 5861               TBLR    >61                 *TRANSFER INTO DM 61
0034 0033 E161               OUT     >61,1               *OUT VALUE CLOCK TO PORT 1
0035 0034 E361               OUT     >61,3               *DUMMY OUTPUT TO PORT 3
0036                    ** LOAD FILTER COEFFICIENTS **
0037 0035 CE04               CNFD                        *CONFIGURE B0 AS D.M.
0038 0036 5588               LARP    AR0                 *SELECT AR0 FOR INDIR ADDR
0039 0037 D000               LRLK    AR0,CH0             *AR0 POINT TO TOP OF BLOCK B0
     0038 0200
0040 0039 CB0A               RPTK    >A                  *REPEAT FOR 11 COEFFICIENTS
0041 003A FCA0               BLKP    H0,*+               *PM >20-2A => DM >200-20A
     003B 0020
0042                    ** MAIN PROCESSING LOOP **
0043 003C FA80     WAIT      BIOZ    MAIN                *NEW SAMPLE WHEN BIO PIN LOW
     003D 0040
0044 003E FF80               B       WAIT                *CONTINUE UNTIL EOC IS DONE
     003F 003C
0045 0040          MAIN
0046 0040 C804               LDPK    4                   *SELECT PAGE 4
0047                    ** INPUT  **
0048 0041 820C               IN      XN,2                *INPUT NEW SAMPLE IN DMA 20C
0049 0042 D000               LRLK    AR0,CH0             *POINT TO FIRST COEFF AT >200
     0043 0200
0050 0044 D100               LRLK    AR1,>216            *AR1=>X(n-10)=X10,LAST SAMPLE
     0045 0216
0051 0046 A000               MPYK    0                   *SET PRODUCT REGISTER TO 0
0052 0047 CA00               ZAC                         *SET ACC TO 0
0053 0048 5589               LARP    AR1                 *SELECT AR1 FOR INDIR ADDR
0054 0049 3C98               LT      *-,0                *TR=X10,AR1=>X9,SELECT AR0 TO
0055               *                                     *POINT AT >200 ,DATA MOVE
0056 004A 38A9               MPY     *+,1                *MULT BY H0,AR0=>H1,SELECT AR1
0057 004B 3F98               LTD     *-,0                *TR=X9,AR1=>X8,DMOV,SEL AR0
0058 004C 38A9               MPY     *+,1                *H1*X9
0059 004D 3F98               LTD     *-,0                *
0060 004E 38A9               MPY     *+,1                *H2*X8
0061 004F 3F98               LTD     *-,0                *
0062 0050 38A9               MPY     *+,1                *H3*X7
0063 0051 3F98               LTD     *-,0                *
0064 0052 38A9               MPY     *+,1                *H4*X6
0065 0053 3F98               LTD     *-,0                *
```

FIGURE 5.6. Lowpass FIR program using LTD/MPY Implementation (LP11RLTD.LST)

```
0066  0054  38A9         MPY      *+,1       *H5*X5
0067  0055  3F98         LTD      *-,0       *
0068  0056  38A9         MPY      *+,1       *H6*X4
0069  0057  3F98         LTD      *-,0       *
0070  0058  38A9         MPY      *+,1       *H7*X3
0071  0059  3F98         LTD      *-,0       *
0072  005A  38A9         MPY      *+,1       *H8*X2
0073  005B  3F98         LTD      *-,0       *
0074  005C  38A9         MPY      *+,1       *H9*X1
0075  005D  3F98         LTD      *-,0       *
0076  005E  38A9         MPY      *+,1       *H10*X0
0077  005F  CE15         APAC                *LAST ACCUMULATE
0078  0060  690B         SACH     YN,1       *SHIFT LEFT 1 DUE TO XTRA BIT
0079              *                          *STORE UPPER 16 BITS
0080              **   OUTPUT   **
0081  0061  E20B         OUT      YN,2       *OUT RESULT y(n) FROM DMA 20B
0082  0062  FF80         B        WAIT       *BRANCH FOR NEXT SAMPLE
      0063  003C
0083                     END
NO ERRORS, NO WARNINGS
```

FIGURE 5.6. (concluded)

3. After the WAIT loop, page 4 with starting address >200 is selected. The newest sample XN is placed in $>20C$ since page 4 is used and XN is OFFSET by $>C$. AR0 is set to point at the first of the 11 coefficients located at DMA >200–$20A$. AR1 is set to point at >216, the bottom of the input samples table, containing $x(n - 10) = X10$.

4. The T Register (TR) is loaded with X10, the content of AR1, AR1 is decremented to point at X9, and AR0 is selected next for indirect addressing. The instruction MPY $* +$,1 multiplies the content of AR0, the first coefficient value H0, with the last sample X10, loaded in the TR, with the result placed in the PR. AR0 is then incremented to point at the second coefficient value H1 and AR1 is selected next for indirect addressing. While the last input sample $x(n - 10)$ need not be data moved in the next higher DM location, each of the other samples $x(n - 9)$, $x(n - 8)$, . . . is moved down, using the LTD instead of the LT instruction, in anticipation of the calculation of $y(n)$ for the next value of n. The input samples, not the coefficients, are loaded in the TR since the delay is associated with the input samples. The LTD instruction combines the DMOV instruction as well as the LT and the APAC instructions. There is no MPYD instruction. Note also that the input samples can be moved down or copied in the next-higher DMA only if they reside in internal DM.

5. Since LTD = LT + APAC + DMOV, the APAC instruction is required after the last multiply in order to accumulate the previous product. The higher 16 bits in the accumulator are then stored in DM specified by YN. Selecting the higher 16 bits with the instruction SACH YN,1 eliminates the extra sign bit as in the previous program. The output $y(n)$ for a specific value of n is at DMA $>20B$ (>200 offset by $>B$). The offset is accomplished by setting YN to $>B$ with the EQU assembler directive and selecting page 4.

FIR Bandpass, Highpass, and Bandstop Filters

Expressions for bandpass, highpass, and bandstop filters can readily be obtained from the expression for a lowpass filter, as in Example 5.1,

$$C_n = \frac{\sin[(f_c/f_N)n\pi]}{n\pi} \tag{5.35}$$

by visualizing the desired filter frequency response in terms of the lowpass filter frequency response. Let us consider the bandpass filter first. The C_n for a bandpass can be obtained by subtracting the C_n of two lowpass filters with cutoff frequencies of f_{c1} and f_{c2}. Figure 5.7 shows how two lowpass filters combine to produce the bandpass filter,

$$C_n = \frac{\sin[(f_{c2}/f_N)n\pi]}{n\pi} - \frac{\sin[(f_{c1}/f_N)n\pi]}{n\pi} \tag{5.36}$$

where f_{c1} is the cutoff frequency of one lowpass filter, f_{c2} is the cutoff frequency of the other lowpass filter, and f_N is one-half the sample frequency.

Next consider the highpass filter. This can be obtained by subtracting a lowpass response from a constant magnitude response, as shown in Figure 5.8. The

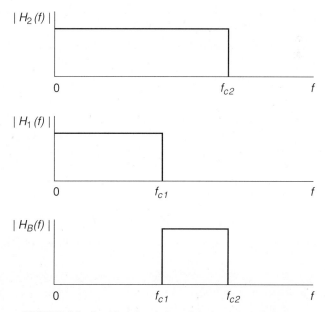

FIGURE 5.7. Graphical description of the bandpass calculation

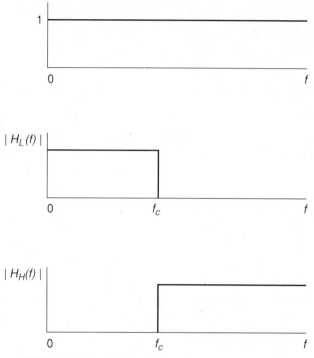

FIGURE 5.8. Graphical description of the highpass calculation

time sequence that gives a constant magnitude frequency response is $\delta(n)$. The subtraction results in

$$C_n = \delta(n) - \frac{\sin[(f_c/f_N)n\,\pi]}{n\,\pi} \tag{5.37}$$

The delta function in (5.37) will only affect the $n = 0$ term of the sequence. Similarly, the coefficients for a bandstop filter can be found using $\delta(n)$ and the resulting coefficients in (5.36).

Exercise 5.1   Bandstop Filter Calculation. Use the graphical method to find the sequence for a bandstop filter.

5.8 IMPROVEMENT OF FIR FILTERS WITH WINDOW FUNCTIONS

A specific class of time-limited weighting functions are referred to as window functions. A finite-duration causal filter was obtained by truncating the infinite-duration impulse response. The truncation of the infinite series, for the implementation of FIR filters, is equivalent to multiplying it with a rectangular window function, which is defined as

$$w_R(n) = \begin{cases} 1 & |n| \le N/2 \\ 0 & \text{otherwise} \end{cases}$$

Hence

$$C_n' = C_n w_R(n) \tag{5.38}$$

where the C_n are the Fourier series coefficients found in (5.29).

The transfer function with window can now be defined as

$$H'(z) = \sum_{i=0}^{N} h_i' z^{-i} \tag{5.39}$$

where

$$h_i' = C_{Q-i}' \qquad 0 \le i \le 2Q \tag{5.40}$$

In general,

$$C_n' = C_n w(n) \tag{5.41}$$

with $C_n' = C_{-n}'$. The C_n's are the Fourier series coefficients from (5.29), C_n' are the coefficients modified by the window sequence, $w(n)$. There are $2Q + 1$ impulse response coefficients, symmetrical such that $h_i' = C_{Q-i}'$ and $C_n' = C_{-n}'$. The duration of the impulse response is $D = 2QT$. The frequency response of an FIR filter with linear phase will be zero at $f = f_s/2$ if the order of the filter is even. Hence one should not chose the order of the filter to be even when designing a highpass or a bandstop filter.

Multiplication in the time domain corresponds to convolution in the frequency domain. The desired window function has a Fourier transform with low-level sidelobes with respect to the mainlobe peak. The abrupt truncation of the infinite series results in poor convergence of the series, causing the oscillations that are present in both the passbands and stopbands, specially near discontinuities. The oscillations can be reduced by multiplying the impulse response coefficients by a window function as in (5.38). As the length of the impulse response is increased, the amplitude of the oscillations will decrease. The Fourier transform $W_R(\omega)$ of the rectangular waveform $w_R(n)$ is proportional to a sinc [a sin (ax/x)] function that has a main lobe and high sidelobes, causing the undesired oscillations.

Several other window functions exist such that their Fourier transform have smaller sidelobes, but with a wider mainlobe. Instead of the abrupt truncation of the infinite series expansion used with the rectangular window sequence, a more gradual truncation is possible.

If a window is used, the frequency response will contain oscillations at each discontinuity of the ideal frequency response, with overshoot equally on either side of the discontinuity of the ideal response. The amplitude of the ripple decreases as you move away from the discontinuity. Certain trade-offs exist since, while on one hand, the oscillations can be reduced to yield a better approximation in the smooth regions, the transition region of the discontinuity would then be increased. Harris [8] provides a very thorough discussion of the trade-offs of various window functions such as the rectangular, Hanning, Hamming, Blackman, Kaiser, and so on. Later in the chapter, an FIR bandpass filter will be implemented using these various windows.

Hanning Window

The Hanning or raised cosine window function is defined as

$$
\begin{aligned}
w_{HA}(n) &= 0.5 + 0.5\cos(2n\,\pi/N) & |n| \le N/2 \\
&= 0 & \text{otherwise}
\end{aligned}
\tag{5.42}
$$

If we multiply the Fourier series coefficients C_n by the Hanning sequence $w_{HA}(n)$, a more gradual truncation of the series expansion will occur, resulting in a much lower sidelobe level compared to the rectangular window. Note that a multiplication of the Hanning sequence with the rectangular sequence,

$$
w_{HA}(n) = w_R(n)[0.5 + 0.5\cos(2n\,\pi/N)]
$$

results in $W_{HA}(\omega)$ as a convolution equation in the frequency domain.

Hamming Window

The Hamming window function is defined as

$$
w_H(n) =
\begin{cases}
0.54 + 0.46\cos(2n\,\pi/N) & \text{for } |n| \le N/2 \\
0 & \text{otherwise}
\end{cases}
\tag{5.43}
$$

which is a modified Hanning window with the highest sidelobe level at approximately -43 dB from the peak of the mainlobe [8].

Blackman Window

The Blackman window function is given by

$$
w_B(n) =
\begin{cases}
0.42 + 0.5\cos(2n\,\pi/N) + 0.08\cos(4n\,\pi/N) & |n| \le N/2 \\
0 & \text{otherwise}
\end{cases}
\tag{5.44}
$$

Its highest sidelobe level is down -58 dB from the peak of the mainlobe as shown in Ref. 8. With the rectangular, Hanning, Hamming, and Blackman windows, increasing the order of the filter results in a decrease of the mainlobe and a decrease in the transition band. The minimum stopband attenuation is independent of the window length. Windows with low-level sidelobes have wider mainlobes.

Kaiser Window

The Kaiser window function provides a trade-off between the amplitude of the ripple, the transition, and the order of the filter. It has a variable parameter to control the level of the sidelobe with respect to the peak of the mainlobe; and as with the previous windows, the width of the mainlobe can be adjusted with the length of the window. The Kaiser window function is defined as

$$w_K(n) = \begin{cases} \dfrac{I_0(b)}{I_0(a)} & |n| \le N/2 \\[2em] 0 & \text{otherwise} \end{cases} \tag{5.45}$$

where

$$b = a[1 - (2n/N)^2]^{1/2}$$

and a is an empirically determined independent variable. $I_0(x)$ is the modified Bessel function of the first kind and can be expressed in terms of a series expansion which can converge quite rapidly,

$$I_{o(x)} = 1 + \frac{0.25x^2}{(1!)^2} + \frac{(0.25x^2)^2}{(2!)^2} + \frac{(0.25x^2)^3}{(3!)^2} + \cdots \tag{5.46}$$

or

$$I_0(x) = 1 + \sum_{n=1}^{\infty} [(1/n!)(x/2)^n]^2$$

Parks–McClellan Window

Parks and McClellan [12] developed the Remez exchange algorithm, which provides equiripple approximation of FIR filters. The impulse response coefficients are varied to obtain the equiripple approximation, while the order of the filter as well as the edges of the passbands and stopbands are fixed. In the Parks–McClellan equiripple FIR design, the frequency spectrum is divided into passbands, stopbands, and transition bands. This design algorithm minimizes the ripple in the passbands and stopbands (not the transition bands) and is still a very efficient one, even though the transition regions, in the design solution, are considered as don't-care regions where the solution may fail.

In Section 5.10, we use a commercially available software package to design FIR filters using Parks–McClellan as well as Kaiser windows.

5.9 FILTER DEVELOPMENT PACKAGE

The filter development package includes a program (FIRDP) that calculates the coefficients to implement lowpass, highpass, bandpass, and bandstop FIR filters using rectangular, Hanning, Hamming, Blackman, and Kaiser windows. The resulting scaled coefficients in hex can be saved and appropriately placed later within a TMS320C25 source code that uses either the MACD or the LTD/MPY implementation. While the MACD implementation requires an external noise source (from a signal analyzer, for example), the LTD/MPY does not, since an input noise is provided by including in the LTD/MPY program the pseudorandom-noise-generator program section of Chapter 3. Appendix D contains the procedure on how to use the FIRDP as well as the listing of support programs written in Basic. In Chapter 6 the filter development package is used again in conjunction with the design of IIR filters. In Section 5.10, a commercially available powerful digital filter design package is used for the design of FIR filters with Kaiser and Parks–McClellan windows.

Example 5.3   Bandpass FIR Filter with 41 Coefficients. A bandpass filter with 41 coefficients is designed with lower and upper cutoff frequencies of 1 kHz and 1.5 kHz, respectively (center frequency of 1250 Hz). Follow the procedure in Appendix

```
0001                     *PSEUDO RANDOM NOISE FEEDING INTO 41 COEFFICIENTS BANDPASS FIR
0002                     *USE OF INTERRUPT, LTD/MPY IMPLEMENTATION, RECTANGULAR WINDOW
0003                     *NO EXTERNAL INPUT REQUIRED SINCE NOISE PROVIDED BY PRN GEN
0004                              IDT      'BP41'
0005              *          DATA MEMORY ALLOCATION/DEFINITION
0006       0060   STAT0     EQU      >60              STATUS REG STORAGE FOR ISR
0007       0061   LH1000    EQU      >61              NOISE GEN POSITIVE SCALER
0008       0062   LHF000    EQU      >62              NOISE GEN NEGATIVE SCALER
0009       0063   RNUMH     EQU      >63              RANDOM NUM UPPER 16-BITS
0010       0064   RNUML     EQU      >64              RANDOM NUM LOWER 16-BITS
0011       0065   MASK15    EQU      >65              BIT-15 MASK
0012       0066   TEMP      EQU      >66              SCRATCH PAD MEMORY AREA
0013       007E   SWDS1     EQU      >7E              RESERVED FOR SWDS USAGE
0014       007F   SWDS2     EQU      >7F              RESERVED FOR SWDS USAGE
0015       0200   CH0       EQU      >0200            FILTER COEFF STORAGE
0016       0229   X0        EQU      >0229            FILTER INPUT SAMPLES
0017              *          MEMORY MAPPED REGISTER LOCATIONS
0018       0002   TIMER     EQU      >0002            INTERVAL TIMER REGISTER
0019       0003   PERIOD    EQU      >0003            PERIOD REGISTER FOR TIMER
0020       0004   INTMSK    EQU      >0004            INTERRUPT MASK REGISTER
0021 0000                            AORG     0
0022 0000 FF80                       B        INIT
     0001 004D
0023 0018                            AORG     >18
0024 0018 FF80                       B        START          TIMER INTERRUPT VECTOR
     0019 0069
0025 0020                            AORG     >20        *****PROGRAM MEMORY DATA STORAGE*****
0026 0020 7E52   SEEDH    ·DATA     >7E52            RANDOM NUM SEED,UPPER 16-BITS
0027 0021 1603   SEEDL     DATA     >1603            RANDOM NUM SEED,LOWER 16-BITS
```

FIGURE 5.9. Bandpass with 41-coefficient FIR program using PRN as input (BP41RLTD.LST)

```
0028 0022 03E7  RATE   DATA    >03E7           OUTPUT DATA RATE (10000 Hz)
0029 0023 0008  IMASK  DATA    >0008           ENABLE TIMER INTERRUPT ONLY
0030                   *
0031                   * Place Coefficients here.
0032                   *
0033 0024 0000  H0     DATA    >0
0034 0025 FF87  H1     DATA    >FF87
0035 0026 0000  H2     DATA    >0
0036 0027 018A  H3     DATA    >18A
0037 0028 02FE  H4     DATA    >2FE
0038 0029 02B7  H5     DATA    >2B7
0039 002A 0000  H6     DATA    >0
0040 002B FC0D  H7     DATA    >FC0D
0041 002C F98B  H8     DATA    >F98B
0042 002D FAD4  H9     DATA    >FAD4
0043 002E 0000  H10    DATA    >0
0044 002F 0653  H11    DATA    >653
0045 0030 09B0  H12    DATA    >9B0
0046 0031 0756  H13    DATA    >756
0047 0032 0000  H14    DATA    >0
0048 0033 F7DA  H15    DATA    >F7DA
0049 0034 F407  H16    DATA    >F407
0050 0035 F748  H17    DATA    >F748
0051 0036 0000  H18    DATA    >0
0052 0037 0904  H19    DATA    >904
0053 0038 0CCD  H20    DATA    >CCD
0054 0039 0904  H21    DATA    >904
0055 003A 0000  H22    DATA    >0
0056 003B F748  H23    DATA    >F748
0057 003C F407  H24    DATA    >F407
0058 003D F7DA  H25    DATA    >F7DA
0059 003E 0000  H26    DATA    >0
0060 003F 0756  H27    DATA    >756
0061 0040 09B0  H28    DATA    >9B0
0062 0041 0653  H29    DATA    >653
0063 0042 0000  H30    DATA    >0
0064 0043 FAD4  H31    DATA    >FAD4
0065 0044 F98B  H32    DATA    >F98B
0066 0045 FC0D  H33    DATA    >FC0D
0067 0046 0000  H34    DATA    >0
0068 0047 02B7  H35    DATA    >2B7
0069 0048 02FE  H36    DATA    >2FE
0070 0049 018A  H37    DATA    >18A
0071 004A 0000  H38    DATA    >0
0072 004B FF87  H39    DATA    >FF87
0073 004C 0000  H40    DATA    >0
0074 004D CE04  INIT   CNFD                    CONFIGURE B0 AS DATA MEMORY
0075 004E C800         LDPK    0               POINT TO DATA PAGE 0(forever)
0076 004F D001         LALK    SEEDH           POINT TO P.M. STORAGE
     0050 0020
0077 0051 5863         TBLR    RNUMH           XFER SEEDH TO RNUMH IN D.M.
0078 0052 CC01         ADDK    1               INCREMENT READ ADDRESS IN ACC
0079 0053 5864         TBLR    RNUML           XFER SEEDL TO RNUML IN D.M.
0080 0054 CC01         ADDK    1               INCREMENT READ ADDRESS IN ACC
0081 0055 5803         TBLR    PERIOD          XFER RATE TO PERIOD REGISTER
0082 0056 CC01         ADDK    1               INCREMENT READ ADDRESS IN ACC
0083 0057 5804         TBLR    INTMSK          XFER IMASK TO MASK REGISTER
0084 0058 D001         LALK    >1000
     0059 1000
0085 005A 6061         SACL    LH1000          INITIALIZE LH1000 TO >1000
0086 005B CE23         NEG
0087 005C 6062         SACL    LHF000          INITIALIZE LHF000 TO >F000
0088 005D D001         LALK    >8000
     005E 8000
0089 005F 6065         SACL    MASK15          INITIALIZE MASK15 TO >8000
0090           * COPY BANDPASS FILTER COEFFS FROM PGMEM TO DMEM
0091 0060 D000         LRLK    0,CH0           LOAD AR0 WITH WRITE ADX
     0061 0200
0092 0062 5588         LARP    0               POINT TO AR0
0093 0063 CB28         RPTK    >28             LOOP COUNT (41 values)
0094 0064 FCA0         BLKP    H0,*+           COPY H0-H40 TO CH0-CH40
     0065 0024
0095 0066 CE00         EINT                    ENABLE INTERRUPT
0096           *
0097 0067 FF80  WAIT   B       WAIT            WAIT HERE UNTIL AN INTERRUPT
     0068 0067
0098           * Modulo 2 sum of the 32-bit noise word bits 31,30,28,and 17
0099           * yields a random bit (1 or 0) shifted into the LSB of the
0100           * noise word, scaled and output to a D/A converter. Since
```

FIGURE 5.9. (continued)

```
0101                            * only the upper 16-bits are involved in determining feedback,
0102                            * the summation is done in lower accumulator for convenience.
0103  0069  2063  START   LAC    RNUMH           GET RANDOM NUM UPPER 16-BITS
0104  006A  4E65          AND    MASK15          MASK BIT-15
0105  006B  0163          ADD    RNUMH,1         ADD BIT-14 TO BIT-15
0106  006C  4E65          AND    MASK15          MASK BIT-15
0107  006D  0363          ADD    RNUMH,3         ADD BIT-12 TO THE SUMMATION
0108  006E  4E65          AND    MASK15          MASK BIT-15
0109  006F  0E63          ADD    RNUMH,14        ADD BIT-1 TO THE SUMMATION
0110  0070  4E65          AND    MASK15          MASK BIT-15
0111  0071  6966          SACH   TEMP,1          SHIFT R 15 (LEFT 1,RIGHT 16)
0112  0072  4063          ZALH   RNUMH           RESTORE RNUMH TO UPPER ACCUM
0113  0073  0064          ADD    RNUML           RESTORE RNUML TO LOWER ACCUM
0114  0074  CE18          SFL                    SHIFT 32-BIT ACCUM LEFT 1
0115  0075  0066          ADD    TEMP            ADD FEEDBACK BIT
0116  0076  6863          SACH   RNUMH           STORE UPPER 16-BIT NOISE WORD
0117  0077  6064          SACL   RNUML           STORE LOWER 16-BIT NOISE WORD
0118  0078  2066          LAC    TEMP            RESTORE RANDOM BIT INTO ACCUM
0119  0079  F680          BZ     MINUS
      007A  007E
0120  007B  2061          LAC    LH1000          SET OUTPUT POSITIVE IF BIT=1
0121  007C  FF80          B      FILT
      007D  007F
0122  007E  2062  MINUS   LAC    LHF000          SET OUTPUT NEGATIVE IF BIT=0
0123                      * THE BANDPASS FILTER STARTS HERE
0124  007F  D000  FILT    LRLK   0,X0            LOAD POINTER TO NEWEST SAMPLE
      0080  0229
0125  0081  5588          LARP   0               POINT TO NEWEST SAMPLE STORAGE
0126  0082  6080          SACL   *               STORE SCALED OUTPUT VALUE
0127  0083  CA00          ZAC
0128  0084  D000          LRLK   0,CH0+>28       LOAD POINTER TO COEFF H40
      0085  0228
0129  0086  D100          LRLK   1,X0+>28        LOAD POINTER TO SAMPLE X40
      0087  0251
0130  0088  5589          LARP   1               POINT TO X40
0131  0089  3F98          LTD    *-,0            LOAD X40
0132  008A  3899          MPY    *-,1            H40*X40
0133  008B  3F98          LTD    *-,0            *
0134  008C  3899          MPY    *-,1            * H39*X39
0135  008D  3F98          LTD    *-,0            *
0136  008E  3899          MPY    *-,1            * H38*X38
0137  008F  3F98          LTD    *-,0            *
0138  0090  3899          MPY    *-,1            * H37*X37
0139  0091  3F98          LTD    *-,0            *
0140  0092  3899          MPY    *-,1            * H36*X36
0141  0093  3F98          LTD    *-,0            *
0142  0094  3899          MPY    *-,1            * H35*X35
0143  0095  3F98          LTD    *-,0            *
0144  0096  3899          MPY    *-,1            * H34*X34
0145  0097  3F98          LTD    *-,0            *
0146  0098  3899          MPY    *-,1            * H33*X33
0147  0099  3F98          LTD    *-,0            *
0148  009A  3899          MPY    *-,1            * H32*X32
0149  009B  3F98          LTD    *-,0            *
0150  009C  3899          MPY    *-,1            * H31*X31
0151  009D  3F98          LTD    *-,0            *
0152  009E  3899          MPY    *-,1            * H30*X30
0153  009F  3F98          LTD    *-,0            *
0154  00A0  3899          MPY    *-,1            * H29*X29
0155  00A1  3F98          LTD    *-,0            *
0156  00A2  3899          MPY    *-,1            * H28*X28
0157  00A3  3F98          LTD    *-,0            *
0158  00A4  3899          MPY    *-,1            * H27*X27
0159  00A5  3F98          LTD    *-,0            *
0160  00A6  3899          MPY    *-,1            * H26*X26
0161  00A7  3F98          LTD    *-,0            *
0162  00A8  3899          MPY    *-,1            * H25*X25
0163  00A9  3F98          LTD    *-,0            *
0164  00AA  3899          MPY    *-,1            * H24*X24
0165  00AB  3F98          LTD    *-,0            *
0166  00AC  3899          MPY    *-,1            * H23*X23
0167  00AD  3F98          LTD    *-,0            *
0168  00AE  3899          MPY    *-,1            * H22*X22
0169  00AF  3F98          LTD    *-,0            *
0170  00B0  3899          MPY    *-,1            * H21*X21
0171  00B1  3F98          LTD    *-,0            *
0172  00B2  3899          MPY    *-,1            * H20*X20
0173  00B3  3F98          LTD    *-,0            *
0174  00B4  3899          MPY    *-,1            * H19*X19
0175  00B5  3F98          LTD    *-,0            *
```

FIGURE 5.9. (continued)

```
0176 00B6 3899          MPY     *-,1        *  H18*X18
0177 00B7 3F98          LTD     *-,0        *
0178 00B8 3899          MPY     *-,1        *  H17*X17
0179 00B9 3F98          LTD     *-,0        *
0180 00BA 3899          MPY     *-,1        *  H16*X16
0181 00BB 3F98          LTD     *-,0        *
0182 00BC 3899          MPY     *-,1        *  H15*X15
0183 00BD 3F98          LTD     *-,0        *
0184 00BE 3899          MPY     *-,1        *  H14*X14
0185 00BF 3F98          LTD     *-,0        *
0186 00C0 3899          MPY     *-,1        *  H13*X13
0187 00C1 3F98          LTD     *-,0        *
0188 00C2 3899          MPY     *-,1        *  H12*X12
0189 00C3 3F98          LTD     *-,0        *
0190 00C4 3899          MPY     *-,1        *  H11*X11
0191 00C5 3F98          LTD     *-,0        *
0192 00C6 3899          MPY     *-,1        *  H10*X10
0193 00C7 3F98          LTD     *-,0        *
0194 00C8 3899          MPY     *-,1        *  H9*X9
0195 00C9 3F98          LTD     *-,0        *
0196 00CA 3899          MPY     *-,1        *  H8*X8
0197 00CB 3F98          LTD     *-,0        *
0198 00CC 3899          MPY     *-,1        *  H7*A7
0199 00CD 3F98          LTD     *-,0        *
0200 00CE 3899          MPY     *-,1        *  H6*X6
0201 00CF 3F98          LTD     *-,0        *
0202 00D0 3899          MPY     *-,1        *  H5*X5
0203 00D1 3F98          LTD     *-,0        *
0204 00D2 3899          MPY     *-,1        *  H4*X4
0205 00D3 3F98          LTD     *-,0        *
0206 00D4 3899          MPY     *-,1        *  H3*X3
0207 00D5 3F98          LTD     *-,0        *
0208 00D6 3899          MPY     *-,1        *  H2*X2
0209 00D7 3F98          LTD     *-,0        *
0210 00D8 3899          MPY     *-,1        *  H1*X1
0211 00D9 3F98          LTD     *-,0        *
0212 00DA 3899          MPY     *-,1        *  H0*X0
0213 00DB CE15          APAC                ACCUMULATE LAST PRODUCT
0214 00DC 6966          SACH    TEMP,1      STORE FILTER OUTPUT
0215 00DD E266          OUT     TEMP,2      OUTPUT TEMP TO D/A
0216 00DE CE00          EINT
0217 00DF CE26          RET
0218 00E0
0219 00E0
NO ERRORS, NO WARNINGS
```

FIGURE 5.9. (concluded)

D to calculate the 41 coefficients. PCWRITE or another editor can be used to insert these coefficients in the source code using a LTD/MPY implementation and a pseudorandom noise generator. Appendix D shows in detail the steps required to implement the 41 coefficients FIR filter with rectangular, Hanning, Hamming, and Blackman, as well as Kaiser windows. Figure 5.9 shows the TMS320C25 listing program, which contains the rectangular window coefficients calculated using the FIRDP. An external input noise source is not needed to obtain the frequency response shown in Figure 5.10, since the pseudorandom-noise-generator program section is incorporated in Figure 5.9. Figures 5.11 through 5.14 compare the output spectrum for the Hanning, Hamming, Blackman, and Kaiser windows with the rectangular window, and Figure 5.15 includes them all.

5.10 DIGITAL FILTER DESIGN PACKAGE

The digital filter design package (DFDP) software is commercially available [11] for the design of FIR filters (as well as IIR). The DFDP allows also for the generation of TMS320C25 source code. A control file directs an AIB initialization file to be linked with a TMS320C25 main program generated using DFDP. The

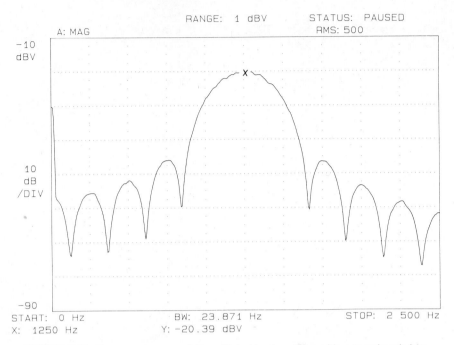

FIGURE 5.10. Frequency response of 41-coefficient bandpass filter with rectangular window

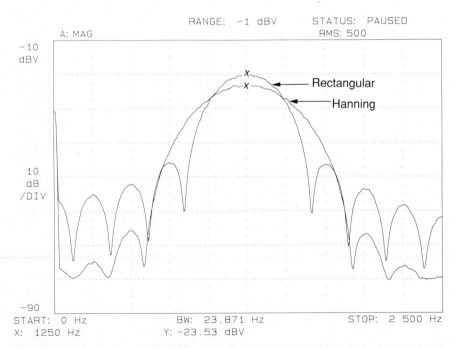

FIGURE 5.11. Frequency response of 41-coefficient bandpass filter comparing Hanning and rectangular windows

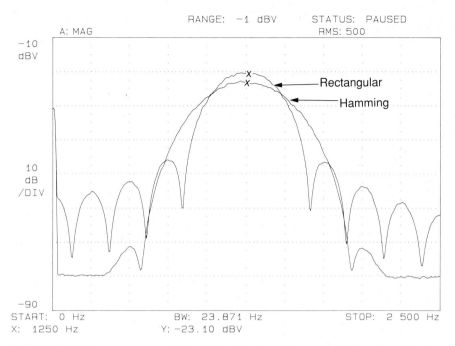

FIGURE 5.12. Frequency response of 41-coefficient bandpass filter comparing Hamming and rectangular windows

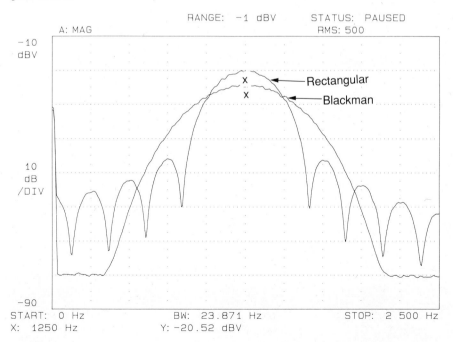

FIGURE 5.13. Frequency response of 41-coefficient bandpass filter comparing Blackman and rectangular windows

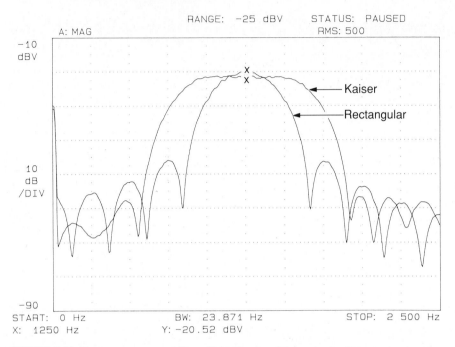

FIGURE 5.14. Frequency response of 41-coefficient bandpas filter comparing Kaiser and rectangular window

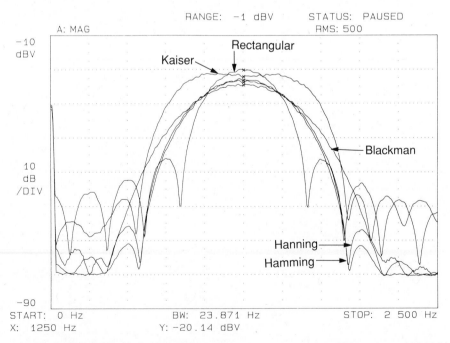

FIGURE 5.15. Frequency responses of 41-coefficient bandpass filter comparing rectangular/Hanning/Hamming/Blackman/Kaiser windows

linking process creates a LOD file that is downloaded into the SWDS (instead of an MPO file). Appendix E contains examples showing the various features available with the DFDP.

Lowpass Filter Example Using COFF

The common object file format (COFF) using an assembler version 5.04 is discussed in Chapter 1 and Appendix A. Figures A.12 to A.14 show the resulting COFF files of the lowpass filter, with 11 coefficients, discussed in Example 5.1.

REFERENCES

[1]   T. W. Parks and C. S. Burrus, *Digital Filter Design*, Wiley, New York, 1987.

[2]   N. Ahmed and T. Natarajan, *Discrete-Time Signals and Systems*, Reston, Reston, Va., 1983.

[3]   DSP Committee, IEEE ASSP (editors), *Selected Papers in Digital Signal Processing II*, IEEE Press, New York, 1976.

[4]   W. D. Stanley, G. R. Dougherty, and R. Dougherty, *Digital Signal Processing*, Reston, Reston, Va., 1975.

[5]   A. V. Oppenheim and R. Schafer, *Discrete-Time Signal Processing*, Prentice-Hall, Englewood Cliffs, N.J., 1989.

[6]   D. J. Defatta, J. G. Lucas, and W. S. Hodgkiss, *Digital Signal Processing: A System Approach*, Wiley, New York, 1988.

[7]   L. R. Rabiner and B. Gold, *Theory and Application of Digital Signal Processing*, Prentice-Hall, Englewood Cliffs, N.J., 1975.

[8]   F. Harris, "On the Use of Windows for Harmonic Analysis with the Discrete Fourier Transform," *Proceedings of the IEEE*, **66**(1), January 1978.

[9]   J. F. Kaiser, "Nonrecursive Digital Filter Design Using the I_0 $-$sinh Window Function," *Proceedings of the IEEE International Symposium on Circuits and Systems*, 1974.

[10]   R. W. Hamming, *Digital Filters*, Prentice-Hall, Englewood Cliffs, N.J., 1983.

[11]   *Digital Filter Design Package (DFDP)*, Atlanta Signal Processors Inc., Atlanta, Ga., 1987.

[12]   T. W. Parks and J. H. McClellan, "Chebychev Approximation for Nonrecursive Digital Filter with Linear Phase," *IEEE Transactions on Circuit Theory*, **CT-19**, March 1972, pp. 189–194.

[13]   L. B. Jackson, *Digital Filters and Signal Processing*, Kluwer Academic, Norwell, Mass., 1986.

[14]   L. C. Ludemen, *Fundamentals of Digital Signal Processing*, Harper & Row, New York, 1986.

[15]   C. S. Williams, *Designing Digital Filters*, Prentice-Hall, Englewood Cliffs, N.J., 1986.

[16] K. S. Lin (editor), *Digital Signal Processing Applications with the TMS320 Family*, Vol. 1, Prentice-Hall, Englewood Cliffs, N.J., 1988.

[17] *TMS320C1x/TMS320C2x Assembly Language Tools User's Guide*, Texas Instruments Inc., Dallas, Tex., 1987.

[18] *TMS320C2x Software Development System User's Guide*, Texas Instruments Inc., Dallas, Tex., 1988.

[19] *TMS320C1x/TMS320C2x Source Conversion Reference Guide*, Texas Instruments Inc., Dallas Tex., 1988.

[20] *TMS32010 Analog Interface Board User's Guide*, Texas Instruments Inc., Dallas, Tex., 1984.

6

Infinite Impulse Response Filters

CONCEPTS AND PROCEDURES

- *Commonly used IIR structures: direct form I and direct form II, parallel, and cascade forms*
- *Design of an IIR filter*
- *Bilinear transformation and design procedure*
- *IIR filter scaling illustrated with an example*
- *The effects of quantization, small-limit-cycle oscillations, and overflow oscillations*

Infinite impulse response (IIR) filters comprise an important fundamental area in digital signal processing [1–8]. The IIR filter is an extension of the FIR filter studied in Chapter 5. This chapter covers three basic IIR filter structures: direct, parallel, and cascade forms. The design procedures covered are limited to conversion of analog filters to discrete filters using the bilinear transformation. Appendix F includes a program (BLT.BAS) to perform the bilinear transformation. A complete example of a Butterworth bandpass filter is given. Scaling and overflow are discussed and illustrated with a laboratory example.

The effects of quantization on the fixed-point calculations are much more pronounced in IIR filters than in FIR filters. A discussion of the effects of quantization errors on the filter performance is given. Small-limit-cycle oscillations and overflow oscillations are described for first- and second-order factors. The student should be able to implement an arbitrary-order IIR filter and understand some of the problems associated with it.

6.1 INTRODUCTION

A general input–output relationship can be written as

$$y(n) = \sum_{k=0}^{N} a_k x(n-k) - \sum_{j=1}^{M} b_j y(n-j) \qquad n \geq 0 \tag{6.1}$$

If all the coefficients b_j are zero, (6.1) becomes the convolution equation in Chapter 4, representing FIR filters. If at least one of the coefficients b_j is nonzero, the equation becomes recursive. The output value at time n depends not only on the input at time n and the previous N inputs, but on the previous M outputs as well. Equation (6.1) represents an infinite impulse response (IIR) filter. To calculate $y(n)$, $y(n-1)$ is needed, which in turn depends on $y(n-2)$, and so on. The impulse response is represented by an infinite number of terms. The input–output equation can be written as

$$y(n) = a_0 x(n) + a_1 x(n-1) + a_2 x(n-2) + \cdots + a_N x(n-N)$$

$$-b_1 y(n-1) - b_2 y(n-2) - \cdots - b_M y(n-M) \tag{6.2}$$

Assuming zero initial conditions and taking the Z-transform of (6.1), we obtain

$$Y(z) = a_0 X(z) + a_1 z^{-1} X(z) + a_2 z^{-2} X(z) + \cdots + a_N z^{-N} X(z)$$

$$-b_1 z^{-1} Y(z) - b_2 z^{-2} Y(z) - \cdots - b_M z^{-M} Y(z)$$

The transfer function, assuming that $N = M$, is

$$H(z) = \frac{Y(z)}{X(z)} = \frac{N(z)}{D(z)} = \frac{a_0 + a_1 z^{-1} + a_2 z^{-2} + \cdots + a_N z^{-N}}{1 + b_1 z^{-1} + b_2 z^{-2} + \cdots + b_N z^{-N}} \tag{6.3}$$

which can also be written as

$$H(z) = \frac{a_0 z^{N} + a_1 z^{N-1} + a_2 z^{N-2} + \cdots + a_N}{z^{N} + b_1 z^{N-1} + b_2 z^{N-2} + \cdots + b_N} = C \prod_{i=1}^{N} \frac{z - z_i}{z - p_i} \tag{6.4}$$

which has N zeros and N poles, hence can be unstable if the magnitude of any pole does not lie inside the unit circle. Note that if all the b_N's are zero, there will be no poles to make the system unstable and $h(n)$ reduces to the nonrecursive equation representing FIR filters. Conditions of stability can be stated as follows:

If $|p_i| < 1$, then $h(n) \rightarrow 0$ as $n \rightarrow \infty$, and the system is stable.

If $|p_i| > 1$, then $h(n) \rightarrow \infty$ as $n \rightarrow \infty$, and the system is unstable.

6.2 REALIZATION FORMS

Various structures will be discussed to represent (6.1): direct forms I and II, cascade, and parallel.

Direct Form I

Equation (6.1) can be realized using the direct form I structure shown in Figure 6.1. This structure shows that $2N$ delays represented by z^{-1} are required. Note that a second-order structure requires four delay elements z^{-1}.

Direct Form II

The direct form II structure is a more practical one, requiring only half as many delay elements. An Nth-order structure can be implemented with N delay elements z^{-1}, as opposed to $2N$ with direct form I. Define an intermediate variable $U(z)$ as

$$U(z) = \frac{X(z)}{D(z)} \tag{6.5}$$

and rewrite (6.3) as

$$Y(z) = N(z)U(z) \tag{6.6}$$

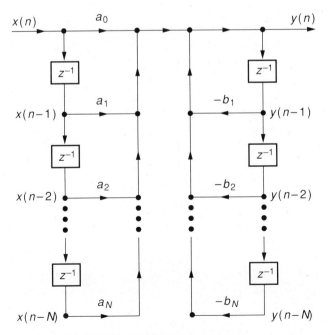

FIGURE 6.1. Direct form I structure

From equation (6.5) we get an expression for the intermediate variable:

$$U(z) = X(z) - b_1 z^{-1} U(z) - b_2 z^{-2} U(z) - \cdots - b_N z^{-N} U(z) \qquad (6.7)$$

or

$$u(n) = x(n) - b_1 u(n-1) - b_2 u(n-2) - \cdots - b_N u(n-N) \qquad (6.8)$$

Equation (6.6) defines the output in terms of the intermediate variable:

$$Y(z) = U(z)(a_0 + a_1 z^{-1} + a_2 z^{-2} + \cdots + a_N z^{-N})$$

or

$$y(n) = a_0 u(n) + a_1 u(n-1) + a_2 u(n-2) + \cdots + a_N u(n-N) \qquad (6.9)$$

The direct form II structure is shown in Figure 6.2, incorporating (6.8) and (6.9), for an Nth-order IIR filter. Note that the structure shows $u(n)$ at the middle top, which satisfies (6.8), and $y(n)$ at the right top, which satisfies (6.9). Only two delay elements would be required for a second-order direct form II structure.

We will see later, in the direct form II implementation, that initially the $u(n-k)$, $k = 1, \ldots, N$ can be set to zero; $u(n_0)$ at time n_0 can be found in terms of $x(n_0)$ and then the output at time n_0 can be found. For example, the sequences are as follows:

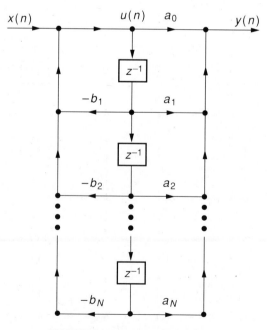

FIGURE 6.2. Direct form II structure

$$u(n_0) = x(n_0) - 0$$
$$u(n_0 + 1) = x(n_0 + 1) - b_1 u(n_0) - 0$$
$$u(n_0 + 2) = x(n_0 + 2) - b_1 u(n_0 + 1) - b_2 u(n_0) - 0$$

.
.
.

and

$$y(n_0) = a_0 u(n_0) + 0$$
$$y(n_0 + 1) = a_0 u(n_0 + 1) + a_1 u(n_0) + 0$$
$$y(n_0 + 2) = a_0 u(n_0 + 2) + a_1 u(n_0 + 1) + a_2 u(n_0) + 0$$

.
.
.

Cascade Form

From (6.4), H(z) can be factored as

$$H(z) = CH_1(z)H_2(z)H_3(z)\cdots H_r(z)$$

as shown in Figure 6.3, with each $H_k(z)$ being a first- or second-order transfer function. The overall cascade form can be efficiently represented in terms of second-order sections, with each individual section composed of the direct form II (or the transpose direct form II discussed in Section 6.6). $H(z)$ can also be written as

$$H(z) = \prod_{i=1}^{N/2} \frac{a_{0i} + a_{1i}z^{-1} + a_{2i}z^{-2}}{1 + b_{1i}z^{-1} + b_{2i}z^{-2}} \tag{6.10}$$

incorporating the constant C into the coefficients a_{0i}, a_{li}, and a_{2i}.

For the fourth-order IIR filter in cascade form, H(z) can be written as

$$H(z) = \frac{(a_{01} + a_{11}z^{-1} + a_{21}z^{-2})(a_{02} + a_{12}z^{-1} + a_{22}z^{-2})}{(1 + b_{11}z^{-1} + b_{21}z^{-2})(1 + b_{12}z^{-1} + b_{22}z^{-2})}$$

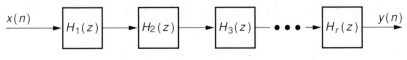

FIGURE 6.3. Cascade form structure

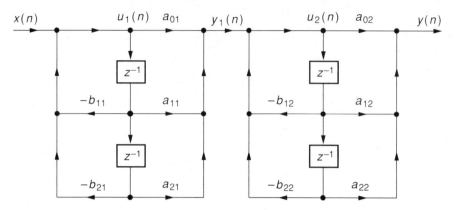

FIGURE 6.4. Fourth-order IIR with two second-order direct form II sections in cascade

Figure 6.4 shows a fourth-order IIR filter cascaded with two second-order sections. In this structure, the output of the first section $y_1(n)$ can be stored in the 32-bit accumulator and used as the input to the second section, avoiding premature truncation of intermediate output results. Although the ordering of the numerator and denominator factors is not unique and does not affect the overall result mathematically, it has been found that with proper ordering of each second-order section, quantization noise can be minimized.

Parallel Form

In the parallel form, $H(z)$ can be expressed as

$$H(z) = C + H_1(z) + H_2(z) + H_3(z) + \cdots + H_r(z)$$

with each $H_k(z)$ being a first- or second-order transfer function as shown in Figure 6.5. In a similar fashion to (6.10), again with $N = M$ in (6.1), $H(z)$ can be written as

$$H(z) = C + \sum_{i=1}^{N/2} \frac{a_{0i} + a_{1i}z^{-1} + a_{2i}z^{-2}}{1 + b_{1i}z^{-1} + b_{2i}z^{-2}} \qquad (6.11)$$

The parallel form can be efficiently represented in terms of second-order sections, with each section composed of the direct form II. Figure 6.6 shows a fourth-order parallel form IIR filter, with two second-order sections, and represents the following transfer function, from (6.11):

$$H(z) = C + \frac{a_{01} + a_{11}z^{-1} + a_{21}z^{-2}}{1 + b_{11}z^{-1} + b_{21}z^{-2}} + \frac{a_{02} + a_{12}z^{-1} + a_{22}z^{-2}}{1 + b_{12}z^{-1} + b_{22}z^{-2}}$$

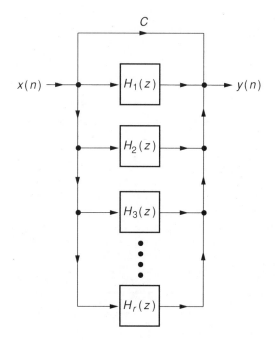

FIGURE 6.5. Parallel form structure

The output $y(n)$ can be found from Figure 6.6 by summing the individual outputs, or

$$y(n) = Cx(n) + \sum_{i=1}^{N/2} y_i(n)$$

6.3 IIR FILTER DESIGN

An IIR filter can be designed using first an analog prototype, taking advantage of well-known analog design results. Then a bilinear transformation is performed in order to map the analog filter design into a corresponding digital filter design. The transfer function of an analog filter $H(s)$ can be used and transformed into a corresponding transfer function $H(z)$ representing a digital filter. The bilinear transformation transforms uniquely the $j\omega$-axis in the s-plane into the unit circle in the z-plane. If $H(s)$ represents a stable system with all its poles in the left-hand side in the s-plane, then $H(z)$ will have all its poles inside the unit circle, which will yield a stable digital filter.

Design of analog-type filters already exists and is well-documented; for example:

1. In the Butterworth design, the magnitude is maximally flat in the passbands.

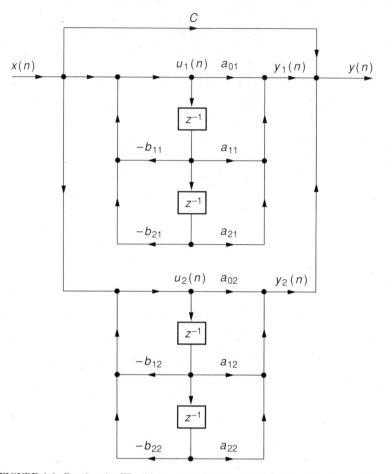

FIGURE 6.6. Fourth-order IIR with two second-order direct form II sections in parallel

2. The Chebychev approximation has lower order than the Butterworth design since it allows for ripple in the passbands.

3. The elliptic approximation yields the lowest order, with equiripple in both passbands and stopbands.

6.4 BILINEAR TRANSFORMATION

The most widely used technique for transforming analog filters to discrete filters is the bilinear transformation (BLT), which is a mapping between the s- and the z-planes. The BLT uses the relationship

$$ s = \frac{z - 1}{z + 1} \tag{6.12} $$

or

$$z = \frac{1 + s}{1 - s} \tag{6.13}$$

The general properties of the BLT are such that:

1. The left-hand side in the s-plane maps *inside* the unit circle in the z-plane.
2. The right-hand side in the s-plane maps *outside* the unit circle in the z-plane.
3. The imaginary or $j\omega$-axis in the s-plane maps *on* the unit circle in the z-plane.

Consider the mapping from the $j\omega$-axis in the s-plane *onto* the unit circle in the z-plane. Let

$$s = j\omega_A \tag{6.14}$$

and

$$z = e^{j\omega_D T} \tag{6.15}$$

where ω_A represents a variable analog frequency which corresponds to a desired or critical digital frequency ω_D. Using (6.14) and (6.15) in (6.12) gives

$$j\omega_A = \frac{e^{j\omega_D T} - 1}{e^{j\omega_D T} + 1} = \frac{e^{j\omega_D T/2}(e^{j\omega_D T/2} - e^{-j\omega_D T/2})}{e^{j\omega_D T/2}(e^{j\omega_D T/2} + e^{-j\omega_D T/2})} \tag{6.16}$$

Solving for ω_A yields

$$\omega_A = \tan \frac{\omega_D T}{2} \tag{6.17}$$

The analog and digital frequencies are such that

$$H(s)\Big|_{s = j\omega_A} = H(z)\Big|_{z = e^{j\omega_D T}} \tag{6.18}$$

The BLT provides a one-to-one mapping over the frequency range of interest. Figure 6.7 illustrates the relationship between positive values for ω_A and ω_D. While the region ω_A between 0 and 1 corresponds to ω_D between 0 and $\omega_s/4$, the entire remainder region of $\omega_A > 1$ is mapped and compressed into the region ω_D between $\omega_s/4$ and $\omega_s/2$. This effect associated with the BLT is referred to as frequency warping. To compensate for this frequency warping, prewarping and scaling are done. Note that the shape of the magnitude and phase responses of the transfer function is preserved.

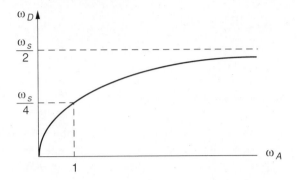

FIGURE 6.7. Transformation showing relationship between ω_A and ω_D

6.5 BILINEAR TRANSFORMATION DESIGN PROCEDURE

The following steps are taken in order to use the BLT and find $H(z)$:

1. Select an appropriate known analog transfer function $H(s)$.

2. Prewarp the critical or desired frequency ω_D in order to obtain the analog frequency ω_A, using (6.17):

$$\omega_A = \tan \frac{\omega_D T}{2}$$

3. Scale the frequency of the selected $H(s)$ in step 1 by ω_A, using

$$H(s)\bigg|_{s=s/\omega_A} \tag{6.19}$$

4. Obtain $H(z)$ using (6.12), or

$$H(z) = H(s/\omega_A)\bigg|_{s=(z-1)/(z+1)} \tag{6.20}$$

We will now illustrate the BLT using an example.

Example 6.1 Lowpass Filter. Design a lowpass filter with bandwidth $B = 1$ r/s at a sampling frequency $f_s = 10$ Hz.

SOLUTION:

1. Choose an appropriate transfer function $H(s)$,

$$H(s) = \frac{1}{s+1}$$

which represents a lowpass filter of bandwidth of 1 r/s.

2. Prewarp ω_D using

$$\omega_A = \tan \frac{\omega_D T}{2}$$

where $\omega_D = B = 1$ r/s $T = 1/f_s = \dfrac{1}{10 \text{ Hz}}$

$$\omega_A = \tan \frac{(1 \times \frac{1}{10})}{2} \simeq 1/20$$

3. Scale $H(s)$ using

$$H(s)\,|_{s=s/\omega_A} = \frac{1}{20s + 1}$$

4. Obtain the desired transfer function $H(z)$ using

$$H(z) = H(s/\omega_A) \Big|_{s=(z-1)/(z+1)} = \frac{1}{\dfrac{20(z-1)}{z+1} + 1} = \frac{z+1}{21z - 19}$$

This procedure can be applied in designing discrete-time (DT) filters using well-documented analog filter functions (Butterworth, Chebychev, etc.). Note that in the case of bandpass or bandstop filters, two critical frequencies ω_{D1} and ω_{D2} are present, representing the lower and upper cutoff frequencies, as illustrated in the following example.

Example 6.2   Butterworth Bandpass Filter.   Design a second-order DT Butterworth bandpass filter with ω_{D1} and ω_{D2} as the lower and upper cutoff frequencies, respectively, and f_s as the sampling frequency.

SOLUTION:

The solution follows from the previous procedure:

1. The transfer function $H(s)$ for a second-order bandpass analog filter is

$$H(s) = \frac{sB}{s^2 + sB + \omega_r^2} \tag{6.21}$$

where the bandwidth $B = \omega_{A2} - \omega_{A1}$ and the center frequency $\omega_r^2 = \omega_{A1}\omega_{A2}$. The transfer function $H(s)$ in (6.21) can be obtained from a first-order lowpass filter $H(s) = 1/(s + 1)$ using the transformation

$$H_{BP}(s) = H_{LP}(s)\Big|_{s=(s^2+\omega_r^2)/sB} \tag{6.22}$$

2. Obtain the corresponding analog frequencies using (6.17), or

$$\omega_{A1} = \tan\frac{\omega_{D1}T}{2} \tag{6.23}$$

and

$$\omega_{A2} = \tan\frac{\omega_{D2}T}{2} \tag{6.24}$$

3. The scaling step is taken care of already using the transformation in (6.22) to obtain (6.21).
4. Map the poles and zeros of $H_{BP}(s)$ into the z-plane using the BLT,

$$H(z) = H_{BP}(s)\big|_{s=(z-1)/(z+1)}$$

where $H(z)$ is the desired transfer function which can be used to implement the bandpass filter.

The following example will be implemented using the TMS320C25.

Example 6.3   TMS320C25 Implementation of Fourth-Order Bandpass IIR Filter Using a Butterworth Design.   Design and implement a fourth-order bandpass IIR filter using direct form II. The lower and upper cutoff frequencies are 1 kHz and 1.5 kHz, respectively, and the sampling frequency is 10 kHz.

SOLUTION:

For the solution, we will first determine a fourth-order bandpass Butterworth $H(s)$ from a second-order lowpass Butterworth transfer function.

1.

$$H(s) = H_{LP}(s)\Big|_{s=(s^2+\omega_r^2)/sB} \tag{6.25}$$

where $H_{LP}(s)$ is the 2nd order Butterworth lowpass transfer function. Substitution yields

$$H(s) = \frac{1}{s^2+\sqrt{2}\,s+1}\Big|_{s=(s^2+\omega_r^2)/sB} \tag{6.26}$$

or

$$H(s) = \frac{s^2 B^2}{s^4 + (\sqrt{2}\,B)s^3 + (2\omega_r^2 + B^2)s^2 + (\sqrt{2}\,B\omega_r^2)s + \omega_r^4} \tag{6.27}$$

2. The analog frequencies ω_{A1} and ω_{A2} can be found using

$$\omega_{A1} = \tan\frac{\omega_{D1}T}{2} = \tan\frac{2\pi \times 1000}{2 \times 10,000} = 0.3249 \tag{6.28}$$

$$\omega_{A2} = \tan\frac{\omega_{D2}T}{2} = \tan\frac{2\pi \times 1500}{2 \times 10,000} = 0.5095 \tag{6.29}$$

3. The center frequency ω_r and the bandwidth B can now be found:

$$\omega_r^2 = (\omega_{A1}) \times (\omega_{A2}) = 0.1655$$

and

$$B = \omega_{A2} - \omega_{A1} = 0.1846$$

Using the calculated values for ω_r and B gives

$$H(s) = \frac{0.03407s^2}{s^4 + 0.26106s^3 + 0.36517s^2 + 0.04322s + 0.0274}$$

4. Determine the desired transfer function H(z) using the BLT,

$$H(z) = H(s)\bigg|_{s=(z-1)/(z+1)} \tag{6.30}$$

The Basic program BLT.BAS in Appendix F is used to apply the bilinear transformation in (6.30) to find $H(z)$, in the form

$$H(z) = \frac{a_0 + a_1 z^{-1} + a_2 z^{-2} + a_3 z^{-3} + a_4 z^{-4}}{1 + b_1 z^{-1} + b_2 z^{-2} + b_3 z^{-3} + b_4 z^{-4}} \tag{6.31}$$

where

$$a_0 = 0.02008$$
$$a_1 = 0$$
$$a_2 = -0.04016$$
$$a_3 = 0$$
$$a_4 = 0.02008 \tag{6.32}$$

and

$$b_1 = -2.5495$$
$$b_2 = 3.2021$$
$$b_3 = -2.0359$$
$$b_4 = 0.64137$$

The transfer function $H(z)$ in (6.31) can then be written as

$$H(z) = \frac{0.02008 - 0.04016z^{-2} + 0.02008z^{-4}}{1 - 2.5495z^{-1} + 3.2021z^{-2} - 2.0359z^{-3} + 0.64137z^{-4}} \tag{6.33}$$

Rewriting (6.8), since it is more convenient to accumulate continuously, we obtain

$$u(n) = x(n) + (-b_1)u(n-1) + (-b_2)u(n-2) + \cdots + (-b_N)u(n-N) \tag{6.34}$$

Equation (6.9) is repeated again,

$$y(n) = a_0 u(n) + a_1 u(n-1) + a_2 u(n-2) + \cdots + a_N u(n-N) \tag{6.35}$$

Equations (6.34) and (6.35) are used to program the fourth-order bandpass IIR filter.

The coefficients a_N converted to hex and scaled by 2^{15} are (see also Appendix F)

$$a_0 => 291$$
$$a_1 = 0$$
$$a_2 => FADD$$
$$a_3 = 0$$
$$a_4 => 291$$

When the coefficients b_N are greater than 1 or less than -1, it is convenient to express them as

$$-b_1 = 2.5495 = (1 + 1 + 0.5495) \quad \text{and} \quad 0.5495 \times 2^{15} = >4656$$

$$-b_2 = -3.2021 = (-1 - 1 - 1 - 0.2021) \quad \text{and} \quad -0.2021 \times 2^{15} = >E622$$

$$-b_3 = 2.0359 = (1 + 1 + 0.0359) \quad \text{and} \quad 0.0359 \times 2^{15} = >498$$

$$-b_4 = -0.64137 \quad \text{and} \quad -0.6417 \times 2^{15} = >ADDD$$

```
0001                  * IIR PROGRAM
0002                  * FOURTH ORDER IIR BUTTERWORTH BANDPASS FILTER, fs=10 KHZ
0003                  * DIRECT FORM II STRUCTURE
0004                  * NO ADD/SUBTRACT , EACH COEFFICIENT SCALED BASED ON VALUE
0005                  *                                        a
0006                  *          x(n)                          0              y(n)
0007                  *           o---->----o---->----o---->----o---->----o---->--o
0008                  *                     |          |  -1     |
0009                  *                     ^     -b   v  z     a
0010                  *                     |      1   |          1   |
0011                  *           o----<----o----o---->----o
0012                  *                     |          |  -1     |
0013                  *                     ^     -b   v  z     a
0014                  *                     |      2   |          2   |
0015                  *           o----<----o----o---->----o
0016                  *                     |          |  -1     |
0017                  *                     ^     -b   v  z     a
0018                  *                     |      3   |          3   |
0019                  *           o----<----o----o---->----o
0020                  *                     |          |  -1     |
0021                  *                     ^     -b   v  z     a
0022                  *                     |      4   |          4   |
0023                  *           o----<----o----o---->----o
0024                  *  STORAGE DEFINITIONS
0025        0000  B1          EQU     >0
0026        0001  B2          EQU     >1
0027        0002  B3          EQU     >2
0028        0003  B4          EQU     >3
0029        0004  A0          EQU     >4
0030        0005  A1          EQU     >5
0031        0006  A2          EQU     >6
0032        0007  A3          EQU     >7
0033        0008  A4          EQU     >8
0034        0009  MODE        EQU     >9          MODE FOR AIB
0035        000A  CLOCK       EQU     >A          RATE
0036        000B  YN          EQU     >B          OUTPUT STORAGE
0037        000C  XN          EQU     >C          INPUT STORAGE
0038        000D  UN          EQU     >D          STORAGE FOR u(n-N)
0039        000E  UNM1        EQU     >E
0040        000F  UNM2        EQU     >F
0041        0010  UNM3        EQU     >10
0042        0011  UNM4        EQU     >11
0043                  *
0044 0000                     AORG    0
0045 0000 FF80                B       START
     0001 002B
0046 0020         TABLE       AORG    >20         START OF COEFFICIENTS-SCALED
0047 0020 5195    MB1         DATA    >5195       2.5495 * 2**13
0048 0021 9989    MB2         DATA    >9989       -3.2021    ..
0049 0022 4126    MB3         DATA    >4126       2.0359     ..
0050 0023 EB78    MB4         DATA    >EB78       -0.6417    ..
0051 0024 0291    AA0         DATA    >291        0.02007 * 2**15
0052 0025 0000    AA1         DATA    >0
0053 0026 FADB    AA2         DATA    >FADB       -0.04016   ..
0054 0027 0000    AA3         DATA    >0
0055 0028 0291    AA4         DATA    >291        0.02007    ..
0056                  * INITIALIZE AIB
0057 0029 000A    MD          DATA    >000A       MODE FOR AIB
0058 002A 03E7    RATE        DATA    >03E7       SAMPLING RATE OF 10 KHZ
0059 002B C806    START       LDPK    6           SELECT PAGE 6, TOP OF B1 >300
0060                  *LOAD FILTER COEFFICIENTS AND OTHER VALUES
0061 002C CE07                SSXM                SET SIGN EXTENSION MODE
0062 002D 5588                LARP    AR0         SELECT AR0 FOR INDIR ADDR
0063 002E D000                LRLK    AR0,>300    AR0 POINTS TO TOP OF B1
```

FIGURE 6.8. Fourth-order IIR bandpass filter program with ADD/SUBTRACT (IIR4BPAS.LST)

```
         002F 0300
0064 0030 CB0A          RPTK      >A           TRANSFER FROM PM >20-2A
0065 0031 FCA0          BLKP      TABLE,*+     TO DM >300-30A
         0032 0020
0066 0033 CA00          ZAC                    SET ACC=0
0067 0034 600D          SACL      UN           INITIALIZE TO 0  u(n-k)
0068 0035 600E          SACL      UNM1
0069 0036 600F          SACL      UNM2
0070 0037 6010          SACL      UNM3
0071 0038 6011          SACL      UNM4
0072 0039 E009          OUT       MODE,0       OUTPUT TO PORT 0
0073 003A E10A          OUT       CLOCK,1      OUTPUT TO PORT 1
0074 003B E30A          OUT       CLOCK,3      DUMMY OUTPUT TO PORT 3
0075 003C FA80  WAIT    BIOZ      MAIN         WAIT UNTIL BIO LOW
         003D 0040
0076 003E FF80          B         WAIT         LOOP BACK
         003F 003C
0077 0040 820C  MAIN    IN        XN,2         BRING IN NEW SAMPLE XN
0078 0041 2F0C          LAC       XN,15        SCALE AND STORE IN ACC
0079 0042 3C0E          LT        UNM1         TR=u(n-1)
0080 0043 3800          MPY       B1           PR= u(n-1) * (-b1 )
0081 0044 0F0E          ADD       UNM1,15      SINCE -b1=2.5495 AND
0082 0045 0F0E          ADD       UNM1,15      WAS SET TO 0.5495
0083 0046 3D0F          LTA       UNM2         TR=u(n-2),ACC PREVIOUS PR
0084 0047 3801          MPY       B2           PR=u(n-2) * (-b2)
0085 0048 1F0F          SUB       UNM2,15      SINCE -b2=-3.2021 AND
0086 0049 1F0F          SUB       UNM2,15      WAS SET TO -0.2021
0087 004A 1F0F          SUB       UNM2,15
0088 004B 3D10          LTA       UNM3         TR=u(n-3),ACC (PR)
0089 004C 3802          MPY       B3           PR=u(n-3) * (-b3)
0090 004D 0F10          ADD       UNM3,15      SINCE -b3=2.0359 AND
0091 004E 0F10          ADD       UNM3,15      WAS SET TO 0.0359
0092 004F 3D11          LTA       UNM4         TR=u(n-4),ACC (PR)
0093 0050 3803          MPY       B4           PR=u(n-4) * (-b4)
0094 0051 CE15          APAC                   ACC LAST PRODUCT
0095 0052 690D          SACH      UN,1         SHIFT 1,STORE UPPER 16 BITS
0096 0053 CA00          ZAC                    START IMPLEMENT ZEROES
0097 0054 3808          MPY       A4           PR=u(n-4)*(a4)
0098 0055 3F10          LTD       UNM3         TR=u(n-3) , ACC , DATA MOVE
0099 0056 3807          MPY       A3           PR=u(n-3)*(a3)
0100 0057 3F0F          LTD       UNM2         TR=u(n-2) , ACC , DATA MOVE
0101 0058 3806          MPY       A2           PR=u(n-2)*(a2)
0102 0059 3F0E          LTD       UNM1         TR=u(n-1) , ACC , DATA MOVE
0103 005A 3805          MPY       A1           PR=u(n-1)*(a1)
0104 005B 3F0D          LTD       UN           TR=u(n) , ACC , DATA MOVE
0105 005C 3804          MPY       A0           PR=u(n)*(a0)
0106 005D CE15          APAC                   ACCUMULATE LAST PRODUCT
0107 005E 690B          SACH      YN,1         SHIFT 1,STORE UPPER 16 BITS
0108 005F E20B          OUT       YN,2         OUTPUT FILTER RESPONSE y(n)
0109 0060 FF80          B         WAIT         GO BACK FOR NEW SAMPLE
```

FIGURE 6.8. (concluded)

since ± 1 can be implemented using add or subtract. The scaled coefficients a_N and the scaled fractional components of $-b_N$ are listed in the program in Figure 6.8. The frequency response of the bandpass filter is shown in Figure 6.9. The Basic program in Appendix F (MAGPHSE.BAS) is used to find the magnitude of (6.33), which is shown in Figure 6.10.

Figure 6.11 shows an alternative method to implement the same fourth-order IIR bandpass filter, without the multiple ADD/SUBTRACT in Figure 6.8, by scaling the coefficients in (6.33) appropriately:

1. While the a_N are scaled by 2^{15}, the b_N are scaled by 2^{13}. This effectively divides the b_N by 4 in (6.34).
2. Since the b_N are divided by 4, the input $x(n)$ is also divided by 4 in (6.34).

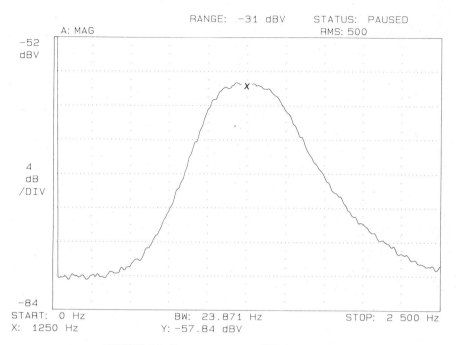

RANGE: −31 dBV   STATUS: PAUSED
A: MAG   RMS: 500

−52 dBV

4 dB /DIV

−84

START: 0 Hz   BW: 23.871 Hz   STOP: 2 500 Hz
X: 1250 Hz   Y: −57.84 dBV

FIGURE 6.9. Frequency response of IIR bandpass filter

3. The resulting $u(n)$ is then multiplied by 4 using SACH UN, 3, which produces a left shift by 3 (instead of by 1 to get rid of the extra sign bit).

Running the program in Figure 6.11 produces an output spectrum similar to that shown in Figure 6.9. A step-by-step procedure is shown in Appendix F for implementing IIR filters using the DFDP [9]. The necessary initialization file is included and can be linked with the main TMS320C25 code generated by the DFDP.

6.6 SCALING AND OVERFLOW

Scaling is a straightforward procedure but without guidelines; students will generally struggle with it. In this section we provide some general rules, a working example to clarify the procedure, and a simple simulation tool to aid in the process. Since cascade second-order sections offer better quantization noise immunity than many of the direct realizations [1], this approach is widely used in implementing IIR filters with fixed-point arithmetic. The cascade method, described previously, breaks higher-order filters into second-order sections which can be realized with the direct form II structure or with the direct form II transpose. General design rules place the sections in ascending Q-order with the highest Q section being placed at the output [8,9]. The input to each section is scaled to prevent overflow or to increase the dynamic range of the input variable. Use of the second-order section in the cascade method makes it a very important filter realization.

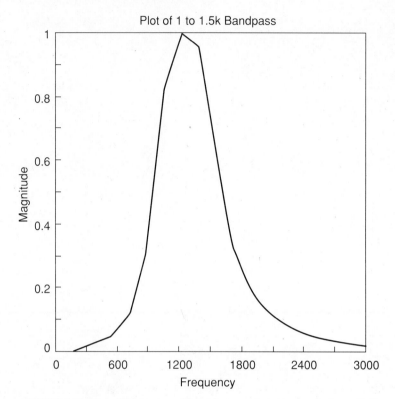

FIGURE 6.10. Plot of IIR bandpass filter using magnitude program in Appendix F

```
0001           *  IIR PROGRAM
0002           *  FOURTH ORDER IIR BUTTERWORTH BANDPASS FILTER, fs=10 KHZ
0003           *  DIRECT FORM II STRUCTURE
0004           *  ADD/SUBTRACT USED WHEN COEFFICIENTS OUTSIDE RANGE(-1==>1)
0005           *                                   a
0006           *      x(n)                          0              y(n)
0007           *          o--->---o---->----->----o---->----o--->---o
0008           *          |                |   -1            |
0009           *                  -b       v z       a
0010           *          |        1       |         1       |
0011           *          o----<----o---->--->----o
0012           *          |                |   -1            |
0013           *                  -b       v z       a
0014           *          |        2       |         2       |
0015           *          o----<----o---->--->----o
0016           *          |                |   -1            |
0017           *                  -b       v z       a
0018           *          |        3       |         3       |
0019           *          o----<----o---->--->----o
0020           *          |                |   -1            |
0021           *                  -b       v z       a
0022           *          |        4       |         4       |
0023           *          o----<----o---->--->----o
```

FIGURE 6.11. Fourth-order IIR bandpass filter program without ADD/SUBTRACT
(IIR4BPSC.LST)

```
0024                    *     STORAGE DEFINITIONS
0025         0000  B1         EQU       >0
0026         0001  B2         EQU       >1
0027         0002  B3         EQU       >2
0028         0003  B4         EQU       >3
0029         0004  A0         EQU       >4
0030         0005  A1         EQU       >5
0031         0006  A2         EQU       >6
0032         0007  A3         EQU       >7
0033         0008  A4         EQU       >8
0034         0009  MODE       EQU       >9        MODE FOR AIB
0035         000A  CLOCK      EQU       >A        RATE
0036         000B  YN         EQU       >B        OUTPUT STORAGE
0037         000C  XN         EQU       >C        INPUT STORAGE
0038         000D  UN         EQU       >D        STORAGE FOR u(n-N)
0039         000E  UNM1       EQU       >E
0040         000F  UNM2       EQU       >F
0041         0010  UNM3       EQU       >10
0042         0011  UNM4       EQU       >11
0043                    *
0044  0000                    AORG      0
0045  0000  FF80              B         START
      0001  002B
0046  0020              TABLE  AORG      >20       START OF COEFFICIENTS-SCALED
0047  0020  4656  FNB1        DATA      >4656     0.5495 * 2**15
0048  0021  E622  FNB2        DATA      >E622     -0.2021   ..
0049  0022  0498  FNB3        DATA      >498      0.0359    ..
0050  0023  ADE8  FNB4        DATA      >ADE8     -0.64137  ..
0051  0024  0291  AA0         DATA      >291      0.02007   ..
0052  0025  0000  AA1         DATA      >0
0053  0026  FADB  AA2         DATA      >FADB     -0.04016  ..
0054  0027  0000  AA3         DATA      >0
0055  0028  0291  AA4         DATA      >291      0.02007   ..
0056                    * INITIALIZE AIB
0057  0029  000A  MD          DATA      >000A     MODE FOR AIB
0058  002A  03E7  RATE        DATA      >03E7     SAMPLING RATE OF 10 KHZ
0059  002B  C806  START       LDPK      6         SELECT PAGE 6,TOP OF B1 >300
0060                    *LOAD FILTER COEFFICIENTS AND OTHER VALUES
0061  002C  CE07              SSXM                SET SIGN EXTENSION MODE
0062  002D  5588              LARP      AR0       SELECT AR0 FOR INDIRECT ADD
0063  002E  D000              LRLK      AR0,>300  AR0 POINTS TO TOP OF B1
      002F  0300
0064  0030  CB0A              RPTK      >A        TRANSFER FROM PM >20-2A
0065  0031  FCA0              BLKP      TABLE,*+  TO DMA >300-30A
      0032  0020
0066  0033  CA00              ZAC                 SET ACC=0
0067  0034  600D              SACL      UN        INITIALIZE TO 0  u(n-N)
0068  0035  600E              SACL      UNM1
0069  0036  600F              SACL      UNM2
0070  0037  6010              SACL      UNM3
0071  0038  6011              SACL      UNM4
0072  0039  E009              OUT       MODE,0    OUTPUT TO PORT 0
0073  003A  E10A              OUT       CLOCK,1   OUTPUT TO PORT 1
0074  003B  E30A              OUT       CLOCK,3   DUMMY OUTPUT TO PORT 3
0075  003C  FA80  WAIT        BIOZ      MAIN      WAIT UNTIL BIO LOW
      003D  0040
0076  003E  FF80              B         WAIT      LOOP BACK
      003F  003C
0077  0040  820C  MAIN        IN        XN,2      BRING IN NEW SAMPLE XN
0078  0041  2D0C              LAC       XN,13     SCALE AND STORE IN ACC
0079  0042  3C0E              LT        UNM1      TR=u(n-1)
0080  0043  3800              MPY       B1        PR=u(n-1) * (-b1/4 )
0081  0044  3D0F              LTA       UNM2      TR=u(n-2) , ACC
0082  0045  3801              MPY       B2        PR=u(n-2) * (-b2/4 )
0083  0046  3D10              LTA       UNM3      TR=u(n-3) , ACC
0084  0047  3802              MPY       B3        PR=u(n-3) * (-b3/4 )
0085  0048  3D11              LTA       UNM4      TR=u(n-4) , ACC
0086  0049  3803              MPY       B4        PR=u(n-4) * (-b4/4 )
0087  004A  CE15              APAC                ACC LAST PRODUCT
```

FIGURE 6.11. (continued)

```
0088 004B 6B0D          SACH      UN,3      SCALE AND MUTIPLY BY 4
0089 004C CA00          ZAC                 START IMPLEMENT ZEROES
0090 004D 3808          MPY       A4        PR=u(n-4)*(a4)
0091 004E 3F10          LTD       UNM3      TR=u(n-3) , ACC , DATA MOVE
0092 004F 3807          MPY       A3        PR=u(n-3)*(a3)
0093 0050 3F0F          LTD       UNM2      TR=u(n-2) , ACC , DATA MOVE
0094 0051 3806          MPY       A2        PR=u(n-2)*(a2)
0095 0052 3F0E          LTD       UNM1      TR=u(n-1) , ACC , DATA MOVE
0096 0053 3805          MPY       A1        PR=u(n-1)*(a1)
0097 0054 3F0D          LTD       UN        TR=u(n) , ACC , DATA MOVE
0098 0055 3804          MPY       A0        PR=u(n)*(a0)
0099 0056 CE15          APAC                ACCUMULATE LAST PRODUCT
0100 0057 690B          SACH      YN,1      SHIFT 1,STORE UPPER 16 BITS
0101 0058 E20B          OUT       YN,2      OUTPUT FILTER RESPONSE y(n)
0102 0059 FF80          B         WAIT      GO BACK FOR NEW SAMPLE
     005A 003C
0103                    END
NO ERRORS, NO WARNINGS
```

FIGURE 6.11. (concluded)

The transfer function of the second-order section is assumed to have the form

$$H(z) = \frac{a_0 + a_1 z^{-1} + a_2 z^{-2}}{1 + b_1 z^{-1} + b_2 z^{-2}} \qquad (6.36)$$

This function will be implemented using the transpose structure shown in Figure 6.12. Realization of a filter in fixed-point arithmetic requires that all the coefficients and the variables be represented in a fixed range. We will use the Q15 format throughout, which has a range of $+0.999\ldots$ to -1. What this means is that all coefficients and variables must be scaled or adjusted to fractions (actually entered as integers in the program) when they are stored in memory. Coefficients greater than 1 are represented by the fraction plus a left shift or an add operation.

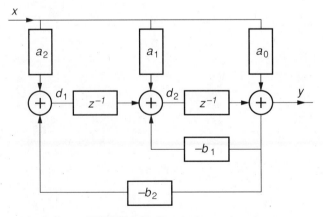

FIGURE 6.12. Transpose structure

Arithmetic overflow is one of the biggest problems in implementing an IIR filter with fixed-point arithmetic. Variable and coefficient values should be as large as possible in order to give the largest possible dynamic range but not so large as to cause overflow. In the case of the transpose structure, overflow can occur at any of the three summing nodes. The input $x(n)$ or the coefficients a_k must be scaled to prevent overflow. When scaling, care must be taken not to change the original $H(z)$. Solving for $y(n)$ in equation (6.36) and scaling $y(n)$ by s yields

$$\frac{y(n)}{s} = \frac{a_0\,x(n)}{s_1\,s_2} + \frac{a_1\,x(n-1)}{s_1\quad s_2} + \frac{a_2\,x(n-2)}{s_1\quad s_2} - \frac{b_1 y(n-1)}{s} - \frac{b_2 y(n-2)}{s}$$
$$(6.37)$$

where $y(n-k)/s$ and $x(n-k)/s_2, k = 0, 1, 2,$ represent the scaled values of $y(n-k)$ and $x(n-k)$, respectively, and $s = s_1 s_2$. The scaling factor s is distributed between the a's and the input signal x(n) as shown in equation (6.37). This equation represents the same relationship as equation (6.36) except that the output is scaled by s. When using the TMS320C25 with a 12-bit analog–digital and digital–analog conversions, the input can be scaled down by a factor of 16 without losing any resolution on the input signal. The next question is: How do we find s? Three commonly used estimates of size are provided by the expressions

$$\sum_{n=0}^{N-1} |h(n)| \qquad\qquad (6.38)$$

$$\left[\sum_{n=0}^{N-1} h(n)^2\right]^{1/2} \qquad\qquad (6.39)$$

$$\max|H(jf)| \qquad\qquad (6.40)$$

where $H(jf)$ is the frequency-domain counterpart of the unit impulse response $h(n)$. The first expression is the upper bound on the response and is the most conservative of the three. Scaling by the first expression is the only one that guarantees no overflow. The scaling estimate in (6.39) is shown for completeness but will not be discussed further. Scaling by expression (6.40) only ensures that no overflow will occur on the steady-state sinusoidal response; transients may still cause overflow.

Estimates on the sizes must be found for the transfer function between the input and the three nodes—in the case of the transpose structure: $Y(z)/X(z) = H(z)$, $D_1(z)/X(z) = H_1(z)$, and $D_2(z)/X(z) = H_2(z)$. These functions can be evaluated by running a simulation similar to the one shown in Figure 6.13. You can use this

```
10       REM **************Filter Simulation Program ***************
30       REM This routine will simulate a transpose 2nd
40       REM iir filter and scale the coefficients
60       REM
70       REM**********************************************************
80       REM
90       REM
100      DIM A(3),B(3)
111      SF=1
112      S=1
115      FLAG=0
122 REM********* Introductory Message************
123      PRINT "          SIMULATION OF 2nd ORDER H(z)"
124      PRINT "          TRANSPOSE DIRECT II FORM"
125      PRINT
126      PRINT "H(z)=[a(0)+a(1)z*-1+a(2)z**-2]/[b(0)+b(1)z*-1+b(2)z**-2]"
127      PRINT
130      REM *******input section*****************
135      IF FLAG=0 THEN GOTO 185
137 REM****reset scale multiplier
140      SF=1
142      PRINT "Filter Coefficients"
150      GOSUB 1200
160      PRINT
170      INPUT "Do you wish to change them (y/n) n";ANS$
180      IF ANS$ <> "y" THEN GOTO 380
185      FLAG=1
186      PRINT "Enter Coefficient value: or <cr> = [default]"
187      PRINT
190      CC=B(0)
195      PRINT "b(0)=[";B(0);"]";
200      INPUT  X$:GOSUB 1290
210      B(0)=CC
220      CC=B(1)
222      PRINT "b(1)=[";B(1);"]";
230      INPUT  X$:GOSUB 1290
240      B(1)=CC
250      CC=B(2)
255      PRINT "b(2)=[";B(2);"]";
260      INPUT  X$:GOSUB 1290
270      B(2)=CC
280      CC=A(0)
285      PRINT "a(0)=[";A(0);"]";
290      INPUT X$:GOSUB 1290
300      A(0)=CC
310      CC=A(1)
315      PRINT "a(1)=[";A(1);"]";
320      INPUT X$:GOSUB 1290
330      A(1)=CC
340      CC=A(2)
345      PRINT "a(2)=[";A(2);"]";
350      INPUT X$:GOSUB  1290
360      A(2)=CC
365      INPUT "Change any of the coefficients[ y/n] n";ANS$
366      PRINT
367      IF ANS$ = "y" GOTO 190
380      REM  Normalize the demoninator
390      IF B(0)=0 THEN PRINT "b(0)=0":INPUT "b(0)=",B(0)
400      B(1)=B(1)/B(0):B(2)=B(2)/B(0)
410      A(0)=A(0)/B(0):A(1)=A(1)/B(0):A(2)=A(2)/B(0)
420      B(0)=1
450      PRINT "Current output scale factor: y-->[y/S] = [y/";S;"]"
462      CC=SF
463      PRINT
490      PRINT "Enter scale factor multiplier [";SF;"]";
500      INPUT ;X$:GOSUB 1290
502      SF=CC
504      S=S*SF
505      PRINT
506      PRINT
510      A(0)=A(0)/SF:A(1)=A(1)/SF:A(2)=A(2)/SF
520      PRINT "          Normalized Coefficients:
525      PRINT "               y-->[y/";S;"]"
```

FIGURE 6.13. Simulation program (IIR2SIM.BAS)

```
530        GOSUB 1200
640        INPUT "Scale the output more [y/n] n";ANS$
645        PRINT
650        IF ANS$="y" THEN GOTO 490
660        PRINT"Select input option:"
670        PRINT "          (1)*impulse"
680        PRINT "          (2)*sinewave"
690        PRINT "          (3)*quit"
700        INPUT  I
702 REM***Call the Indicated Routine
705        ON I GOSUB 1440,5010,750
730        PRINT :INPUT "Quit [y /n] y";ANS$
732        IF ANS$ = "n" GOTO 140
733        PRINT"          Coefficients:"
734        PRINT"          y=[y/";S;"]"
735        GOSUB 1200
736        SK=2^15
737        B(1)=SGN(B(1))*INT(ABS(B(1)*SK)+.5)
738        B(2)=SGN(B(2))*INT(ABS(B(2)*SK)+.5)
739        A(0)=SGN(A(0))*INT(ABS(A(0)*SK)+.5)
740        A(1)=SGN(A(1))*INT(ABS(A(1)*SK)+.5)
741        A(2)=SGN(A(2))*INT(ABS(A(2)*SK)+.5)
750        PRINT "          Quantized Coefficients:"
752        PRINT "          y=[y/";S;"]"
754        GOSUB 1200
755        END
756 REM******************End of Main*********************
757        REM
758        REM
760        REM********CALC SUBROUTINE***************************
770        REM  Function:  this routine calculates a 2nd order section
780        REM             with a transpose structure.
790        REM             It saves the sum of |h(n)| for the impulse input
800        REM             It saves the Max H(jw) for the sinewave input
810        REM Inputs:
820        REM     I--the type of input to the filter
825        REM     W--the frequency for the sine option
830        REM     KLIM--the # of time samples
841        REM Outputs:
842        REM     Impulse response--
843        REM          |h(t)|,|h1(t)|,|h2(t)|
844        REM     Sine response--
845        REM          Max H(jw), Max H1(jw), Max H2(jw)
850        REM  *************************************************************
860        REM
870        REM
900        REM
910        REM Initialize storage
920        Y=0:Y2=0:Z1=0:Z2=0
940        REM Each loop equal one sample time
960        FOR K=0 TO KLIM
970        REM     input a value--x
975             IF I=1 THEN GOSUB 1130 ELSE X=COS (W*K)
990        REM     filter calculation
1000            Y=Z1+A(0)*X
1020            Z1=A(1)*X-B(1)*Y+Z2
1030            Z2=A(2)*X-B(2)*Y
1032            YMOLD=YM
1034            IF ABS(Y)> YM THEN YM=ABS(Y)
1040       REM     accumulate H,H1,H2
1042         IF I = 2 THEN GOTO 1100
1050            H=H+ABS(Y)
1060            H1=H1+ABS(Z1)
1070            H2=H2+ABS(Z2)
1080          GOTO 1106
1100            IF ABS (Y) > YMAX THEN YMAX = ABS(Y)
1101            IF ABS (Z1) >Z1MAX THEN Z1MAX = ABS(Z1)
1102            IF ABS (Z2) > Z2MAX THEN Z2MAX = ABS (Z2)
1106       NEXT K
1108       RETURN
1110       REM
1120       REM***********subroutine x*****************************
1130       REM Input function--impulse
```

FIGURE 6.13. (continued)

```
1140   REM returns an x value
1150   REM
1160   REM *****************************************************
1170        X=0
1180        IF K=0 THEN X=1
1190   RETURN
1200   REM **************Coefficient Print*********************
1210   REM ********Displays the coefficients******************
1220   REM
1230   PRINT
1240   PRINT "                 b(0)=";B(0)
1242   PRINT "                 b(1)=";B(1)
1244   PRINT "                 b(2)=";B(2)
1250   PRINT "                 a(0)=";A(0)
1252   PRINT "                 a(1)=";A(1)
1254   PRINT "                 a(2)=";A(2)
1260   PRINT
1270   RETURN
1280   REM
1290   REM**********String to Decimal Convert****************
1300   REM
1310   FLAG=0
1320   L=LEN(X$)
1330   IF L= 0 THEN GOTO 1420
1340   T$=LEFT$(X$,1)
1350   IF T$<>"+" OR T$<>"-" THEN GOTO 1390
1360   IF T$="-" THEN FLAG=1
1370   X$=RIGHT$(L-1)
1380   T$=LEFT$(X$,1)
1390   IF T$="." THEN X$="0"+X$
1400   CC=VAL(X$)
1410   IF FLAG=1 THEN CC=-CC
1420   RETURN
1430 REM End of String Convert*******************
1435 REM
1436 REM
1440 REM****Impulse Subroutine*****************
1460 REM*************************************
1461 REM******* Initialize Accumulators
1462   H=0:H1=0:H2=0
1463   KLIM=25
1464   CC=KLIM
1465   PRINT "Time Samples=[";KLIM;"]";
1466   INPUT X$:GOSUB 1290
1467   KLIM=CC
1468   PRINT
1470   GOSUB 760
1475   PRINT "          IMPULSE RESPONSES"
1476   PRINT
1480   PRINT "        S |h(t) |=";H
1490   PRINT "        S |h1(t)|=";H1
1500   PRINT "        S |h2(t)|=";H2
1510   PRINT
1515   RETURN
5010 REM *************Sine Wave Response************
5020 REM Function: Finds the maximum sinusoidal responses
5021 REM              at each nodes
5030 REM Routines CAlled:
5031 REM          Calculate
5032 REM                      returns Max |H(jw)|'s
5040 REM Inputs:
5041 REM          Keyboard--
5042 REM                   Sampling Frequency
5043 REM                   # of time samples
5044 REM                   # of frequency points
5046 REM          Display--
5047 REM                   Max |H(jw)|'s
5050 REM*************************************************
5120   PI = 3.14159
5125 REM *************input section ****************
5130 REM  input sample frequency
5140   CC=FS
5142   PRINT "Enter Values: or <cr> = [default]"
5143   PRINT_____
```

FIGURE 6.13. (continued)

```
5150    PRINT "Sample Frequency=[";FS;"Hz]";
5162    INPUT X$:GOSUB 1290
5170    FS= CC
5172    PRINT
5180    I$="s"
5182 REM input # of frequency points
5184    NN=50
5185    CC=NN
5190    PRINT "Frequency Points=[";NN;"]";
5200    INPUT X$:GOSUB 1290
5210    NN=CC
5212    PRINT
5214 REM # of time samples
5220       KLIM=NN/2
5225    CC=KLIM
5230     PRINT "Time Samples =[";KLIM;"]";
5240      INPUT X$:GOSUB 1290
5241    KLIM=CC
5242    PRINT
5250 REM   frequency increment
5260 REM         f0 = fs/(2*NN)
5270 REM         stf = f0/fs
5280    STF= 1!/(2*NN)
5290 REM   Set maxima to zero
5310    Z1MAX = 0
5320    Z2MAX = 0
5350 REM ********frequency loop
5360    FOR N=1 TO  NN
5370 REM******set initial conditions
5380           Z1 = 0
5390           Z2 = 0
5400           Y = 0
5410           Y2 = 0
5420 REM*******set a new frequency
5430           W = 2*PI* N* STF
5450 REM******Calculate the response at this frequency
5460           GOSUB 760
5470           NEXT N
5472           PRINT "          MAX SINE RESPONSES"
5473           PRINT
5480           PRINT "          ymax=";YMAX
5482           PRINT "          z1max="; Z1MAX
5483           PRINT "          z2max="; Z2MAX
5490           RETURN
```

FIGURE 6.13. (concluded)

program as a template or write your own from scratch. The bulk of the program in Figure 6.13 is devoted to the input of the coefficients, scaling, and menus. If you omit these features, a simple simulation can easily be written. To do a different structure, all that must be changed is the CALC subroutine which calculates the expression

$$\sum_{n=0}^{N-1} |h(n)|$$

or the maximum steady-state sinusoidal response, $\max|H(jf)|$. The scaling procedure is best clarified by considering an example of the implementation of a second-order IIR filter.

Example 6.4   Scaling a Transpose Second-Order Function.   Write a TMS320C25 program that implements the following transfer function:

$$H(z) = \frac{9.6603 + 2.0z^{-1} - 7.6603z^{-2}}{4.6928 - 4.0z^{-1} + 3.3072z^{-2}} = \frac{2.0585 + 0.4262z^{-1} - 1.6324z^{-2}}{1 - 0.8524z^{-1} + 0.7047z^{-2}}$$

$$(6.41)$$

SOLUTION:

This transfer function has a dc value of 1 and a peak value of approximately 12 at 10 Hz with a sampling frequency of 60 Hz. In order to measure the transfer function in the laboratory, it will be frequency scaled by changing the sample rate from 60 Hz to 10 kHz—thus scaling the peak to 1667 Hz. Frequency scaling works as follows: Scaling the sampling frequency by a factor of k shifts the frequency response by a factor of k on the frequency axis. In this example, the sample frequency is multiplied by a factor of 10 kHz/60 Hz; therefore, the peak frequency will be shifted to (10 kHz/60) ×(10 Hz)= 1667 Hz.

Running the simulation program with the set of inputs,

$$a_0 = 2.0585 \qquad a_1 = 0.4262 \qquad a_2 = -1.6324$$
$$b_0 = 1.0 \qquad b_1 = -0.8524 \qquad b_2 = 0.7047$$

gives the following outputs:

$$\sum_{n=0}^{N-1} |h(n)| = 16.5$$

$$\sum_{n=0}^{N-1} |h_1(n)| = 14.4$$

$$\sum_{n=0}^{N-1} |h_2(n)| = 13.3$$

$$\max|H(jf)| = 12.3$$

$$\max|H_1(jf)| = 10.5$$

$$\max|H_2(jf)| = 10.3$$

The simulation indicates a scale factor of 16 (the nearest power of 2). The input can absorb the entire scale of 16 without any loss in resolution since the I/O ports use 12-bit converters. To represent the a coefficients easily, they will be scaled by 2, and the input x(n) will be scaled by 8. Using the simulation program to scale the output by 2, results in the a coefficients being divided by 2 and thus the following list of integer decimal coefficients:

$$a_0 = 33,728 \qquad a_1 = 6983 \qquad a_2 = -26,744 \qquad b_1 = -27,930 \qquad b_2 = 23,093$$

Notice that a_0 will not fit into 16 bits; therefore, it will be divided by 2, resulting in $a_0 = 16,864$ and entered along with the other coefficients into the program memory table in the TMS320C25 program shown in Figure 6.14. The actual value of a_0 is twice what is entered into the program table. When a_0 is used in the calculation, it will be doubled (on line 77 of Figure 6.14) to give the correct value. The program in Figure 6.14 was executed on the SWDS using the AIB for I/O. The measured frequency response is shown in Figure 6.15. Notice that the peak is very close to 1667 Hz and the dynamic range from dc to the peak is about 1 to 12.

The delay variables d_1 and d_2 are stored as 16-bit quantities in the program of Figure 6.14. The 32-bit accumulated sum for each is truncated on every pass through the calculation. If these two delay variables are stored as 32-bit quantities, the intermediate overflows at these summing nodes will not affect the value $y(n)$ as long as it is in the range between $+1$ and -1 [5,10,11]. With this variation, the filter need only be scaled to ensure that $y(n)$ does not overflow, even though overflows on d_1 and d_2 are allowed. The TMS320C25 has an overflow saturation mode which will make digital overflow behave as overflow does in analog circuits: namely, saturate to the limit in either direction. If this mode is set, intermediate overflows will cause errors in the final answer, $y(n)$.

The following are general guidelines for scaling IIR filters:

1. Simulate the transfer function to get estimates on the node variables.
2. Scale the input or the input coefficients by the largest estimate to prevent overflow. The variables and coefficients should be as large as possible to maximize the accuracy and dynamic range of the transfer.
3. Coefficients greater than 1 can be represented by dividing it by a power of 2 and when using it, multiply by a power of 2 to produce the correct result. Addition of ± 1 can also be used to accomplish this for numbers slightly over 1.
4. Intermediate overflows will not affect the correctness of the final answer provided that the following conditions are met:
   a. The overflow saturation mode is inactive.
   b. The final result of the sum is in the range from $+1$ to -1.

Although simulation of a filter is an important step in successfully realizing a filter, it is frequently omitted in a student's laboratory procedure. Detecting overflows is much easier with a simulation than in a real-time operation. The unit impulse and the frequency response together give a balanced estimate of the magnitude of the node variables in a structure and can be used to scale the filter.

6.7 QUANTIZATION EFFECTS

Quantization errors occur in digital filters because digital systems must represent signal amplitudes and coefficient values in discrete steps. The size of each step

```
0001                         IDT      'IIRSC'
0002             *****************************************************
0003             * 2ND ORDER TRANSPOSE DIRECT FORM II REALIZATION*
0004             *****************************************************
0005             *SCALE x(n)  BY 8
0006             *SCALE B'S BY 2
0007             * FILTER DESCRIPTION:
0008             *
0009             *  x(n) ------+---->-----+----->----+
0010             *             |A2        |A1        |A0
0011             *             |          |          |         y(n)
0012             *             O---Z*-1---O---Z*-1---O---+-->--
0013             *             ^                 |
0014             *             |                 |    -B1     |
0015             *             |                 +------------+
0016             *             |                    -B2       |
0017             *             +------------<------------+
0018 0000                      AORG 0
0019 0000 FF80                 B        START
     0001 0007'
0020 0000                      PSEG
0021 0000             FCNST
0022             *FILTER CONSTANTS IN PROGRAM MEMORY
0023 0000 41E0                 DATA     16864          A0/2
0024 0001 1B47                 DATA     6983           A1
0025 0002 9788                 DATA     -26744         A2
0026 0003 6D1A                 DATA     27930          -B1
0027 0004 A5CB                 DATA     -23093         -B2
0028 0005
0029             *AIB SETUP PARAMETERS
0030 0005 00FA                 DATA     >FA            TRANSPARENT MODE
0031 0006 03E7                 DATA     999            SAMPLE RATE=10k
0032 0007                      PEND
0033             *DATA MEMORY
0034 0000                      DSEG                    START OF DATA SEGMENT
0035             *FILTER CONSTANTS IN DATA MEMORY
0036 0000        A0      BSS      1                     A0/2
0037 0001        A1      BSS      1                     A1
0038 0002        A2      BSS      1                     A2
0039 0003        B1      BSS      1                     -B1
0040 0004        B2      BSS      1                     -B2
0041 0005
0042 0005        MODE    BSS      1                     TRANSPARENT MODE
0043 0006        SRATE   BSS      1                     SAMPLE RATE=10K
0044 0007
0045             * VARIABLES
0046 0007        X       BSS      1                     INPUT
0047 0008        D1      BSS      1                     FIRST DELAY VARIABLE
0048 0009        Y       BSS      1                     OUTPUT
0049 000A        D2      BSS      1                     SECOND DELAY VARIABLE
0050 000B                DEND
0051 0007                PSEG                           START OF PROGRAM SEGMENT
0052             * INITIALIZE THE PROCESSOR
0053 0007 CE09   START   SPM      1                     SET AUTOMATIC LEFT SHIFT ON P OU
0054 0008 CE07           SSXM                           SET SIGN EXTENSION MODE
0055             * INITIALIZE DATA MEMORY
0056 0009 5588           LARP     AR0                    SELECT AR0
0057 000A D000           LRLK     AR0,A0                 POINT TO START OF   DATA TABLE
     000B 0000"
0058 000C CB06           RPTK     6                      LOAD REPEAT COUNTER
0059 000D FCA0           BLKP     FCNST,*+               MOVE THE CONSTANTS TO D MEMORY
     000E 0000'
0060 000F CA00           ZAC
0061 0010 CB03           RPTK     3                      LOAD REPEAT COUNTER
0062 0011 68A0           SACH     *+                     ZERO VARIABLES
0063             * INITIALIZE AIB
0064 0012 E005"          OUT      MODE,PA0               TRANSPARENT MODE
0065 0013 E106"          OUT      SRATE,PA1              FS=10k HZ
0066 0014 E306"          OUT      SRATE,PA3              DUMMY WRITE
0067 0015
0068             * MAIN PROCESSING LOOP
0069 0015 FA80   WAIT    BIOZ     IIR                    WAIT FOR SAMPLE
     0016 0019'
```

FIGURE 6.14. Second-order transpose filter program (IIR2TR.LST)

```
0070 0017 FF80         B      WAIT
     0018 0015'
0071 0019 8207" IIR     IN     X,PA2             INPUT IT
0072 001A 2D07"         LAC    X,13              SCALE INPUT BY 8
0073 001B 6807"         SACH   X                 X= X/8
0074 001C 3C07"         LT     X                 T=X
0075 001D 3800"         MPY    A0                P=(A0*X)/2
0076 001E 4008"         ZALH   D1                ACC=D1
0077 001F CE15          APAC                     DOUBLE A0
0078 0020 3A01"         MPYA   A1                ACC=D1+A0*X : P=(A1*X)/2
0079 0021 6809"         SACH   Y                 STORE OUTPUT
0080 0022 400A"         ZALH   D2                ACC=D2
0081 0023 3D09"         LTA    Y                 ACC=D2+A1*X
0082 0024 3803"         MPY    B1                P=B1*Y
0083 0025 3A04"         MPYA   B2                ACC=D2+A1*X+B1*Y : P=(B2*Y)/2
0084 0026 6808"         SACH   D1                SAVE 1ST ORDER DELAY
0085 0027 3E07"         LTP    X                 ACC=B2*Y
0086 0028 3802"         MPY    A2                P=(A2*X)/2
0087 0029 CE15          APAC                     ACC=A2*X
0088 002A 680A"         SACH   D2                STORE 2ND ORDER DELAY
0089 002B E209"         OUT    Y,PA2             OUTPUT Y
0090 002C FF80          B      WAIT              DO IT AGAIN
     002D 0015'
0091 002E              PEND
0092                   END
NO ERRORS, NO WARNINGS
```

FIGURE 6.14. (concluded)

is governed by the number of bits used to represent the values. The effects of quantization are most noticeable on short-word-length microprocessors or special-purpose digital systems and can be minimized by increasing the number of bits of resolution. Increasing the bit resolution may not be within the time or memory budget. As a result, other methods of minimizing the effects must be sought.

There are two sources of quantization error in digital filters: (1) coefficient quantization and (2) signal quantization. Coefficient quantization causes an error in the value of the coefficients because of the finite number of bits available to represent the value. This error, in turn, causes the transfer function $H(z)$ to be perturbed from the target; thus the desired $H(z)$ goes to some function, $H(z) + e$, where e is the error introduced by coefficient quantization. Since the coefficient is truncated only once, during the design of the filter, this error is constant. After the design, the filter can be simulated and redesigned if the filter does not meet the specifications due to the truncation of the filter coefficients.

Signal quantization errors occur along the calculation path whenever a product or sum is truncated or rounded. If the signal amplitude and spectrum are sufficiently large, these errors can usually be modeled as uncorrelated random signals at the point of truncation. Thus the noise analysis is treated as a linear problem.

Signal and quantization noise occurs in both floating-point and fixed-point systems. Quantization is usually less of a problem in floating-point representations because of the large number of bits of resolution usually present and its dynamic scaling ability. Because of this, coupled with the fact that most of our work with the TMS320C25 is with fixed-point arithmetic, we will confine our discussion to fixed-point arithmetic.

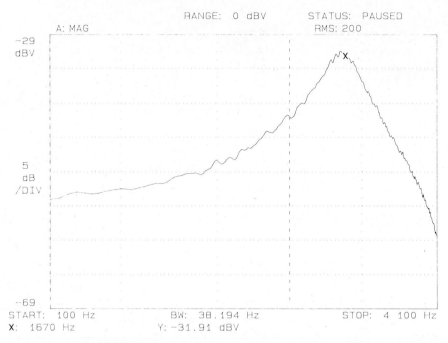

FIGURE 6.15. Frequency response of second-order transpose filter

Coefficient Quantization

Coefficient quantization errors occur because there is insufficient word length to represent the filter coefficients accurately. Pole placement in some parts of the z-plane is more sensitive to coefficient quantization than in other parts. To gain some insight into the effects of coefficient quantization on pole placement, let us consider first a one-pole IIR filter and then a two-pole direct form IIR filter.

The one-pole transfer function is

$$H(z) = \frac{1}{1 - bz^{-1}} \tag{6.42}$$

and its pole, p, is located at $z = b$. If we have two fractional bits, then b, and therefore p, can take on only the values $(0, \pm 0.25, \pm 0.5, \pm 0.75)$ along the real axis of the z-plane and still remain stable. Pole placement anywhere else inside the unit circle is impossible unless the number of fractional bits is increased.

The two-pole direct form transfer function is given by

$$H(z) = \frac{1}{(1 + b_1 z^{-1} + b_2 z^{-2})} \tag{6.43}$$

where $z^2 + b_1 z + b_2 = (z - p_1)(z - p_2) = 0$ defines the poles. If the coefficients are real and the poles are complex, they must occur in conjugates, $p_1 = p$ and $p_2 = p^*$. Furthermore, $b_2 = p_1 p_2 = |p|^2$ and $b_1 = -2 \operatorname{Re}(p)$. The discrete values of b_1 constrain the real part of the poles, as in the case of the one-pole transfer function, and the discrete values of b_2 constrain the radial spacing of the poles to discrete values. Therefore, the poles of this transfer function can only lie at the intersection of the vertical lines representing the real values and the concentric circles representing the radial values. These intersecting lines are shown in Figure 6.16 for the case of two fractional bits. Three values of b_2 will give a radius inside the unit circle. They are (0.25, 0.5, 0.75) and the corresponding radii are (0.5, 0.707, 0.866). The values of b_1 that will give real parts inside the first-quadrant unit circle are (2.0, 1.75, 1.50, 1.25, 1.0, 0.75, 0.5, 0.25), and the corresponding values of the real part of p are (1.0, 0.875, 0.75, 0.625, 0.5, 0.375, 0.25, 0.125). Poles can only be placed at the intersection of a circle and a vertical line. Examination of Figure 6.16 shows that the area around $z = \pm 1$ has fewer possible pole locations than does the area around $\pm j$. This indicates that poles cannot be placed as accurately in the region of $z = \pm 1$ as in other regions of the circle.

The sparseness of the pole locations near ± 1 indicate that whenever poles must be placed in that region, they will be very sensitive to coefficient quantization. This means that narrow lowpass and highpass filters are sensitive to coefficient quantization because their poles must lie in the regions of $+1$ and -1, respectively.

Coefficient quantization is also affected by oversampling. While oversampling tends to decrease aliasing and more accurately approximate a signal, it also tends to drive the poles nearer to $z = +1$ and therefore cause greater coefficient sensitivity.

The state equation formulation yields a structure that has uniform spacing of the poles in the z-plane:

$$\begin{bmatrix} u_1(n+1) \\ u_2(n+1) \end{bmatrix} = \begin{bmatrix} a_1 & a_2 \\ -a_2 & a_1 \end{bmatrix} \begin{bmatrix} u_1(n) \\ u_2(n) \end{bmatrix} + [b_1 \quad b_2] \begin{bmatrix} x_1(n) \\ x_2(n) \end{bmatrix} \tag{6.44}$$

or in matrix notation,

$$\bar{u}(n+1) = \mathbf{A}\bar{u}(n) + \mathbf{B}\bar{x} \tag{6.45}$$

The poles of the system are given by

$$\det|\mathbf{I}z - \mathbf{A}| = 0$$

or

$$\det \begin{vmatrix} z - a_1 & -a_2 \\ a_2 & z - a_1 \end{vmatrix} \tag{6.46}$$

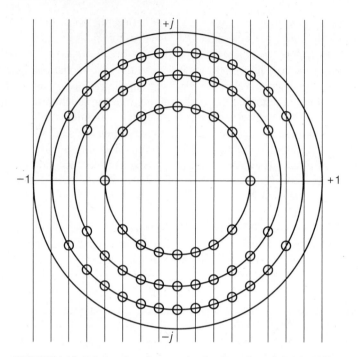

FIGURE 6.16. Pole locations for second-order direct form (2 fractional bits)

The solution to (6.46) is

$$z_1 = -a_1 + ja_2 \qquad \text{and} \qquad z_2 = -a_1 - ja_2 \qquad (6.47)$$

which are the poles of the system. The flow graph of equation (6.44) is shown in Figure 6.17. Notice that a_1 and a_2 are the coefficients needed to calculate the output and that they are also the real and imaginary parts of the poles for the second-order system. As a result, a uniformly space, rectangular grid is obtained and is shown in Figure 6.18 with the 2 fractional bits.

As the order of the equation of the direct form denominator increases, so does the sensitivity of its poles to coefficients quantization [1, 4, 12]. A polynomial representing the denominator can be expressed in a power series or in a factored form, as

$$P(z) = 1 - \sum_{n=1}^{N-1} b(n)z^{-n} = z^{-N} \prod_{n=1}^{N} (z - z_n) \qquad (6.48)$$

The movement of the ith pole is given by

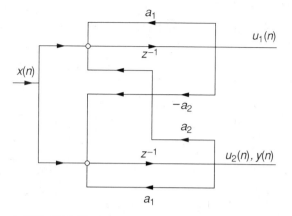

FIGURE 6.17. Second-order state-variable flow graph

$$\Delta z_i = \sum_{n=1}^{N} \frac{\partial z_i}{\partial b(n)} \Delta b(n) \qquad i = 1, 2, 3, \cdots, N \qquad (6.49)$$

The sensitivity of the *i*th pole to the *k*th coefficient can be defined by

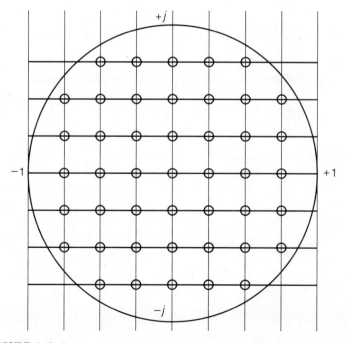

FIGURE 6.18. Pole locations for second-order state-variable filter (2 fractional bits)

$$S_{ik} = \frac{\partial z_i}{\partial b(k)} \tag{6.50}$$

The sensitivity can be expressed in terms of the partial derivatives of the polynomial of equation (6.48).

$$\frac{\partial P}{\partial z_i}\bigg|_{z=z_i} \frac{\partial z_i}{\partial b(k)} = \frac{\partial P(z)}{\partial b(k)}\bigg|_{z=z_i} \tag{6.51}$$

Taking the derivative of the left-hand side of equation (6.48) with respect to $b(k)$ yields

$$\frac{\partial P(z)}{\partial b(k)}\bigg|_{z=z_i} = -z_i^{-k} \tag{6.52}$$

Taking the derivative of the right-hand side of equation (6.48) with respect to the ith pole, z_i, yields

$$\frac{\partial P(z)}{\partial z_i}\bigg|_{z=z_i} = -z^{-N} \prod_{n=1}^{N}(z_i - z_n) \qquad n \neq i \tag{6.53}$$

Substituting (6.52) and (6.53) into equation (6.51) and solving for S_{ik} gives

$$S_{ik} = \frac{\partial z_i}{\partial b(k)} = \frac{z_i^{N-k}}{\prod_{n=1}^{N}(z_i - z_n)} \qquad n \neq i \tag{6.54}$$

From equation (6.54) several observations can be made:

1. The terms in the denominator of (6.54) can be represented by vectors drawn from each pole to the ith pole. The numerator is a vector drawn from the origin to the ith pole. The shorter the denominator vectors, the greater is the sensitivity.
2. If the poles are close together, the terms in the denominator are small and the sensitivity becomes large. Therefore, filters with the poles clustered together, such as narrowband filters, are very sensitive to coefficient quantization.
3. Poles near the unit circle tend to be more sensitive because of the z term in the numerator is larger.

Since the direct form of a higher-order filter results in a denominator similar to the one just discussed, it is very sensitive to coefficient quantization and is usually not used in filter designs above second order. Instead, the cascade or parallel form made up of second-order direct forms is normally used. Its complex conjugate poles are realized separately, thus making the quantization errors independent of the pole placement in other sections.

The next topic to discuss is the movement of the zeros of the numerator. The FIR filter is an example of a filter with only zeros; therefore, it will be discussed before taking up the zeros associated with the IIR filter. Most FIR filters are designed to have a linear phase, which implies the following symmetry:

$$a(i) = \pm a(N - i)$$

Perturbing each side of the equation with a quantization error still preserves the basic symmetry, and hence the linear phase property. The zeros of an FIR filter with linear phase must lie on the unit circle or they must be reciprocal radii pairs [2]. Therefore, a zero on the unit circle will remain on the unit circle unless the quantization causes it to move far enough to form a reciprocal radii pair.

Exercise 6.1   Coefficient Quantization and FIR Zeros.   Design an FIR lowpass filter with a Kaiser window using the ASPI DFDP. Observe and record the location of the zeros in the stopband with no quantization (this can be done by plotting the frequency response of the filter within the package). Next, quantize the coefficients to 9 bits. Observe the movement of the stopband zeros. Have they been retained? How much did the stopband amplitude change?

Now let us return to the zeros of the IIR filter. Since the higher-order direct form was ruled out because of sensitivity to coefficient quantization, we are now left with the cascade and parallel forms, both of whose poles can be placed independent of other second-order section poles. The same is not true of the zeros of the parallel form. Since the zeros do not appear explicitly in the parallel form, they must be calculated by recombining the parallel terms into one term, the numerator which specifies the zeros. Therefore, using the same argument as was used for the direct form poles, namely (6.54), we can show that the zeros of the parallel form are very sensitive to coefficient quantization.

The zeros in the cascade form, on the other hand, can be specified completely independent of the other section's zeros; and therefore they are relatively insensitive to coefficient quantization. The equation

$$a(2)z^{-2} + a(1)z^{-1} + 1 = 0 \qquad (6.55)$$

defines the zeros of a second-order cascade section. If the zeros are on the unit circle, $a(2) = 1$, there is no quantization involved with $a(2)$; thus the zeros must remain on the unit circle under quantization. The other coefficient, $a(1)$, will suffer from quantization, but this coefficient will only move the poles along the unit circle. The only case where the zeros can move off the circle is if they become real, in which case they would have reciprocal radii. The IIR digital filters that are designed from continuous analog filters, such as the Butterworth, the Chebyshev

I and II, and the elliptic, have all their zeros on the $j\omega$ axis or at infinity. If the bilinear transformation is used, the zeros of these filters will always be on the unit circle. Since one of the coefficients of the cascade form is usually a 1, this cuts down on coefficient storage and multiplication.

Exercise 6.2   Coefficient Quantization and IIR Zeros. Zeros are very important in creating stopbands in filters. Zeros are generally placed on the unit circle in the stopband region, thus driving the signal to very low levels at frequencies near them.

   a. Using the DFDP package, design a stopband filter. Select the elliptic type to keep the order small. Observe and record the zero crossings of the unquantized filter in the stopband (this can be accomplished by plotting from the package environment). Note that the ASPI package uses the cascade second-order form.
   b. Next, quantize the coefficients to 9 bits and observe and record the zero crossings in the stopband (this can be done in the laboratory or in the package environment).
   c. Using the same filter transfer function as was found in part (a), break it into parallel sections and implement it. Test it in the laboratory and check the zero crossings in the stopband.

Signal Quantization

Signal quantization occurs when the input is converted from analog to digital form, when the output signal is truncated to fewer bits than used for arithmetic, and during multiplications involving truncation or rounding. Since the input/output quantization errors are usually governed by the existing hardware, we shall concentrate our efforts on the truncation and round-off errors. Round-off errors can usually be modeled as random noise sources which are uncorrelated with each other and with the signal. This model is usually good if the amplitude is sufficiently large and the signal has a reasonably broad spectrum.

The two's-complement representation has a zero mean on the quantization noise if the signal is rounded, but has bias on the mean if the signal is truncated. This is usually of no consequence, but in some applications it may be important to have no dc on the output. The error introduced by rounding is between $\pm Q/2$, where Q is the quantization width. The variance of the rounded error signal is the same as the mean square of the error signal (since its mean is zero) and is obtained by assuming that the quantization error is evenly distributed over the quantization interval:

$$\sigma^2 = \frac{1}{Q} \int_{-Q/2}^{Q/2} e^2 de = \frac{Q^2}{12} \tag{6.56}$$

where Q is the quantization level and is 2^{-15} for 15 fractional bits.

Consider a filter whose ith summing node has several multiplications feeding into it, as shown in Figure 6.19(a). Assume that rounding is performed after the summation, which is the usual case with the TMS320C25. The rounding operation will create one noise source $e(n)$ with a variance of σ^2 which can be fed into the summing node as shown in Figure 6.19(b). Assuming that the transfer function from the ith summing node to the output is $H_i(j\omega)$, the noise variance at the output from the round-off at the ith summing node is

$$N_i(j\omega) = \sigma^2 |H_i(j\omega)|^2 \tag{6.57}$$

Since the noise sources are uncorrelated, the variances add, thus for K branches:

$$N(j\omega) = \sigma^2 \sum_{i=1}^{K} |H_i(j\omega)|^2 \tag{6.58}$$

Assuming that the mean noise component is zero, the total average power at the output is

$$e^2(n) = \frac{1}{\omega_n} \int_0^{\omega_n} N(\omega)d\omega \tag{6.59}$$

where ω_n is the upper limit of the noise band. Substituting (6.58) into (6.59) yields

$$e^2(n) = \sigma^2 \sum_{i=1}^{K} \frac{1}{\omega_n} \int_0^{\omega_n} |H_i(j\omega)|^2 d\omega \tag{6.60}$$

Using Parseval's theorem, (6.60) can be written in terms of the unit sample response from each node to the output:

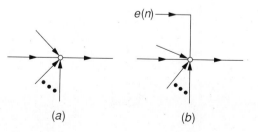

(a) (b)

FIGURE 6.19. Round-off error at a node: (a) several multiply branches into a node; (b) round-off model for the node

$$e^2(n) = \sigma^2 \sum_{i=1}^{K} \sum_{n=0}^{\infty} |h_i(n)|^2 \qquad (6.61)$$

Exercise 6.3   Simulation of Round-off Noise.   Simulate the filter shown in Figure 6.20 in a higher-level language. Find $e^2(n)$ for the filter assuming 15 fractional bits.

To evaluate $e^2(n)$ in (6.61) can be very tedious. Instead, the filter can be simulated with a high-level language, and each h_i can be found by superposition, squared, and then summed to give the total effect of round-off on the output signal. The superposition can be done as follows:

1. A unit impulse $\delta(n)$ can be injected into the ith node of the simulation with all other sources set to zero. Measure $h_i(n)$ and $|h_i(n)|^2$.

2. Repeat step 1 for all the nodes, accumulating the result each time to give $e^2(n)$.

Small Limit Cycles

Signal quantization can be modeled as uncorrelated white noise if the signal amplitude is large compared to the quantization noise and if the signal spectrum is reasonably broad. An example in which the white noise model does not apply is the small-limit-cycle oscillation, which can occur in filters with zero inputs. Filters with poles close to the unit circle are susceptible to this problem. What happens is that the round-off error produces an effective coefficient which places the pole(s) on the unit circle. The small limit cycle can be detected as a hum when the input to the filter is reduced to zero.

Consider the following first-order difference equation as a example of a system that can support small-limit-cycle oscillations:

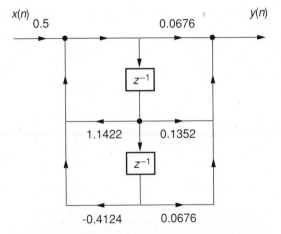

FIGURE 6.20. IIR filter for simulation problem

$$y(n) = by(n-1) + x(n) \tag{6.62}$$

This equation has a pole at $-b$ in the z-plane. For illustration, a 4-bit sign magnitude word will be used (1 sign bit plus 3 fractional bits) and b will be set to -0.5. With $x(n) = -\delta(n)$, the following solutions are obtained:

No Rounding	With Rounding	Resulting Bit Pattern
$y(0) = 0.50$	$y(0) = 0.50$	0. 1 0 0 ∣ 0
		0. 1 0 0 (rounded)
$y(1) = -0.25$	$y(1) = -0.25$	1. 0 1 0 ∣ 0
		1. 0 1 0 (rounded)
$y(2) = 0.125$	$y(2) = 0.125$	0. 0 0 1 ∣ 0
		0. 0 0 1 (rounded)
$y(3) = -0.0625$	$y(3) = -0.125$	1. 0 0 0 ∣ 1
		1. 0 0 1 (rounded)
$y(4) = 0.03125$	$y(4) = 0.125$	0. 0 0 0 ∣ 1
		0. 0 0 1 (rounded)
.	.	.
.	.	.
.	.	.
$y(9) = 0.0009$	$y(9) = 0.125$	0. 0 0 0 ∣ 1
		0. 0 0 1 (rounded)
$y(10) = 0.00005$	$y(10) = -0.125$	1. 0 0 0 ∣ 1
		0. 0 0 1 (rounded)

The solution done on a calculator with virtually no rounding decays toward zero. The solution rounded to 3 bits is oscillating between ± 0.125 with no input and is an example of a small limit cycle caused by rounding. This system will not oscillate if truncation is used instead of rounding, since truncation is less susceptible to small-limit-cycle oscillations. When the solution reaches its limit cycle, $y(n) = y(n-1)$, and the effect of rounding has made the system behave as if $b = -1$ with the pole on the unit circle.

The rounding operation in (6.62) can be expressed as

$$|R[by(n-1)] - by(n-1)| \le 0.5 \times 2^{-B} \tag{6.63}$$

where $R[\cdot]$ represents the rounding operation and B is the number of fractional bits [4]. Moreover, when the system is in its limit cycle,

$$|R[by(n-1)]| = |y(n-1)| \tag{6.64}$$

which indicates that the effective value of b is 1 and the pole is on the unit circle.

Substitute (6.64) into (6.63) and use the distributive property of the inequality/absolute operation to get:

$$|y(n-1)| - |by(n-1)| \leq 0.5 \times 2^{-B} \qquad (6.65)$$

which can be solved for $|y(n-1)|$,

$$|y(n-1)| \leq \frac{0.5 \times 2^{-B}}{1 - |b|} \qquad (6.66)$$

Equation (6.66) defines the range of $y(n-1)$ when the system is in a limit cycle. Whenever the $y(n-1)$ enters the region defined by (6.66), it will go into the limit-cycle mode and remain there until a sufficiently large signal on the input brings it out of this mode. This frequency of oscillation is one-half the sample frequency, $f_s/2$. The region defined by (6.66) is also called the dead band. Notice that for $b = -0.5$, the limit cycle is ± 0.125. Also, notice that if b in equation (6.62) approaches 1, the amplitude of the limit cycle increases. The largest fraction we can make b in our example is $\frac{7}{8}$, which gives a limit cycle of 0.5.

Second-order difference equations can also have limit cycles similar to the one just discussed. The situation is more complicated in that at least two different modes have been analyzed and a third mode reported [2, 13, 14]. We will confine ourselves to only one of these modes, the type that is similar to the one which occurs in the first-order equation just discussed. Consider the difference equation

$$y(n) = b_1 y(n-1) + b_2 y(n-2) + x(n) \qquad (6.67)$$

whose poles are defined by the equation

$$1 - b_1 z^{-1} - b_2 z^{-2} = 0 \qquad (6.68)$$

If $b_2 = -1$, the poles will be on the unit circle and the system will oscillate. There are two rounding operations, one with b_1 and one with b_2. The rounding operation $R[b_2 y(n-2)]$ can place the poles on the unit circle and make b_2 effectively -1, if

$$R[b_2 y(n-2)] = y(n-2) \qquad (6.69)$$

From the rounding operation (6.63), we have the relationship

$$|R[b_2 y(n-2)] - b_2 y(n-2)| \leq 0.5 \times 2^{-B} \qquad (6.70)$$

Substitute (6.69) into (6.70) to get the oscillation limit or the dead band,

$$|y(n - 1)| \leq \frac{0.5 \times 2^{-B}}{1 - |b_2|} \qquad (6.71)$$

If the input is zero and $y(n - 1)$ enters the dead band region defined by (6.71), the coefficient b_2 will effectively be -1 due to the round-off operation, thus placing the poles on the unit circle and causing the system to oscillate. The frequency of oscillation will be controlled by b_1, not by the sample rate as was the case with the first-order equation.

Exercise 6.4 Small-Cycle Oscillation. Implement the following filter on the TMS320C25,

$$H(z) = \frac{1}{1 - 1.73z^{-1} + 0.99z^{-2}}$$

with $f_s = 10$ kHz. Observe the output on an oscilloscope. First excite the filter by driving it with a sine wave for a few seconds. Then remove the sine wave source. If a small limit cycle exists, a small oscillation will remain indefinitely on the output.

Overflow Oscillations

On machines that use two's-complement arithmetic, it is possible to get large-scale oscillations caused by overflow. When overflow occurs, the amplitude of the number goes from a largest positive number to a large negative number, or vice versa. Consider the following scenario. A filter's output first overflows in the positive direction, which causes the output to jump to a large negative value. This negative value after overflow is also a large number which is fed back into the filter and will probably cause overflow again—this time in the negative direction, causing another jump back to the positive limit, and so on, thus setting up a large-scale oscillation.

The TMS320C25 can operate in a saturation mode, which will cause the over-flows to saturate in a fashion similar to the way in which amplifiers saturate, thus avoiding the large-scale jump and the overflow oscillation. This mode will eliminate overflow oscillation only if the overflow occurs out of the accumulator. The saturation mode does not cause saturation if the overflow resulted from a left shift of the parallel shifter.

Exercise 6.5 Overflow Oscillation. Implement the following filter with the transpose direct form on the TMS320C25:

$$H(z) = \frac{1}{1 - z^{-1} + 1/2z^{-2}}$$

with $f_s = 10$ kHz. Set the input equal to 0.5 with zero initial conditions. Do you get a large-scale limit cycle? Try other combinations of initial conditions and inputs to see if a large-scale limit cycle can be set up. Set the saturation mode and test to see how the performance of the filter is changed.

REFERENCES

[1]   T. W. Parks and C. S. Burrus, *Digital Filter Design*, Wiley, New York, 1987.

[2]   L. B. Jackson, *Digital Filters and Signal Processing*, Kluwer Academic, Norwell, Mass., 1986.

[3]   W. D. Stanley, G. R. Dougherty, and R. Dougherty, *Digital Signal Processing*, Reston, Reston, Va., 1984.

[4]   A. V. Oppenheim and R. Schafer, *Discrete-Time Signal Processing*, Prentice-Hall, Englewood Cliffs, N.J., 1989.

[5]   L. R. Rabiner and B. Gold, *Theory and Application of Digital Signal Processing*, Prentice-Hall, Englewood Cliffs, N.J., 1975.

[6]   N. Ahmed and T. Natarajan, *Discrete-Time Signals and Systems*, Reston, Reston, Va., 1983.

[7]   DSP Committee, IEEE ASSP (editors), *Selected Papers in Digital Signal Processing II*, IEEE Press, New York, 1976.

[8]   L. B. Jackson, "Roundoff Noise Analysis for Fixed-Point Digital Filters Realized in Cascade or Parallel Form," *IEEE Transactions on Audio and Electroacoustics*, **Au-18**, June 1970, pp. 107–122.

[9]   *Digital Filter Design Package (DFDP)*, Atlanta Signal Processors Inc., Atlanta, Ga., 1987.

[10] S. Waser and M. Flynn, *Introduction to Arithmetic for Digital Systems Designers*, Holt, Rinehart and Winston, New York, 1982.

[11] H. L. Garner, "Theory of Computer Addition and Overflows," *IEEE Transactions on Computers*, **C-27** (4), April 1978, pp. 297–301.

[12] J. F. Kaiser, "Some Practical Considerations in the Realization of Linear Digital Filters," *Proceedings of the 3rd Allerton Conference on Circuit and System Theory*, October 1965, pp. 621–633.

[13] L. B. Jackson, "An Analysis of Limit Cycles due to Multiplicative Rounding in Recursive Digital Filters," *Proceedings of the 7th Allerton Conference on Circuit and System Theory*, 1969, pp. 69–78.

[14] V. B. Lawrence and K. V. Mina, "A New and Interesting Class of Limit Cycles in Recursive Digital Filters," *Proceedings of the IEEE International Symposium on Circuit and Systems*, April 1977, pp. 191–194.

[15] DSP Committee, IEE ASSP (editors), *Programs for Digital Signal Processing*, IEEE Press, New York, 1979.

[16] D. L. Jones and T. W. Parks, *A Digital Signal Processing Laboratory*, Prentice-Hall, Englewood Cliffs, N.J., 1988.

[17] K. S. Lin (editor), *Digital Signal Processing Applications with the TMS320 Family*, Vol. 1, Prentice-Hall, Englewood Cliffs, N.J., 1988.

[18] D. J. DeFatta, J. G. Lucas, and W. S. Hodgkiss, *Digital Signal Processing: A System Approach*, Wiley, New York, 1988.

[19] *Second-Generation TMS320 User's Guide*, Texas Instruments Inc., Dallas, Tex., 1987.

[20] *TMS320C1x/TMS320C2x Assembly Language Tools User's Guide*, Texas Instruments Inc., Dallas, Tex., 1987.

[21] *TMS320C2x Software Development System User's Guide*, Texas Instruments Inc., Dallas, Tex., 1988.

[22] *TMS320C1x/TMS320C2x Source Conversion Reference Guide*, Texas Instruments Inc., Dallas, Tex., 1988.

[23] *TMS32010 Analog Interface Board User's Guide*, Texas Instruments Inc., Dallas, Tex., 1984.

7

Fast Fourier Transform

CONCEPTS AND PROCEDURES

- *Computation of the discrete Fourier transform*
- *Examples covering both the decimation-in-frequency and the decimation-in-time, with radix-2 and radix-4*
- *Use of the indirect addressing instruction, with reverse carry, to perform bit reversal*

The discrete Fourier transform (DFT) is widely used in digital signal processing for the analysis of discrete-time signals [1–9]. It is used to transform a discrete-time sequence into a discrete-frequency representation. The direct computation of the DFT requires approximately N^2 complex operations, N being the number of samples.

A class of efficient DFT calculations are called fast Fourier transforms (FFT's). FFTs reduce considerably the computational requirements of the DFT, from approximately N^2 to $N(\log N)$ for a basic FFT. Two different procedures for arriving at an FFT are covered: decimation-in-frequency (DIF) and decimation-in-time (DIT).

We discuss the FFT for computing an N-point DFT, with both radix-2 and radix-4 algorithms. Examples for both radices are presented: a radix-2 FFT for $N = 8$ and $N = 16$, and a radix-4 FFT for $N = 16$. The procedure used to calculate the FFT causes the set of data to become scrambled. A special instruction, indirect addressing with reverse carry, is available with the TMS320C25 to perform an unscrambling operation on the data [10].

7.1 INTRODUCTION

The Fourier transform is used in DSP to convert data or information from the time domain to the frequency domain. Data available in the frequency domain

149

can be converted back to the time domain using the inverse Fourier transform. A very efficient algorithm based on the discrete Fourier transform is known as the fast Fourier transform (FFT). In recent years, special cases and modifications of the FFT as well as other transform procedures, such as the Hartley transform and the Winograd transform, to convert data from the time domain to the frequency domain, or vice versa, have evolved. Speed is a very important factor in the FFT process since real-time computation of the FFT is required in many applications. Digital signal processors such as the TMS320C25 are well suited for this application.

7.2 DISCRETE FOURIER TRANSFORM ALGORITHM

The Fourier transform of an analog signal $x(t)$ is defined as

$$X(\omega) = \int_{-\infty}^{\infty} x(t) e^{-j\omega t} dt \qquad (7.1)$$

Similarly, the discrete Fourier transform (DFT) of a sampled signal $x(nT)$ is defined as

$$X(k) = \sum_{n=0}^{N-1} x(n) W_N^{nk} \qquad k = 0, 1, \cdots, N - 1 \qquad (7.2)$$

or

$$X(k) = x(0) + x(1) W_N^k + x(2) W_N^{2k} + \cdots + x(N - 1) W_N^{(N-1)k}$$

where $x(nT)$ is replaced by $x(n)$ with the sampling period T implied and the frame has N samples; $W_N = e^{-j2\pi/N}$ is the phase term and is known as the *twiddle factor*. The computation required by the DFT can be intensive, especially for large N. To compute each $X(k)$, we need approximately N complex additions and N complex multiplications. Since there are N DFT coefficients, we would then need a total of approximately N^2 complex additions and N^2 complex multiplications. The FFT is an efficient algorithm for the calculation of (7.2). It takes advantage of the periodicity and symmetry of the twiddle factor W_N:

$$W_N^k = W_N^{k+N} = -W_N^{k+N/2} \qquad (7.3)$$

as shown in Figure 7.1. The twiddle factor W_N will also be written as W, with the number of samples N implied.

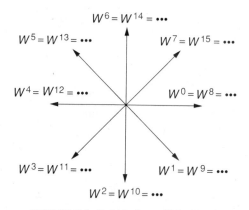

FIGURE 7.1. Twiddle factor W for $N = 8$

7.3 DEVELOPMENT OF THE FFT ALGORITHM: RADIX-2

The FFT reduces significantly the amount of computation required by decomposing an N-point DFT into successively smaller transforms. The radix of the FFT is determined by the size of the smallest transform. For example, for a radix-2 FFT, N must be a power of 2 and the smallest transform is the 2-point DFT. Initially, an N-point DFT can be decomposed into two $(N/2)$-point DFTs. Each $(N/2)$-point DFT can be further decomposed into two $(N/4)$-point DFTs, and so on, until a set of $(N/2)$ 2-point DFTs is obtained. Before proceeding, we must find a rule for dividing an N-point FFT into two $(N/2)$-point FFTs. Once we have this rule, we can apply it repeatedly until we arrive at $(N/2)$ 2-point FFTs. Let us start by considering an N-point time sequence $x(n)$. The sequence $x(n)$ can be decomposed into two halves:

$$x(0), x(1), \cdots, x(N/2 - 1)$$

and

$$x(N/2), x(N/2 + 1), \cdots, x(N - 1)$$

Also, equation (7.2) can be divided into two summations,

$$X(k) = \sum_{n=0}^{(N/2)-1} x(n) W^{nk} + \sum_{n=N/2}^{N-1} x(n) W^{nk} \tag{7.4}$$

By letting $n = n + N/2$, in the second summation, both summation limits can be made the same,

$$X(k) = \sum_{n=0}^{(N/2)-1} x(n) W^{nk} + W^{kN/2} \sum_{n=0}^{(N/2)-1} x(n + N/2) W^{nk} \tag{7.5}$$

Since $W^{kN/2} = e^{-j\pi k} = (-1)^k$, equation (7.5) becomes

$$X(k) = \sum_{n=0}^{(N/2)-1} [x(n) + (-1)^k x(n + N/2)]W^{nk} \tag{7.6}$$

Equation (7.6) can be separated into an equation for even k and an equation for odd k,

$$X(k) = \sum_{n=0}^{(N/2)-1} [x(n) + x(n + N/2)]W^{nk} \qquad \text{for } k \text{ even} \tag{7.7}$$

and

$$X(k) = \sum_{n=0}^{(N/2)-1} [x(n) - x(n + N/2)]W^{nk} \qquad \text{for } k \text{ odd} \tag{7.8}$$

Substituting $k = 2k$ for even k and $k = 2k + 1$ for odd k yields

$$X(2k) = \sum_{n=0}^{(N/2)-1} [x(n) + x(n + N/2)]W^{2nk} \qquad k = 0, 1, \cdots, N - 1 \tag{7.9}$$

and

$$X(2k + 1) = \sum_{n=0}^{(N/2)-1} [x(n) - x(n + N/2)]W^n W^{2nk} \qquad k = 0, 1, \cdots, N - 1 \tag{7.10}$$

To write (7.9) and (7.10) as $(N/2)$-point FFTs, one can replace W_N^2 by $W_{N/2}$. Furthermore, let

$$a(n) = x(n) + x(n + N/2) \tag{7.11}$$

and

$$b(n) = x(n) - x(n + N/2) \qquad n = 0, 1, \cdots, N/2 - 1 \tag{7.12}$$

to give two separate $N/2$ FFTs:

$$X(2k) = \sum_{n=0}^{(N/2)-1} a(n)W_{N/2}^{nk} \tag{7.13}$$

and

$$X(2k + 1) = \sum_{n=0}^{(N/2)-1} b(n) W_N^n W_{N/2}^{nk} \qquad (7.14)$$

The sum of $X(2k)$ and $X(2k + 1)$ is the N-point FFT,

$$X(k) = X(2k) + X(2k + 1)$$

or

$$X(k) = \sum_{n=0}^{(N/2)-1} a(n) W_{N/2}^{nk} + \sum_{n=0}^{(N/2)-1} b(n) W_{N/2}^{nk} W_N^n \qquad (7.15)$$

Equation (7.15) is the rule for rearranging the sequence so that two $(N/2)$-point FFTs can be added to produce an N-point FFT. Figure 7.2 graphically depicts this rule for $N = 8$. Note that the even X's are on the upper half and the odd X's on the lower half. By splitting the N-point DFT into two $(N/2)$-point DFTs, the total number of complex multiplications and additions is now reduced to approximately $N + 2(N/2)^2$. Additional reduction in the number of computations is possible when both the symmetry and periodicity in (7.3) are utilized to find the X's. Repeating the decimation process, each of the $(N/2)$-point DFTs can be computed using two $(N/4)$-point DFTs and so on until we have $(N/2)$ 2-point DFTs. The 2-point DFT is as far as the decomposition can go. The DFT of a 2-point sequence can be obtained from equation (7.2).

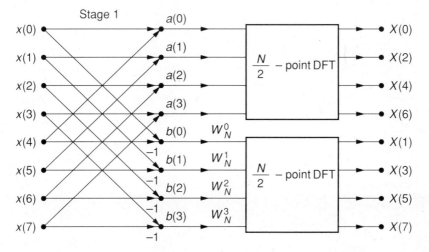

FIGURE 7.2. Decomposition of N-point DFT into two $(N/2)$-point DFTs

$$X(k) = \sum_{n=0}^{1} x(n) W_2^{nk} \tag{7.16}$$

Expanding equation (7.16) gives the individual terms. There are no multiplications required and only two additions.

$$X(0) = x(0)W_2^0 + x(1)W_2^0 = x(0) + x(1) \tag{7.17}$$

$$X(1) = x(0)W_2^0 + x(1)W_2^1 = x(0) - x(1) \tag{7.18}$$

Figure 7.3 shows a flow graph for a 2-point FFT referred to sometimes as a butterfly. The butterfly is equivalent to equations (7.17) and (7.18). The value at the top output node of the graph is equal to the sum of the values flowing into it. The value at the bottom output node is equal to the value at the top input node minus the value directly across from it. Multiplicative factors (twiddle factors) are written next to the flow path (for the 2-point butterfly, they are all unity).

In preparation for the next decimation, the outputs in the upper half in Figure 7.2 are numbered $X(0)$, $X(4)$, as the even values and $X(2)$, $X(6)$, as the odd values. Similarly, the outputs in the lower half are reordered $X(1)$, $X(5)$, as the even and $X(3)$, $X(7)$, as the odd values.

The two-stage decomposition process is shown in Figure 7.4. The complete decomposition of an 8-point DFT is shown in Figure 7.5. This algorithm is referred to as decimation-in-frequency (DIF). In this process, the output sequence $X(k)$, which is the discrete Fourier transform (DFT) of $x(n)$, is broken down successively into smaller subsequences. Note that the decimation process goes through M iterations or stages, where $N = 2^M$. Figure 7.5 shows an in-place transform; that is, the data memory locations used for the input samples are also used for the output results. Since the computations are complex, $2N$ data memory locations are required to store real and imaginary data values. Note that the same data memory locations can also be used for storing intermediate results. The data memory requirements for storing the twiddle factor W can be reduced if the 13-bit MPYK instruction is used in lieu of the 16-bit MPY instruction. This gain in data storage is made at the expense of precision in the sine and cosine terms of the twiddle factor W.

A very important consideration in the FFT implementation is scaling. Overflow is caused not only by the input data, but also by intermediate output results

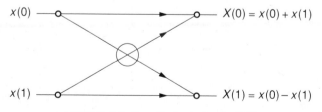

FIGURE 7.3. Butterfly for the 2-point FFT

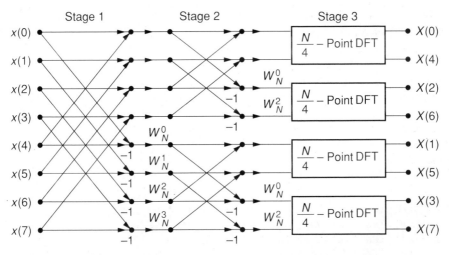

FIGURE 7.4. Decomposition of N-point DFT into four $(N/4)$-point DFTs

after each iteration. After an iteration, the intermediate values should be less than 1. Scaling can be accomplished by dividing by 2 before each iteration. The resulting output would then be $(1/2)^M$, which is one-eighth of the actual output value for a three-stage 8-point FFT. Scaling can be accomplished using the shifting capabilities of the TMS320C25.

The final order of the DFT sequence is $X(0)$, $X(4)$, $X(2)$, $X(6)$, $X(1)$, $X(5)$, $X(3)$, and $X(7)$. This order is a direct result of the rearrangement of the x's in the algorithm. The output sequence obtained is referred to as *scrambled*. A *bit-reversal* procedure is used to unscramble the output sequence. The bit reversal is implemented by swapping the order of the bits in the following fashion: 100 is

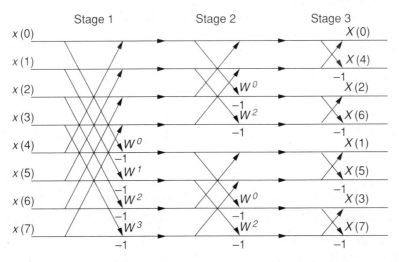

FIGURE 7.5. An 8-point DFT decimation-in-frequency

replaced by 001 by swapping the first and third bits. By using bit reversal on $X(4)$, $X(6)$, $X(1)$, and $X(3)$, the output sequence will become unscrambled. Note that the bit-reversal process is also applicable for larger values of N. For example, for 5 bits, ($N = 32$) the first bit would be swapped with the fifth, and the second bit would be swapped with the fourth.

Bit reversal can be done by swapping data and generally requires one temporary storage location for swapping. However, with the double precision of the accumulator, bit reversal can be performed without the use of a temporary storage location, using instead, instructions that can address the lower and higher 16 bits of the accumulator independently. For example, to swap the contents of data memory locations DM1 and DM2, the following steps can be performed:

1. ZALH   DM1   to load content of DM1 in upper 16 bits of ACC
2. ADDS   DM2   to load content of DM2 in lower 16 bits of ACC
3. SACL   DM1   to store lower ACC in DM1
4. SACH   DM2   to store upper ACC in DM2

Later in the chapter we will show that by scrambling the input sequence, the output sequence will be ordered (unscrambled).

Bit Reversal with Indirect Addressing Mode

An alternative method for proper sequencing of data is to take advantage of the indirect addressing mode with bit reversal, introduced in Chapter 2 [10,11]. The instruction

```
IN *BR0+,2
```

inputs the data from port 2 and transfers them directly into the data memory location specified by the current auxiliary register, ARX. After using the content of ARX as the address, the content of AR0 is added to the content of ARX with reverse carry propagation. For example, if AR0 contains 1000 and ARX contains 1000001000, the addition with reverse carry will be executed as follows:

$$
\begin{array}{r}
1000001000 \\
1000 \\
\hline
1000000100
\end{array}
$$

This single instruction combines an input instruction with a bit-reversal function. The following illustrates the bit-reversal procedure for swapping data addresses.

Swapping of Addresses

Given is the following ordered sequence of data samples of length $N = 8$ for an 8-point FFT, with their associated even memory addresses:

Data	Binary Representation	Memory Address
X0	000	512
X1	001	514
X2	010	516
X3	011	518
X4	100	520
X5	101	522
X6	110	524
X7	111	526

These data values represent the real components of the data samples (the imaginary components would be located in the odd memory addresses 513, 515, · · ·, 527). It is desired to scramble or reorder the sequence as follows:

Data	Bit-Reversal Representation	Memory Address (Decimal)	Memory Address (Binary)
X0	000	512	0000 0010 0000 0000
X4	100	520	0000 0010 0000 1000
X2	010	516	0000 0010 0000 0100
X6	110	524	0000 0010 0000 1100
X1	001	514	0000 0010 0000 0010
X5	101	522	0000 0010 0000 1010
X3	011	518	0000 0010 0000 0110
X7	111	526	0000 0010 0000 1110

Note that the sequence of data X0, X4, . . . , X7 follows from a bit reversal on the binary sequence.

The following program segment will perform bit reversal on the 8-length sequence just shown:

```
LRLK      AR1,512
LARK      AR0,8
LARP      AR1
RPTK      7
IN        *BRO+,2
```

The program segment loads AR1 with the base address 512, loads AR0 with the length of the FFT, reads the input samples X0, X1, . . . , X7 and transfers them directly into memory addresses 512, 520, 516, . . . , 526, in the scrambled or desired order. As the content of AR0 = 1000 (binary) is subsequently added, with reverse carry, to the address location contained in AR1, the desired addresses are obtained. The bit reversal with reverse carry procedure can also be used to resequence a scrambled output. In this case the instruction OUT *BRO–,2, with the current auxiliary register loaded with the last (highest) address, performs a similar operation as the IN *BR0+, 2.

Example 7.1   8-Point FFT Radix-2, Decimation-in-Frequency. Figure 7.5 is used to compute an 8-point FFT. The input data samples shown in Figure 7.6 represent a rectangular waveform. Consider the intermediate results, after the first stage, in Figure 7.6. $x(0)$ is replaced by $[x(0) + x(4)]$, $x(1)$ by $[x(1) + x(5)]$, and so on, with $x(7)$ replaced by $[x(3) - x(7)]\, W^3$. For $N = 8$, the values for the twiddle factors W_8 are

$$W^0 = 1$$

$$W^1 = e^{-j2\pi/8} = e^{-j\pi/4} = \cos(\pi/4) - j\sin(\pi/4) = 0.707 - j0.707$$

$$W^2 = -j$$

$$W^3 = -0.707 - j0.707$$

$$W^4 = -1 = -W^0$$

$$W^5 = -0.707 + j0.707 = -W^1$$

$$W^6 = +j = -W^2$$

$$W^7 = 0.707 + j0.707 = -W^3$$

Use the diagram in Figure 7.6 to calculate $X(3)$:

1. After stage 1:

$$x(4) \rightarrow [x(0) - x(4)]$$

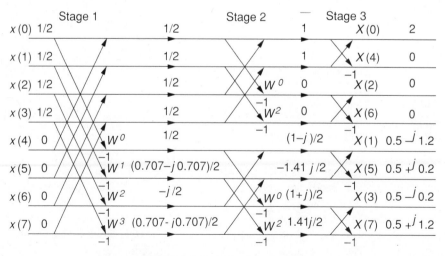

FIGURE 7.6. Example of 8-point FFT

$$x(5) \rightarrow [x(1) - x(5)]W$$
$$x(6) \rightarrow [x(2) - x(6)]W^2$$
$$x(7) \rightarrow [x(3) - x(7)]W^3$$

2. After stage 2:

$$x(6) \rightarrow [x(4) - x(6)] \quad \text{with } x(4) \text{ and } x(6) \text{ obtained from stage 1}$$
$$x(6) \rightarrow [x(0) - x(4)] - [x(2) - x(6)]W^2$$
$$x(7) \rightarrow [x(5) - x(7)]W^2 \quad \text{with } x(5) \text{ and } x(7) \text{ obtained from stage 1}$$
$$x(7) \rightarrow \{[x(1) - x(5)]W - [x(3) - x(7)]W^3\}W^2$$
$$x(7) \rightarrow [x(1) - x(5)]W^3 + [x(3) - x(7)]W$$

3. After stage 3:

$$X(3) = x(6) + x(7) = [x(0) - x(4)] - [x(2) - x(6)]W^2$$
$$+ [x(1) - x(5)]W^3 + [x(3) - x(7)]W$$

with $x(6)$ and $x(7)$ obtained from stage 2. This result can be verified using (7.2) or (7.14):

$$X(3) = [x(0) - x(4)] + [x(1) - x(5)]WW^2 + [x(2) - x(6)]W^2W^4$$
$$+ [x(3) - x(7)]W^3(-W^2)$$

using $W^4 = -1$. Substituting values for the input sequence and the twiddle factors gives,

$$X(3) = \frac{1 - 0.414j}{2}$$

Bit reversal can then be used to reorder the output sequence X's. Note that to obtain a plot of the output, the magnitude must be taken, since each of the X's may be complex.

7.4 DECIMATION-IN-TIME

The previous implementation which used the decomposition of the output sequence into smaller subsequences was referred to as decimation-in-frequency (DIF). If instead, the input sequence is decomposed into smaller subsequences, the process is referred to as decimation-in-time (DIT). Let us proceed by decomposing $x(n)$

into an even sequence $[x(0), x(2), \ldots, x(2n)]$ and an odd sequence $[x(1), x(3), \ldots, x(2n + 1)]$.

This time, we need a rule for combining two $(N/2)$-point FFTs into an N-point FFT,

$$X(k) = \sum_{n=0}^{(N/2)-1} x(2n) W^{2nk} + \sum_{n=0}^{(N/2)-1} x(2n + 1) W^{(2n+1)k} \tag{7.19}$$

Using $W_N^2 = W_{N/2}$ yields

$$X(k) = \sum_{n=0}^{(N/2)-1} x(2n) W_{N/2}^{nk} + W_N^k \sum_{n=0}^{(N/2)-1} x(2n + 1) W_{N/2}^{nk} \tag{7.20}$$

Equation (7.20) is such that each of the summations is an $(N/2)$-point DFT. Let

$$C(k) = \sum_{n=0}^{(N/2)-1} x(2n) W_{N/2}^{nk} \tag{7.21}$$

and

$$D(k) = \sum_{n=0}^{(N/2)-1} x(2n + 1) W_{N/2}^{nk} \tag{7.22}$$

Equation (7.16) can be rewritten as

$$X(k) = C(k) + W_N^k D(k) \qquad k = 0, 1, \cdots, N - 1 \tag{7.23}$$

which shows that $X(k)$ is a linear combination of two $(N/2)$-point DFTs. Notice that for $C(k)$ and $D(k)$, the range on k is 0 to $(N/2)$ -1, but the range on k for $X(k)$ is from 0 to $N - 1$; therefore, equation (7.23) must be interpreted for $k > (N/2)$ $- 1$. Since $W_N^{N/2+k} = -W_N^k$ and the coefficients, $C(k)$ and $D(k)$, are periodic in $N/2$, equation (7.23) can be rewritten for arguments greater than $N/2$:

$$X(k + N/2) = C(k) - W_N^k D(k) \qquad k = 0, 1, \cdots, N/2 - 1 \tag{7.24}$$

Substituting $N = 8$ in equations (7.23) and (7.24) gives

$$X(k) = C(k) + W_8^k D(k) \qquad 0 \le k \le 3 \tag{7.25}$$

and

$$X(k + 4) = C(k) - W_8^k D(k) \qquad 0 \le k \le 3 \tag{7.26}$$

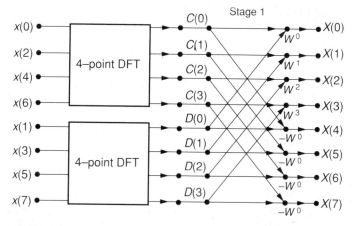

FIGURE 7.7. An 8-point DFT into two 4-point DFTs (decimation-in-time)

The results in equations (7.25) and (7.26) represent Figure 7.7, which shows the combination of two 4-point sequences into one 8-point sequence. Repeating the process, each ($N/4$)-point time sequence is further divided into two 2-point sequences as shown in Figure 7.8. Now, we are ready to take the DFT of each 2-point time sequence and use the flow graph in Figure 7.7 to combine them into two 4-point sequences, then into one 8-point sequence. The complete decomposition, which is a decimation-in-time, is shown in Figure 7.9. Note that the input sequence appears scrambled in a fashion similar to the output sequence $X(k)$ in the decimation-in-frequency process. The output sequence is properly ordered and need not be unscrambled. Computationally, there is no difference between the two decimation procedures.

Another way to look at the 8-point decimation-in-time algorithm is the following:

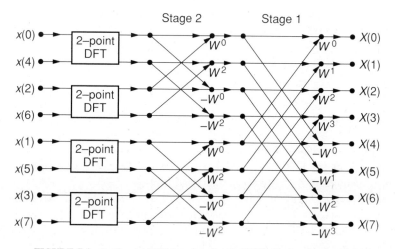

FIGURE 7.8. An 8-point DFT into four 2-point DFTs (decimation-in-time)

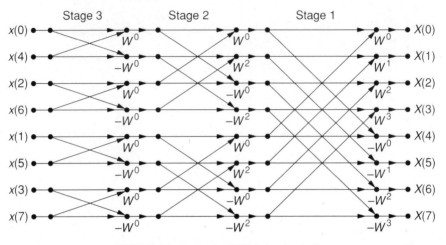

FIGURE 7.9. An 8-point DFT decimation-in-time

1. The 8-point time sequence is successively divided by 2 until we get four 2-point sequences. This is as far as the division process can go.

2. Take the DFT of each 2-point sequence. This is easily done; it requires just two additions. We now have transformed the time sequence into the frequency domain.

3. The remainder of the procedure deals with how to combine these N frequency samples into an 8-point DFT. Equations (7.23) and (7.24) give us the rule for going from two ($N/2$)-point sequences to one N-point sequence. This rule must be applied at each stage as shown in the flow graph in Figure 7.9.

7.5 INVERSE FAST FOURIER TRANSFORM

The procedure used to develop the FFT algorithm can be extended to find the inverse discrete Fourier transform (IDFT) defined as

$$x(n) = \frac{1}{N} \sum_{n=0}^{N-1} X(k) W_N^{-nk} \qquad n = 0, 1, \cdots, N - 1 \qquad (7.27)$$

where $W_N = e^{-j2\pi/N}$, as before. If we compare (7.27) with (7.2), we see that the same FFT algorithm can also be used to compute the IDFT, by noting two differences: a scaling factor of $1/N$, and W_N^{-nk} instead of W_N^{nk}. Hence the same flow graphs for the FFT can be used if we replace the twiddle factor W_N by its complex conjugate W_N^* and indicate a scaling by $1/N$.

A more practical way of doing the IFFT will now be explained. If we multiply both sides of equation (7.27) by N and take the complex conjugate, the expression on the right is the DFT of $X^*(k)$ and can be calculated using the FFT algorithms just discussed:

$$Nx^*(n) = \sum_{n=0}^{N-1} X^*(k) W^{nk} \tag{7.28}$$

writing W_N as W. Solving equation (7.28) results in a rule for doing the IDFT:

$$x(n) = \frac{1}{N} \left[\sum_{n=0}^{N-1} X^*(k) W^{nk} \right]^* \tag{7.29}$$

The rule that equation (7.29) implies is: Conjugate the input $X(k)$, take the DFT, then take the conjugate of the summation and divide it by N. Normally, $x(n)$ is real; therefore, the second conjugation is unnecessary. As equation (7.27) demonstrates, only one FFT program is necessary to do both the FFT DFT and the IDFT.

7.6 RADIX-4 DECIMATION-IN-FREQUENCY

In a radix-4 FFT, each butterfly has four inputs and outputs instead of two and the number of terms in the sequence must be a power of 4. The radix-4 algorithm provides an additional reduction in the amount of computation. Consider a DIF decomposition with (7.2) broken into four components instead of two:

$$X(k) = \sum_{n=0}^{(N/4)-1} x(n) W^{nk} + \sum_{n=N/4}^{(N/2)-1} x(n) W^{nk} + \sum_{n=N/2}^{(3N/4)-1} x(n) W^{nk} + \sum_{n=3N/4}^{N-1} x(n) W^{nk} \tag{7.30}$$

With appropriate substitution of n in (7.30), $X(k)$ can be written as

$$X(k) = \sum_{n=0}^{(N/4)-1} x(n) W^{nk} + W^{kN/4} \sum_{n=0}^{(N/4)-1} x(n + N/4) W^{nk}$$

$$+ W^{kN/2} \sum_{n=0}^{(N/4)-1} x(n + N/2) W^{nk} + W^{3kN/4} \sum_{n=0}^{(N/4)-1} x(n + 3N/4) W^{nk} \tag{7.31}$$

Using $W^{kN/4} = (-j)^k$, $W^{kN/2} = (-1)^k$, and $W^{3kN/4} = (j)^k$, the equation (7.31) becomes

$$X(k) = \sum_{n=0}^{(N/4)-1} [x(n) + (-j)^k x(n + N/4) + (-1)^k x(n + N/2)$$

$$+ (j)^k x(n + 3N/4)] W^{nk} \tag{7.32}$$

Substituting $W_N^4 = W_{N/4}$, we can decompose (7.32) into four sequences, each as an $(N/4)$-point FFT:

$$X(4k) = \sum_{n=0}^{(N/4)-1} [x(n) + x(n + N/4) + x(n + N/2)$$

$$+ x(n + 3N/4)]W_{N/4}^{nk} \qquad (7.33)$$

$$X(4k + 1) = \sum_{n=0}^{(N/4)-1} \{[x(n) - jx(n + N/4) - x(n + N/2)$$

$$+ jx(n + 3N/4)]W_N^n\}W_{N/4}^{nk} \qquad (7.34)$$

$$X(4k + 2) = \sum_{n=0}^{(N/4)-1} \{[x(n) - x(n + N/4) + x(n + N/2)$$

$$- x(n + 3N/4)]W_N^{2n}\}W_{N/4}^{nk} \qquad (7.35)$$

$$X(4k + 3) = \sum_{n=0}^{(N/4)-1} \{[x(n) + jx(n + N/4) - x(n + N/2)$$

$$- jx(n + 3N/4)]W_N^{3n}\}W_{N/4}^{nk} \qquad (7.36)$$

Equations (7.33) through (7.36) each represent an $(N/4)$-point DFT and they show how to arrange an N-point sequence into four groups which have DFTs that can be summed to give an N-point FFT.

Example 7.2   16-Point FFT, Radix-4, Decimation-In-Frequency. Figure 7.10 shows the flow graph representing equations (7.33) through (7.36). Let $N = 16$ and the input samples $x(0) = x(1) = \cdots = x(7) = 1$ and $x(8) = x(9) = \cdots = x(15) = 0$. The twiddle factors needed $W_L^m = e^{-j2m\pi/L}$ are as shown:

m	W_N^m	$W_{N/4}^m$
0	1	1
1	$0.9238 - j0.3826$	$-j$
2	$0.707 - j0.707$	-1
3	$0.3826 - j0.9238$	$+j$
4	$0 - j$	1
5	$-0.3826 - j0.9238$	$-j$
6	$-0.707 - j0.707$	-1
7	$-0.9238 - j0.3826$	$+j$

Note that $W_N^8 = -W_N^0$, $W_N^9 = -W_N^1$, and so on. Similarly, the coefficients $W_{N/4}^4$ are periodic in $\frac{N}{4}$ (i.e., $W_{N/4}^4 = W_{N/4}^0$, $W_{N/4}^5 = W_{N/4}^1$, etc.).

The resulting sequence, after stage 1, can be obtained using the flow graph in Figure 7.10. Note that -1, $\pm j$ are not shown in Figure 7.10.

$$x(0) \rightarrow [x(0) + x(4) + x(8) + x(12)]W^0 = 1 + 1 + 0 + 0 = 2$$

$$\cdot$$
$$\cdot$$
$$\cdot$$

$$x(4) \rightarrow [x(0) - jx(4) + x(8) + x(12)]W^0 = 1 - j + 0 + 0 = 1 - j$$

$$\cdot$$
$$\cdot$$
$$\cdot$$

$$x(12) \rightarrow [x(0) + jx(4) - x(8) - jx(12)]W^0 = 1 + j - 0 - 0 = 1 + j$$

$$\cdot$$
$$\cdot$$
$$\cdot$$

$$x(15) \rightarrow [x(3) + jx(7) - x(11) - jx(15)]W^9 = [1 + j - 0 - 0](-W^1)$$
$$= (1 + j)(-0.9238 + j0.3826) = -1.307 - j0.541$$

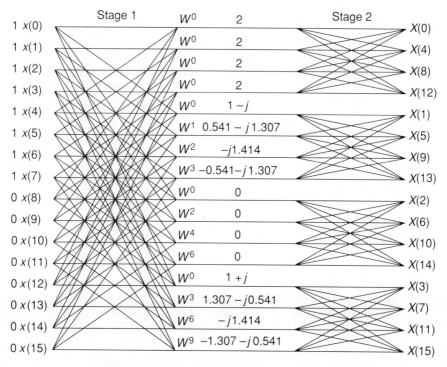

FIGURE 7.10. Radix-4 decimation-in-frequency for $N = 16$

For example, $X(5)$ can be found from the flow graph and verified using (7.31) with $k = 1$,

$$X(5) = (1-j)W_{16}^0 + (1-j)W_{16}^1(-j) + (1-j)W_{16}^2(-1)$$

$$+ (1-j)W_{16}^3(j)$$

$$= (1-j) + (1-j)(0.9238 - j0.3826)(-j)$$

$$+ (1-j)(0.707 - j0.707)(-1) + (1-j)(0.3826 - j0.9238)(j)$$

$$= (1-j) + (0.5412 - j1.3064)(-j) + (-j1.414)(-1)$$

$$+ (-0.5412 - j1.3064)(j) = 1 - j0.6684$$

The output sequence is listed in Table 7.1 along with the magnitude values. Note that from the flow graph, the constants $W_{N/4}$ are implemented with ± 1 and $\pm j$. The order of the output sequence is scrambled. If the indices of the outputs are represented in base 4 (which uses digits 0,1,2,3), they can be unscrambled using digit reversal. For example, $X(8)$ would be swapped with $X(2)$, since $(8)_D$ in decimal represents $(20)_4$ in base 4. Digits 0 and 1 would be reversed to yield $(02)_4$, which is $(02)_D$ in decimal. Similarly, $X(4)$ would be swapped with $X(1)$, and so on, as shown in Table 7.2.

The output magnitude is shown in Figure 7.11 and represents a sinc function, the transform of the rectangular waveform specified by the input sequence in Figure 7.10.

Figures 7.12 and 7.13 show a radix-2, 16-point FFT, using decimation-in-frequency and decimation-in-time, respectively (included for comparison). With the same values for the input samples $x(n)$ as with the 16-point FFT, radix-

TABLE 7.1 Output Sequence for $N = 16$, Radix -4

		Magnitude
$X(0) =$	8	8
$X(1) =$	$1 - j5.028$	5.1255
$X(2) =$	0	0
$X(3) =$	$1 - j1.496$	1.8
$X(4) =$	0	0
$X(5) =$	$1 - j0.668$	1.2028
$X(6) =$	0	0
$X(7) =$	$1 - j0.2$	1.0195
$X(8) =$	0	0
$X(9) =$	$1 + j0.2$	1.0195
$X(10) =$	0	0
$X(11) =$	$1 + j0.668$	1.2028
$X(12) =$	0	0
$X(13) =$	$1 + j1.496$	1.8
$X(14) =$	0	0
$X(15) =$	$1 + j5.028$	5.1264

TABLE 7.2   Radix-4 Digit Reversal for $N = 16$

Scrambled	Ordered
$X(0)$	$X(0)$
$X(4)$	$X(1)$
$X(8)$	$X(2)$
$X(12)$	$X(3)$
$X(1)$	$X(4)$
$X(5)$	$X(5)$
$X(9)$	$X(6)$
$X(13)$	$X(7)$
$X(2)$	$X(8)$
$X(6)$	$X(9)$
$X(10)$	$X(10)$
$X(14)$	$X(11)$
$X(3)$	$X(12)$
$X(7)$	$X(13)$
$X(11)$	$X(14)$
$X(15)$	$X(15)$

4, of Figure 7.10, the same output results are obtained in all three cases. While higher radices can provide a reduction in arithmetic, addressing and programming considerations become more complex. As a result, the radix-2 and radix-4 FFTs are the most widely used. Figure 7.14 shows a flow graph for a 32-point FFT, with radix-2, using decimation-in-frequency.

Several FFT programs are available on a diskette provided by Texas Instruments included with the SWDS package. In Chapter 9 we discuss several FFT-related projects, including the implementation of a real-time FFT board.

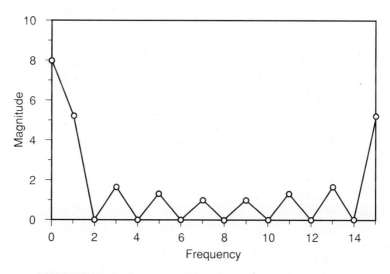

FIGURE 7.11. Output magnitude for a rectangular input $N = 16$, radix-4

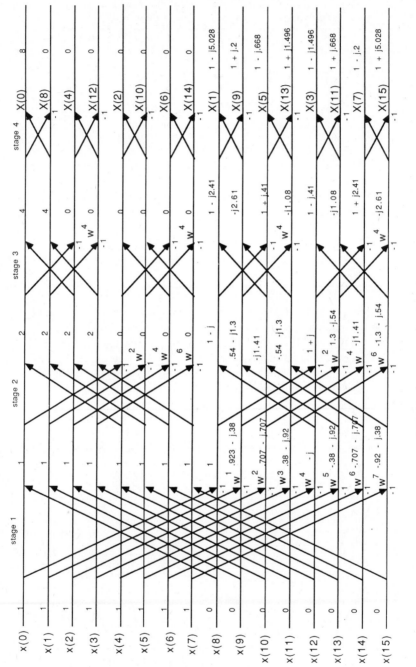

FIGURE 7.12. A 16-point DIF flowchart, radix-2

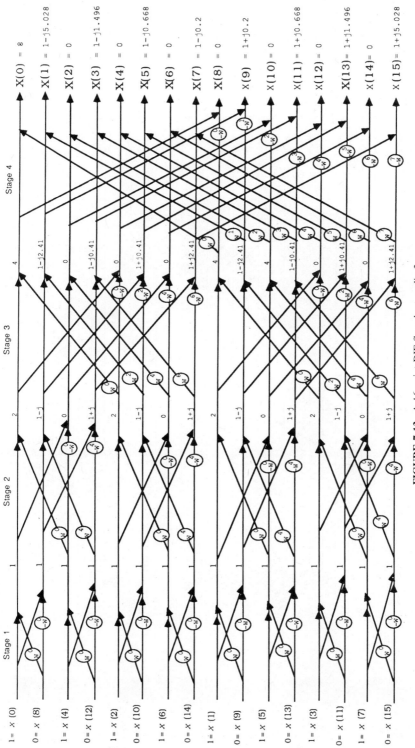

FIGURE 7.13. A 16-point DIF flowchart, radix-2

169

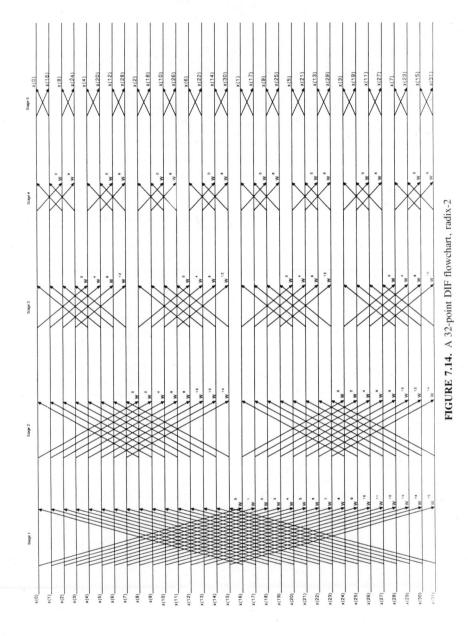

FIGURE 7.14. A 32-point DIF flowchart, radix-2

REFERENCES

[1] J. W. Cooley and J. W. Tukey, "An Algorithm for Machine Computation of Complex Fourier Series," *Mathematics of Computation*, **19**, 1965.

[2] E. O. Brigham, *The Fast Fourier Transform*, Prentice-Hall, Englewood Cliffs, N.J., 1974.

[3] DSP Committee, IEEE ASSP (editors), *Selected Papers in Digital Signal Processing II*, IEEE Press, New York, 1976.

[4] DSP Committee, IEEE ASSP (editors), *Programs for Digital Signal Processing*, IEEE Press, New York, 1979.

[5] G. D. Bergland, "A Guided Tour of the Fast Fourier Transform," *IEEE Spectrum*, **6**, July 1969.

[6] N. Ahmed and T. Natarajan, *Discrete-Time Signals and Systems*, Reston, Reston, Va., 1983.

[7] L. R. Rabiner and B. Gold, *Theory and Application of Digital Signal Processing*, Prentice-Hall, Englewood Cliffs, N.J., 1975.

[8] K. S. Lin (editor), *Digital Signal Processing Applications with the TMS320 Family*, Vol. 1, Prentice-Hall, Englewood Cliffs, N.J., 1988.

[9] C. S. Burrus and T. W. Parks, *DFT/FFT and Convolution Algorithms Theory and Implementation*, Wiley, New York, 1988.

[10] C. S. Burrus, "Unscrambling for Fast DFT Algorithms," *IEEE Transactions on Acoustics, Speech, and Signal Processing*, **ASSP-36**, July 1988.

[11] *Second Generation User's Guide*, Texas Instruments Inc., Dallas, Tex., 1987.

[12] A. V. Oppenheim and R. Schafer, *Discrete-Time Signal Processing*, Prentice-Hall, Englewood Cliffs, N.J., 1989.

[13] D. J. DeFatta, J. G. Lucas, and W. S. Hodgkiss, *Digital Signal Processing: A System Approach*, Wiley, New York, 1988.

[14] H. F. Silverman, "An Introduction to Programming the Winograd Fourier Transform Algorithm (WFTA)," *IEEE Transactions on Acoustics, Speech, and Signal Processing*, **ASSP-25**, April 1977.

[15] R. N. Bracewell, "The Fast Hartley Transform," *Proceedings of the IEEE*, **72** (8), August 1984.

[16] A. Zakhor and A. V. Oppenheim, "Quantization Errors in the Computation of the Discrete Hartley Transform," *IEEE Transactions on Acoustics, Speech, and Signal Processing*, **ASSP-35** (2), October 1987.

[17] H. V. Sorensen, D. L. Jones, C. S. Burrus, and M. T. Heideman, "On Computing the Discrete Hartley Transform," *IEEE Transactions on Acoustics, Speech, and Signal Processing*, **ASSP-33** (4), October 1985.

[18]   M. Vetterli and P. Duhamel, "Split-Radix Algorithms for Length-p^m DFT's," *IEEE Transactions on Acoustics, Speech, and Signal Processing*, **ASSP-37**, January 1989.

[19]   G. Goertzel, "An Algorithm for the Evaluation of Finite Trigonometric Series," *American Mathematical Monthly*, **65**, January 1958.

[20]   S. Kay and R. Sudhaker, "A Zero Crossing Spectrum Analyzer," *IEEE Transactions on Acoustics, Speech, and Signal Processing*, **ASSP-34**, February 1986.

8

Adaptive Filters

CONCEPTS AND PROCEDURES

- *Introduction to four basic adaptive structures*
- *Presentation of the linear adaptive combiner*
- *Development of the LMS algorithm*
- *Several examples show how to apply adaptive techniques*

Adaptive filters are best used in situations where the statistics of a signal are slowly changing and the filter must adjust to compensate. In this chapter we describe the basic concept of adaptive filtering and four structures that have been successful in the past [1–4]. The discussion is limited to a very simple but powerful filter called the linear adaptive combiner, which is nothing more than an adjustable FIR filter. The search algorithm, which provides the strategy for adjusting the filter constants, is limited to the LMS algorithm. The chapter concludes with several laboratory examples of adaptive filters. The aim of this chapter is to give the student a basic, intuitive understanding of adaptive filters reinforced through the laboratory examples at the end of the chapter.

8.1 INTRODUCTION

An adaptive system is a system whose coefficients can automatically adjust to a changing environment or input signal. The transition from implementing digital filters with DSP microprocessors to implementing adaptive digital filters is a natural extension. With all the decision-making capability of the processor already present, it seems a shame to build a rigid system that cannot adjust to changing conditions. Adaptive filters are used to their greatest advantage when there is an uncertainty about the characteristics of a signal or when the characteristics change during

173

the filter's operation. An adaptive system can learn the signal characteristics and track slow changes. Analysis of adaptive systems is more difficult than analysis of nonadaptive systems; therefore, in this chapter we rely on intuitive ideas and laboratory examples to develop a basic understanding of adaptive systems.

8.2 STRUCTURES

Although many different configurations are possible for an adaptive system, we examine four basic ones that have worked well with a number of applications in the past. Conceptually the adaptive scheme is fairly simple. Most of the adaptive schemes can be described by the structure shown in Figure 8.1. In this figure the output y is compared with a desired output d to produce an error signal e. The error signal is input to the adaptive algorithm that adjusts the variable filter to satisfy some predetermined criteria or rules. The desired signal is usually the most difficult to obtain. One of the first questions that probably comes to mind is: Why are we trying to generate the desired signal at y if we already know it? Surprisingly, in many applications the desired signal does exist somewhere in the system or is known a priori. The challenge in applying adaptive techniques is to figure out where to get the desired signal d, what to make the output y, and what to make the error e. The following paragraphs will show how y, d, and e are chosen for some situations where adaptive techniques can be applied.

System Identification

The adaptive approach is very useful in system modeling or identification, and the structure used is shown in Figure 8.2. The response of an unknown system to an input x is the desired signal in this scheme. The response of the adaptive filter y, to the same input x, is compared to the unknown system response d, to create the error signal e, which is used to adjust the adaptive block so that it matches the unknown system. The filter will be adjusted until the error signal reaches approximately zero, at which time the adaptation is complete. If the unknown system is linear, and not time varying, then after the adaptation is complete, the

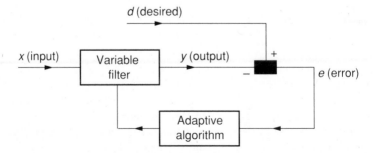

FIGURE 8.1. General adaptive structure

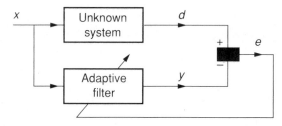

FIGURE 8.2. System identification

filter's characteristics no longer change. The input/output behavior matches that of the unknown system; hence it is a model. When the system has adapted, y matches d and e is approximately zero. The input should be bandlimited white noise with a spectrum broad enough to excite the poles of the unknown system.

Output Noise Cancellation

A second application of an adaptive system is the elimination of output distortion or noise in a system whose output must follow the input. Examples of such a system are an amplifier or a servomechanism. The diagram for this situation is shown in Figure 8.3. The output noise is modeled as additive noise inserted at the output of the system. The corrupted output provides the input to the adaptive filter. This input is filtered to produce a clean output at y. The output y is compared with the desired signal, which is the delayed input. The delay is introduced to match the delays of the system plus the adaptive filter. Again, after the system has adapted, y matches d and the error e tends to zero.

Additive Input Noise Cancellation

A third example is the noise canceler [1,2] shown in Figure 8.4. Here the signal x is contaminated with additive noise n. The filter input in this case is n', which is correlated to the noise n in the signal. This usually means that the noise comes from the same source but has been modified by the environment in some way. When this system has adapted, y approaches the additive noise n, and the error signal approaches the input signal x. If x is uncorrelated with n, the strategy is to minimize $E(e^2)$, where $E()$ is the expected value. The expected value is generally

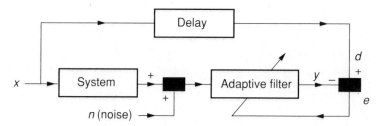

FIGURE 8.3. Output noise cancellation

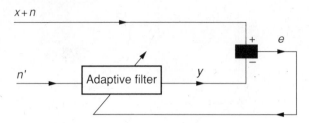

FIGURE 8.4. Additive input noise cancellation

unknown; therefore, it is usually approximated with a running average or with the instantaneous function itself. Its signal component, $E(x^2)$, will be unaffected and only its noise component $E[(n - y)^2]$ will be minimized, thus giving an e signal whose noise component is minimized. A more complete discussion of how this works is given in Widrow and Stearns [1].

8.3 ADAPTIVE LINEAR COMBINER

We will consider one of the most useful adaptive filter structures—the linear adaptive combiner. Two cases occur when using the linear combiner: (1) multiple inputs and (2) a single input.

Multiple Inputs

The case of multiple inputs is described in Figure 8.5. The configuration consists of K independent input signals each of which is weighted by $w(k)$ and combined to form the output,

$$y(n) = \sum_{k=0}^{K} w(k, n)x(k, n) \tag{8.1}$$

The input can be represented as a $(K + 1)$-dimensional vector,

$$\mathbf{X}(n) = [x(0,n) \quad x(1, n) \quad \cdots \quad x(K, n)]^{\mathrm{T}} \tag{8.2}$$

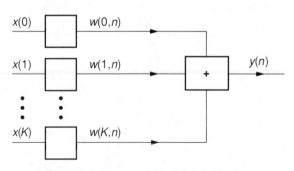

FIGURE 8.5. Linear combiner with multiple inputs

where n is the time index and the transpose T is used so that the vector can be written on one line.

Single Input

In the case of a single input, the structure reduces to a $(K + 1)$-tap FIR filter with adjustable coefficients as shown in Figure 8.6. Each delayed input is weighted and summed to produce the output,

$$y(n) = \sum_{k=0}^{K} w(k, n)x(n - k) \qquad (8.3)$$

The single input and the weights can also be written as vectors,

$$\mathbf{X}(n) = [x(n) \quad x(n - 1) \quad \cdots \quad x(n - K)]^{\mathrm{T}} \qquad (8.4)$$

$$\mathbf{W}(n) = [w(0, n) \quad w(1, n) \quad w(2, n) \quad \cdots \quad w(K, n)]^{\mathrm{T}} \qquad (8.5)$$

where n is the time index, which will frequently be dropped from the notation for both W and x.

Using the vector notation, (8.3) is cast as

$$y(n) = \mathbf{X}^{\mathrm{T}}(n)\mathbf{W}(n) = \mathbf{W}^{\mathrm{T}}(n)\mathbf{X}(n) \qquad (8.6)$$

Equations (8.1), (8.3), and (8.6) as well as Figures 8.5 and 8.6 all contain the same information. To become more familiar with the notation, let us examine a filter with two weights and a single input:

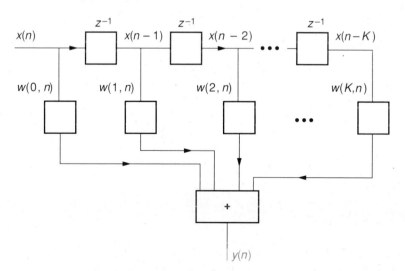

FIGURE 8.6. Adaptive linear combiner with single input

Example 8.1   Two Weights.   Verify that equations (8.3) and (8.6) and Figure 8.7 give the same y for a two-weight filter.

SOLUTION:

For $K = 1$, equation (8.3) reduces to

$$y(n) = \sum_{k=0}^{1} w(k,n)x(n-k) = w(0,n)x(n) + w(1,n)x(n-1)$$

or with the time index n implied on the weights,

$$y(n) = w(0)x(n) + w(1)x(n-1)$$

The equation above can also be obtained using (8.6),

$$y(n) = \begin{bmatrix} x(n) & x(n-1) \end{bmatrix} \begin{bmatrix} w(0) \\ w(1) \end{bmatrix} = \begin{bmatrix} w(0) & w(1) \end{bmatrix} \begin{bmatrix} x(n) \\ x(n-1) \end{bmatrix}$$

which reduces to

$$y(n) = x(n)w(0) + x(n-1)w(1)$$

which can also be obtained by summing the signals at the node of the two-weight diagram shown in Figure 8.7.

As can be seen in Figure 8.7, the linear combiner with a single input is just an

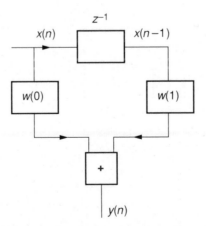

FIGURE 8.7. Two-weight linear combiner

FIR filter with adjustable coefficients. Although this is a very simple configuration, it can handle many of the adaptive applications.

8.4 PERFORMANCE FUNCTION

In the preceding section we provided a structure for the filter whose characteristics may be changed by adjusting the weights. However, we still need a way to judge how well the filter is operating—a performance measure is needed. The performance function will be based on the error, which is obtained from the block diagram in Figure 8.1, with the time index incorporated:

$$e(n) = d(n) - y(n) \tag{8.7}$$

The square of this function is

$$e^2(n) = d^2(n) - 2d(n)y(n) + y^2(n) \tag{8.8}$$

which is the instantaneous squared-error function. In terms of the weights, it becomes

$$e^2(n) = d^2(n) - 2d(n)\mathbf{X}^T(n)\mathbf{W} + \mathbf{W}^T\mathbf{X}(n)\mathbf{X}^T(n)\mathbf{W} \tag{8.9}$$

where the time index on the $\mathbf{W}$ has been dropped. Equation (8.9) represents a quadratic surface in $\mathbf{W}$, which means that the highest power of the weights is the squared power. The strategy will be to adjust the weights so that the squared-error function will be a minimum.

To understand the performance surface equation (8.9), consider the case of one weight. The error surface then becomes

$$e^2(n) = d^2(n) - 2d(n)x(n)w(0) + x^2(n)w^2(0) \tag{8.10}$$

which is a second-order function in $w(0)$. To find the minimum, set the derivative of (8.10) with respect to $w(0)$ equal to zero, or

$$\frac{de^2(n)}{dw(0)} = -2d(n)x(n) + 2x^2(n)w(0) = 0 \tag{8.11}$$

resulting in

$$w(0) = \frac{d(n)}{x(n)} \tag{8.12}$$

which is the value of $w(0)$ that yields the desired minimum.

Since the signals d and x are functions of time, the minimum and the performance surface also fluctuate with the signals. This is not desirable; we would feel more comfortable with a rigid performance function. To eliminate this problem we can take the expected value of the squared-error function, which for the one weight becomes

$$E[e^2(n)] = E[d^2(n)] - 2E[d(n)x(n)]w(0) + E[x^2(n)]w^2(0) \qquad (8.13)$$

This performance function is called the mean-squared error.

Note that the expected value of any sum is the sum of the expected values. The expected value of a product is the product of the expected values only if the variables are statistically independent. The signals $d(n)$ and $x(n)$ are generally not statistically independent. If the signals d and x are statistically time invariant, the expected values of the signal products of d and x are constants, and equation (8.13) is rewritten as

$$E[e^2(n)] = A - 2Bw(0) + Cw^2(0) \qquad (8.14)$$

where A, B, and C are constants.

Using equation (8.13) as the performance function for one weight results in a fixed minimum point on a rigid performance function,

$$w(0) = B/C \qquad (8.15)$$

A plot of the one-dimensional error function with respect to $w(0)$ is shown in Figure 8.8. This is a simple second-order curve in two dimensions ($E[e^2]$, $w(0)$) with a single minimum at $w(0) = B/C$. If we examine two weights, a three-dimensional second-order surface that resembles a bowl will result. With more weights, a higher-dimensional second-order surface will result which cannot be

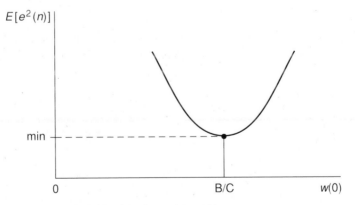

FIGURE 8.8. One-weight performance curve

visualized by humans. In practice, the weights (the weight in this case) will start at some initial value w_i and are adjusted in increments toward the minimum value of the performance function. The procedure for adjusting the weights is a subject of the next section.

Taking the mean of the general squared-error function, equation (8.9), results in a general mean-squared-error performance function:

$$E[e^2(n)] = E[d^2(n)] - 2E[d(n)\mathbf{X}^T(n)]\mathbf{W} + \mathbf{W}^T E[\mathbf{X}(n)\mathbf{X}^T(n)]\mathbf{W} \qquad (8.16)$$

Again notice that the mean value of any sum is the sum of the mean values. The product values of d and $\mathbf{X}$ and $\mathbf{X}$ with X^T cannot be further reduced since the mean value of a product is the product of mean values only when the two variables are statistically independent; d and $\mathbf{X}$ are generally not independent. This is still the same second-order performance surface as before, but now it is not fluctuating with d and $\mathbf{X}$ but is rigid. However, if d and $\mathbf{X}$ are statistically time varying, the error surface will wiggle as the statistics of d and $\mathbf{X}$ change.

8.5 SEARCHING FOR THE MINIMUM

In this section we deal with how the weights should be adjusted to find the minimum in a reasonably efficient fashion. Of course, the weights could be adjusted randomly, but life is too short. Since we will be dealing with real-time events and changes that must be tracked, we need a relatively fast way of reaching the minimum.

Consider the one-weight system again to get an idea of how this search can be conducted. Initially, the weight will equal some arbitrary value $w(0, n)$, and it will be adjusted in a stepwise fashion until the minimum is reached (Figure 8.9). The size and direction of the step are the two things that must be chosen when making a step. Each step will consist of adding an increment to $w(0, n)$. Notice that if the current value of $w(0, n)$ is to the right of the minimum, the step must be negative (but the derivative of the curve is positive); similarly, if the current value is to the left of the minimum, the increment must be positive (but the derivative is negative). This observation leads to the conclusion that the negation of the derivative indicates the proper direction of the increment. Since the derivative vanishes at the minimum, it can also be used to adjust the step size. With these observations we conclude that the step size and direction can be made proportional to the negative of the derivative and the iteration for the weights can be expressed as

$$w(0, \; n + 1) = w(0, \; n) - \beta \frac{dE[e^2]}{dw(0)} \qquad (8.17)$$

where β is an arbitrary positive constant usually chosen between .1 and 1. As

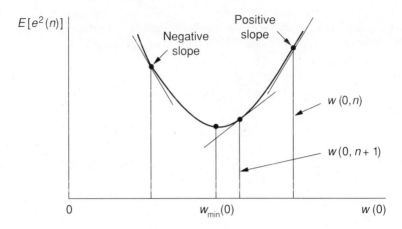

FIGURE 8.9. Minimum search on one weight

shown in Figure 8.9, repeated application of equation (8.17) will cause $w(0)$ to move by steps from its initial value until it reaches the minimum.

The derivative of the function used in the one-dimensional search can be extended to an N-dimensional surface by replacing it with the gradient of the function. The gradient is a vector of first derivatives with respect to each of the weights:

$$\text{grad}\{E[e^2]\} = \text{grad}\{P\} = \left[\frac{\partial P}{\partial w(0)} \quad \frac{\partial P}{\partial w(1)} \cdots \frac{\partial P}{\partial w(K)} \right]^{\text{T}} \tag{8.18}$$

The gradient points in the direction in which the function, in this case P, increases most rapidly. Therefore, the step size and direction can be made proportional to the gradient of the performance function.

Similarly, the minimum of the N-dimensional performance curve occurs when the gradient vanishes,

$$\text{grad}\{P\} = 0 \tag{8.19}$$

or when the partial derivative with respect to each weight vanishes,

$$\frac{\partial P}{\partial w(0)} = 0, \quad \frac{\partial P}{\partial w(1)} = 0, \quad \cdots \quad \frac{\partial P}{\partial w(K)} = 0 \tag{8.20}$$

Replacing the single weight with a vector of weights and the derivative with the gradient in equation (8.17) gives the multiple weight iteration rule,

$$\mathbf{W}(n + 1) = \mathbf{W}(n) - \beta \,\text{grad}\{P\} \tag{8.21}$$

The only issue left to resolve is how to find grad $\{P\}$. To get a simple, yet practical way to find grad $\{P\}$, we will use an estimate for it rather than the exact gradient. Instead of using the gradient of the expected squared error, we will approximate it with the grad $\{e^2\}$:

$$\text{grad}\{P\} \simeq \text{grad}\{e^2\} \qquad (8.22)$$

To get a workable expression, let us perform the gradient operation on the squared-error function,

$$\text{grad}\{e^2\} = 2e\ \text{grad}\{e\} \qquad (8.23)$$

where

$$e(n) = [d(n) - \mathbf{X}^{\mathrm{T}}(n)\mathbf{W}(n)] \qquad (8.24)$$

Substitution yields

$$\text{grad}\{e^2\} = 2e\ \text{grad}[d(n) - \mathbf{X}^{\mathrm{T}}(n)\mathbf{W}(n)] \qquad (8.25)$$

Expanding the gradient term gives

$$\text{grad}\{e^2\} = 2e\begin{bmatrix} \dfrac{\partial e}{\partial w(0)} \\ \dfrac{\partial e}{\partial w(1)} \\ \vdots \\ \dfrac{\partial e}{\partial w(K)} \end{bmatrix} = -2e\begin{bmatrix} x(0) \\ x(1) \\ \vdots \\ x(K) \end{bmatrix} \qquad (8.26)$$

and

$$\text{grad}\{e^2(n)\} = -2e(n)\mathbf{X}(n) \qquad (8.27)$$

Substituting this result for grad $\{P\}$ in equation (8.21) results in

$$\mathbf{W}(n+1) = \mathbf{W}(n) + 2\beta e(n)\mathbf{X}(n) \qquad (8.28)$$

The time index n has been included in the last two equations, implying that e will be updated every sample time. Notice that if e goes to zero, then $\mathbf{W}(n+1) = \mathbf{W}(n)$ and the weights remain constant.

Equation (8.28) forms the single most important result of this chapter, and it is the basis for the least mean squared (LMS) algorithm. This equation allows the weights to be updated without squaring, averaging, or differentiating, yet it is powerful and efficient. This equation will be used in the following examples.

8.6 APPLICATIONS OF THE LMS ALGORITHM

Several examples will be used to illustrate various applications of adaptive signal processing using the least mean squared criteria discussed previously [1–4].

Example 8.2   Adaptation with BASIC.   This example illustrates the simulation of the adaptive structure shown in Figure 8.1. The adaptive filter is a single-input, 21-weight filter; the weights are adjusted using the LMS algorithm. As an example, the simulation program in Figure 8.10 adapts a cosine signal of amplitude 2 to a sine signal of amplitude 1. The simulation program adjusts the components of coefficients vector W of length 21 using,

$$w_k(n + 1) = w_k(n) + \beta e(n)x(n - k) \qquad (8.29)$$

which expresses the same relationship as (8.28), with the constant 2β replaced by β. The input signal $x(n)$ is chosen as $\sin[(2n\pi(f/f_s)]$ with frequency 1 kHz, β determines the speed and accuracy of the adaptation process, and $e(n) = d(n) - y(n)$ is the error between a desired signal $d(n) = 2\cos[(2n\pi(f/f_s)]$ and the output of a 21-coefficients filter,

$$y(n) = \sum_{k=0}^{20} w(k)x(n - k) \qquad (8.30)$$

```
10                    BASIC PROGRAM FOR ADAPTIVE FILTERING USING THE LMS
20                    ADAPTATION OF COEFFICIENTS W TO HAVE OUTPUT Y EQUAL
30                    DESIRED SIGNAL D. X SIMULATED AS NOISE.
40  N=20                          'ORDER OF FILTER = N + 1
50  BETA=.05                      'GAIN FACTOR FOR COEFF. UPDATE
60  NS=40                         'NUMBER OF SAMPLES TO BE CALCULATED
70  FS=8000                       'SAMPLE FREQ. = 8 KHZ.
80  DIM W(20),X(20)               'ARRAYS FOR COEFF.AND NOISE SAMPLES
90  FOR J=0 TO N                  'INITIALIZE TO ZERO
100 W(J)=0                         'ALL COEFFICIENTS
110 X(J)=0                         'AND NOISE SAMPLES
120 NEXT J
130 PI=3.1415926#
140 REM OPEN FILES TO PLOT PARAMETERS USING A PLOT (PCPLOT) PROGRAM
150 OPEN "O",#1,"DESIRED"         'DESIRED SAMPLES
160 OPEN "O",#2,"Y_OUT"           'FILTERED OUTPUT SAMPLES
170 OPEN "O",#3,"ERROR"           'ERROR SAMPLES
180 OPEN "O",#4,"NOISE"           'NOISE SAMPLES
190 FOR T=0 TO NS                 'START MAIN PROGRAM - (NS+1) TIME SAMPLES
200 X(0)=SIN(2*PI*T*1000/FS)      'NEW NOISE SAMPLE (1 KHZ SINE)
210 D=2*COS(2*PI*T*1000/FS)       'DESIRED SAMPLE (1 KHZ 2*COSINE)
220 Y=0                           'CLEAR FILTER OUTPUT
230 FOR I=0 TO N                  'CONVOLVE THE COEFF. WITH NOISE SAMPLES
240 Y=Y+W(I)*X(I)                 'OUTPUT FILTER CALCULATION
250 NEXT I
260 E=D-Y                         'ERROR = DESIRED - FILTER OUTPUT
270 FOR I=N TO 0 STEP -1
280 W(I)=W(I)+2*BETA*E*X(I)       'UPDATE COEFFICIENTS
290 IF I <> 0 THEN X(I)=X(I-1)    'READY FOR NEXT SAMPLE (DATA MOVE)
300 NEXT I
310 PRINT#1,T/FS;",","D
320 PRINT#2,T/FS;",","Y
330 PRINT#3,T/FS;",","E
340 PRINT#4,T/FS;",","X(0)
350 NEXT T
```

FIGURE 8.10. BASIC program for adaptation (ADAPT.BAS)

Figure 8.11 shows the output y of the adaptive filter approaching the desired input signal d.

Example 8.3   Adaptive Echo Cancellation. This example demonstrates an adaptive echo cancellation scheme, as shown in Figure 8.12, using the least mean squared (LMS) algorithm to update the coefficients. The input to the A/D is random noise, 0 to 5 kHz. The noise source is effectively bandlimited by the 4.7-kHz lowpass filter on the AIB and was sampled at 10 kHz. The output of a 41-coefficient FIR bandpass filter *(ECHO)*, centered at one-fourth of the sampling frequency, simulates the echo signal from a transmission line used in a telephone system. With the current telephone lines, a significant portion of the transmitted signal returns as an echo that is detectable to the talker. The adaptive filter or echo cancellation filter (ECF) is an FIR filter. Both the adaptive ECF and the echo simulation filter are implemented with 41 coefficients. The error signal $e(n)$ is obtained as the difference between the output $y(n)$ of the echo bandpass filter and the output $p(n)$ of the echo cancellation filter. The 41 ECF coefficients are updated using

$$w(n + 1,\ k) = w(n,\ k) + \beta x(n,\ k)e(n) \qquad (8.31)$$

where the value of β can be adjusted to speed the convergence/adaptation process.

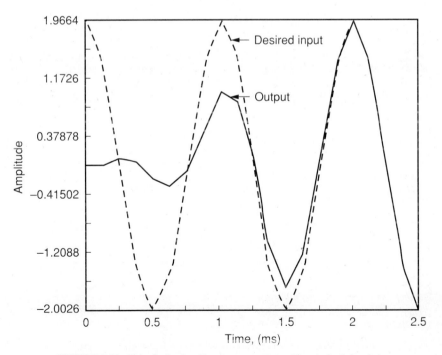

FIGURE 8.11. Plot of adaptive filter output y approaching desired signal d

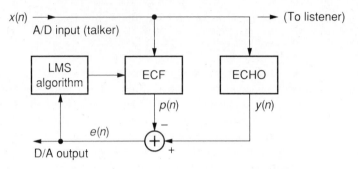

FIGURE 8.12. Adaptive echo cancellation block diagram

For a particular time n, each coefficient $w(k)$ is replaced by a new coefficient, unless $e(n)$ is zero. The following steps describe the operation of the system using continuous adaptation:

1. First, the random noise source is the input, through the AIB, to the ECHO filter. Figure 8.13 displays the output of the echo filter *before* adaptation (the upper graph), which represents the echo signal that we wish to attenuate using the ECF. This result can be obtained by running the program listed in Figure 8.14, which implements a 41-coefficient bandpass FIR filter, centered

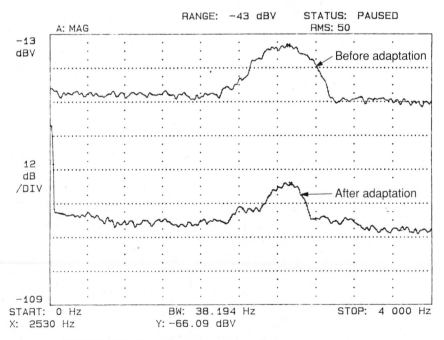

FIGURE 8.13. Frequency response of output before and after adaptation

```
0001                    *ECHO CANCELLATION SIMULATION
0002                    *SIMULATED ECHO y(n) IS OUTPUT
0003                    *FROM 41-COEFFICIENTS BANDPASS FIR FILTER-WITH NO ADAPTATION
0004                    *CENTER FREQUENCY OF FILTER APPROX. 1/4 THE SAMPLING FREQUENCY
0005 0000                       AORG    0
0006 0000 FF80                  B       START
     0001 004B
0007                    *
0008 0020               TABLE   AORG    >20                     *P.M. START AT >20
0009 0020 FFD4          H0      DATA    >FFD4                   *FIRST COEFFICIENT   H0=H40
0010 0021 FFF0          H1      DATA    >FFF0                   *SECOND COEFFICIENT  H1=H39
0011 0022 FF87          H2      DATA    >FF87                   *THIRD COEFFICIENT   H2=H38
0012 0023 0073          H3      DATA    >0073                   *H3=H37
0013 0024 0130          H4      DATA    >0130                   *
0014 0025 FF6D          H5      DATA    >FF6D                   *  .
0015 0026 FF01          H6      DATA    >FF01                   *  .
0016 0027 000D          H7      DATA    >000D                   *  .
0017 0028 FE8A          H8      DATA    >FE8A                   *
0018 0029 0140          H9      DATA    >0140                   *
0019 002A 0757          H10     DATA    >0757                   *
0020 002B FCFD          H11     DATA    >FCFD                   *
0021 002C EF1F          H12     DATA    >EF1F                   *
0022 002D 0483          H13     DATA    >0483                   *
0023 002E 1CE5          H14     DATA    >1CE5                   *
0024 002F FB0E          H15     DATA    >FB0E                   *
0025 0030 D71B          H16     DATA    >D71B                   *
0026 0031 03DE          H17     DATA    >03DE                   *
0027 0032 31CE          H18     DATA    >31CE                   *
0028 0033 FE89          H19     DATA    >FE89                   *
0029 0034 CAE6          H20     DATA    >CAE6                   *
0030 0035 FE89          H21     DATA    >FE89                   *
0031 0036 31CE          H22     DATA    >31CE                   *
0032 0037 03DE          H23     DATA    >03DE                   *
0033 0038 D71B          H24     DATA    >D71B                   *
0034 0039 FB0E          H25     DATA    >FB0E                   *
0035 003A 1CE5          H26     DATA    >1CE5                   *
0036 003B 0483          H27     DATA    >0483                   *
0037 003C EF1F          H28     DATA    >EF1F                   *
0038 003D FCFD          H29     DATA    >FCFD                   *
0039 003E 0757          H30     DATA    >0757                   *
0040 003F 0140          H31     DATA    >0140                   *
0041 0040 FE8A          H32     DATA    >FE8A                   *
0042 0041 000D          H33     DATA    >000D                   *
0043 0042 FF01          H34     DATA    >FF01                   *
0044 0043 FF6D          H35     DATA    >FF6D                   *
0045 0044 0130          H36     DATA    >0130                   *
0046 0045 0073          H37     DATA    >0073                   *
0047 0046 FF87          H38     DATA    >FF87                   *
0048 0047 FFF0          H39     DATA    >FFF0                   *H39=H1
0049 0048 FFD4          H40     DATA    >FFD4                   *H40=H0
0050                    *INITIALIZE AIB
0051 0049 000A          MD      DATA    >000A                   *MODE FOR AIB
0052 004A 03E7          RATE    DATA    >03E7                   *SAMPLING FREQUENCY fs=10 KHZ
0053                    *ASSIGNMENT OF VARIABLES-DATA PAGE 6 ( >300 )
0054      0000          MODE    EQU     >0                      *MODE FOR AIB
0055      0001          CLOCK   EQU     >1                      *SAMPLE FREQ
0056      0002          YN      EQU     >2                      *FIR OUTPUT
0057      0003          XN      EQU     >3                      *CURRENT SAMPLE
0058                    *INITIALIZATION OF VARIABLES
0059 004B C806          START   LDPK    6                       *SELECT PAGE 6,TOP OF B1 >300
0060 004C CA49                  LACK    MD                      *LOAD ACC WITH (MD)
0061 004D 5800                  TBLR    MODE                    *TRANSFER INTO DMA 300
0062 004E E000                  OUT     MODE,0                  *OUT VALUE MODE TO PORT 0
0063 004F CA4A                  LACK    RATE                    *LOAD ACC WITH (RATE)
0064 0050 5801                  TBLR    CLOCK                   *TRANSFER INTO DMA 301
0065 0051 E101                  OUT     CLOCK,1                 *OUT VALUE CLOCK TO PORT 1
0066 0052 E301                  OUT     CLOCK,3                 *DUMMY OUTPUT TO PORT 3
0067                    * LOAD ECHO FILTER COEFFICIENTS
0068 0053 5588                  LARP    AR0                     *SELECT AR0 FOR INDIR ADDRESS
0069 0054 D000                  LRLK    AR0,>200                *AR0 POINT TO TOP OF BLOCK B0
     0055 0200
0070 0056 CB28                  RPTK    >28                     *REPEAT FOR 41 COEFFICIENTS
0071 0057 FCA0                  BLKP    TABLE,*+                *PMA 20-48 DMA 200-248
     0058 0020
0072                    *INPUT NEW SAMPLE (XN)
0073 0059 FA80          WAIT    BIOZ    MAIN                    *NEW SAMPLE WHEN BIO PIN LOW
```

FIGURE 8.14. Program for implementation of echo without adaptation (ECHOWA.LST)

```
          005A 005D
0074 005B FF80               B       WAIT            *CONTINUE UNTIL EOC IS DONE
          005C 0059
0075 005D CE05    MAIN       CNFP                     *CONFIGURE BLOCK B0 AS P.M.
0076 005E 8203               IN      XN,2            *INPUT NEW SAMPLE IN DMA 303
0077 005F D100               LRLK    AR1,>32B        *AR1=>x(n-40)   LAST SAMPLE
          0060 032B
0078 0061 5589               LARP    AR1             *SELECT AR1 FOR INDIR ADDRESS
0079 0062 A000               MPYK    0               *SET PRODUCT REGISTER TO 0
0080 0063 CA00               ZAC                     *SET ACC TO 0
0081              * ECHO FIR BANDPASS FILTER
0082 0064 CB28               RPTK    >28             *DO 41 MULTIPLY
0083 0065 5C90               MACD    >FF00,*-        *H0*x(n-40) + H1*x(n-39) +...
          0066 FF00
0084 0067 CE15               APAC                    *LAST ACCUMULATE
0085 0068 DE02               ADLK    1,14            *ROUND RESULT
          0069 0001
0086 006A 6902               SACH    YN,1            *STORE IN OUTPUT y(n)
0087              * OUTPUT (SIMULATED ECHO WITHOUT CANCELLATION ROUTINE)
0088 006B E202               OUT     YN,2            *OUTPUT y(n) TO PORT 2
0089 006C FF80               B       WAIT            *BRANCH FOR NEW SAMPLE
          006D 0059
0090                         END
NO ERRORS, NO WARNINGS
```

FIGURE 8.14. (concluded)

at one-fourth of the sampling frequency. Figure 8.15 shows the time-domain representation of the noise source.

2. The program listed in Figure 8.16 performs a continuous adaptation. A relatively small value of β (FF) is used so that one can "see" the adaptation process. The rate of adaptation can be modified by choosing other values for β, which is stored into data memory at PMA >82. The error signal $e(n)$, output from the D/A to an oscilloscope, can be observed to decrease slowly toward zero, over a period of approximately 10 seconds, after the adaptation begins. The initial ECF coefficients reside in DMA >229–251 until transferred into PMA starting at >FF29. The coefficients updating process takes place with the instruction starting at PMA >80.

During the initial adaptation of the coefficients, note that AR1 points at DMA >333 (not DMA >332), which contains $x(n-40)$, due to the "data move" (done previously) associated with the MACD instruction at PMA >78. After each of

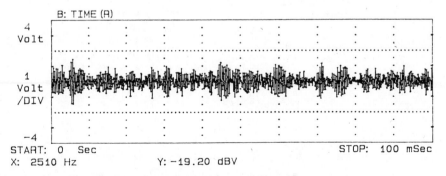

FIGURE 8.15. Time-domain representation of noise source

```
0001                    *ADAPTIVE ECHO CANCELLATION
0002                    *CONTINUOUS ADAPTATION USING THE LMS ALGORITHM
0003                    *41 COEFFICIENTS BANDPASS (FIR) FILTERS FOR BOTH
0004                    *SIMULATED ECHO AND ADAPTIVE ECHO CANCELLATION FILTERS
0005 0000                         AORG    0
0006 0000 FF80                    B       START
     0001 004D
0007 0020          TABLE          AORG    >20              *P.M. START AT PMA 20
0008 0020 FFD4     H0             DATA    >FFD4            *FIRST COEFFICIENT   H0=H40
0009 0021 FFF0     H1             DATA    >FFF0            *SECOND COEFFICIENT  H1=H39
0010 0022 FF87     H2             DATA    >FF87            *THIRD COEFFICIENT   H2=H38
0011 0023 0073     H3             DATA    >0073            *H3=H37
0012 0024 0130     H4             DATA    >0130            *
0013 0025 FF6D     H5             DATA    >FF6D            *    .
0014 0026 FF01     H6             DATA    >FF01            *    .
0015 0027 000D     H7             DATA    >000D            *    .
0016 0028 FE8A     H8             DATA    >FE8A            *
0017 0029 0140     H9             DATA    >0140            *
0018 002A 0757     H10            DATA    >0757            *
0019 002B FCFD     H11            DATA    >FCFD            *
0020 002C EF1F     H12            DATA    >EF1F            *
0021 002D 0483     H13            DATA    >0483            *
0022 002E 1CE5     H14            DATA    >1CE5            *
0023 002F FB0E     H15            DATA    >FB0E            *
0024 0030 D71B     H16            DATA    >D71B            *
0025 0031 03DE     H17            DATA    >03DE            *
0026 0032 31CE     H18            DATA    >31CE            *
0027 0033 FE89     H19            DATA    >FE89            *
0028 0034 CAE6     H20            DATA    >CAE6            *
0029 0035 FE89     H21            DATA    >FE89            *
0030 0036 31CE     H22            DATA    >31CE            *
0031 0037 03DE     H23            DATA    >03DE            *
0032 0038 D71B     H24            DATA    >D71B            *
0033 0039 FB0E     H25            DATA    >FB0E            *
0034 003A 1CE5     H26            DATA    >1CE5            *
0035 003B 0483     H27            DATA    >0483            *
0036 003C EF1F     H28            DATA    >EF1F            *
0037 003D FCFD     H29            DATA    >FCFD            *
0038 003E 0757     H30            DATA    >0757            *
0039 003F 0140     H31            DATA    >0140            *
0040 0040 FE8A     H32            DATA    >FE8A            *
0041 0041 000D     H33            DATA    >000D            *
0042 0042 FF01     H34            DATA    >FF01            *
0043 0043 FF6D     H35            DATA    >FF6D            *
0044 0044 0130     H36            DATA    >0130            *
0045 0045 0073     H37            DATA    >0073            *
0046 0046 FF87     H38            DATA    >FF87            *
0047 0047 FFF0     H39            DATA    >FFF0            *H39=H1
0048 0048 FFD4     H40            DATA    >FFD4            *H40=H0
0049              *INITIALIZE AIB
0050 0049 000A     MD             DATA    >000A            *MODE FOR AIB
0051 004A 03E7     RATE           DATA    >03E7            *SAMPLING FREQUENCY fs=10 KHZ
0052              *INITIALIZE COUNTERS, TIMERS
0053 004B 0000     TIMER          DATA    0                *ADAPTATION TIMER
0054 004C 0029     COUNT          DATA    >29              *ECF COEFF UPDATE COUNTER
0055              *ASSIGNMENT OF VARIABLES
0056      0000     MODE           EQU     >0               *MODE FOR AIB(NO OFFSET)
0057      0001     CLOCK          EQU     >1               *SAMPLE FREQ (10 KHZ)
0058      0002     YN             EQU     >2               *ECHO FIR OUTPUT
0059      0003     PN             EQU     >3               *ADAPTIVE ECF OUTPUT
0060      0004     EN             EQU     >4               *ERROR OUTPUT e(n)= y(n)-p(n)
0061      0005     B              EQU     >5               *ADAPTATION GAIN FACTOR
0062      0006     BE             EQU     >6               *B*ERROR
0063      0007     CNTR           EQU     >7               *COEFF UPDATE LOOP COUNTER
0064      000A     XN             EQU     >A               *CURRENT INPUT SAMPLE
0065              *INITIALIZATION OF VARIABLES
0066 004D CE07     START          SSXM                     *SET SIGN EXTENSION MODE
0067 004E C806                    LDPK    6                *SELECT PAGE 6,TOP OF B1 >300
0068 004F CA49                    LACK    MD               *LOAD ACC WITH (MD)
0069 0050 5800                    TBLR    MODE             *TRANSFER INTO DMA 300
0070 0051 E000                    OUT     MODE,0           *OUT VALUE MODE TO PORT 0
0071 0052 CA4A                    LACK    RATE             *LOAD ACC WITH (RATE)
0072 0053 5801                    TBLR    CLOCK            *TRANSFER INTO DMA 301
0073 0054 E101                    OUT     CLOCK,1          *OUT VALUE CLOCK TO PORT 1
0074 0055 E301                    OUT     CLOCK,3          *DUMMY OUTPUT TO PORT 3
```

FIGURE 8.16. Program for adaptive echo cancellation: continuous adaptation (ECHOADPT.LST)

```
0075 0056 CA4C              LACK    COUNT           *LOAD ACC WITH COUNT
0076 0057 5807              TBLR    CNTR            *TRANSFER INTO DMA 307
0077                 * LOAD ECHO FILTER AND ECHO CANCELLATION FILTER(ECF)COEFF
0078 0058 5588              LARP    AR0             *SELECT AR0 FOR INDIR ADDRESS
0079 0059 D000              LRLK    AR0,>200        *AR0 POINT TO TOP OF BLOCK B0
     005A 0200
0080 005B CB28              RPTK    >28             *REPEAT FOR 41 COEFFICIENTS
0081 005C FCA0              BLKP    TABLE,*+        *PMA 20-48 INTO DMA 200-228
     005D 0020
0082 005E CA00              ZAC                     *ACC=0
0083 005F CB28              RPTK    >28             *INIT TO ZERO ECF COEFFICIENTS
0084 0060 60A0              SACL    *+              *ZERO FILL DMA 229-251
0085                 *INPUT NEW SAMPLE (XN)
0086 0061 FA80      WAIT    BIOZ    MAIN            *NEW SAMPLE WHEN BIO PIN LOW
     0062 0065
0087 0063 FF80              B       WAIT            *CONTINUE UNTIL EOC IS DONE
     0064 0061
0088 0065 CE05      MAIN    CNFP                    *CONFIGURE BLOCK B0 AS P.M.
0089 0066 820A              IN      XN,2            *INPUT NEW SAMPLE IN DMA 30A
0090 0067 D100              LRLK    AR1,>332        *AR1=>x(n-40)  LAST SAMPLE
     0068 0332
0091 0069 5589              LARP    AR1             *SELECT AR1 FOR INDIR ADDRESS
0092 006A A000              MPYK    0               *SET PRODUCT REGISTER TO 0
0093 006B CA00              ZAC                     *SET ACC TO 0
0094                 *FIR (ECHO) BANDPASS FILTER OUTPUT Y(n)
0095 006C CB28              RPTK    >28             *DO 41 MULTIPLY
0096 006D 5D90              MAC     >FF00,*-        *H0*x(n-40) + H1*x(n-39) +...
     006E FF00
0097 006F CE15              APAC                    *LAST ACCUMULATE
0098 0070 DE02              ADLK    1,14            *ROUND RESULT
     0071 0001
0099 0072 6902              SACH    YN,1            *STORE FIR ECHO OUTPUT y(n)
0100                 * ECHO CANCELLATION FILTER (ECF) OUTPUT p(n)
0101 0073 A000              MPYK    0               *ZERO=>PR
0102 0074 CA00              ZAC                     *ZERO=>ACC
0103 0075 D100              LRLK    AR1,>332        *AR1=>x(n-40) LAST SAMPLE
     0076 0332
0104 0077 CB28              RPTK    >28             *DO 41 MULT WITH ECF COEFF W
0105 0078 5C90              MACD    >FF29,*-        *W(0)*x(n-40)+W(1)*x(n-39)+...
     0079 FF29
0106 007A CE15              APAC                    *LAST ACCUMULATE
0107 007B DE02              ADLK    1,14            *ROUND RESULT
     007C 0001
0108 007D 6903              SACH    PN,1            *STORE IN ECF OUTPUT p(n)
0109                 * CALCULATE ERROR e(n)= y(n)-p(n)
0110 007E 2002              LAC     YN              *FIR ECHO OUTPUT=>ACC
0111 007F 1003              SUB     PN              *y(n)-p(n)
0112 0080 6004              SACL    EN              *STORE RESULT AS ERROR e(n)
0113                 *UPDATE ECF COEFF W(n+1,k)=W(n,K)+B*e(n)*x(n-k)
0114 0081 CAFF              LACK    >FF             *LOAD ADAPT CONSTANT
0115 0082 6005              SACL    B               *STORE IT
0116 0083 CE04      LMS     CNFD                    *BLOCK B0 IS NOW DATA MEMORY
0117 0084 CA00              ZAC                     *ZERO THE ACC
0118 0085 3C05              LT      B               *B => TREG
0119 0086 3804              MPY     EN              *B*ERROR
0120 0087 CE14              PAC                     *PUT B*ERROR INTO ACC
0121 0088 DE02              ADLK    1,14            *ROUND THE RESULT
     0089 0001
0122 008A 6906              SACH    BE,1            *STORE ROUNDED RESULT IN BE
0123                 * THE BEEF
0124 008B 3C06      ADPT    LT      BE              *LOAD TREG WITH BE
0125 008C D000              LRLK    0,>229          *AR0=>W0 FIRST ECF COEFF
     008D 0229
0126 008E D100              LRLK    1,>333          * SINCE DMOV(FROM MACD)
     008F 0333
0127 0090 5588              LARP    0               *ARP => AR0
0128 0091 2F89      NXT     LAC     *,15,1          *COEFF=>ACC,LSHIFT 15,ARP=>AR1
0129 0092 3898              MPY     *-,0            *BE*x(n-k)+W(n) TO UPDATE W
0130 0093 CE15              APAC                    *ACCUMULATE PREVIOUS RESULT
0131 0094 DE02              ADLK    1,14            *ROUND RESULT
     0095 0001
0132 0096 69A0              SACH    *+,1            *UPDATE ECF COEFF W
```

FIGURE 8.16. (continued)

```
0133                    *COEFFICIENT UPDATE LOOP COUNTER
0134 0097 2007              LAC     CNTR        *ACC=CNTR
0135 0098 CD01              SUBK    >1          *DECREMENT COUNTER
0136 0099 6007              SACL    CNTR,0      *LOW ACC=>CNTR, NO SHIFT
0137 009A F580              BNZ     NXT         *GO TO NXT IF CNTR NE 0
     009B 0091
0138 009C CA29              LACK    >29         *RESET COEFF UPDATE COUNTER
0139 009D 6007              SACL    CNTR
0140 009E E204    OUT       OUT     EN,2        *OUTPUT ERROR e(n)=y(n)-p(n)
0141 009F FF80              B       WAIT        *BRANCH FOR NEW SAMPLE
     00A0 0061
0142                        END
: NO ERRORS, NO WARNINGS
```

FIGURE 8.16. (concluded)

the coefficients $w_0, w_1, \cdots, w_{40}$ is updated for a specific time n, a new sample is acquired from the A/D and the process repeats.

While the program of Figure 8.16 performs the adaptation process continuously, a modified version of this program is shown in Figure 8.17 which illustrates an "Adapt and Freeze" procedure. The following list highlights some of the features:

1. The echo spectrum $y(n)$ before and also after adaptation is shown in Figure 8.13. A 40- to 50-dB reduction of the echo signal is shown within a frequency range of 0 to 4 kHz. Figure 8.18 displays the echo cancellation filter coefficients before and after the adaptation, which shows that the adaptive ECF models its coefficients after the fixed echo filter such that the impulse responses from both filters are almost the same.

2. A variable flag is introduced such that no adaptation takes place if flag = 0 (PMA >8D).

3. A larger value of adaptation constant is used (C00) in PMA >4E, which enables the adaptation to be done in a much shorter time period, and still converge accurately.

4. A software timer is used to stop the adaptation after a predetermined time interval. In a practical situation, there may be only a limited amount of time during which adaptation can be performed.

```
0001                    *ADAPTIVE ECHO CANCELLATION
0002                    *ADAPT AND FREEZE- USING THE LMS ALGORITHM
0003                    *41 COEFFICIENTS BANDPASS (FIR) FILTERS FOR BOTH
0004                    *SIMULATED ECHO AND ADAPTIVE ECHO CANCELLATION FILTERS
0005 0000                   AORG    0
0006 0000 FF80              B       START
     0001 0050
0007                    *   .
0008 0020        TABLE      AORG    >20         *P.M. START AT >20
0009 0020 FFD4   H0         DATA    >FFD4       *FIRST COEFFICIENT  H0=H40
0010 0021 FFF0   H1         DATA    >FFF0       *SECOND COEFFICIENT H1=H39
0011 0022 FF87   H2         DATA    >FF87       *THIRD COEFFICIENT  H2=H38
0012 0023 0073   H3         DATA    >0073       *H3=H37
0013 0024 0130   H4         DATA    >0130       *
0014 0025 FF6D   H5         DATA    >FF6D       * .
0015 0026 FF01   H6         DATA    >FF01       * .
0016 0027 000D   H7         DATA    >000D       * .
0017 0028 FE8A   H8         DATA    >FE8A       *
0018 0029 0140   H9         DATA    >0140       *
```

FIGURE 8.17. Program for adaptive echo cancellation: adapt and freeze (ECHOADFR.LST)

```
0019 002A 0757   H10    DATA   >0757      *
0020 002B FCFD   H11    DATA   >FCFD      *
0021 002C EF1F   H12    DATA   >EF1F      *
0022 002D 0483   H13    DATA   >0483      *
0023 002E 1CE5   H14    DATA   >1CE5      *
0024 002F FB0E   H15    DATA   >FB0E      *
0025 0030 D71B   H16    DATA   >D71B      *
0026 0031 03DE   H17    DATA   >03DE      *
0027 0032 31CE   H18    DATA   >31CE      *
0028 0033 FE89   H19    DATA   >FE89      *
0029 0034 CAE6   H20    DATA   >CAE6      *
0030 0035 FE89   H21    DATA   >FE89      *
0031 0036 31CE   H22    DATA   >31CE      *
0032 0037 03DE   H23    DATA   >03DE      *
0033 0038 D71B   H24    DATA   >D71B      *
0034 0039 FB0E   H25    DATA   >FB0E      *
0035 003A 1CE5   H26    DATA   >1CE5      *
0036 003B 0483   H27    DATA   >0483      *
0037 003C EF1F   H28    DATA   >EF1F      *
0038 003D FCFD   H29    DATA   >FCFD      *
0039 003E 0757   H30    DATA   >0757      *
0040 003F 0140   H31    DATA   >0140      *
0041 0040 FE8A   H32    DATA   >FE8A      *
0042 0041 000D   H33    DATA   >000D      *
0043 0042 FF01   H34    DATA   >FF01      *
0044 0043 FF6D   H35    DATA   >FF6D      *
0045 0044 0130   H36    DATA   >0130      *
0046 0045 0073   H37    DATA   >0073      *
0047 0046 FF87   H38    DATA   >FF87      *
0048 0047 FFF0   H39    DATA   >FFF0      *H39=H1
0049 0048 FFD4   H40    DATA   >FFD4      *H40=H0
0050                    *INITIALIZE AIB
0051 0049 000A   MD     DATA   >000A      *MODE FOR AIB
0052 004A 03E7   RATE   DATA   >03E7      *SAMPLING FREQUENCY fs=10 KHZ
0053                    *INITIALIZE COUNTERS, TIMERS
0054 004B 0000   TIMER  DATA   0          *ADAPTATION TIMER
0055 004C 0029   COUNT  DATA   >29        *ECF COEFF UPDATE COUNTER
0056 004D 0001   FLAG   DATA   >1         **ADAPT IF FLAG=1
0057 004E 0C00   BETA   DATA   >C00       **GAIN ADAPTATION CONSTANT
0058 004F 0004   LOOP   DATA   >4         **FOR DURATION OF ADAPTATION
0059                    *ASSIGNMENT OF VARIABLES
0060      0000   MODE   EQU    >0         *MODE FOR AIB(NO OFFSET)
0061      0001   CLOCK  EQU    >1         *SAMPLE FREQ (10 KHZ)
0062      0002   YN     EQU    >2         *ECHO FIR OUTPUT
0063      0003   PN     EQU    >3         *ADAPTIVE ECF OUTPUT
0064      0004   EN     EQU    >4         *ERROR OUTPUT e(n)= y(n)-p(n)
0065      0005   B      EQU    >5         *ADAPTATION GAIN FACTOR
0066      0006   BE     EQU    >6         *B*ERROR
0067      0007   CNTR   EQU    >7         *COEFF UPDATE LOOP COUNTER
0068      0008   ACOUNT EQU    >8         *ADAPT COUNTER INIT TO 0
0069      0009   ADAPT  EQU    >9         *ADAPT FLAG,1=ADAPT,0=FROZEN
0070      000A   XN     EQU    >A         *CURRENT INPUT SAMPLE
0071      0035   TIME   EQU    >35        *ADAPTATION TIME COUNTER
0072                    *INITIALIZATION OF VARIABLES
0073 0050 CE07   START  SSXM              *SET SIGN EXTENSION MODE
0074 0051 C806          LDPK   6          *SELECT PAGE 6,TOP OF B1 >300
0075 0052 CA49          LACK   MD         *LOAD ACC WITH (MD)
0076 0053 5800          TBLR   MODE       *TRANSFER INTO DMA 300
0077 0054 E000          OUT    MODE,0     *OUT VALUE MODE TO PORT 0
0078 0055 CA4A          LACK   RATE       *LOAD ACC WITH (RATE)
0079 0056 5801          TBLR   CLOCK      *TRANSFER INTO DMA 301
0080 0057 E101          OUT    CLOCK,1    *OUT VALUE CLOCK TO PORT 1
0081 0058 E301          OUT    CLOCK,3    *DUMMY OUTPUT TO PORT 3
0082 0059 CA4C          LACK   COUNT      *LOAD ACC WITH COUNT
0083 005A 5807          TBLR   CNTR       *TRANSFER INTO DMA 307
0084 005B CA4B          LACK   TIMER      **LOAD ACC WITH TIMER
0085 005C 5808          TBLR   ACOUNT     **TRANSFER INTO DMA 308
0086 005D CA4D          LACK   FLAG       **LOAD ACC WITH FLAG
0087 005E 5809          TBLR   ADAPT      **TRANSFER INTO DMA 309
0088 005F CA4E          LACK   BETA       *LOAD ACC WITH BETA
0089 0060 5805          TBLR   B          *TRANSFER INTO DMA 305
0090 0061 CA4F          LACK   LOOP       *LOAD ACC WITH LOOP
0091 0062 5835          TBLR   TIME       *TRANSFER INTO DMA 335
0092              * LOAD FIR (ECHO) FILTER AND ECF COEFFICIENTS
0093 0063 5588          LARP   AR0        *SELECT AR0 FOR INDIR ADDR
0094 0064 D000          LRLK   AR0,>200   *AR0 POINT TO TOP OF BLK B0
     0065 0200
```

```
0095 0066 CB28              RPTK    >28             *REPEAT FOR 41 COEFFICIENTS
0096 0067 FCA0              BLKP    TABLE,*+        *PMA 20-48 INTO DMA 200-228
     0068 0020
0097 0069 CA00              ZAC                     *ACC=0
0098 006A CB28              RPTK    >28             *INIT TO ZERO ECF COEFFICIENTS
0099 006B 60A0              SACL    *+              *ZERO FILL DMA 229-251
0100               *INPUT NEW SAMPLE (XN)
0101 006C FA80     WAIT     BIOZ    MAIN            *NEW SAMPLE WHEN BIO PIN LOW
     006D 0070
0102 006E FF80              B       WAIT            *CONTINUE UNTIL EOC IS DONE
     006F 006C
0103 0070 CE05     MAIN     CNFP                    *CONFIGURE BLOCK B0 AS P.M.
0104 0071 820A              IN      XN,2            *INPUT NEW SAMPLE IN DMA 30A
0105 0072 D100              LRLK    AR1,>332        *AR1=>x(n-40)   LAST SAMPLE
     0073 0332
0106 0074 5589              LARP    AR1             *SELECT AR1 FOR INDIR ADDRESS
0107 0075 A000              MPYK    0               *SET PRODUCT REGISTER TO 0
0108 0076 CA00              ZAC                     *SET ACC TO 0
0109               *FIR (ECHO) BANDPASS FILTER OUTPUT y(n)
0110 0077 CB28              RPTK    >28             *DO 41 MULTIPLY
0111 0078 5D90              MAC     >FF00,*-        *H0*x(n-40) + H1*x(n-39) +...
     0079 FF00
0112 007A CE15              APAC                    *LAST ACCUMULATE
0113 007B DE02              ADLK    1,14            *ROUND RESULT
     007C 0001
0114 007D 6902              SACH    YN,1            *STORE FIR ECHO OUTPUT y(n)
0115               * ECHO CANCELLATION FILTER (ECF) OUTPUT p(n)
0116 007E A000              MPYK    0               *ZERO => PR
0117 007F CA00              ZAC                     *ZERO => ACC
0118 0080 D100              LRLK    AR1,>332        *AR1=>x(n-40) LAST SAMPLE
     0081 0332
0119 0082 CB28              RPTK    >28             *DO 41 MULT WITH ECF COEFF W
0120 0083 5C90              MACD    >FF29,*-        *W(0)*x(n-40)+W(1)*x(n-39)+...
     0084 FF29
0121 0085 CE15              APAC                    *LAST ACCUMULATE
0122 0086 DE02              ADLK    1,14            *ROUND RESULT
     0087 0001
0123 0088 6903              SACH    PN,1            *STORE IN ECF OUTPUT p(n)
0124               * CALCULATE ERROR e(n)= y(n)-p(n)
0125 0089 2002              LAC     YN              *FIR (ECHO) OUTPUT=>ACC
0126 008A 1003              SUB     PN              *y(n)-p(n)
0127 008B 6004              SACL    EN              *STORE RESULT AS ERROR e(n)
0128               * UPDATE ECF COEFF W(n+1,k)=W(n,k)+B*e(n)*x(n-k)
0129 008C 2009              LAC     ADAPT           **CHECK ADAPTATION FLAG
0130 008D F680              BZ      OUT             **BRANCH OUT IF ADAPT=0
     008E 00C6
0131 008F D001              LALK    >72C0           **INITIALIZE ADAPTATION TIMER
     0090 72C0
0132 0091 1008              SUB     ACOUNT          **TIMER - ACOUNT => ACC
0133 0092 F280              BLEZ    SMALL           **IF ACC LE ZERO GOTO SMALL
     0093 0096
0134 0094 FF80              B       LMS             **OTHERWISE BRANCH TO LMS
     0095 0098
0135 0096 CA05     SMALL    LACK    >05             *LOAD ACC WITH SMALL BETA VALUE
0136 0097 6005              SACL    B               *STORE IN B
0137 0098 CE04     LMS      CNFD                    *BLOCK B0 IS NOW DATA MEMORY
0138 0099 CA00              ZAC                     *ZERO THE ACC
0139 009A 3C05              LT      B               *BETA => TREG
0140 009B 3804              MPY     EN              *B*ERROR
0141 009C CE14              PAC                     *PUT B*ERROR INTO ACC
0142 009D DE02              ADLK    1,14            *ROUND THE RESULT
     009E 0001
0143 009F 6906              SACH    BE,1            *STORE ROUNDED RESULT IN BE
0144               *THE BEEF
0145 00A0 3C06     ADPT     LT      BE              *LOAD TREG WITH BE
0146 00A1 D000              LRLK    0,>229          *AR0=>W0 FIRST ECF COEFF
     00A2 0229
0147 00A3 D100              LRLK    1,>333          *DMA 333 SINCE DMOV(FROM MACD)
     00A4 0333
0148 00A5 5588              LARP    0               *ARP => AR0
0149 00A6 2F89     NXT      LAC     *,15,1          *COEFF=>ACC,LSHFT 15,ARP=>AR1
0150 00A7 3898              MPY     *-,0            *BE*x(n-k)+W(n)
0151 00A8 CE15              APAC                    *ACCUMULATE PREVIOUS RESULT
0152 00A9 DE02              ADLK    1,14            *ROUND RESULT
     00AA 0001
```

FIGURE 8.17. (continued)

```
0153 00AB 69A0        SACH    *+,1          *UPDATE ECF COEFF W
0154              *COEFFICIENT UPDATE LOOP COUNTER
0155 00AC 2007        LAC     CNTR          *ACC=CNTR
0156 00AD CD01        SUBK    >1            *DECREMENT COUNTER
0157 00AE 6007        SACL    CNTR,0        *LOW ACC=>CNTR, NO SHIFT
0158 00AF F580        BNZ     NXT           *GOTO NXT IF CNTR NE 0
     00B0 00A6
0159 00B1 CA29        LACK    >29           *RESET COEFF UPDATE COUNTER
0160 00B2 6007        SACL    CNTR          *
0161 00B3 D001        LALK    >7750         **FINE CONVERGENCE TIMER
     00B4 7750
0162 00B5 1008        SUB     ACOUNT        **DECREMENT TIMER
0163 00B6 F380        BLZ     DONE          **GO TO DONE IF TIMED OUT
     00B7 00BD
0164 00B8 2008        LAC     ACOUNT        **
0165 00B9 CC01        ADDK    >1            **INCREMENT ACOUNT
0166 00BA 6008        SACL    ACOUNT        **NEW ACOUNT
0167 00BB FF80        B       OUT           **
     00BC 00C6
0168              *ADAPT TIMER
0169 00BD CA00   DONE ZAC                   **ZERO THE ACC
0170 00BE 6008        SACL    ACOUNT        **INITIALIZE ACOUNT
0171 00BF 2035        LAC     TIME          **UPDATE TIME
0172 00C0 CD01        SUBK    >1            **
0173 00C1 6035        SACL    TIME          **TIME-1 => TIME
0174 00C2 F580        BNZ     OUT           **CONTINUE WITH ADAPT IF NE 0
     00C3 00C6
0175 00C4 CA00        ZAC                   **OTHERWISE  ADAPT FLAG = 0
0176 00C5 6009        SACL    ADAPT         **(STOP ADAPTATION)
0177 00C6 E204   OUT  OUT     EN,2          *OUTPUT ERROR e(n)=y(n)-p(n)
0178 00C7 FF80        B       WAIT          *BRANCH FOR NEW SAMPLE
     00C8 006C
0179                  END
NO ERRORS, NO WARNINGS
```

FIGURE 8.17. (concluded)

n	ECHO FIR	ECF Before	ECF After
0	FFD4	0000	FFD0
1	FFF0	.	FFE8
2	FF87	.	FF8E
3	0073	.	0072
4	0130	.	013B
5	FF6D	.	FF5E
6	FF01	.	FF12
7	000D	.	FFF0
8	FE8A	.	CE70
9	0140	.	0129
10	0757	.	0767
11	FCFD	.	FCE7
12	EF1F	.	EF3C
13	0483	.	0485
14	1CE5	.	1CD0
15	FB0E	.	FB16
16	D71B	.	D741
17	03DE	.	03E1
18	31CE	.	31A9
19	FE89	.	FE86
20	CAE6	.	CB0F
21	FE89	.	FE94
22	31CE	.	31A2

23	0 3DE	.	0 3E 6
24	D7 1 B	.	D7 4 3
25	FB 0 E	.	FB 1 A
26	1 CE 5	.	1 CC 6
27	0 4 8 3	.	0 4 8 2
28	EF 1 F	.	EF 2 E
29	FCFD	.	FD0 4
30	0 7 5 7	.	0 7 5 4
31	0 1 4 0	.	0 1 4 1
32	FE 8 A	.	FE 8 B
33	0 0 0D	.	0 0 1 0
34	FF 0 1	.	FEFE
35	FF 6D	.	FF 7 3
36	0 1 3 0	.	0 1 3 4
37	0 0 7 3	.	0 0 8 0
38	FF 8 7	.	FF 8 6
39	FFF0	.	FFF 4
40	FFD4	0 0 0 0	FFD3

FIGURE 8.18. Echo cancellation filter coefficients before and after adaptation

Example 8.4   Signal Identification. This example demonstrates an application of adaptive signal processing for the identification of a signal. For this example, the analog interface chip (AIC) described in Chapter 3 is essential since this application requires two separate inputs. The system in Figures 8.19 and 8.20 is implemented, with the summer shown in Figure 8.21. The desired signal d is fed to the input IN of the AIC. The output of the summer, which contains the desired signal d with added noise n, is fed to the second input, AUX IN, of the AIC. This composite signal with additive noise $(d + n)$ is the input to a 51-tap FIR adaptive filter. Its coefficients are adapted so that the output y is the same as the desired signal d (making $e = 0$), thus removing the random noise component in the composite signal. This type of filter design has possible applications in an environment where a template or reference signal d is known or can be derived and is present in the received composite signal $(d + n)$. The program listed in Figure 8.22 uses the LMS algorithm to update the 51 coefficients in order to achieve the smallest amount of error between the filtered output y and the desired signal d. Figure 8.23 shows an alternative program incorporating the use of MACROS, discussed in Chapter 3.

The signal d is obtained from a function generator and the input noise n from a bandlimited random noise source generator. The pseudorandom noise generator discussed in Chapters 3 and 5 is used as the input bandlimited random noise centered at 1 kHz. Figure 8.24 shows the input signal and input noise as well as the output signal and output noise after adaptation. The noise is seen to be reduced by approximately 30 to 40 dB after adaptation.

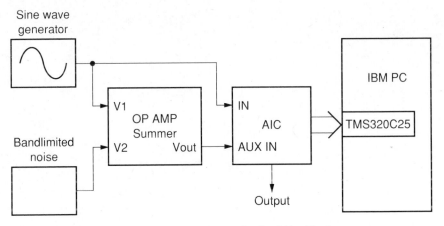

FIGURE 8.19. Block diagram for signal identification

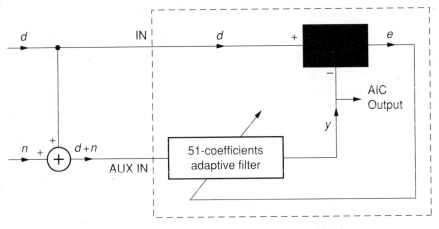

FIGURE 8.20. Adaptive filter diagram for signal identification

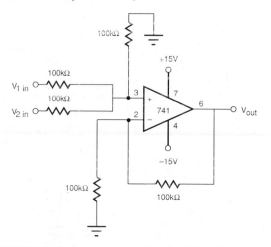

FIGURE 8.21. Summer circuit used for signal identification system

```
0001                    * ADAPTIVE FILTER PROGRAM
0002                    * USES AIC. OUTPUT AT y
0003                    * CONTINUOUS ADAPTION
0004          0061      OUTPUT   EQU     >61          STORAGE FOR OUTPUT SAMPLES
0005 0000                        AORG    0            INITIALIZE AIC AND THE
0006 0000 FF80                   B       INIT         MAIN PROGRAM PARAMETERS
     0001 002B
0007      FF00      WPROG    EQU     >FF00        START OF COEFFICIENTS IN P.M.
0008      0200      WDATA    EQU     >0200        START OF COEFFICIENTS IN D.M.
0009      0068      ONE      EQU     >68          STORAGE FOR 1 USED IN ROUNDING OFF
0010      0069      BETA     EQU     >69          STORAGE FOR ERROR WEIGHTING TERM
0011      006A      ERR      EQU     >6A          ERROR
0012      006B      ERRF     EQU     >6B          ERROR FUNCTION ( ERROR * BETA )
0013      006D      DPLUSN   EQU     >6D          NEW D+N SAMPLE STORAGE
0014      006E      Y        EQU     >6E          OUTPUT OF FILTER STORAGE (AIC OUT)
0015      006F      D        EQU     >6F          DESIRED SAMPLE STORAGE
0016      0332      LASTDN   EQU     >0332        LAST D+N SAMPLE IN LIST OF 51
0017 001C                    AORG    >1C          TRANSMIT INTERRUPT (XMIT) VECTOR
0018 001C FF80                B       XINT         BRANCH TO XMIT INTERRUPT ROUTINE
     001D 0028
0019 0020                    AORG    >20
0020                    * SECONDARY COMMUNICATION DATA TABLE ( FOR AIC 'S REGISTERS )
0021 0020 0003      TABLE    DATA    >03,>2448,>03,>244A,>03,>0063
     0021 2448
     0022 0003
     0023 244A
     0024 0003
     0025 0063
0022 0026 0001      DONE     DATA    >01          ROUNDING DATA ( 1 )
0023 0027 2000      DBETA    DATA    >2000        ERROR WEIGHTING FACTOR
0024 0028 2061      XINT     LAC     OUTPUT       LOAD ACC WITH OUTPUT SAMPLE
0025 0029 6001               SACL    >01          STORE IT IN DXR, DMA 1
0026 002A CE26               RET                  RETURN
0027                    * INITIALIZE PROCESSOR
0028 002B D001      INIT     LALK    >07E0        INIT ST1, FRAME SYNC PULSES AS INPUT
     002C 07E0
0029 002D 6067               SACL    >67          TXM=0,XF=0(LOW) TO AIC (RESET) PIN 2
0030 002E 5167               LST1    >67          FSM=1 FRAME SYNC PULSES NOT IGNORED
0031 002F CA00               ZAC                  SET ACC=0
0032 0030 6001               SACL    >01          CLEAR DXR (DATA XMIT REG) IN DMA 01
0033 0031 6061               SACL    OUTPUT       INIT OUTPUT TO 0
0034 0032 CA20               LACK    >20          LOAD IMR (INTER MASK REG)= 0010 0000
0035 0033 6004               SACL    >04          MASK ALL INTER EXCEPT 'XINT',DMA 4
0036 0034 5588               LARP    0            SELESCT AR0
0037 0035 C062               LARK    AR0,>62      TRANSFER XMIT AND MAIN PROGAM DATA
0038 0036 CB07               RPTK    >07          FROM PMA 20-28
0039 0037 FCA0               BLKP    TABLE,*+     INTO DMA 62-69
     0038 0020
0040                    * INITIALIZE AIC REGISTERS    ( SECONDARY COMMUNICATION )
0041 0039 C105               LARK    AR1,05       SELECT AR1 AS LOOP COUNTER
0042 003A C062               LARK    AR0,>62      AR0 POINTS AT >62
0043 003B CE0D               SXF                  DISABLE AIC RESET WITH EXT FLAG XF=1
0044 003C CE1F      SCND     IDLE                 ENABLE AND WAIT FOR XMIT ( INTM=0 )
0045 003D 20A9               LAC     *+,0,AR1     TRANSFER DATA STARTING AT DMA >62
0046 003E 6061               SACL    OUTPUT       OUTPUT TO THE AIC
0047 003F FB98               BANZ    SCND,*-,AR0  BRANCH IF AR1 NOT 0,DEC AR1
     0040 003C
0048 0041 CE1F               IDLE                 WAIT FOR DATA TO BE STORED IN DXR
0049 0042 CE1F      START    IDLE                 WAIT FOR XMIT OF DATA TO THE AIC
0050 0043 206E               LAC     Y            LOAD OUTPUT SAMPLE (Y) IN ACC
0051 0044 D005               ORK     >03          OR IT FOR SECONDARY COMM
     0045 0003
0052 0046 6001               SACL    >01          SEND TO DXR (UPPER 14 BITS TO D/A)
0053 0047 CA73               LACK    >73          ACC = DATA FOR SEC COMM
0054 0048 6061               SACL    OUTPUT       READY FOR CHANGE TO 'AUX IN'
0055 0049 CE1F               IDLE                 ENABLE AND WAIT FOR A XMIT INTER
0056 004A 2000               LAC     >0           ACC = AUX INPUT
0057                    * MAIN PROGRAM
0058 004B 606D               SACL    DPLUSN       STORE D+N SAMPLE
0059 004C CE05      ADPFIR   CNFP                 B0 AS P.M. (FOR 'MACD')
0060 004D A000               MPYK    0            CLEAR PRODUCT REG
0061 004E 2E68               LAC     ONE,14       SET UP TO ROUND OFF
0062 004F 558B               LARP    AR3          POINT TO AR3
0063 0050 D300               LRLK    AR3,LASTDN   AR3 = LAST D+N SAMPLE
     0051 0332
0064 0052 CB32      FIR      RPTK    >32          CONVOLVE THE D+N SAMPLES
0065 0053 5C90               MACD    WPROG,*-     WITH THE COEF
     0054 FF00
```

FIGURE 8.22. TMS320C25 program for signal identification using the AIC (ADAPTYO.LST)

```
0066 0055 CE04          CNFD              B0 AS D.M.
0067 0056 CE15          APAC              ADD LAST MULT
0068 0057 696E          SACH    Y,1       Y = OUTPUT OF FILTER
0069 0058 CE1F          IDLE              WAIT FOR XMIT OF DATA TO AIC
0070 0059 206E          LAC     Y         ACC = OUTPUT SAMPLE (Y)
0071 005A D005          ORK     >03       OR IT SECONDARY COMM
     005B 0003
0072 005C 6001          SACL    >01       SEND TO DXR (UPPER 14 BITS TO D/A)
0073 005D CA63          LACK    >63       ACC = DATA FOR SEC COMM
0074 005E 6061          SACL    OUTPUT    READY FOR CHANGE TO PRI 'IN'
0075 005F CE1F          IDLE              ENABLE AND WAIT FOR XMIT IMTER
0076 0060 2000          LAC     >0        ACC = INPUT
0077 0061 606F          SACL    D         STORE IN D
0078 0062 206E          LAC     Y         ACC = Y
0079 0063 CE23          NEG               NEGATE
0080 0064 006F          ADD     D         ADD DESIRED
0081 0065 606A          SACL    ERR       STORE ERROR (E = D + N - Y)
0082 0066 3C6A          LT      ERR       LOAD TR WITH ERROR
0083 0067 3869          MPY     BETA      ERROR * BETA
0084 0068 CE14          PAC
0085 0069 0E68          ADD     ONE,14    ROUND OFF
0086 006A 696B          SACH    ERRF,1    USE AS EEROR FUNCTION
0087 006B 55A0          MAR     *+        POINT FIRST D+N SAMPLE
0088 006C 206D          LAC     DPLUSN    ADD NEW D+N SAMPLE
0089 006D 6080          SACL    *         TO LIST OF 51
0090 006E D200          LRLK    AR2,WDATA AR2 = TOP OF 51 COEFF
     006F 0200
0091 0070 D300          LRLK    AR3,LASTDN AR3 = LAST D+N SAMPLE
     0071 0332
0092 0072 C131          LARK    AR1,>31   AR1 = COUNTER (50 TIMES)
0093 0073 3C6B          LT      ERRF      TR = ERROR FUNCTION
0094 0074 389A          MPY     *-,AR2    NEW COEFF= ERRF * D+N(50) + OLD COEF
0095 0075 7B8B  ADAPT   ZALR    *,AR3     READY TO ADD OLD COEFF
0096 0076 3A9A          MPYA    *-,AR2    ERRF * NEXT D+N, ADD PREV
0097 0077 68A9          SACH    *+,0,AR1  STORE NEW COEFF
0098 0078 FB9A          BANZ    ADAPT,*-,AR2 CHECK FOR LAST COEFF
     0079 0075
0099 007A 7B80          ZALR    *         ROUND
0100 007B CE15          APAC              ADD LAST OLD COEFF
0101 007C 6880          SACH    *         STORE LAST NEW COEFF
0102 007D FF80          B       START     BRANCH TO START OVER.
     007E 0042
0103                    END
NO ERRORS, NO WARNINGS
```

FIGURE 8.22. (concluded)

Example 8.5   Amplitude and Phase Adaptation. This example is a real-time TMS320C25 implementation of the program of Example 8.2. It implements the system in Figure 8.25. A desired sinusoidal signal d is used as the input to the AIC IN. This signal, d is also the input to an op amp. The output of the op amp is the signal d delayed by $180°$ and attenuated by 50%. The op amp output n is the input to a 51-tap adaptive FIR filter through the AIC input AUX IN. The signal identification program listed in Figure 8.22 or 8.23 (with macros) which uses a continuous adaptation can also be used here to produce an output y that has the same phase and amplitude as the desired signal d.

Example 8.6   Noise Cancellation. This example considers the system in Figure 8.26. The noise n, used as the input to the AIC AUX IN, is the input to an adaptive FIR filter with 51 coefficients. This noise n is also the input to an op amp, the output of which is added to the signal d. This composite input $(d + n_0)$ is fed to the AIC "IN." The program listed in Figure 8.27 performs the adaptation for a specified duration with the signal d set to zero. The filter produces an output, y, adapted

```
* ADAPTIVE FILTER PROGRAM -  OUTPUT AT y
* USES AIC AND MACROS. CONTINUOUS ADAPTATION
* DESIRED TO AIC " IN ", DESIRED AND NOISE TO AIC " AUX IN "
          MLIB      'A:'
CSR       EQU       >0073         SELECT AUX. INPUT,SYNC MODE
*                                 FULL SCALE = +3 VOLTS TO -3 VOLTS
SRATE     EQU       >244A         SAMPLE RATE
FLTCLK    EQU       >2448         SWITCHED CAPACITOR CLK RATE
AUX       EQU       >73           SELECT AUX INPUT
PRI       EQU       >63           SELECT PRIMARY INPUT
WPROG     EQU       >FF00         START OF COEFFICIENTS IN P.M.
WDATA     EQU       >0200         START OF COEFFICIENTS IN D.M.
LASTDN    EQU       >0332         LAST D+N SAMPLE IN LIST OF 51
ONE       EQU       >01           1 USED IN ROUNDING OFF
BETA      EQU       >0FFF         ERROR WEIGHTING FACTOR
*             STORAGE
          DSEG
ERR       BSS 1                   ERROR
ERRF      BSS 1                   ERROR FUNCTION ( ERROR * BETA )
DPLUSN    BSS 1                   NEW D+N SAMPLE STORAGE
Y         BSS 1                   OUTPUT OF FILTER STORAGE (AIC OUT)
D         BSS 1                   DESIRED SAMPLE STORAGE
          DEND
          AORG      0             INITIALIZE AIC AND THE
          B         START         MAIN PROGRAM PARAMETERS
          PSEG
*       INITIALIZE PROCESSOR AND AIC
START     INIT CSR,SRATE,FLTCLK,TABLE
MAIN      IDLE                    WAIT FOR XMIT OF DATA TO THE AIC
          INN       DPLUSN        INPUT D+N SAMPLE FROM AUX. IN
          OUTS      Y,PRI         OUTPUT AND SWITCH TO PRI INPUT
*   MAIN PROCESS
ADPFIR    CNFP                    B0 AS P.M. (FOR 'MACD')
          MPYK      0             CLEAR PRODUCT REG
          LALK      ONE,14        SET UP TO ROUND OFF
          LARP      AR3           POINT TO AR3
          LRLK      AR3,LASTDN    AR3 = LAST D+N SAMPLE
FIR       RPTK      >32           CONVOLVE THE D+N SAMPLES
          MACD      WPROG,*-      WITH THE COEFF
          CNFD                    B0 AS D.M.
          APAC                    ADD LAST MULT
          SACH      Y,1           Y = OUTPUT OF FILTER
          IDLE                    WAIT FOR XMIT OF DATA TO THE AIC
          INN       D             INPUT D SAMPLE FROM PRI INPUT
          OUTS      Y,AUX         OUTPUT AND SWITCH TO AUX INPUT
          LAC       Y             ACC = OUTPUT SAMPLE (Y)
          NEG                     NEGATE
          ADD       D             ADD DESIRED
          SACL      ERR           STORE ERROR (E = D + N -Y)
          LT        ERR           LOAD TR WITH ERROR
          MPYK      BETA          ERROR * BETA
          PAC
          ADLK      ONE,14        ROUND OFF
          SACH      ERRF,1        USE AS ERROR FUNCTION
          MAR       *+            POINT TO FIRST D+N SAMPLE
          LAC       DPLUSN        ADD NEW D+N SAMPLE
          SACL      *             TO LIST OF 51
          LRLK      AR2,WDATA     AR2 = TOP OF 51 COEFF
          LRLK      AR3,LASTDN    AR3 = LAST D+N SAMPLE
          LARK      AR1,>31       AR1 = COUNTER (50 TIMES)
          LT        ERRF          TR = ERROR FUNCTION
          MPY       *-,AR2        NEW COEFF= ERRF * D+N(50) = OLD COEF
ADAPT     ZALR      *,AR3         READY TO ADD OLD COEFF
          MPYA      *-,AR2        ERRF * NEXT D+N, ADD PREV
          SACH      *+,0,AR1      STORE NEW COEFF
          BANZ      ADAPT,*-,AR2  CHECK FOR LAST COEFF
          ZALR      *             ROUND
          APAC                    ADD LAST OLD COEFF
          SACH      *             STORE THE LAST NEW COEFF
          B         MAIN          BRANCH TO START OVER
          PEND
          END
```

FIGURE 8.23. TMS320C25 program for signal identification using the AIC-WITH MACROS (ADAPTYOM.ASM)

to the noise n_0 which is correlated with n. After the coefficients are adapted, the signal d is added and the program runs to minimize the system output e, to equal the desired signal d. The output e is the difference between the filter output y and the input $(d + n_0)$.

The system was tested with a sinusoidal input as noise. Figure 8.28 shows a

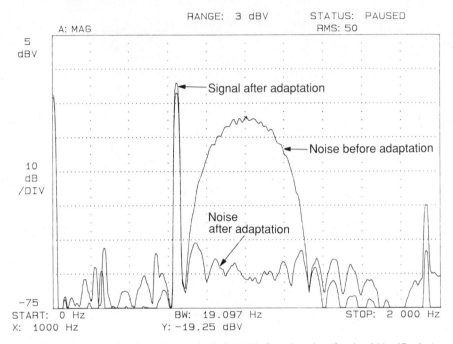

FIGURE 8.24. Spectrum of signal and noise before and after adaptation (for signal identification)

plot of the noise before and after adaptation, a reduction of approximately 30 dB. Figure 8.29 shows the input signal and noise (before adaptation), and Figure 8.30 shows the output signal and the noise after adaptation.

Next, the system was tested with a bandlimited noise centered at 1 kHz. Figure 8.31 shows the input signal and noise before and after adaptation, with the input noise reduced by approximately 20 to 30 dB after adaptation. Figure 8.31 also

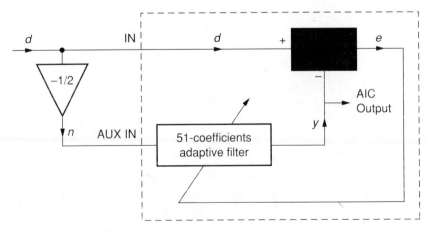

FIGURE 8.25. Amplitude and phase adaptation block diagram

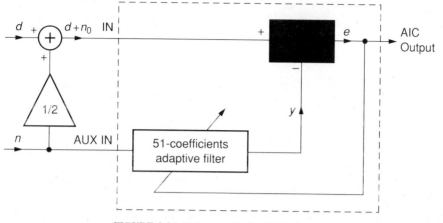

FIGURE 8.26. Noise cancellation block diagram

shows the output signal. Note that the signal after adaptation is superimposed on the input signal.

Example 8.7   Noise Cancellation in Speech Signals. This example also implements the block diagram in Figure 8.26. It is similar to the last example; however, it is designed with the consideration that the desired signal, d, is a speech signal. The following strategy is used to eliminate additive random noise from the speech:

l. First, the coefficients are adjusted without speech present but with random noise n as input. In the absence of the speech signal ($d = 0$), the filter will adapt to the noise n_0; thus the output e will go to zero and the filter output will match the correlated noise n_0.

```
0001                        * ADAPTIVE FILTER PROGRAM
0002                        * USES AIC . NOISE IN AIC 'AUX IN' IS ADAPTED TO CANCEL NOISE
0003                        * IN AIC 'IN'. OUTPUT AT e
0004                        * AFTER ADAPTATION,DESIRED SIGNAL AND NOISE APPLIED AT AIC 'IN'
0005         0061  OUTPUT   EQU    >61     STORAGE FOR OUTPUT SAMPLES
0006 0000                   AORG   0       INITIALIZE AIC AND THE
0007 0000 FF80              B      INIT    MAIN PROGRAM PARAMETERS
     0001 002C
0008      FF00  WPROG       EQU    >FF00   START OF COEFFICIENTS IN P.M.
0009      0200  WDATA       EQU    >0200   START OF COEFFICIENTS IN D.M.
0010      0068  ONE         EQU    >68     STORAGE FOR 1 USED IN ROUNDING OFF
0011      0069  BETA        EQU    >69     STORAGE FOR ERROR WEIGHTING TERM
0012      006A  NADAPT      EQU    >6A     NUMBER OF TIMES COEFFICIENTS ADAPTED
0013      006B  ERR         EQU    >6B     ERROR
0014      006C  ERRF        EQU    >6C     ERROR FUNCTION ( ERROR * BETA )
0015      006D  N           EQU    >6D     NEW NOISE SAMPLE STORAGE
0016      006E  Y           EQU    >6E     OUTPUT OF FILTER STORAGE
0017      0332  LASTN       EQU    >0332   LAST NOISE SAMPLE IN LIST OF 51
0018      0071  NEWDN       EQU    >71     NEWEST D+N SAMPLE STORAGE
0019      0072  OLDDN       EQU    >72     OLD D+N SAMPLE
0020 001C                   AORG   >1C     TRANSMIT INTERRUPT (XMIT) VECTOR
0021 001C FF80              B      XINT    BRANCH TO XMIT INTERRUPT ROUTINE
     001D 0029
```

FIGURE 8.27. TMS320C25 program for noise cancellation (ADAPTEO.LST)

```
0022 0020                    AORG     >20
0023              * SECONDARY COMMUNICATION XMIT DATA TABLE FOR AIC REGISTERS
0024 0020 0003   TABLE  DATA     >03,>2448,>03,>244A,>03,>0063
     0021 2448
     0022 0003
     0023 244A
     0024 0003
     0025 0063
0025 0026 0001   DONE   DATA     >01           ROUNDING DATA ( 1 )
0026 0027 0FFF   DBETA  DATA     >0FFF         ERROR WEIGHTING FACTOR
0027 0028 2000   DNDAPT DATA     >2000         NUMBER OF ADAPTIONS
0028              * TRANSMIT INTERRUPT ROUTINE
0029 0029 2061   XINT   LAC      OUTPUT        LOAD ACC WITH OUTPUT SAMPLE
0030 002A 6001          SACL     >01           STORE IT IN DXR, DMA 1
0031 002B CE26          RET                    RETURN
0032              * INITIALIZE PROCESSOR
0033 002C D001   INIT   LALK     >07E0         INIT ST1, FRAME SYNC PULSES AS INPUT
     002D 07E0
0034 002E 6067          SACL     >67           TXM=0,XF=0(LOW) TO AIC (RESET) PIN 2
0035 002F 5167          LST1     >67           FSM=1 FRAME SYNC PULSES NOT IGNORED
0036 0030 CA00          ZAC                    SET ACC=0
0037 0031 6001          SACL     >01           CLEAR DXR (DATA XMIT REG) IN DMA 01
0038 0032 6061          SACL     OUTPUT        INIT OUTPUT TO 0
0039 0033 CA20          LACK     >20           LOAD IMR (INTER MASK REG)= 0010 0000
0040 0034 6004          SACL     >04           MASK ALL INTER EXCEPT 'XINT',DMA 4
0041 0035 5588          LARP     0             SELECT AR0
0042 0036 C062          LARK     AR0,>62       TRANSFER XMIT AND MAIN PROGRAM DATA
0043 0037 CB08          RPTK     >08           FROM PMA 20-28
0044 0038 FCA0          BLKP     TABLE,*+      INTO DMA 62-6A
     0039 0020
0045              * INITIALIZE AIC REGISTERS    ( SECONDARY COMMUNICATION )
0046 003A C105          LARK     AR1,05        SELECT AR1 AS LOOP COUNTER
0047 003B C062          LARK     AR0,>62       AR0 POINTS AT >62
0048 003C CE0D          SXF                    DISABLE AIC RESET WITH EXT FLAG XF=1
0049 003D CE1F   SCND   IDLE                   ENABLE AND WAIT FOR XMIT (INTM=0)
0050 003E 20A9          LAC      *+,0,AR1      TRANSFER DATA STARTING AT DMA >62
0051 003F 6061          SACL     OUTPUT        FOR OUTPUT TO THE AIC
0052 0040 FB98          BANZ     SCND,*-,AR0   BRANCH BACK IF AR1 NOT 0,DEC AR1
     0041 003D
0053 0042 CE1F          IDLE                   WAIT FOR DATA TO BE STORED IN DXR
0054 0043 CE1F   OUT    IDLE                   WAIT FOR XMIT OF DATA TO THE AIC
0055 0044 206B          LAC      ERR           LOAD OUTPUT SAMPLE (ERROR) IN ACC
0056 0045 D005          ORK      >03           OR IT FOR SECONDARY COMM
     0046 0003
0057 0047 6001          SACL     >01           SEND TO DXR (UPPER 14  BITS TO D/A )
0058 0048 CA73          LACK     >73           ACC = DATA FOR SEC COMM
0059 0049 6061          SACL     OUTPUT        READY FOR CHANGE TO 'AUX IN'
0060 004A CE1F          IDLE                   ENABLE AND WAIT FOR A XMIT INTER
0061 004B 2000          LAC      >0            ACC = AUX INPUT
0062              * MAIN PROGRAM
0063 004C 606D          SACL     N             STORE NOISE SAMPLE
0064 004D CE05   ADPFIR CNFP                   B0 AS P.M. (FOR 'MACD')
0065 004E A000          MPYK     0             CLEAR PRODUCT REG
0066 004F 2E68          LAC      ONE,14        SET UP TO ROUND OFF
0067 0050 558B          LARP     AR3           POINT TO AR3
0068 0051 D300          LRLK     AR3,LASTN     AR3 = LAST N SAMPLE
     0052 0332
0069 0053 CB32   FIR    RPTK     50            CONVOLVE THE N SAMPLES
0070 0054 5C90          MACD     WPROG,*-      WITH THE COEF
     0055 FF00
0071 0056 CE04          CNFD                   B0 AS D.M.
0072 0057 CE15          APAC                   ADD LAST MULT
0073 0058 696E          SACH     Y,1           Y = OUTPUT OF FILTER
0074 0059 CE1F          IDLE                   WAIT FOR XMIT OF DATA TO AIC
0075 005A 206B          LAC      ERR           ACC = OUTPUT SAMPLE (ERROR)
0076 005B D005          ORK      >03           OR IT FOR SECONDARY COMM
     005C 0003
0077 005D 6001          SACL     >01           SEND TO DXR (UPPER 14 BITS TO D/A)
0078 005E CA63          LACK     >63           ACC = DATA FOR SEC COMM ·
0079 005F 6061          SACL     OUTPUT        READY FOR CHANGE TO PRI 'IN'
0080 0060 CE1F          IDLE                   ENABLE AND WAIT FOR XMIT INTER
0081 0061 2000          LAC      >0            ACC = INPUT
0082 0062 6071          SACL     NEWDN         STORE AS NEW D+N
0083 0063 206E          LAC      Y             ACC = Y
0084 0064 CE23          NEG                    NEGATE
0085 0065 0072          ADD      OLDDN         ADD OLD D+N
```

FIGURE 8.27. (continued)

```
0086 0066 606B              SACL    ERR         STORE ERROR (E = D + N -Y)
0087 0067 3C6B              LT      ERR         LOAD TR WITH ERROR
0088 0068 3869              MPY     BETA        ERROR * BETA
0089 0069 CE14              PAC
0090 006A 0E68              ADD     ONE,14      ROUND OFF
0091 006B 696C              SACH    ERRF,1      USE AS ERROR FUNCTION
0092 006C 55A0              MAR     *+          POINT TO FIRST N SAMPLE
0093 006D 206D              LAC     N           ADD NEW N SAMPLE
0094 006E 6080              SACL    *           TO LIST OF 51
0095 006F 2071              LAC     NEWDN       ACC = NEW D+N
0096 0070 6072              SACL    OLDDN       OLD D+N = NEW D+N
0097 0071 206A              LAC     NADAPT      ACC = NO. OF ADAPTIONS
0098 0072 F680              BZ      OUT         BRANCH BACK IF ADAPTED
     0073 0043
0099 0074 CD01              SUBK    01          ELSE SUB 1 FROM COUNTER
0100 0075 606A              SACL    NADAPT      STORE NO. OF ADAPTS LEFT
0101 0076 D200              LRLK    AR2,WDATA   AR2 = TOP OF 51 COEFF
     0077 0200
0102 0078 D300              LRLK    AR3,LASTN   AR3 = LAST N SAMPLE
     0079 0332
0103 007A C131              LARK    AR1,49      AR1 = COUNTER (50 TIMES)
0104 007B 3C6C              LT      ERRF        TR  = ERROR FUNCTION
0105 007C 389A              MPY     *-,AR2      NEW COEFF = ERRF * N(50) + OLD COEFF
0106 007D 7B8B    ADAPT     ZALR    *,AR3       READY TO ADD OLD COEFF
0107 007E 3A9A              MPYA    *-,AR2      ERRF * NEXT N(K), ADD PREV
0108 007F 68A9              SACH    *+,0,AR1     STORE NEW COEFF
0109 0080 FB9A              BANZ    ADAPT,*-,AR2 CHECK FOR LAST COEFF
     0081 007D
0110 0082 7B80              ZALR    *           ROUND
0111 0083 CE15              APAC                ADD LAST OLD COEFF
0112 0084 6880              SACH    *           STORE LAST NEW COEFF
0113 0085 FF80              B       OUT         BRANCH TO START OVER
     0086 0043
0114                        END
NO ERRORS, NO WARNINGS
```

FIGURE 8.27. (concluded)

2. Next, speech, mixed with noise correlated with the filter input noise n, is introduced as the reference signal, $d + n_0$.

3. Upon detecting the presence of speech, the algorithm stops the adaptation. It then uses the coefficients based on the input filter noise, n, and the correlated reference noise n_0, with no speech present, to cancel the additive noise in the speech.

4. When the speech drops below a certain threshold, the adaptation is restarted after a specified delay.

Since voice has many intervals where there is a low-level signal, this scheme will allow the coefficients to be updated continually, thus allowing the system to adapt to noise with a changing statistics.

The speech detector is implemented in the program listed in Figure 8.32. To detect when the speech signal is present, the program uses the power level of both the noise, n, and the signal corrupted by noise, $d + n_0$. The noise inputs are scanned to save the highest absolute value, within a window, selected from the last 36 noise samples. Knowing that the noise corrupting the speech is $n/2$ or less, the signal plus noise $(d + n_0)$ is compared to a threshold $T = 1/2[\text{abs}(n_{max})]$. If $(d + n_0)$ is greater than this threshold T, the speech signal is detected and the adaptation process is halted.

To avoid adapting with the signal present, a delay feature in the program is used. The adaptation calculation is delayed by 10 samples, thus allowing the adaptation

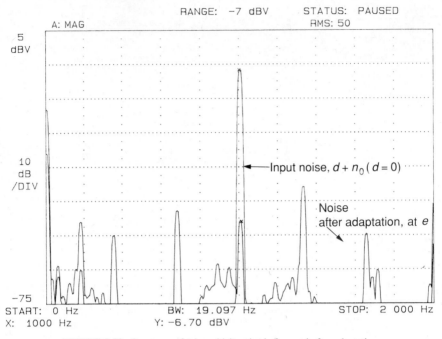

FIGURE 8.28. Spectrum of (sinusoidal) noise before and after adaptation

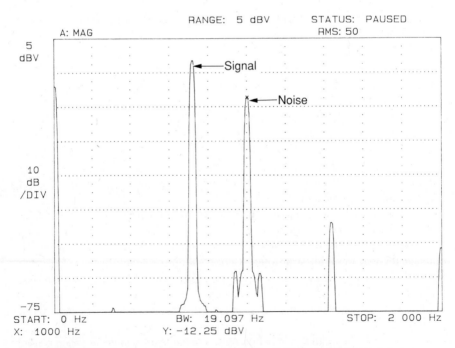

FIGURE 8.29. Spectrum of (sinusoidal) signal and noise before adaptation at $d + n_o$

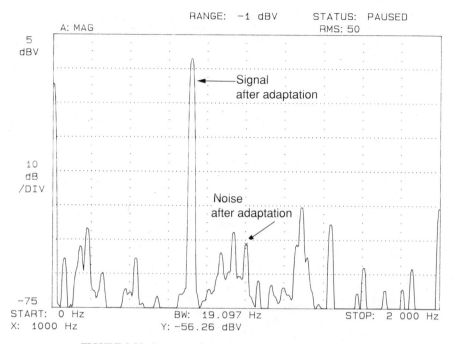

RANGE: −1 dBV   STATUS: PAUSED
A: MAG   RMS: 50

FIGURE 8.30. Spectrum of signal and noise after adaptation (at e)

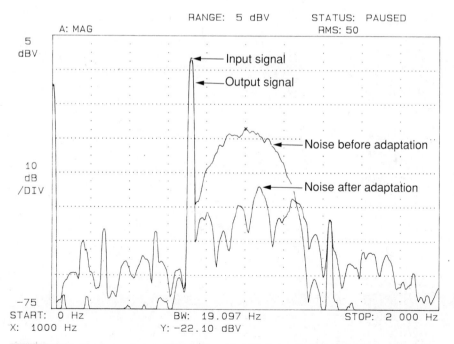

RANGE: 5 dBV   STATUS: PAUSED
A: MAG   RMS: 50

FIGURE 8.31. Spectrum of sinusoidal signal and bandlimited noise before (at d + n_o) and after adaptation (at e)

```
0001                  * ADAPTIVE FILTER PROGRAM FOR SPEECH SIGNAL - USES THE AIC
0002        0061  OUTPUT EQU    >61         STORAGE FOR OUTPUT SAMPLES
0003 0000              AORG   0            INITIALIZE AIC AND THE
0004 0000 FF80         B      INIT         MAIN PROGRAM PARAMETERS
     0001 002B
0005      FF00  WPROG  EQU    >FF00        START OF COEFFICIENTS IN P.M.
0006      0200  WDATA  EQU    >0200        START OF COEFFICIENTS IN D.M.
0007      0068  ONE    EQU    >68          STORAGE FOR 1 USED IN ROUNDING OFF
0008      0069  BETA   EQU    >69          STORAGE FOR ERROR WEIGHTING TERM
0009      006A  FLAG   EQU    >6A          FLAG TO STOP AND RESTART ADAPTING
0010      006B  ERR    EQU    >6B          ERROR
0011      006C  ERRF   EQU    >6C          ERROR FUNCTION ( ERROR * BETA )
0012      006D  N      EQU    >6D          NEWEST NOISE SAMPLE STORAGE
0013      006E  Y      EQU    >6E          OUTPUT OF FILTER STORAGE
0014      0032  WN     EQU    >32          NUMBER OF COEFFICIENTS - 1
0015      033C  LASTN  EQU    >030A+WN     LAST NOISE SAMPLE IN LIST OF 51
0016      0380  LASTDN EQU    >0380        SAMPLE AT END OF D+N DELAY
0017      0071  DN     EQU    >71          NEWEST D+N SAMPLE STORAGE
0018      0073  MAXCOM EQU    >73          MAX VALUE OF OLD N SAMPLES
0019      0074  MAXABS EQU    >74          MAX VALUE N UPDATED
0020 001C             AORG   >1C          TRANSMIT INTERRUPT (XMIT) VECTOR
0021 001C FF80        B      XINT         BRANCH TO XMIT INTERRUPT ROUTINE
     001D 0028
0022 0020             AORG   >20
0023                  * SECONDARY COMMUNICATION XMIT DATA TABLE FOR AIC REGISTERS
0024 0020 0003  TABLE DATA   >03,>2448,>03,>244A,>03,>0063
     0021 2448
     0022 0003
     0023 244A
     0024 0003
     0025 0063
0025 0026 0001  DONE  DATA   >01          ROUNDING DATA ( 1 )
0026 0027 05FF  DBETA DATA   >05FF        ERROR WEIGHTING FACTOR
0027                  * TRANSMIT INTERRUPT ROUTINE
0028 0028 2061  XINT  LAC    OUTPUT       LOAD ACC WITH OUTPUT SAMPLE
0029 0029 6001        SACL   >01          STORE IT IN DXR, DMA 1
0030 002A CE26        RET                 RETURN
0031                  * INITIALIZE PROCESSOR
0032 002B D001  INIT  LALK   >07E0        INIT ST1, FRAME SYNC PULSES AS INPUT
     002C 07E0
0033 002D 6067        SACL   >67          TXM=0,XF=0(LOW) TO AIC (RESET) PIN 2
0034 002E 5167        LST1   >67          FSM=1 FRAME SYNC PULSES NOT IGNORED
0035 002F CA00        ZAC                 SET ACC=0
0036 0030 6001        SACL   >01          CLEAR DXR (DATA XMIT REG) IN DMA 01
0037 0031 6061        SACL   OUTPUT       INIT OUTPUT TO 0
0038 0032 606A        SACL   FLAG         CLEAR FLAG TO ALLOW ADAPTATION
0039 0033 6073        SACL   MAXCOM       CLEAR MAX N OLD
0040 0034 6074        SACL   MAXABS       CLEAR MAX N UPDATED
0041 0035 C418        LARK   AR4,24       AR4 = MAX N UPDATE COUNTER
0042 0036 CA20        LACK   >20          LOAD IMR (INTER MASK REG)= 0010 0000
0043 0037 6004        SACL   >04          MASK ALL INTER EXCEPT 'XINT',DMA 4
0044 0038 5588        LARP   0            SELECT AR0
0045 0039 C062        LARK   AR0,>62      TRANSFER XMIT AND MAIN PROGRAM DATA
0046 003A CB07        RPTK   >07          FROM PMA 20-27
0047 003B FCA0        BLKP   TABLE,*+     INTO DMA 62-69
     003C 0020
0048                  * INITIALIZE AIC REGISTERS   ( SECONDARY COMMUNICATION )
0049 003D C105        LARK   AR1,05       SELECT AR1 AS LOOP COUNTER
0050 003E C062        LARK   AR0,>62      AR0 POINTS AT >62
0051 003F CE0D        SXF                 DISABLE AIC RESET WITH EXT FLAG XF=1
0052 0040 CE1F  SCND  IDLE                ENABLE AND WAIT FOR XMIT (INTM=0)
0053 0041 20A9        LAC    *+,0,AR1     TRANSFER DATA STARTING AT DMA >62
0054 0042 6061        SACL   OUTPUT       FOR OUTPUT TO THE AIC
0055 0043 FB98        BANZ   SCND,*-,AR0  BRANCH BACK IF AR1 NOT 0,DEC AR1
     0044 0040
0056 0045 CE1F        IDLE                WAIT FOR DATA TO BE STORED IN DXR
0057 0046 CE1F  OUT   IDLE                WAIT FOR XMIT OF DATA TO THE AIC
0058 0047 206B        LAC    ERR          LOAD OUTPUT SAMPLE (ERROR) IN ACC
0059 0048 D005        ORK    >03          OR IT FOR SECONDARY COMM
     0049 0003
0060 004A 6001        SACL   >01          SEND TO DXR (UPPER 14  BITS TO D/A )
0061 004B CA73        LACK   >73          ACC = DATA FOR SEC COMM
0062 004C 6061        SACL   OUTPUT       READY FOR CHANGE TO 'AUX IN'
0063 004D CE1F        IDLE                ENABLE AND WAIT FOR A XMIT INTER
0064 004E 2000        LAC    >0           ACC = AUX INPUT
0065                  * MAIN PROGRAM
0066 004F 606D        SACL   N            STORE NOISE SAMPLE
0067 0050 CE1B        ABS                 ACC = ABSOLUTE VALUE OF N
0068 0051 1074        SUB    MAXABS       COMPARE TO LAST UPDATED MAX N
0069 0052 F280        BLEZ   SKIP1        BRANCH IF MAXABS < NEW N SAMPLE
     0053 0057
0070 0054 206D        LAC    N            OTHERWISE ACC = NEW N
0071 0055 CE1B        ABS                 ACC = ABSOLUTE VALUE OF N
0072 0056 6074        SACL   MAXABS       MAX N IS UPDATED
0073 0057 558C  SKIP1 LARP   AR4          ARP = 4
0074 0058 FB90        BANZ   ADPFIR       BRANCH IF NOT 0, AR4 = AR4-1
     0059 005F
```

FIGURE 8.32. Program listing for adaptation with speech signal (ADAPTS.LST)

```
0075 005A C418            LARK    AR4,24       OTHERWISE RELOAD COUNTER
0076 005B 2F74            LAC     MAXABS,15    ACC = UPDATED N DIVIDED BY 2
0077 005C 6873            SACH    MAXCOM       CHANGE VALUE USED TO COMPARE
0078 005D CA00            ZAC                  CLEAR ACC
0079 005E 6074            SACL    MAXABS       CLEAR TO GET A NEW UPDATED N MAX
0080 005F CE05    ADPFIR  CNFP                 B0 AS P.M. (FOR 'MACD')
0081 0060 A000            MPYK    0            CLEAR PRODUCT REG
0082 0061 2E68            LAC     ONE,14       SET UP TO ROUND OFF
0083 0062 558B            LARP    AR3          POINT TO AR3
0084 0063 D300            LRLK    AR3,LASTN    AR3 = LAST N SAMPLE
     0064 033C
0085 0065 CB32    FIR     RPTK    WN           CONVOLVE THE N SAMPLES
0086 0066 5C90            MACD    WPROG,*-     WITH THE COEF
     0067 FF00
0087 0068 CE04            CNFD                 B0 AS D.M.
0088 0069 CE15            APAC                 ADD LAST MULT
0089 006A 696E            SACH    Y,1          Y = OUTPUT OF FILTER
0090 006B CB09            RPTK    >09          REPEAT 10 TIMES
0091 006C 5690            DMOV    *-           MOVE DATA IN DELAY LINE
0092 006D 55A0            MAR     *+           AR3 = TOP OF N SAMPLES(300H)
0093 006E 206D            LAC     N            ACC = NEWEST N SAMPLE
0094 006F 608D            SACL    *,0,AR5      ADD TO TOP OF DELAY
0095 0070 CE1F            IDLE                 WAIT FOR XMIT OF DATA TO AIC
0096 0071 206B            LAC     ERR          ACC = OUTPUT SAMPLE (ERROR)
0097 0072 D005            ORK     >03          OR IT FOR SECONDARY COMM
     0073 0003
0098 0074 6001            SACL    >01          SEND TO DXR (UPPER 14 BITS TO D/A)
0099 0075 CA63            LACK    >63          ACC = DATA FOR SEC COMM
0100 0076 6061            SACL    OUTPUT       READY FOR CHANGE TO PRI 'IN'
0101 0077 CE1F            IDLE                 ENABLE AND WAIT FOR XMIT INTER
0102 0078 2000            LAC     >0           ACC = INPUT
0103 0079 6071            SACL    DN           STORE AS NEW D+N
0104 007A D500            LRLK    AR5,LASTDN   AR5 = BOTTOM OF DELAY
     007B 0380
0105 007C 206E            LAC     Y            ACC = Y
0106 007D CE23            NEG                  NEGATE
0107 007E 0090            ADD     *-           ADD OLD D+N
0108 007F 606B            SACL    ERR          STORE ERROR (E = D + N -Y)
0109 0080 3C6B            LT      ERR          LOAD TR WITH ERROR
0110 0081 3869            MPY     BETA         ERROR * BETA
0111 0082 CE14            PAC
0112 0083 0E68            ADD     ONE,14       ROUND OFF
0113 0084 696C            SACH    ERRF,1       USE AS ERROR FUNCTION
0114 0085 CB08            RPTK    >08          REPEAT 9 TIMES
0115 0086 5690            DMOV    *-           MOVE DATA IN D+N DELAY LINE
0116 0087 55A0            MAR     *+           AR3 = TOP OF D+N SAMPLES(377H)
0117 0088 2071            LAC     DN           ACC = NEWEST D+N SAMPLE
0118 0089 6080            SACL    *            ADD TO TOP OF DELAY
0119 008A CE1B            ABS                  ACC = ABSOLUTE VALUE OF NEW D+N
0120 008B 1073            SUB     MAXCOM       COMPARE TO MAX N OLD
0121 008C F280            BLEZ    SKIP2        BRANCH IF NO SPEECH IS DETECTED
     008D 0092
0122 008E CA14            LACK    >14          OTHERWISE SET FLAG TO COUNT
0123 008F 606A            SACL    FLAG         20 TIMES AFTER DETECTION
0124 0090 FF80            B       OUT          DO NOT ADAPT FOR 20 CYCLES & BRANCH
     0091 0046
0125 0092 206A    SKIP2   LAC     FLAG         ACC = FLAG COUNT
0126 0093 F680            BZ      INITA        BRANCH TO START ADAPT IF FLAG CLEAR
     0094 0099
0127 0095 CD01            SUBK    01           OTHEWISE DEC ACC
0128 0096 606A            SACL    FLAG         TO DEC FLAG COUNT
0129 0097 FF80            B       OUT          DO NOT ADAPT ( STILL IN 20 CYCLES )
     0098 0046
0130 0099 D200    INITA   LRLK    AR2,WDATA    AR2 = TOP OF 51 COEFF
     009A 0200
0131 009B D300            LRLK    AR3,LASTN    AR3 = LAST N SAMPLE
     009C 033C
0132 009D C131            LARK    AR1,WN-1     AR1 = COUNTER (50 TIMES)
0133 009E 3C6C            LT      ERRF         TR  = ERROR FUNCTION
0134 009F 389A            MPY     *-,AR2       NEW COEFF = ERRF * N(50) + OLD COEFF
0135 00A0 7B8B    ADAPT   ZALR    *,AR3        READY TO ADD OLD COEFF
0136 00A1 3A9A            MPYA    *-,AR2       ERRF * NEXT N(K), ADD PREV
0137 00A2 68A9            SACH    *+,AR1       STORE NEW COEFF
0138 00A3 FB9A            BANZ    ADAPT,*-,AR2 CHECK FOR LAST COEFF
     00A4 00A0
0139 00A5 7B80            ZALR    *            ROUND
0140 00A6 CE15            APAC                 ADD LAST OLD COEFF
0141 00A7 6880            SACH    *            STORE LAST NEW COEFF
0142 00A8 FF80            B       OUT          BRANCH TO START OVER
     00A9 0046
0143                      END
NO ERRORS, NO WARNINGS
```

FIGURE 8.32. (concluded)

MEASUREMENT PAUSED

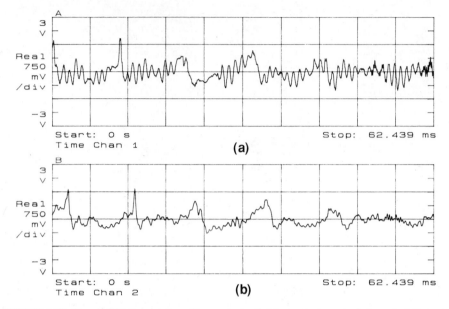

FIGURE 8.33. Plot of adaptation process with speech signal: (*a*) speech plus noise before adaptation; (*b*) speech after adaptation

to stop before the threshold is reached. Conversely, when the speech detector senses that the voice is below the threshold, adaptation does not restart until a specified delay (20 samples) has occurred.

The system was tested with a bandlimited noise centered at 1 kHz. The noise reduction after adaptation is approximately the same as in Example 8.6 (20 to 30 dB). Figure 8.33(*a*) shows a time plot of the $d + n_0$ before adaptation, and Figure 8.33(*b*) shows a time plot of the system's output e after adaptation, with the noise eliminated.

REFERENCES

[1] B. Widrow and S. D. Stearns, *Adaptive Signal Processing*, Prentice-Hall, Englewood Cliffs, N.J., 1985.

[2] B. Widrow et al., "Adaptive Noise Cancelling: Principles and Applications," *Proceedings of the IEEE*, **63** (12), December 1975.

[3] J. R. Treichler, C. R. Johnson, Jr., and M. G. Larimore, *Theory and Design of Adaptive Filters*, Wiley, New York, 1987.

[4] K. S. Lin (editor), *Digital Signal Processing Applications with the TMS320 Family*, Vol. 1, Prentice-Hall, Englewood Cliffs, N.J., 1988.

[5]   M. L. Honig and D. G. Messershmidt, *Adaptive Filters: Structures, Algorithms and Applications*, Kluwer Academic, Norwell, Mass., 1984.

[6]   D. G. Messerschmitt, "Echo Cancellation in Speech and Data Transmission," *IEEE Journal on Selected Topics in Communications*, **SAC-2** (2), March 1984, pp. 283–303.

[7]   S. T. Alexander, *Adaptive Signal Processing Theory and Applications*, Springer-Verlag, New York, 1986.

[8]   S. T. Alexander, "Fast Adaptive Filters: A Geometrical Approach," *IEEE Transactions on Acoustics, Speech, and Signal Processing*, **ASSP-34**, October 1986.

[9]   S. Haykin, *Adaptive Filter Theory*, Prentice-Hall, Englewood Cliffs, N.J., 1986.

[10]   J. G. Proakis and D. G. Manolakis, *Introduction to Digital Signal Processing*, Macmillan, New York, 1988.

[11]   C. F. Cowan and P. F. Grant, *Adaptive Filters*, Prentice-Hall, Englewood Cliffs, N.J., 1985.

[12]   *Second-Generation TMS320 User's Guide*, Texas Instruments Inc., Dallas, Tex., 1987.

[13]   *TLC32040I, TLC32040C, TLC32041I, TLC32041C Analog Interface Circuits, Advanced Information*, Texas Instruments Inc., Dallas, Tex., September 1987.

9

Digital Signal Processing Applications: Student Projects and Laboratory Exercises

HIGHLIGHTS

• *Twenty-seven student-developed projects are described.*

• *Several of the project descriptions include TMS320C25 code and test results.*

• *One- to two-week laboratory exercises conclude the chapter.*

In this chapter, we describe 27 projects developed and performed by our students. We owe a special debt to all the students who have made this chapter possible, in particular: J. Banna, J. Bazinet, K. Benevides, S. Farnham, R. Fugere, P. Furze, J. Gauthier, N. Goyette, R. Irey, E. Jackson, B. Joyce, R. Lemieux, K. Lindell, J. Nery, W. Peterson, K. Spencer, N. Vlamis, and K. Zemlok.

The projects include applications in filtering, communications, and control. Students can use the projects as a source of ideas for their projects, or as examples of implementation of various DSP algorithms. Digital Signal Processing Applications with the TMS320 Family *[1] and Chapter 8 are also good sources of project ideas. Adaptive filtering projects are described in Chapter 8, with TMS320C25 code included.*

9.1 MULTIRATE FILTERING: A SHAPED PSEUDORANDOM NOISE GENERATOR

In this section we discuss a multirate filtering project which is a six-week to one-semester project.

Problem Statement

The objective of this project is to design an efficient programmable random-noise generator, controllable in $\frac{1}{3}$-octave bands, with a frequency range over nine octaves from approximately 0 to 5000 Hz.

Design Considerations

The approach used is to create a binary random signal, then feed it into a bank of bandpass filters that shape the spectrum to the desired form, as shown in Figure 9.1. The frequency range is divided into nine octave regions, and each octave region is controlled with three filters, whose coefficients are combined to give one set of filter coefficients for each octave. Each of the $\frac{1}{3}$-octave filters that make up the composite filter has a bandwidth of approximately 23% of its center frequency and a stopband rejection of greater than 45 dB. The amplitude of each $\frac{1}{3}$-octave filter is individually controlled to provide flexibility in shaping the pseudorandom noise spectrum. To meet these requirements, 41 coefficients are needed for the highest $\frac{1}{3}$-octave filter. Thus the corresponding composite filter also has 41 coefficients. The sample rate of the output is chosen to be 16,393 Hz, in order to make it easily twice the highest band. To meet the filter specifications in each region with a constant sample frequency, the filter length must be doubled in going from one octave filter to the next-lower one. Therefore, the lowest octave filter would need 41×2^8 coefficients. With nine filters ranging from 41 coefficients to 41×256 coefficients, the computational effort is considerable; as a result, a multirate approach, shown in Figure 9.1, is used to decrease the amount of computation needed to produce the output [8–16]. The sample rate of the highest-octave band filter is processed at 16,393 samples per second.

The output of the noise generator provides uncorrelated noise inputs to each of the nine sets of bandpass filters. The coefficients are designed such that the center frequency and bandwidth are determined by the sampling frequency. Each stage, controllable in $\frac{1}{3}$ octave, has an individual scaling factor which adjusts that specific $\frac{1}{3}$-octave band. Overall, 27 levels can be individually controlled. The output of each of the octave bandpass filters (except the ninth stage) is provided as an input to an interpolation filter (see Figure 9.1). To minimize the ripple in the output spectrum, each adjacent $\frac{1}{3}$-octave band filter has crossover frequencies at the 3-dB points. A double buffered output was used, since the outputs to the D/A need to be evenly spaced in time, and the processing required cannot provide evenly spaced samples within the bandwidth of the TMS320C25. Figure 9.2 shows the magnitude of the first FIR bandpass filter (sample frequency = 64) with a center frequency of approximately one-fourth the sampling frequency.

The sample rate of the highest octave band filter is processed as 16,393 samples per second and each successively lower band is processed at half the rate of the next-higher band, interpolated up by a factor of 2 and summed with the next-higher band output, as shown in the multirate filter functional block diagram in Figure 9.1. The decimation and interpolation operations are explained in Appendix H.

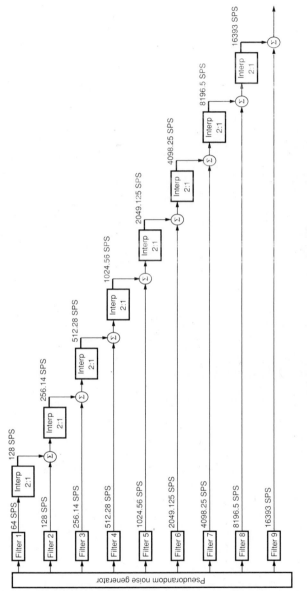

FIGURE 9.1. Multirate filter functional block diagram

213

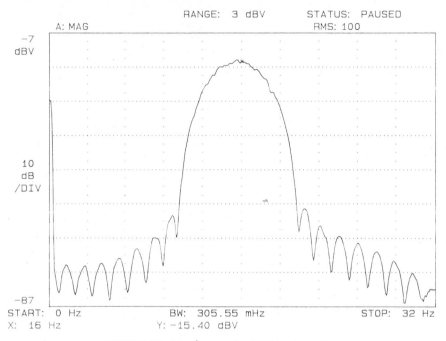

FIGURE 9.2. First $\frac{1}{3}$-octave band FIR bandpass filter

The coefficients of each set of three successive $\frac{1}{3}$-octave filters are combined into a set of coefficients representing one octave band. As a result, only nine octave band filters are implemented, while maintaining individual level control for twenty-seven $\frac{1}{3}$-octave bands. Coefficient scaling and normalization is used to increase the dynamic range of the noise generator.

Each $\frac{1}{3}$-octave filter was implemented as a 41st-order finite impulse response (FIR) filter to realize the necessary filter bandwidth and stopband rejection. Since the center frequency and bandwidth of an FIR bandpass filter are directly proportional to the sampling frequency, only three unique sets of 41 coefficients are required. For each octave band, three separate sets of coefficients are used for the lower, middle, and higher $\frac{1}{3}$-octave bands. The three sets of coefficients are combined as follows, for each of the nine bands:

First band: $A_0 = (AL_0)(L_1) + (AM_0)(L_2) + (AH_0)(L_3)$

$$\vdots$$

$$A_{40} = (AL_{40})(L_1) + (AM_{40})(L_2) + (AH_{40})(L_3)$$

$$\vdots$$

Ninth band: $A_0 = (AL_0)(L_{25}) + (AM_0)(L_{26}) + (AH_0)(L_{27})$

$$\vdots$$

$$A_{40} = (AL_{40})(L_{25}) + (AM_{40})(L_{26}) + (AH_{40})(L_{27})$$

or

$$A_{ij} = (AL_j)(L_{3i-2}) + (AM_j)(L_{3i-1}) + (AH_j)(L_{3i})$$

where $i = 1, \ldots, 9$ bands, $j = 0, \ldots, 40$ coefficients, and AL_j, AM_j, and AH_j represent the jth coefficient of the lower, middle, and higher octave band, respectively. $L_1, L_2, \ldots, L_{27}$ represent the level of each $\frac{1}{3}$-octave filter.

Efficiencies of Multirate Filtering

To increase the efficiency of the system, lower-frequency octave bands are processed at a lower sample rate and interpolated up to a higher sample rate to be summed with the next-higher octave band filter output, as shown in Figure 9.1. Each one of the eight interpolation filters is a 21st-order lowpass FIR filter which provides a 2:1 data rate increase and has a cutoff frequency at approximately one-fourth of the sampling frequency. For each input, the interpolation filter provides two outputs, implemented as follows:

$$Y_{1n} = X_0I_0 + 0I_1 + X_1I_2 + 0I_3 + \cdots + X_{10}I_{20} = \sum_{n=0}^{10} X_nI_{2n}$$

$$Y_{2n} = 0I_0 + X_0I_1 + 0I_2 + X_1I_3 + \cdots + X_9I_{19} = \sum_{n=0}^{10} X_nI_{2n+1}$$

where Y_{1n} and Y_{2n} are the first and second interpolated outputs, respectively, X_n are the filter inputs, and I_n are the interpolation filter coefficients.

Implementation of this bank of twenty-seven $\frac{1}{3}$-octave filters by utilizing multiple sample rates requires approximately 1.7 million multiply/accumulate per second, which is well within the 10-MHz processing bandwidth of the TMS320C25. Consider a multirate nine-band structure, using 41 coefficients for each of the nine octave bandpass filters and 21 coefficients for each of the eight interpolation filters (no interpolation filter required for band 9). The number of multiply/accumulate per second (MAC/S) for this structure is given by:

$$\text{MAC}/S = (O_n + I_n)SR_n + 0_9SR_9$$

where SR_n is the sample rate of the nth filter section, O_n is the number of coefficients in the nth band octave filter and I_n is the number of coefficients in the nth interpolation filter. Evaluating this expression using the rounded sampling frequencies from Table 9.1 gives

$$\text{MAC}/S = (41 + 21)(64 + 128 + 256 + 512 + 1024 + 2048 + 4096$$
$$+ 8192) + (41)(16, 393) \approx 1.68 \times 10^{+6}$$

The efficiency of the multirate filter comes from the fact that each octave band is processed at half the sampling rate of the next stage. The output of each stage is up-sampled or interpolated by 2 (to the next-higher sample rate) and is summed with the output of the next stage. The interpolator is processed in two sections to provide the data rate increase by a factor of 2. The alternative approach, with each stage running at the same (highest) sample rate, would require each preceding filter stage to have twice the number of coefficients as the next-higher stage, in order to obtain the same impulse response as the multirate approach.

The overall program, including the pseudorandom noise generation, utilizes approximately 63% of the available processor time and occupies 1368 words of

TABLE 9.1   Characteristics of $\frac{1}{3}$- Octave Bands

Sampling Frequency	Octave Band	Center Frequency	Bandwidth
64.035	1(Low)	12.5	2.9
	(Medium)	15.75	3.8
	(High)	20.0	4.5
128.07	2(Low)	25.0	5.8
	(Medium)	31.5	7.6
	(High)	40.0	9.3
256.14	3(Low)	50.0	11.2
	(Medium)	63.0	15.1
	(High)	80.0	17.8
512.28	4(Low)	100.0	23.2
	(Medium)	125.0	30.2
	(High)	160.0	35.6
1024.56	5(Low)	200.0	46.3
	(Medium)	250.0	60.5
	(High)	315.0	71.2
2049.125	6(Low)	400.0	92.6
	(Medium)	500.0	121.0
	(High)	630.0	142.4
4098.25	7(Low)	800.0	185.0
	(Medium)	1000.0	242.0
	(High)	1260.0	285.0
8196.5	8(Low)	1600.0	370.0
	(Medium)	2000.0	484.0
	(High)	2520.0	570.0
16393	9(Low)	3200.0	741.0
	(Medium)	4000.0	967.0
	(High)	5040.0	1140.0

data memory and 568 words of program memory. If a single sample rate of 16,393 Hz was used for all the filter stages, approximately 41K words of data memory would be required, an increase by a factor of 30. Such an approach would cause severe processor loading, since a large amount of external memory is required which is not accessed as quickly as on-chip memory. Furthermore, if the increased processor loading and memory requirements could be tolerated, a single rate filter would require approximately 335 million multiply/accumulate per second, which cannot be realized with the TMS320C25. The required MAC/S for the single-rate case follows:

$$t = (N - 1)T$$

where t is the impulse response duration, N the number of filter coefficients, and T the sampling period. Then

$$t_s = (N_s - 1)T_s \qquad (9.1)$$

$$t_m = (N_m - 1)T_m \qquad (9.2)$$

where T_s and T_m are, respectively, the single- and multirate sampling period, and N_s and N_m are, respectively, the number of single and multirate coefficients. For a single-rate alternative with impulse response t_s to yield the same multirate impulse response t_m, (9.1) = (9.2) and

$$(N_s - 1)T_s = (N_m - 1)T_m$$

or

$$N_s = (N_m - 1)(F_s/F_m) + 1 \qquad (9.3)$$

where F_s and F_m are the corresponding sampling frequencies. Applying the expression $N_s - 1$ to the filter structure yields a geometric series with a ratio of two:

$$(N_s - 1)_1 = (40)(16,393/64) = 10,246,$$

$$\text{for band 1 } (F_m = 64)$$

$$(N_s - 1)_2 = (40)(16,393/128) = 5,123$$

$$\vdots$$

$$((N_s - 1)_9 = (40)(16,393/16,393) = 40,$$

$$\text{for band 9 } (F_m = 16,393)$$

The summation of series allows the overall N_s to be easily found,

$$N_s = 20,449$$

Multiplying N_s times the sample rate gives the MAC/S for an equivalent single rate filter:

$$\text{MAC}/S = (N_s)(16,393) \approx 335 \times 10^{+6}$$

The previously calculated multirate filter multiply requirement is .503% of this single-rate requirement. Compared with the single-rate approach, the multirate approach provides a considerable reduction in processing time, due to the significantly lower multiply/accumulate rate.

A further discussion of the multirate filter is included in Appendix H.

Results and Conclusions

The performance of the multirate filter structure is evaluated and demonstrated by stimulating each octave band filter with pseudorandom noise (PRN). The noise generator is a software-based implementation of a maximal-length-sequence technique for generating pseudorandom numbers, as shown in Figure 9.3. Figure 9.4 shows the frequency responses of the three $\frac{1}{3}$-octave bandpass filters for band 8, with the resulting eight-octave bandpass filter frequency response shown in Figure 9.5. Figure 9.6 shows the frequency response of each of the nine center $\frac{1}{3}$-octave bandpass filters. Table 9.1 shows the center frequencies and bandwidth of each $\frac{1}{3}$-octave filter.

Figure 9.7 shows a program (with only a few coefficients), which includes the pseudorandom-noise-generator routine as well as the bandpass filter section, but excludes the interpolator filter. Figure 9.7 also includes the codes necessary for the communication with the AIC [17,18].

Programming considerations include the use of subroutine and in-line coding to maximize processing time while minimizing the memory requirements.

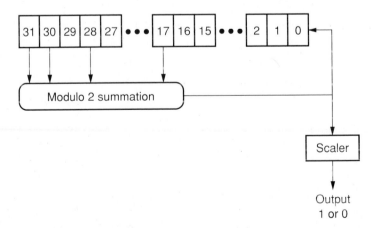

FIGURE 9.3. A 32-bit noise generator

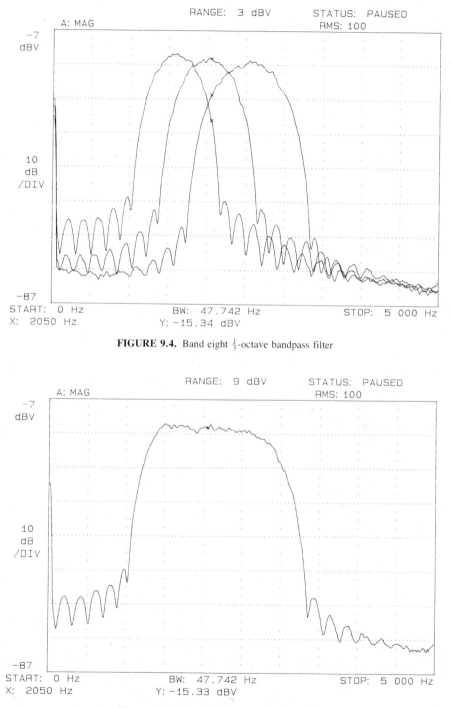

FIGURE 9.4. Band eight $\frac{1}{3}$-octave bandpass filter

FIGURE 9.5. Band eight octave bandpass filter

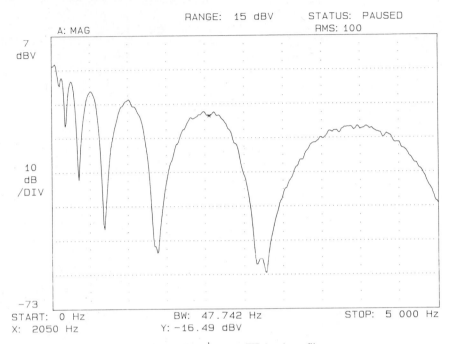

RANGE: 15 dBV   STATUS: PAUSED
A: MAG   RMS: 100

7
dBV

10
dB
/DIV

−73
START: 0 Hz   BW: 47.742 Hz   STOP: 5 000 Hz
X: 2050 Hz   Y: −16.49 dBV

FIGURE 9.6. Nine $\frac{1}{3}$-octave FIR bandpass filters

```
*32-BIT MAXIMUM LENGTH SHIFT REGISTER PSEUDO-RANDOM NOISE
*GENERATOR FEEDING INTO A BANDPASS FILTER WITH
*CENTER FREQUENCY APPROX 1/4 SAMPLE RATE OF 16,393 HZ
            IDT      'BP41'
*        DATA MEMORY ALLOCATION/DEFINITION
OUTPUT  EQU      >60              OUTPUT SAMPLE STORAGE
LH1000  EQU      >61              NOISE GEN POSITIVE SCALER
LHF000  EQU      >62              NOISE GEN NEGATIVE SCALER
RNUMH   EQU      >63              RANDOM NUM UPPER 16-BITS
RNUML   EQU      >64              RANDOM NUM LOWER 16-BITS
MASK15  EQU      >65              BIT-15 MASK
TEMP    EQU      >66              SCRATCH PAD MEMORY AREA
SWDS1   EQU      >7E              RESERVED FOR SWDS USAGE
SWDS2   EQU      >7F              RESERVED FOR SWDS USAGE
CH0     EQU      >0200            FILTER COEFF STORAGE
X0      EQU      >0229            FILTER INPUT SAMPLES
INTMSK  EQU      >0004            INTERRUPT MASK REGISTER
        AORG     0
        B        INIT             BRANCH TO INIT TMS320C25
        AORG     >1A              RECEIVE INTER(RINT) VECTOR
        B        RINT             BRANCH TO RECEIVE INTER
        AORG     >1C              TRANSMIT INTER(XMIT) VECTOR
XINT    LAC      OUTPUT           LOAD ACC WITH OUTPUT SAMPLE
        SACL     >01              STORE IN DXR,DMA 1
        RET                       RETURN FOR MORE SEC DATA
RINT    LAC      OUTPUT           LOAD ACC WITH OUT SAMPLE
        ANDK     >FFFC            MASK 2 LSB SINCE 14 BIT D/A
        SACL     >01              STORE IN DXR
        RET
        AORG     >40      *****PROGRAM MEMORY DATA STORAGE*****
SEEDH   DATA     >7E52            RANDOM NUM SEED,UPPER 16-BITS
SEEDL   DATA     >1603            RANDOM NUM SEED,LOWER 16-BITS
IMASK   DATA     >0010            ENABLE AIC RECEIVE INTER ONLY
H0      DATA     >FFD4            BANDPASS FILTER COEFF 0 TO 40
H1      DATA     >FFF0
.
.
.
H40     DATA     >FFD4
*        AIC BOX - TMS320C25 COMMUNICATION INITIALIZATION
```

FIGURE 9.7. Partial program incorporating the PRN and using the AIC (AICBP41)

```
* SECONDARY COMMUNICATION DATA TABLE FOR AIC REGISTERS
TABLE   DATA    >03,>67,>03,>448A,>03,>1224
* INITIALIZE PROCESSOR
INIT    LALK    >03E0       INIT ST1,FRAME SYNC PULSES AS INPUT
        SACL    >67         TXM=0(LOW) CONN TO AIC(RESET) PIN 2
        LST1    >67         FSM=1 FRAME SYNC PULSES NOT IGNORED
        ZAC                 SET ACC TO 0
        SACL    >01         CLEAR DXR(DATA XMIT REG),IN DMA 01
        SACL    OUTPUT      INIT OUTPUT TO 0
        LACK    >20         LOAD IMR(INTER MASK REG)=0010 0000 TO
        SACL    INTMSK      MASK ALL INTER EXCEPT 'XINT',IN DMA 4
        LARP    0           SELECT AR0
        LARK    AR0,>62     TRANSMIT DATA FROM PMA >20 TO >25
        RPTK    >05         INTO DMA >62 TO >67
        BLKP    TABLE,*+
* INITIALIZE AIC REGISTERS (SECONDARY COMMUNICATION)
        LARK    AR1,05      SELECT AR1 AS LOOP COUNTER
        LARK    AR0,>62     DMA START LOCATION OF XMIT DATA IN AR0
        SXF                 DISABLE AIC RESET WITH EXT FLAG XF=1
SCND    IDLE                ENABLE AND WAIT FOR XMIT INTER(INTM=0)
        LAC     *+,0,AR1    TRANSFER DATA STARTING AT >62
        SACL    OUTPUT      FOR OUTPUT TO AIC
        BANZ    SCND,*-,AR0 BRANCH BACK IF AR0 # 0,DEC AR1
        IDLE                WAIT FOR LAST XMIT DATA
* END OF AIC BOX INITIALIZATION
        CNFD                CONFIGURE B0 AS DATA MEMORY
        LDPK    0           POINT TO DATA PAGE 0(forever)
        TBLR    RNUML       XFER SEEDL TO RNUML IN D.M.
        ADDK    1           INCREMENT READ ADDRESS IN ACC
        TBLR    INTMSK      XFER IMASK TO MASK REGISTER
        LALK    >1000
        SACL    LH1000      INITIALIZE LH1000 TO >1000
        NEG
        SACL    LHF000      INITIALIZE LHF000 TO >F000
        LALK    >8000
        SACL    MASK15      INITIALIZE MASK15 TO >8000
* COPY BANDPASS FILTER COEFFS FROM PGMEM TO DMEM
        LRLK    0,CH0       LOAD AR0 WITH WRITE ADX
        LARP    0           POINT TO AR0
        RPTK    >28         LOOP COUNT (41 values)
        BLKP    H0,*+       COPY H0-H40 TO CH0-CH40
*
WAIT    IDLE                WAIT HERE UNTIL AN INTERRUPT
* Modulo 2 sum of the 32-bit noise word bits 31,30,28,and 17
* yields a random bit (1 or 0) shifted into the LSB of the
* noise word, scaled and output to a D/A converter. Since
* only the upper 16-bits are involved in determining feedback,
* the summation is done in lower accumulator for convenience.
START   LAC     RNUMH       GET RANDOM NUM UPPER 16-BITS
        AND     MASK15      MASK BIT-15
        ADD     RNUMH,1     ADD BIT-14 TO BIT-15
        AND     MASK15      MASK BIT-15
        ADD     RNUMH,3     ADD BIT-12 TO THE SUMMATION
        AND     MASK15      MASK BIT-15
        ADD     RNUMH,14    ADD BIT-1 TO THE SUMMATION
        AND     MASK15      MASK BIT-15
        SACH    TEMP,1      SHIFT R 15 (LEFT 1,RIGHT 16)
        ZALH    RNUMH       RESTORE RNUMH TO UPPER ACCUM
        ADD     RNUML       RESTORE RNUML TO LOWER ACCUM
        SFL                 SHIFT 32-BIT ACCUM LEFT 1
        ADD     TEMP        ADD FEEDBACK BIT
        SACH    RNUMH       STORE UPPER 16-BIT NOISE WORD
        SACL    RNUML       STORE LOWER 16-BIT NOISE WORD
        LAC     TEMP        RESTORE RANDOM BIT INTO ACCUM
        BZ      MINUS
        LAC     LH1000      SET OUTPUT POSITIVE IF BIT=1
        B       FILT
MINUS   LAC     LHF000      SET OUTPUT NEGATIVE IF BIT=0
* THE BANDPASS FILTER STARTS HERE
FILT    LRLK    0,X0        LOAD POINTER TO NEWEST SAMPLE
        LARP    0           POINT TO NEWEST SAMPLE STORAGE
        SACL    *           STORE SCALED OUTPUT VALUE
        ZAC
        LRLK    0,CH0+>28   LOAD POINTER TO COEFF H40
        LRLK    1,X0+>28    LOAD POINTER TO SAMPLE X40
        LARP    1           POINT TO X40
        LTD     *-,0        LOAD X40
        MPY     *-,1        H40*X40
        LTD     *-,0        *
        MPY     *-,1        * H39*X39

        .
        .
        .
        LTD     *-,0        *
        MPY     *-,1        * H0*X0
        APAC
        SACH    OUTPUT,1    STORE FILTER OUTPUT
        B       WAIT
```

FIGURE 9.7. (concluded)

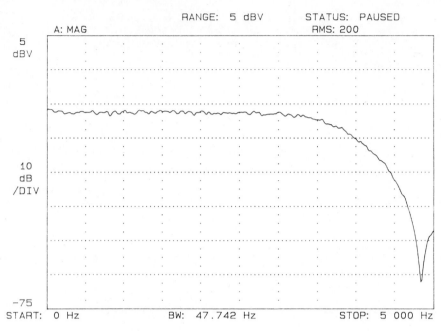

FIGURE 9.8. Interpolator (lowpass) filter

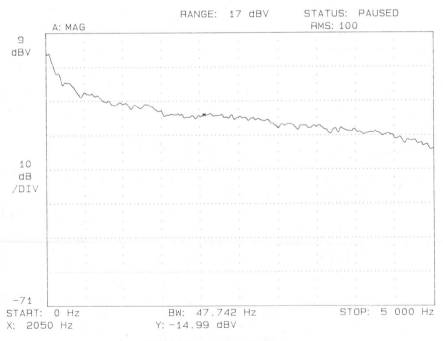

FIGURE 9.9. Shaped output noise

The powerful pair of instructions RPTK/MACD were not used because interrupts are held off during the execution of the repeat instruction. Figure 9.8 shows the interpolator (lowpass) filter, obtained without the bandpass filters. Figure 9.9 shows the nine bandpass filters with octave-band coefficients, with equal scaling, providing the shaped noise spectrum output.

9.2 MODULATION TECHNIQUES

Frequency hopping is a technique in which the carrier frequency is changed randomly at predetermined time intervals. By randomly changing the carrier frequency, a level of security in the transmission of information via unsecure media (radio waves) can be accomplished. To capture the transmitted information, the listener (receiver) must possess knowledge such as the number of modulating carriers and their frequency values, the sequence used to select them and the time intervals at which they are changed. A receiver matched to the transmitter is capable of demodulating the signal, thus providing a relatively secure medium.

Problem Statement

This project considers the implementation of a digital modulator capable of frequency hopping. The oscillator example of Chapter 4, and the pseudo-random-noise-generator program explained in Chapters 3 and 5, provide a good background for this project. Both AM and FM and frequency-hopping techniques are implemented.

Design and Implementation Considerations

Figure 9.10 shows the hardware configuration for both the AM and FM frequency-hopping modulator.

Amplitude Modulator (AM)

Figure 9.11 shows a functional block diagram of an amplitude modulator as well as a sinusoid generated with the oscillator codes and the resulting output AM waveform. The input signal $s(T)$ for the A/D is scaled to ensure that all values are positive, preventing distortions associated with amplitude-modulating input signals that contain both positive and negative values. The output of the scaler is multiplied with the output of a digital oscillator that generates a sinusoidal function whose frequency is determined by the frequency selector. The digital oscillator equation is repeated here:

$$y(n) = Ay(n-1) + By(n-2) + Cx(n-1)$$

where $A = 2 \cos \omega T$, $B = -1$, $C = \sin \omega T$ and T is the sampling period, which

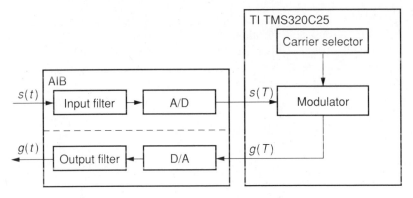

FIGURE 9.10. Functional block diagram of frequency hopping modulator

is equal to the execution time of the calculation loop. Based on the programmed constants, A, B, and C, the oscillator generates a frequency with period proportional to T. The output of the modulator is then fed to the D/A to obtain an analog signal.

Frequency Modulator (FM)

Figure 9.12 shows a functional block diagram of a frequency modulator with the resulting FM output waveform (top). The input signal $s(T)$, from the A/D, is multiplied by a predetermined constant K whose range specifies the bandwidth of the modulator output. The resulting product (fi), which is always less than the value of K due to the scaling of the input, is then added to A, which controls the desired operating frequency of the digital oscillator. When the input signal is zero, the oscillator operates at its base frequency, which is determined by the constants A, B, C in the oscillator equation and the program period, T, identical to the amplitude modulator. When the input is at its maximum negative value, the output frequency is the base frequency plus half of K (range). When the output is at its maximum positive value, the output frequency is the base frequency minus half of K (range). The output of the modulator is then fed to the D/A to obtain an analog signal.

Carrier Selector

Figure 9.13 shows the functional block diagram of a carrier selector. This selector determines the frequency of the digital oscillator, which is the carrier frequency of the AM or FM modulator. The carrier frequencies were restricted to 8, 12, and 16 kHz for AM and 11, 16, and 20 kHz for FM.

The carrier selector, or the frequency-hopping function, consists of the major elements shown in Figure 9.13: the timer, the pseudorandom number generator, and the AND functions. The timer determines the rate at which the carrier is changed.

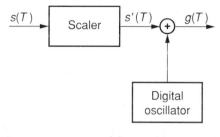

(a)

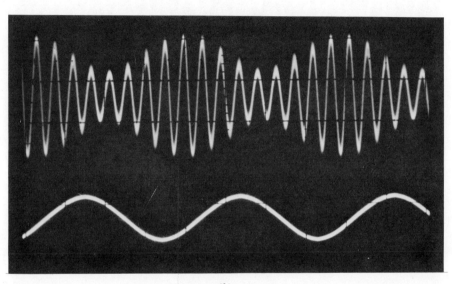

(b)

FIGURE 9.11. Amplitude modulator: (a) functional block diagram; (b) carrier (bottom) and modulated waveform (top)

The pseudorandom number generator, upon activation by the timer, calculates a random 2-bit number, of which three values are valid for the three carriers (0 is not valid). The output of the number generator becomes the input to the AND functions, which generate a selection index, Fc.

A complete listing of the programs for AM, FM, and AM/FM hopping are shown in Figures 9.14 through 9.17. Since the output frequency of the digital oscillator is dependent on the program period T, a delay loop was incorporated as part of T. This delay loop, controlled by the variable STALL (appearing in Figures 9.16 and 9.17), is used to determine the carrier frequency. The oscillator frequency increases as the value of STALL is decreased.

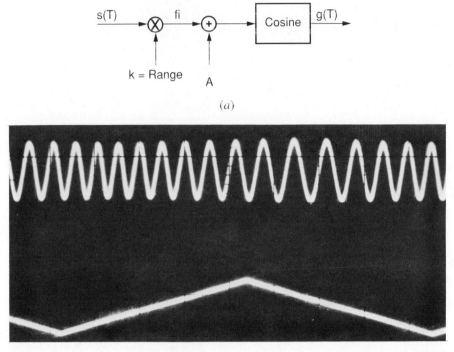

(a)

(b)

FIGURE 9.12. Frequency modulator: (a) functional block diagram; (b) input and modulated waveform

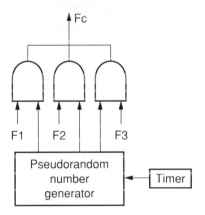

FIGURE 9.13. Carrier selector

On-board the AIB is a 4.7-kHz lowpass output filter. To support the higher-frequency outputs of the modulator, a lowpass output filter with a cutoff frequency of 20 kHz was constructed. This required changing some of the values of the resistors and capacitors of the AIB filter.

Conclusions

To receive information properly, the receiving station's timers, carriers, and pseudorandom number generators must be synchronized with the transmitting station, thus providing a desired level of encryption. Incorporation of additional modulating techniques, such as pulse-code and pulse-width modulation, and using frequency-division multiplexing to transmit data bits in parallel over a signal telephone line, can enhance this project.

```
0001                 * AM PROGRAM
0002                 *DIGITAL OSCILLATOR AND MODULATOR PROGRAM    Fs=10 KHZ
0003                 *Y(n)=A Y(n-1) + B Y(n-2) + C X(n-1)
0004                 * Y0 =A Y1 + B Y2    FOR N>1   X(n-1)=1 FOR n=1 (= 0 OTHERWISE)
0005      0060  MODE         EQU    >60    *TRANSPARENT MODE FOR AIB
0006      0061  RATE         EQU    >61    *RATE=(10 MHZ / Fs) -1=999=03E7
0007      0062  Y0           EQU    >62    *INITIALLY Y0=0, OUTPUT DMA
0008      0063  Y1           EQU    >63    *Y(1)=C=SIN(2x3.14xFd/Fs)
0009      0064  Y2           EQU    >64    *Y(2)=0
0010      0065  A            EQU    >65    *A=2 COS(WT)
0011      0066  B            EQU    >66    *B=-1
0012      0067  SIG          EQU    >67    *INPUT SIGNAL
0013      0068  OPS          EQU    >68    *OUTPUT SIGNAL
0014            *                          *A,B,C DIV BY 2 (MAX VALUE <1 )
0015            *                          *A=678E,B=C000,C=259E    Fd=1KHZ
0016            *                          *A=278E,B=C000,C=3CDE    Fd=2KHZ
0017 0000                    AORG   >0
0018 0000 FF80               B      START
     0001 0027
0019 0020       TABLE        AORG   >20    *PM START AT >20
0020 0020 00FA  MDE          DATA   >00FA
0021 0021 01F3  RTE          DATA   >01F3
0022 0022 0000  Y00          DATA   >0000
0023 0023 2998  Y01          DATA   >2998
0024 0024 0000  Y02          DATA   >0000
0025 0025 5F1F  AA           DATA   >5F1F
0026 0026 C000  BB           DATA   >C000
0027            *
0028 0027 5588  START        LARP   0
0029 0028 D000               LRLK   0,>0060
     0029 0060
0030 002A CB06               RPTK   6
0031 002B FCA0               BLKP   TABLE,*+
     002C 0020
0032 002D E060               OUT    MODE,0 *OUTPUT MODE = >FA TO PORT 0
0033 002E E161               OUT    RATE,1 *OUTPUT RATE =03E7 TO PORT 1
0034 002F FA80  WAIT         BIOZ   MAIN   *WAIT TIL BIO=0(EOC OF A/D LOW)
     0030 0033
0035           *                           *BRANCH TO MAIN AFTER EOC
0036 0031 FF80               B      WAIT   *A/D PROVIDES TIMING
     0032 002F
0037           *MAIN SECTION
0038 0033 C064  MAIN         LARK   0,Y2   *LOAD AUX REG AR0 WITH Y2
0039 0034 C166               LARK   1,B    *LOAD AUX REG AR1 WITH B
0040 0035 CA00               ZAC           *ZERO THE ACCUMULATOR
0041 0036 5588               LARP   0      *AR0 FOR INDIRECT ADDRESSING
0042 0037 3C99               LT     *-,1   *TR=(AR0),DEC TO Y1,SELECT AR1
0043 0038 3898               MPY    *-,0   * (AR1)x(TR)=BxY2,DEC AR1,=>AR0
0044 0039 3F89               LTD    *,1    *TR=(AR0),ADD PR ,MOVE"DOWN"
0045           *                           *LTD=LT+APAC+DMOV
0046 003A 3888               MPY    *,0    * (AR1)x(TR)=AxY1
0047 003B CE15               APAC          *ADD PREVIOUS (PR) TO ACC
0048 003C 6962               SACH   Y0,1   *SHIFT 1 DUE TO EXTRA BIT,STORE
0049           *                           *UPPER 16 BITS(DIVIDE BY 2**15)
0050           *    THIS SECTION DIVIDES THE INPUT SIGNAL IN HALF AND ADDS A
```

FIGURE 9.14. Program listing for AM (AM.LST)

```
0051                          *        CONSTANT TO IT TO ENSURE ALL VALUES ARE POSITIVE
0052 003D 8267                         IN      SIG,2       *READ INPUT, STORE IN DMA >67
0053 003E 2067                         LAC     SIG         *LOAD ACC WITH INPUT SIGNAL
0054 003F CE19                         SFR                 *SHIFT RIGHT TO DIVIDE BY 2
0055 0040 D002                         ADLK    >4000       *ADD 16384 TO ENSURE THAT ALL
     0041 4000
0056                          *                            *ALL VALUES OF SIG ARE POSITIVE
0057 0042 6067                         SACL    SIG         *STORE NEW VALUE IN DMA >67
0058 0043 E267                         OUT     SIG,2       *OUTPUT SIG
0059                          *
0060                          *        THIS NEXT SECTION MULTIPLIES THE CONDITIONED INPUT WITH
0061                          *        THE CARRIER TO GIVE A MODULATED OUTPUT
0062                          *
0063 0044 3C62                         LT      Y0          *PUT VALUE OF CARRIER IN TR
0064 0045 3867                         MPY     SIG         *MULTIPLY WIYH INPUT
0065 0046 CE15                         APAC                *PUT PRODUCT IN ACC
0066 0047 6968                         SACH    OPS,1       *SHIFT 1 FOR EXTRA SIGN BIT,
0067                          *                            *STORE UPPER 16 BITS IN DMA >68
0068                          *
0069                          *        CORRECT FOR DIVISION OF A, B, AND C
0070                          *
0071 0048 2062                         LAC     Y0          *SINCE A,B,C WAS DIVIDED BY 2
0072 0049 0062                         ADD     Y0          *DOUBLE RESULT
0073 004A 6062                         SACL    Y0          *LOWER 16 BITS(STORE IN DMA 62)
0074 004B 5662                         DMOV    Y0          *MOVE"DOWN"IN DM FOR NEXT N
0075 004C 2068                         LAC     OPS         *SINCE Y0 WAS DIVIDED BY 2
0076 004D 0068                         ADD     OPS         *DOUBLE OUTPUT
0077 004E 6068                         SACL    OPS         *STORE MODULATED OUTPUT
0078 004F FF80                         B       WAIT        *BRANCH BACK FOR NEXT  N
     0050 002F
0079                                   END
NO ERRORS, NO WARNINGS
```

FIGURE 9.14. (concluded)

```
0001                          * FM PROGRAM
0002                          *DIGITAL OSCILLATOR AND FREQUENCY MODULATOR PROGRAM    Fs=30 KHZ
0003                          *Y(n)=A Y(n-1) + B Y(n-2) + C X(n-1)
0004                          * Y0 =A Y1 + B Y2    FOR N>1    X(n-1)=1 FOR n=1 (= 0 OTHERWISE)
0005      0060 MODE           EQU     >60         *TRANSPARENT MODE FOR AIB
0006      0061 RATE           EQU     >61         *RATE=(10 MHZ / Fs) -1=332=14C
0007      0062 Y0             EQU     >62         *INITIALLY Y0=0, OUTPUT DMA
0008      0063 Y1             EQU     >63         *Y(1)=C=SIN(2x3.14xFd/Fs)
0009      0064 Y2             EQU     >64         *Y(2)=0
0010      0065 A              EQU     >65         *A=2 COS(WT)
0011      0066 B              EQU     >66         *B=-1
0012      0067 RNG            EQU     >67         *INPUT SIGNAL
0013      0068 SIG            EQU     >68         *FREQUENCY RANGE
0014                          *                   *A,B,C DIV BY 2 (MAX VALUE <1 )
0015                          *                   *A=74EE,B=C000,C=259E Fd=2KHZ
0016                          *                   *A=55A6,B=C000,C=3CDE Fd=4KHZ
0017                          *                   *A=5F1F,B=C000,C=2998 Fd=3.5KHZ
0018 0000                                AORG    >0
0019 0000 FF80                           B       START
     0001 0028
0020 0020      TABLE           AORG     >20        *PM START AT >20
0021 0020 00FA MDE             DATA     >00FA
0022 0021 014C RTE             DATA     >014C
0023 0022 0000 Y00             DATA     >0000
0024 0023 2998 Y01             DATA     >2998
0025 0024 0000 Y02             DATA     >0000
0026 0025 278E AA              DATA     >278E
0027 0026 C000 BB              DATA     >C000
0028 0027 0979 RG              DATA     >0979
0029                          *
0030 0028 5588 START           LARP     0
0031 0029 D000                 LRLK     0,>0060
     002A 0060
0032 002B CB07                 RPTK     7
0033 002C FCA0                 BLKP     TABLE,*+
     002D 0020
0034 002E E060                 OUT      MODE,0     *OUTPUT MODE = >FA TO PORT 0
0035 002F E161                 OUT      RATE,1     *OUTPUT RATE =03E7 TO PORT 1
0036 0030 CE07                 SSXM                *SET SIGN EXTESION MODE
0037 0031 FA80 WAIT            BIOZ     MAIN       *WAIT TIL BIO=0(EOC OF A/D LOW)
     0032 0035
0038                          *                    *BRANCH TO MAIN AFTER EOC
0039 0033 FF80                 B        WAIT       *A/D PROVIDES TIMING
     0034 0031
0040                          *
0041                          *        THIS SECTION READS THE INPUT, MULTIPLIES IT WITH THE
0042                          *        RANGE VALUE, AND ADDS THE PRODUCT TO THE VALUE OF A THAT
0043                          *        GENERATES A 3.5KHZ FREQUENCY
0044                          *
0045 0035 CA00 MAIN            ZAC                 *ZERO ACCUMULATOR
```

FIGURE 9.15. Program listing for FM (FM.LST)

```
0046 0036 8268                    IN      SIG,2    *READ INPUT
0047 0037 3C67                    LT      RNG      *PUT RANGE VALUE IN TR
0048 0038 3868                    MPY     SIG      *MULTIPLY RANGE X INPUT
0049 0039 CE15                    APAC             *PUT PRODUCT IN ACC
0050 003A 6965                    SACH    A,1      *SHIFT 1 FOR EXTRA SIGN BIT
0051 003B 2065                    LAC     A        *STORE AND RETRIEVE TO GET HIGH
0052 003C D002                    ADLK    >5F1F    *ADD VALUE OF A FOR 3.5KHZ
     003D 5F1F
0053 003E 6065                    SACL    A        *STORE VALUE IN DMA >65
0054             *
0055             *         THIS SECTION IS AN OSCILLATOR IN WHICH THE FREQUENCY IS
0056             *         CONTROLLED BY THE VALUE OF A
0057             *
0058 003F C064                    LARK    0,Y2     *LOAD AUX REG AR0 WITH Y2
0059 0040 C166                    LARK    1,B      *LOAD AUX REG AR1 WITH B
0060 0041 CA00                    ZAC              *ZERO THE ACCUMULATOR
0061 0042 5588                    LARP    0        *AR0 FOR INDIRECT ADDRESSING
0062 0043 3C99                    LT      *-,1     *TR=(AR0),DEC TO Y1,SELECT AR1
0063 0044 3898                    MPY     *-,0     *(AR1)x(TR)=BxY2,DEC AR1,=>AR0
0064 0045 3F89                    LTD     *,1      *TR=(AR0),ADD PR ,MOVE"DOWN"
0065             *                                 *LTD=LT+APAC+DMOV
0066 0046 3888                    MPY     *,0      *(AR1)x(TR)=AxY1
0067 0047 CE15                    APAC             *ADD PREVIOUS (PR) TO ACC
0068 0048 6962                    SACH    Y0,1     *SHIFT 1 DUE TO EXTRA BIT,STORE
0069             *                                 *UPPER 16 BITS(DIVIDE BY 2**15)
0070             *         CORRECT FOR DIVISION OF A, B, AND C BY 2
0071             *
0072 0049 2062                    LAC     Y0       *SINCE A,B,C WERE DIVIDED BY 2
0073 004A 0062                    ADD     Y0       *DOUBLE RESULT
0074 004B 6062                    SACL    Y0       *LOWER 16 BITS(STORE IN DMA 62)
0075 004C 5662                    DMOV    Y0       *MOVE"DOWN"IN DM FOR NEXT N
0076 004D E262                    OUT     Y0,2      *OUTPUT DMA >62
0077 004E FF80                    B       WAIT     *BRANCH BACK FOR NEXT  N
     004F 0031
0078                              END
NO ERRORS, NO WARNINGS
```

FIGURE 9.15. (concluded)

```
0001                  *                AMJUMP
0002                  *DIGITAL OSCILLATOR AND AMPLITUDE MODULATOR PROGRAM THAT
0003                  *SELECTS ITS CARRIER FREQUENCY USING A RANDOM NUMBER GENERATOR
0004                  *Fs=30 KHz   PROGRAM CYCLE RATE = 333 KHz
0005                  *Y(n)=A Y(n-1) + B Y(n-2) + C X(n-1)
0006                  * Y0 =A Y1 + B Y2   FOR N>1   X(n-1)=1 FOR n=1 (= 0 OTHERWISE)
0007      0060   MODE        EQU     >60      *TRANSPARENT MODE FOR AIB
0008      0061   RATE        EQU     >61      *RATE=(10 MHZ / Fs)  -1=332=014C
0009      0062   Y0          EQU     >62      *INITIALLY Y0=0, OUTPUT DMA
0010      0063   Y1          EQU     >63      *Y(1)=C=SIN(2x3.14xFd/Fs)
0011      0064   Y2          EQU     >64      *Y(2)=0
0012      0065   A           EQU     >65      *A=2 COS(WT) WORKING VALUE
0013      0066   B           EQU     >66      *B=-1
0014      0067   STLL        EQU     >67      *OFFSET FOR VALUE OF "A"
0015      0068   STPH        EQU     >68      *RANDOM NUM HIGH BYTE
0016      0069   STPL        EQU     >69      *RANDOM NUM LOW BYTE
0017      006A   A1          EQU     >6A      *REFERANCE VALUE OF "A"
0018      006B   TEMP        EQU     >6B      *TEMP. STORAGE
0019      006C   OPS         EQU     >6C      *OUTPUT SIGNAL
0020      006D   SIG         EQU     >6D      *INPUT SIGNAL
0021             *                            *A,B,C DIV BY 2 (MAX VALUE <1 )
0022             *                            *A=5F1F,B=C000,C=2998
0023 0000                    AORG    >0
0024 0000 FF80               B       START
     0001 002B
0025 0020        TABLE       AORG    >20      *PM START AT >20
0026 0020 00FA   MDE         DATA    >00FA
0027 0021 014C   RTE         DATA    >014C
0028 0022 0000   Y00         DATA    >0000
0029 0023 2998   Y01         DATA    >2998
0030 0024 0000   Y02         DATA    >0000
0031 0025 5F1F   AA          DATA    >5F1F
0032 0026 C000   BB          DATA    >C000
0033 0027 0030   ST          DATA    >0030
0034 0028 A72E   STH         DATA    >A72E
0035 0029 D572   STL         DATA    >D572
0036 002A 5F1F   AA1         DATA    >5F1F
0037             *
0038 002B 5588   START       LARP    0
0039 002C D000               LRLK    0,>0060
     002D 0060
0040 002E CB0A               RPTK    10
0041 002F FCA0               BLKP    TABLE,*+
     0030 0020
```

FIGURE 9.16. Program listing for AM hopping (AMJUMP.LST)

```
0042 0031 E060          OUT     MODE,0    *OUTPUT MODE = >FA TO PORT 0
0043 0032 E161          OUT     RATE,1    *OUTPUT RATE =014C TO PORT 1
0044 0033 CE07          SSXM              *SET SIGN EXTESION MODE
0045 0034 D200          LRLK    2,60      *SET VALUE FOR FIRST DELAY
     0035 003C
0046 0036 D300          LRLK    3,3000    *SET VALUE FOR SECOND DELAY
     0037 0BB8
0047 0038 D400          LRLK    4,32      *SET STALL COUNT
     0039 0020
0048              *
0049              *      THIS SECTION READS THE INPUT DIVIDES IT IN HALF AND
0050              *      ADDS A CONSTANT TO IT TO ENSURE THAT ALL VALUES ARE
0051              *      POSITIVE
0052              *
0053 003A 826D INPT     IN      SIG,2     *READ INPUT
0054 003B 206D          LAC     SIG       *LOAD ACC WITH INPUT SIGNAL
0055 003C CE19          SFR               *SHIFT RIGHT TO DIVIDE BY 2
0056 003D D002          ADLK    >4000     *ADD 16384 TO ENSURE THAT
     003E 4000
0057              *                        *ALL VALUES ARE POSITIVE
0058 003F 606D          SACL    SIG       *STORE VALUE IN DATA MEM
0059              *
0060              *      THIS SECTION IS A TIME DELAY THAT DETERMINES WHEN
0061              *      TO CHANGE THE CARRIER FREQUENCY
0062              *
0063 0040 558A MAIN     LARP    2         *SET ARP TO FIRST DELAY
0064 0041 FB90          BANZ    BEGIN     *BRANCH IF COUNT IS NOT 0
     0042 0050
0065              *                        *OTHERWISE DECREMENT COUNT
0066 0043 D200          LRLK    2,60      *RESET COUNTER (AR2) AFTER LOOP
     0044 003C
0067 0045 558B          LARP    3         *SET ARP TO SECOND DELAY
0068 0046 FB90          BANZ    BEGIN     *BRANCH IF COUNT NOT ZERO
     0047 0050
0069 0048 D300          LRLK    3,3000    *RESET COUNTER (AR3) AFTER LOOP
     0049 0BB8
0070              *
0071              *      THIS SECTION SETS THE VALUE OF "STLL" ,WHICH DETERMINES
0072              *      HOW MUCH CARRIER FREQUENCY IS DECREASED
0073              *
0074 004A 2069          LAC     STPL      *GET RANDOM NUMBER
0075 004B D004          ANDK    >30       *AND WITH 2 BITS TO ALLOW 3
     004C 0030
0076              *                        *POSSIBLE FREQUENCIES
0077 004D F680          BZ      GO        *DON'T LET STLL BE ZERO OR
     004E 0054
0078              *                        *FREQUENCY WILL BE TOO HIGH
0079 004F 6067          SACL    STLL      *STORE NEW VALUE OF "STLL"
0080              *
0081              *      THIS LOOP CONTROLS THE CARRIER FREQUENCY BY EXTENDING
0082              *      THE PROGRAM AND SLOWING IT DOWN
0083              *
0084 0050 558C BEGIN    LARP    4         *SET ARP TO STALL COUNTER
0085 0051 FB90 STALL    BANZ    STALL     *DELAY PROGRAM TO DECREASE
     0052 0051
0086              *                        *CARRIER FREAQUENCY
0087 0053 3467          LAR     4,STLL    *RESET STALL COUNTER
0088              *
0089              *      THIS SECTION GENERATES A RANDOM VALUE STPH
0090              *
0091 0054 2068 GO       LAC     STPH      *PUT STPH IN ACC
0092 0055 D004          ANDK    >8000     *MASK BIT 15
     0056 8000
0093 0057 0168          ADD     STPH,1    *SHIFT 1, ADD BIT 15 TO BIT 14
0094 0058 D004          ANDK    >8000     *MASK BIT 15
     0059 8000
0095 005A 0368          ADD     STPH,3    *ADD BIT 12 TO SUMMATION
0096 005B D004          ANDK    >8000     *MASK BIT 15
     005C 8000
0097 005D 0E68          ADD     STPH,14   *ADD BIT 1 TO SUMMATION
0098 005E D004          ANDK    >8000     *MASK BIT 15
     005F 8000
0099 0060 696B          SACH    TEMP,1    *STORE SUMMATION
0100 0061 4068          ZALH    STPH      *RESTORE STPH TO HIGH ACC
0101 0062 0069          ADD     STPL      *RESTORE STPL TO LOW ACC
0102 0063 CE18          SFL               *SHIFT 32 BIT ACC LEFT 1
0103 0064 006B          ADD     TEMP      *ADD FEED BACK BIT
0104 0065 6868          SACH    STPH      *STORE NEW STPH
0105 0066 6069          SACL    STPL      *STORE NEW STPL
0106              *
0107              *      THIS SECTION GENERATES THE CARRIER WITH THE FREQUENCY
0108              *      SET BY THE VALUE OF "A"
0109              *
0110 0067 C064          LARK    0,Y2      *LOAD AUX REG AR0 WITH Y2
0111 0068 C166          LARK    1,B       *LOAD AUX REG AR1 WITH B
0112 0069 CA00          ZAC               *ZERO THE ACCUMULATOR
0113 006A 5588          LARP    0         *AR0 FOR INDIRECT ADDRESSING
0114 006B 3C99          LT      *-,1      *TR=(AR0),DEC TO Y1,SELECT AR1
```

FIGURE 9.16. (continued)

```
0115 006C 3898                    MPY      *-,0       *(AR1)x(TR)=BxY2,DEC AR1,=>AR0
0116 006D 3F89                    LTD      *,1        *TR=(AR0),ADD PR ,MOVE"DOWN"
0117                  *                                *LTD=LT+APAC+DMOV
0118 006E 3888                    MPY      *,0        *(AR1)x(TR)=AxY1
0119 006F CE15                    APAC                *ADD PREVIOUS (PR) TO ACC
0120 0070 6962                    SACH     Y0,1       *SHIFT 1 DUE TO EXTRA BIT,STORE
0121                  *                                *UPPER 16 BITS(DIVIDE BY 2**15)
0122                  *
0123                  *   THIS NEXT SECTION MULTIPLIES THE CONDITIONED INPUT WITH
0124                  *   THE CARRIER TO GIVE A MODULATED OUTPUT
0125                  *
0126 0071 CA00                    ZAC                 *ZERO ACC
0127 0072 3C62                    LT       Y0         *PUT VALUE OF CARRIER IN TR
0128 0073 386D                    MPY      SIG        *MULTIPLY WITH INPUT
0129 0074 CE15                    APAC                *PUT PRODUCT IN ACC
0130 0075 696C                    SACH     OPS,1      *SHIFT 1 FOR EXTRA SIGN BIT,
0131                  *                                *STORE UPPER 16 BITS IN "OPS"
0132                  *
0133                  *   CORRECT FOR DIVISION OF A, B, AND C BY 2
0134                  *
0135 0076 2062                    LAC      Y0         *SINCE A,B,C WAS DIVIDED BY 2
0136 0077 0062                    ADD      Y0         *DOUBLE RESULT
0137 0078 6062                    SACL     Y0         *LOWER 16 BITS(STORE IN DMA 62)
0138 0079 5662                    DMOV     Y0         *MOVE"DOWN"IN DM FOR NEXT N
0139 007A 206C                    LAC      OPS        *SINCE Y0 WAS DIVIDED BY 2
0140 007B 006C                    ADD      OPS        *DOUBLE OUTPUT
0141 007C 606C                    SACL     OPS        *STORE MODULATED OUTPUT
0142 007D E26C                    OUT      OPS,2      *OUTPUT (DM >62) TO PORT 2
0143 007E FA80                    BIOZ     INPT       *READ INPUT IF BIO=0
     007F 003A
0144 0080 FF80                    B        MAIN       *BRANCH BACK FOR NEXT  N
     0081 0040
0145                              END
NO ERRORS, NO WARNINGS
```

FIGURE 9.16. (concluded)

```
0001                  * FMJUMP
0002                  *DIGITAL OSCILLATOR AND FREQUENCY MODULATOR PROGRAM
0003                  *THAT SELECTS ITS CARRIER FREQUENCY USING A RANDOM NUMBER
0004                  *GENERATOR
0005                  *Y(n)=A Y(n-1) + B Y(n-2) + C X(n-1)
0006                  * Y0 =A Y1 + B Y2    FOR N>1   X(n-1)=1 FOR n=1 (= 0 OTHERWISE)
0007      0060 MODE             EQU      >60        *TRANSPARENT MODE FOR AIB
0008      0061 RATE             EQU      >61        *RATE=(10 MHZ / Fs) -1=332=14C
0009      0062 Y0               EQU      >62        *INITIALLY Y0=0, OUTPUT DMA
0010      0063 Y1               EQU      >63        *Y(1)=C=SIN(2x3.14xFd/Fs)
0011      0064 Y2               EQU      >64        *Y(2)=0
0012      0065 A                EQU      >65        *A=2 COS(WT)
0013      0066 B                EQU      >66        *B=-1
0014      0067 RNG              EQU      >67        *FREQ RANGE, SETS BANDWIDTH
0015      0068 RNUMH            EQU      >68        *RANDOM NUMBER (HIGH BYTE)
0016      0069 RNUML            EQU      >69        *RANDOM NUMBER (LOW BYTE)
0017      006A RSET             EQU      >6A        *FREQUENCY SELECT
0018      006B TEMP             EQU      >6B        *STORAGE FOR FEEDBACK BIT
0019      006C SIG              EQU      >6C        *INPUT SIGNAL
0020      006D A1               EQU      >6D        *REFERENCE VALUE FOR "A"
0021                  *
0022                  *                                *A,B,C DIV BY 2 (MAX VALUE <1 )
0023                  *                                *A=5F1F,B=C000,C=2998
0024                  *
0025 0000                       AORG     >0
0026 0000 FF80                  B        START
     0001 002C
0027 0020         TABLE         AORG     >20        *PM START AT >20
0028 0020 00FA    MDE           DATA     >00FA
0029 0021 014C    RTE           DATA     >014C      *Fs=30 KHz
0030 0022 0000    Y00           DATA     >0000
0031 0023 2998    Y01           DATA     >2998
0032 0024 0000    Y02           DATA     >0000
0033 0025 5F1F    AA            DATA     >5F1F
0034 0026 C000    BB            DATA     >C000
0035 0027 0900    RG            DATA     >0900
0036 0028 18DC    RNMH          DATA     >18DC
0037 0029 4587    RNML          DATA     >4587
0038 002A 0030    RST           DATA     >30
0039 002B 5F1F    AA1           DATA     >5F1F
0040                  *
0041                  *  LOADS INITIAL VALUES INTO AUXILIARY DATA REGISTERS
0042                  *
0043 002C 5588    START         LARP     0
0044 002D D000                  LRLK     0,>0060
     002E 0060
```

FIGURE 9.17. Program listing for FM hopping (FMJUMP.LST)

```
0045 002F CB0B          RPTK    11
0046 0030 FCA0          BLKP    TABLE,*+
     0031 0020
0047 0032 E060          OUT     MODE,0   *OUTPUT MODE = >FA TO PORT 0
0048 0033 E161          OUT     RATE,1   *OUTPUT RATE =014C TO PORT 1
0049 0034 D200          LRLK    2,87     *PRESET STRETCH COUNTER
     0035 0057
0050 0036 D300          LRLK    3,60     *PRESET FIRST DELAY
     0037 003C
0051 0038 D400          LRLK    4,5000   *PRESET SECOND DELAY
     0039 1388
0052 003A D500          LRLK    5,50     *PRESET FREQUENCY SELECTOR
     003B 0032
0053 003C CE07          SSXM             *SET SIGN EXTESION MODE
0054             *
0055             * THIS SECTION READS THE INPUT, MULTIPLIES IT WITH THE
0056             * RANGE VALUE, AND ADDS THE PRODUCT TO THE VALUE "A1",
0057             * THAT GENERATES A CARRIER WAVE, AND STORES THE SUM IN
0058             * "A". CHANGING VALUE OF "A" CHANGES THE CARRIER FREQUENCY
0059             *
0060 003D 826C  INPT    IN      SIG,2    *READ INPUT
0061 003E CA00  MAIN    ZAC              *ZERO ACC
0062 003F 3C67          LT      RNG      *PUT RANGE VALUE IN TR
0063 0040 386C          MPY     SIG      *MULTIPLY (RANGE X INPUT)
0064 0041 CE15          APAC             *STORE PRODUCT IN ACC
0065 0042 6965          SACH    A,1      *SHIFT 1 AND STORE
0066 0043 2065          LAC     A        *PUT RNG X SIG IN ACC
0067 0044 006D          ADD     A1       *ADD REF VALUE
0068 0045 6065          SACL    A        *STORE IN "A"
0069             *
0070             * OSCILLATOR IN WHICH THE FREQUENCY IS CONTROLLED BY   "A"
0071             * AND THE EXECUTION TIME OF THE PROGRAM
0072             *
0073 0046 C064          LARK    0,Y2     *LOAD AUX REG AR0 WITH Y2
0074 0047 C166          LARK    1,B      *LOAD AUX REG AR1 WITH B
0075 0048 CA00          ZAC              *ZERO THE ACCUMULATOR
0076 0049 5588          LARP    0        *AR0 FOR INDIRECT ADDRESSING
0077 004A 3C99          LT      *-,1     *TR=(AR0),DEC TO Y1,SELECT AR1
0078 004B 3898          MPY     *-,0     *(AR1)x(TR)=BxY2,DEC AR1,=>AR0
0079 004C 3F89          LTD     *,1      *TR=(AR0),ADD PR ,MOVE"DOWN"
0080             *                       *LTD=LT+APAC+DMOV
0081 004D 3888          MPY     *,0      *(AR1)x(TR)=AxY1
0082 004E CE15          APAC             *ADD PREVIOUS (PR) TO ACC
0083 004F 6962          SACH    Y0,1     *SHIFT 1 DUE TO EXTRA BIT,STORE
0084             *                       *UPPER 16 BITS(DIVIDE BY 2**15)
0085             *   CORRECT FOR DIVISION OF A, B, AND C BY 2
0086             *
0087 0050 2062          LAC     Y0       *SINCE A,B,C WERE DIVIDED BY 2
0088 0051 0062          ADD     Y0       *DOUBLE RESULT
0089 0052 6062          SACL    Y0       *LOWER 16 BITS(STORE IN DMA 62)
0090 0053 5662          DMOV    Y0       *MOVE"DOWN"IN DM FOR NEXT N
0091 0054 E262          OUT     Y0,2     *OUTPUT DMA >62
0092             *
0093             *     THIS SECTION OF THE PROGRAM GENERATES A RANDOM NUMBER
0094             *     THAT IS USED TO VARY THE LENGTH OF THE PROGRAM
0095             *
0096 0055 2068          LAC     RNUMH    *GET RANDOM NUMBER HI BYTE
0097 0056 D004          ANDK    >8000    *MASK BIT 15
     0057 8000
0098 0058 0168          ADD     RNUMH,1  *SHIFT 1,ADD BIT 14 TO BIT 15
0099 0059 D004          ANDK    >8000    *MASK BIT 15
     005A 8000
0100 005B 0368          ADD     RNUMH,3  *ADD BIT 12 TO THE SUMMATION
0101 005C D004          ANDK    >8000    *MASK BIT 15
     005D 8000
0102 005E 0E68          ADD     RNUMH,14 *ADD BIT 1 TO THE SUMMATION
0103 005F D004          ANDK    >8000    *MASK BIT 15
     0060 8000
0104 0061 696B          SACH    TEMP,1   *SHIFT LEFT AND STORE
0105 0062 4068          ZALH    RNUMH    *RESTORE RNUMH TO HIGH ACC
0106 0063 0069          ADD     RNUML    *RESTORE RNUML TO LOW ACC
0107 0064 CE18          SFL              *SHIFT 32 BIT ACC LEFT 1
0108 0065 006B          ADD     TEMP     *ADD FEEDBACK BIT
0109 0066 6868          SACH    RNUMH    *STORE UPPER BYTE
0110 0067 6069          SACL    RNUML    *STORE LOWER BYTE
0111             *
0112             *     THESE TWO LOOPS EXTEND THE LENGTH OF THE PROGRAM
0113             *     TO DECREASE THE CARRIER FREQUENCY
0114 0068
0115 0068 558A          LARP    2        *SET ARP TO STALL COUNTER
```

FIGURE 9.17. (continued)

```
0116 0069 FB90  STALL        BANZ    STALL   *STALL PROGRAM FOR LOWER FREQS
     006A 0069
0117 006B D200               LRLK    2,00    *RESET STALL COUNT
     006C 0000
0118 006D 558D               LARP    5       *SET ARP TO FREQ SELECT
0119 006E FB90  SET          BANZ    SET     *STALL VARIABLE TO SET FREQ.
     006F 006E
0120 0070 356A               LAR     5,RSET  *RESET FREQ SELECT COUNTER
0121            *
0122            * THIS SECTION OF THE PROGRAM IS A TIME DELAY THAT SETS
0123            * THE RATE AT WHICH THE CARRIER FREQUENCY IS RANDOMLY CHANGED
0124            *
0125 0071 558B               LARP    3       *SET ARP TO FIRST DELAY
0126 0072 FB90               BANZ    GO      *CONTINUE PROGRAM IF NOT 0
     0073 0081
0127 0074 D300               LRLK    3,60    *RESET FIRST DELAY
     0075 003C
0128 0076 558C               LARP    4       *SET ARP TO 2ND DELAY
0129 0077 FB90               BANZ    GO      *CONTINUE PROGRAM IF NOT 0
     0078 0081
0130 0079 D400               LRLK    4,5000  *RESET 2ND DELAY
     007A 1388
0131            *
0132            *            THIS SECTION CALCULATES THE VALUE OF "RSET", THE RANDOM
0133            *            VALUE THAT SETS THE CARRIER FREQUENCY
0134            *
0135 007B 2069               LAC     RNUML   *GET RANDOM NUMBER
0136 007C D004               ANDK    >30     *DELAY 16,32, OR 48
     007D 0030
0137 007E F680               BZ      GO      *DON'T LET RSET BE ZERO OR THE
     007F 0081
0138            *                            *FREQUENCY WILL BE TOO HIGH
0139 0080 606A               SACL    RSET    *STORE NEW VALUE OF RSET
0140            *
0141            *            BRANCH BACK TO BEGINING OF PROGRAM, READ INPUT IF
0142            *            BIO=0
0143            *
0144 0081 FA80  GO           BIOZ    INPT    *BRANCH TO READ INPUT IF BIO=0
     0082 003D
0145 0083 FF80               B       MAIN    *BRANCH BACK FOR NEXT  N
     0084 003E
0146                         END
NO ERRORS, NO WARNINGS
```

FIGURE 9.17. (concluded)

9.3 PRESSURE MEASUREMENT USING A RESONANT WIRE TECHNIQUE

Problem Statement

The objective of this project is to measure a wire's resonant frequency as a function of pressure. A vibrating wire sensor assembly with bellows, or a similar pressure source, is needed to implement this project. The system configuration is shown in Figure 9.18.

Design Considerations

Sine-wave generators are frequently used in many applications, such as instrumentation, with both speed and accuracy being major concerns. One such application is a differential pressure transmitter that measures the difference between two pressures via the use of the resonant wire technique. The natural frequency of a wire loaded in tension is proportional to the square root of the tensile force. When the wire is placed in a magnetic field and an oscillating current is passed through it, the magnetic field induced by the current reacts with the permanent magnetic field causing the wire to vibrate. As the wire vibrates within the magnetic field, a voltage is induced in phase with the oscillating drive signal and peaks at the resonant frequency. Peak detection schemes can be utilized to determine the resonant frequency of the wire.

A sensor assembly, shown in Figure 9.19, with a resonant frequency range from 1700 to 3000 Hz, is used in this project. As pressure is applied and the wire is loaded, the resonant frequency increases. A precision sine wave is generated, with the oscillator program given in Chapter 4, and is swept through the operating range of the sensor assembly. The method of peak detection is used to determine the resonant frequency. The voltage across the wire can also be observed on an oscilloscope. When the output sweep range is halted, the resonant frequency waveform is then displayed on the oscilloscope.

Although the sensor used has a range of about 1500 Hz, a much smaller sweep range (1770 to 1850 Hz) is utilized to limit the amount of code required. Higher ranges are possible and can be implemented using the available 4K words of maskable program ROM.

Implementation and Testing

As the output frequency of the AIB is swept through the operating range of the sensor assembly, the resonant point is detected by the peak detection software and can be observed on the oscilloscope. When a peak in the signal is detected, the (A) coefficient value from the digital oscillator (using the notation of Chapter 4) is saved. This value is then used in a lookup table to get the corresponding (C) coefficient for the resonant frequency. The two coefficients are then copied into their respective data memory locations for use in the steady frequency oscillator

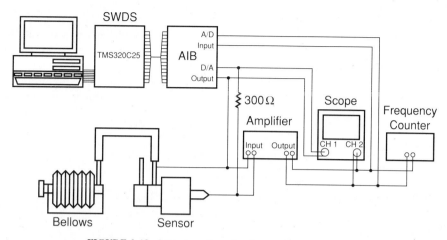

FIGURE 9.18. System configuration for pressure measurement

portion of the program. If the pressure is changed via the bellows, a new resonant frequency must be determined. A complete listing and description of the program used are included in Appendix I.

Conclusions

The use of the TMS320C25 for the creation of a digital sine wave to drive the vibrating wire, and simultaneously to detect its resonant frequency, exhibits the processor's operational speed even though a crude but effective method of peak amplitude detection is used. This project can be modified to incorporate Texas Instruments' TMS320/E15 EPROMs to store the lookup tables used for the sine and cosine parameters associated with the frequency range. This ROM space would

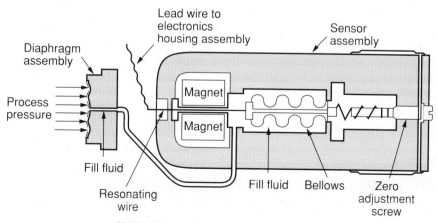

FIGURE 9.19. Vibrating wire sensor assembly

allow for larger frequency ranges, thus giving greater flexibility. Adaptive filtering may be an effective way to detect the original resonance of the wire, as well as to ensure that the program tracks any changes in the resonance of the wire when the input pressure is varied. The digital signal processor, along with a similar type of sensing device, can be used in any number of pressure-sensing applications requiring accuracy and a rapid response time (i.e., depth or altitude sensing, or storage tank fluid level detection).

9.4 PRESSURE-TO-VOLTAGE CONVERSION

This project requires some special hardware: a compressor or pressure source similar to the one used in the preceding project, and a pressure-to-frequency transducer.

Problem Statement

The objective of this project [19] is to design an efficient pressure-to-voltage conversion system, a block diagram of which is shown in Figure 9.20.

Design Considerations

The system in Figure 9.20 includes a compressor and calibrated pressure source, and a pressure-to-frequency transducer. Since the transducer can be in an industrial environment, with electrical noise present, its output signal is filtered to produce a compensated output voltage. The transducer must be selected so that the output signal is within the sampling range of the AIB. The input pressure has a range of 0 to 700 inches of water (in. H_2O). The transducer has a range of output resonant frequencies from 1806.45 Hz at atmospheric pressure (0 in. H_2O), to 2973.53 Hz at 700 in. H_2O. The TMS320 is used for signal filtering, frequency counting, and output compensation.

The TMS320 implements an FIR bandpass filter, of order 61, with a Parks–McClellan window. The resonant frequency of the transducer circuit is determined using a zero-crossing technique, by counting the number of sign changes over a set period of time. The transducer has a certain amount of nonlinearity, with a maximum error at midrange and tapering to zero at the usable extremes. A second-order Taylor series approximation is used to compensate for this nonlinearity. Figures 9.21 and 9.22 show the frequencies deviation and uncompensated voltage deviation versus input pressure.

Results and Conclusions

The compensation technique produces a shifted frequency that is applied to output a corrected voltage to the D/A. As a result, a reduction from an 11% nonlinearity to approximately 2% is obtained, as shown in Figure 9.23. With an input pressure from 0 to 700 inches of water, the corresponding output voltage is linear within 2% over the range from 0.98 to 6.72 V.

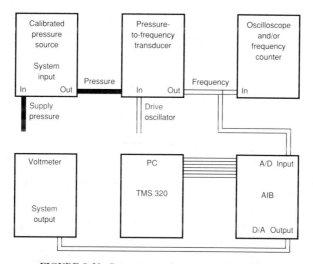

FIGURE 9.20. Pressure-to-voltage conversion system

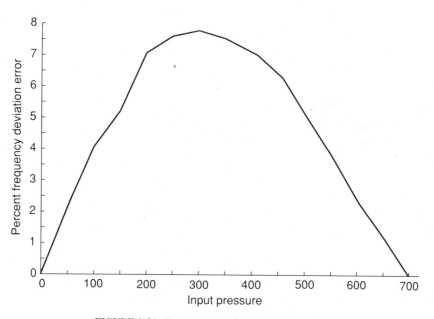

FIGURE 9.21. Frequency versus pressure of transducer

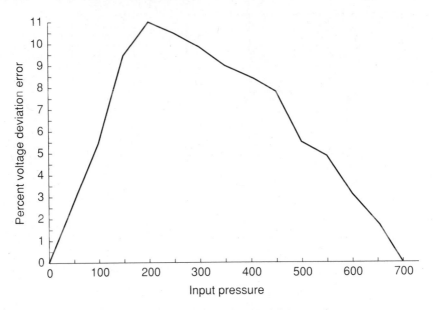

FIGURE 9.22. Uncompensated output voltage deviation versus input pressure

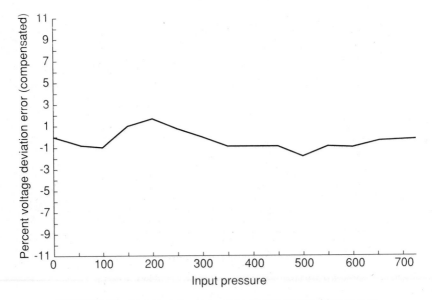

FIGURE 9.23. Compensated output voltage deviation versus input pressure

9.5 ADDITIONAL INPUT CHANNEL FOR THE AIB

Problem Statement

Many applications, such as the adaptive projects in Chapter 8, require two input channels: one for the input signal plus noise and one for reference information. The objective of this project is to add an additional input channel to the AIB.

Design Considerations

The AIB provides three ports for expansion. Ports 4 and 5 are for extended memory and port 3 is an I/O expansion port with processor interface buffers already in place. The existing sample rate clock can be shared between the existing A/D unit and the expansion A/D unit. Thus all that needs to be added is a sample-and-hold unit, an A/D unit, and an antialiasing filter.

Since the AIB was designed, the cost-performance ratio on A/D units has gone down considerably. For example, the AD7870 used in this design (see Figure 9.24) does a 12-bit conversion with a maximum sample rate of 100 kHz, has an on-chip sample-and-hold, and costs only about $33.00 in small quantities. The AD7870 is the centerpiece of this design and it is supported by several additional parts:

1. A -5-Volt voltage regulator. The AIB supplies $+5$ and ± 12 V., but the AD7870 needs $+5$ and -5 V.
2. An LM318 amplifier to buffer the input signal.
3. Two JK flip-flops to provide the proper End-Of-Conversion (EOC) pulse to the AIB and a BUSEN$\sim$ signal for the AD7870.
4. Two inverters and several capacitors, registers, and connectors.

The circuit schematic without the antialiasing and reconstruction filters is shown in Figure 9.25. The output data lines of the A/D connect to P1 on the AIB, which has data lines for a D/A unit. The circuitry shown is mounted on a separate board with all connections to the AIB coming through a cable from P1. Space was allowed for a D/A unit and for the antialiasing and reconstruction filters. A clock signal, which is routed via pin 40 of P1, is needed to run the flip-flops. To free pin 40, the existing foil that is connected to it must be cut, disconnecting an unneeded signal from P1. The clock is then taken from pin 6 of U49 (on the AIB) to pin 40 on P1 (on the AIB) by connecting a wire between them. This is the only alteration necessary on the AIB. The two flip-flops generate the EOC pulse, used to generate the BIO signal on the AIB, and the BUSEN$\sim$ (BUS ENABLE), needed by the A/D chip to enable the data on its outputs. Although the BUSEN$\sim$ can be generated by the processor "read signals," doing it this way allows the unit to work with both the TMS320C25 and the TMS32010. A timing diagram is shown in Figure 9.26. The start of conversion is initiated by $\sim$SOC2 from the AIB.

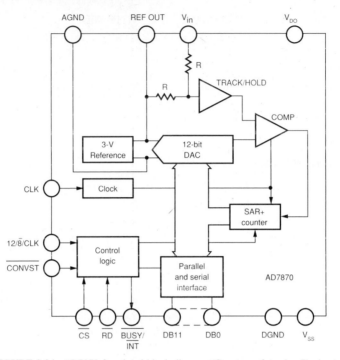

FIGURE 9.24. AD7870 functional block diagram (Courtesy of Analog Devices)

Upon completion of the conversion, approximately 8 μs later, ~INT is brought low by the AD7870 chip, signaling that the conversion is complete. On the next rising clock edge, BUSEN~ is brought low by the flip-flop circuit. This enables the tri-state outputs of the A/D and presents valid data on the output lines of the chip. On the next rising edge, the flip-flop U4B generates EOC~, which latches the A/D converter data into the AIB expansion port. If both A/D channels are used simultaneously, either channel may cause the interrupt or the BIO signal; therefore, the program must check the two bits AD1S and AD2S in the status register to find out which unit is done.

Testing the Circuitry

Two partial loop programs, shown in Figures 9.27 and 9.28, were used to test the new channel. For the first, a sine-wave source with peak-to-peak voltage less than 6 V is connected to the input of the A/D. With the program running, the same waveform delayed is observed on the output of the D/A channel. The second test program inputs a sine wave of frequency f and then outputs a sine wave with twice the frequency, $2f$. The channel is able to handle signals up to 50 kHz which is the Nyquist frequency for a 100 kHz sampling frequency.

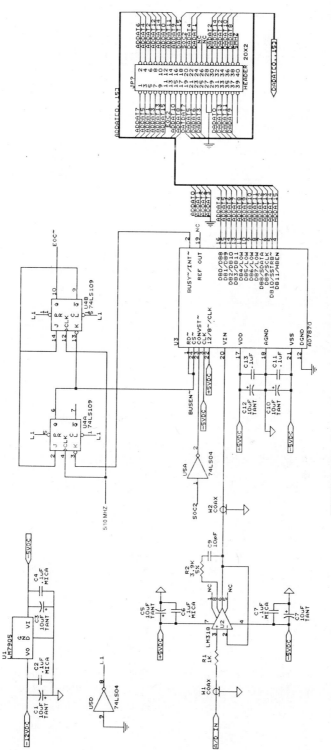

FIGURE 9.25. Schematic for additional AIB input channel

241

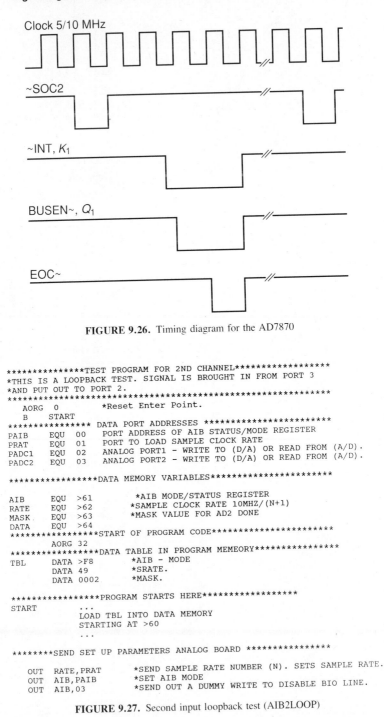

FIGURE 9.26. Timing diagram for the AD7870

```
***************TEST PROGRAM FOR 2ND CHANNEL*****************
*THIS IS A LOOPBACK TEST. SIGNAL IS BROUGHT IN FROM PORT 3
*AND PUT OUT TO PORT 2.
*************************************************************
      AORG   0           *Reset Enter Point.
      B      START
**************** DATA PORT ADDRESSES ***********************
PAIB    EQU    00    PORT ADDRESS OF AIB STATUS/MODE REGISTER
PRAT    EQU    01    PORT TO LOAD SAMPLE CLOCK RATE
PADC1   EQU    02    ANALOG PORT1 - WRITE TO (D/A) OR READ FROM (A/D).
PADC2   EQU    03    ANALOG PORT2 - WRITE TO (D/A) OR READ FROM (A/D).

****************DATA MEMORY VARIABLES**********************

AIB     EQU    >61       *AIB MODE/STATUS REGISTER
RATE    EQU    >62       *SAMPLE CLOCK RATE 10MHZ/(N+1)
MASK    EQU    >63       *MASK VALUE FOR AD2 DONE
DATA    EQU    >64
***************START OF PROGRAM CODE**********************
      AORG 32
****************DATA TABLE IN PROGRAM MEMEORY***************
TBL     DATA   >F8       *AIB - MODE
        DATA   49        *SRATE.
        DATA   0002      *MASK.

***************PROGRAM STARTS HERE******************
START          ...
               LOAD TBL INTO DATA MEMORY
               STARTING AT >60
               ...

********SEND SET UP PARAMETERS ANALOG BOARD ***************

      OUT  RATE,PRAT     *SEND SAMPLE RATE NUMBER (N). SETS SAMPLE RATE.
      OUT  AIB,PAIB      *SET AIB MODE
      OUT  AIB,03        *SEND OUT A DUMMY WRITE TO DISABLE BIO LINE.
```

FIGURE 9.27. Second input loopback test (AIB2LOOP)

```
******************** MAIN PROGRAM LOOP *********************
LOOPA    BIOZ LOOPB          WAIT FOR AN A/D TO FINISH.
         B    LOOPA
*
LOOPB    IN   DATA,PAIB *MAKE SURE IT IS THE CORRECT A/D.
         LAC  DATA
         AND  MASK
         BZ   LOOPA          *TEST BIT 1 IN STATUS REGISTER
*
         IN   DATA,PADC2     *READ A VALUE
         OUT  DATA,PADC1     *OUTPUT IT
         B    LOOPA
         END
```

FIGURE 9.27. (concluded)

```
****A SECOND TEST PROGRAM FOR THE 2ND INPUT CHANNEL*****
*************DOUBLE FREQUENCY LOOPBACK TEST ************
*INPUTS A SINEWAVE AND OUTPUTS A SINEWAVE OF DOUBLE
* THE FREQUENCY
*********************************************************
     AORG 000          *Reset Enter Point.
     B    START
**************** DATA PORT ADDRESSES ********************
PAIB    EQU  00    PORT ADDRESS OF AIB MODE/STATUS REG
PRAT    EQU  01    PORT TO LOAD SAMPLE CLOCK RATE
PADC1   EQU  02    ANALOG PORT1 - WRITE TO (D/A) OR READ FROM (A/D).
PADC2   EQU  03    ANALOG PORT2 - WRITE TO (D/A) OR READ FROM (A/D).
*************DATA MEMORY VARIABLES*********************
AIB     EQU  >61        *EXTENDED MEMORY NO COUNT.
RATE    EQU  >62        *SAMPLE CLOCK RATE 5E6/(N+1)
MASK    EQU  >63        *MASK VALUE FOR AD2 DONE.
DATA    EQU  >64        *TEMP DATA STORAGE
        AORG 32         *Origin - Start Of Code.
**************** DATA TABLE ************************
TBL     DATA >F8        *AIB - NO COUNT
        DATA 99         *SRATE
        DATA 0002       *MASK.
****************PROGRAM STARTS HERE********************
START    ...
         LOAD TBL INTO DATA MEMORY
         STARTING AT >60
         ...

****** SEND SET UP PARAMETERS ANALOG BOARD **************
    OUT  RATE,PRAT      *SEND SAMPLE RATE NUMBER (N). SETS SAMPLE RATE.
    OUT  AIB,PAIB       *SEND AIB DATA TO PORT AIB.
    OUT  AIB,03         *SEND OUT A DUMMY WRITE TO DISABLE BIO LINE.
***************** MAIN PROGRAM LOOP **************************
LOOPA    BIOZ LOOPB          WAIT FOR AN A/D TO FINISH.
         B    LOOPA
*
LOOPB    IN   DATA,PAIB *MAKE SURE IT IS THE CORRECT A/D.
         LAC  DATA
         AND  MASK
         BZ   LOOPA          *TEST STATUS REGISTER BIT 1
         IN   DATA,PADC2
         LT   DATA           *SQUARE DATA
         MPY  DATA           *(ASIN(wt))^2 = 1/2 + (1/2)*SIN(2wt)
         PAC
         SACH DATA,1         *STORE IN DATA.
*
         OUT  DATA,PADC1     *DISPLAY RESULT.
         B    LOOPA
         END
```

FIGURE 9.28. Second input double frequency loopback test (AIB2FREQ)

9.6 FAST FOURIER TRANSFORM IMPLEMENTATION

In Chapter 7 we described in detail the radix-2 and radix-4 fast Fourier transform (FFT) and included several flow graphs [20].

Problem Statement

This project describes an 8-point and a 32-point radix-2 FFT and provides a good background for implementing higher-order FFTs as well as FFTs of higher radices.

Design Considerations

The FFT algorithm can be implemented using three major operations:

1. One that does not involve computations (additions and subtractions) with the twiddle complex factor, $W = e^{-j2\pi/N}$, where N is the number of sample points
2. One that does involve computations with W
3. One that performs a bit reversal

We will describe in detail these three major operations with an 8-point FFT and show how higher-order FFTs follow.

8-Point FFT

Consider the partial program, shown in Figure 9.29, which implements an 8-point, radix-2 FFT, using decimation-in-frequency. Before describing the three major operations stated above, we will discuss the following:

1. *Constants declaration.* For an 8-point ($N = 8$) FFT, only $W^0, W^1, \cdots,$ $W^{(N/2)-1}$ are needed, because of the symmetry of W,

$$W^{N/2} = -W^0, \quad W^{(N/2)+1} = -W^1, \quad \text{etc.}$$

and the periodicity of W, $W^N = W^0$, $W^{N+1} = W^1$, and so on. Using $W^k = e^{-j2\pi k/N}$,

$$W^k = \cos(2\pi k/N) - j\,\sin(2\pi k/N) \qquad k = 0, 1, \cdots, (N/2) - 1$$

For example, with $k = 1$,

$$W^1 = \cos(\pi/4) - j\sin(\pi/4) = WR + jWI$$

and $WR = (0.707)(2^{15}) = 23,170$ and $WI = (-0.707)(2^{15}) = -23,170$.

2. *Variables declaration.* We need to declare 16 memory locations for x, 8 for the real part (XR0, XR1, . . . , XR7) and 8 for the imaginary part (XI0, XI1, . . . , XI7). Note that the intermediate results are complex quantities. Two temporary memory locations are used for unscrambling the output sequence using a bit-reversal procedure. Additional (program or data) memory locations are needed for storing the cosine and sine values associated with W.

For larger FFTs, both program and data memory spaces are used for storing the sample and coefficient values.

Macro Definition

Since the same types of operations are repeated often in an FFT algorithm, the use of macros is effective. One can pass different variables to the same macro. The use of macros was introduced in association with the analog interface chip (AIC) in Chapter 3.

Macro for Bit Reversal (BITR)

This macro is used to arrange the output sequence in proper order, since each $(N/2)$-point decomposition after each stage produces a sequence of even-and odd-ordered outputs. In this macro, the two real variables passed as R1 and R2 are swapped: for example, X0 and X4. This swap is also done with the imaginary components, passed along as I1 and I2.

As an alternative, the bit-reversal procedure could have also have been implemented, using the accumulator, in lieu of a temporary storage location. This procedure addresses separately the accumulator's lower and higher 16 bits. It can also be implemented in a more efficient fashion by taking advantage of the indirect addressing with bit-reversal instruction, available with the TMS320C25, discussed in Chapter 2, and in Chapter 7.

Macro for Computations without Sine or Cosine Terms (FREEBF)

In this section we add and subtract two variables, such as the first stage calculations, X0 = X0 + X4 and X4 = X0 − X4. Note that after stage 1, X0 is the sum of the original values for X0 and X4.

Macro for Computation with Sine and Cosine Terms (CONSBF)

The first part of this macro is similar to the previous macro (FREEBF) to add and subtract two variables. However, the second part multiplies two complex quantities. For example, with the real and imaginary components of X and W as arguments, the following operation is performed:

$$[(XR + jXI)(WR + jWI)] = [(XR)(WR) - (XI)(WI)] + j[(XI)(WR) + (XR)(WI)]$$

```
**PARTIAL 8-POINT FFT PROGRAM, DECIMATION IN FREQUENCY
*        .
*        .
* TWIDDLE CONSTANTS W
COS0    DATA    32767
COS1    DATA    23170
COS2    DATA    0
COS3    DATA    -23170
SIN0    DATA    0
SIN1    DATA    -23170
SIN2    DATA    -32767
SIN3    DATA    -23170
* SAMPLES-REAL PART
XR0     EQU     512
XR1     EQU     514
XR2     EQU     516
XR3     EQU     518
XR4     EQU     520
XR5     EQU     522
XR6     EQU     524
XR7     EQU     526
* SAMPLES-IMAGINARY PART
XI0     EQU     513
XI1     EQU     515
XI2     EQU     517
XI3     EQU     519
XI4     EQU     521
XI5     EQU     523
XI6     EQU     525
XI7     EQU     527
* TEMPORARY STORAGE LOCATION
TR1     EQU     528
TI1     EQU     529
* STORAGE FOR COSINE AND SINE (W)
C0      EQU     530
C1      EQU     531
C2      EQU     532
C3      EQU     533
S0      EQU     534
S1      EQU     535
S2      EQU     536
S3      EQU     537
***** MACRO DEFINITIONS
* BIT REVERSAL MACRO
BITR    $MACRO R1,I1,R2,I2
        LAC     :R1.S:
        SACL    TR1
        LAC     :R2.S:
        SACL    :R1.S:
        LAC     TR1
        SACL    :R2.S:
        LAC     :I1.S:
        SACL    TR1
        LAC     :I2.S:
        SACL    :I1.S:
        LAC     TR1
        SACL    :I2.S:
        $END    BITR
* BUTTERFLY WITHOUT SINE & COSINE TERMS
FREEBF  $MACRO R1,I1,R2,I2
        LAC     :R1.S:
        SUB     :R2.S:
        SACL    TR1
        LAC     :I1.S:
        SUB     :I2.S:
        SACL    TI1
        LAC     :R1.S:
        ADD     :R2.S:
        SACL    :R1.S:
        LAC     :I1.S:
        ADD     :I2.S:
        SACL    :I1.S:
        LAC     TR1
        SACL    :R2.S:
        LAC     TI1
        SACL    :I2.S:
        $END    FREEBF
* BUTTERFLY WITH COSINE (K1) AND SINE (K2) TERMS
CONSBF  $MACRO R1,I1,R2,I2,K1,K2
        LAC     :R1.S:
        SUB     :R2.S:
        SACL    TR1
        LAC     :I1.S:
        SUB     :I2.S:
        SACL    TI1
        LAC     :R1.S:
        ADD     :R2.S:
        SACL    :R1.S:
```

FIGURE 9.29. Partial program to implement an 8-point FFT (FFT8)

```
                   LAC    :I1.S:
                   ADD    :I2.S:
                   SACL   :I1.S:
        ****

                   ZAC
                   LT     TR1
                   MPY    :K1.S:
                   LTA    TI1
                   MPY    :K2.S:
                   SPAC
                   SACH   :R2.S:,1
                   ZAC
                   LT     TR1
                   MPY    :K2.S:
                   LTA    TI1
                   MPY    :K1.S:
                   APAC
                   SACH   :I2.S:,1
                   $END   CONSBF
        ****** THE BEEF ******
        *FIRST STAGE (ITERATION)
        FIR FREEBF XR0,XI0,XR4,XI4
            CONSBF XR1,XI1,XR5,XI5,C0,S0
            CONSBF XR2,XI2,XR6,XI6,C1,S1
            CONSBF XR3,XI3,XR7,XI7,C2,S2
        ***********************************
        *SECOND STAGE (ITERATION)
            FREEBF XR0,XI0,XR2,XI2
            CONSBF XR1,XI1,XR3,XI3,C1,S1
            FREEBF XR4,XI4,XR6,XI6
            CONSBF XR5,XI5,XR7,XI7,C1,S1
        ***********************************
        * THIRD (LAST) STAGE (ITERATION)
            FREEBF XR0,XI0,XR1,XI1
            FREEBF XR2,XI2,XR3,XI3
            FREEBF XR4,XI4,XR5,XI5
            FREEBF XR6,XI6,XR7,XI7
        ***********************************
        * BIT REVERSAL
            BITR XR1,XI1,XR4,XI4
            BITR XR3,XI3,XR6,XI6
        ***********************************
        *GET POWER SPECTRUM
        *          .
        *          .
                   END
```

FIGURE 9.29. (concluded)

Macro Calls

This section calls the three macros to perform their tasks.

1. After the SECOND STAGE (iteration), for example (see the 8-point FFT flow graph in Figure 7.5), FREEBF is called to add and subtract the real and imaginary parts of X0 and X2.

2. CONSBF is called, with the sample values X1 and X3 as well as the cosine and sine values passed along to CONSBF. Note the similarity of a macro to a Fortran call statement.

3. Steps (1) and (2) are repeated to handle the third stage (iteration).

4. The last two statements call the bit-reversal macro.

Note that codes for initialization and the transfer of data from program to data memory are not included. To plot the results, an additional section of code (a square-root macro) would be needed to compute the power spectrum, essentially computing the magnitude of $(XR + jXI)$. Figure 9.30 shows a similar partial program to implement a 32-point FFT. In this case, 64 memory locations are used for the real and imaginary components of X and 32 locations $(W^0, W^1, \ldots, W^{15})$ for the real and imaginary components of W. Figure 9.31 shows the power spectrum of a 256-point FFT. This was not a real-time implementation, since the input samples were assigned values to simulate a rectangular waveform [i.e., $x(0) = x(1) = \cdots x(127) = 1$ and $x(128) = x(129) = \cdots x(255) = 0$].

An extension of this project would be a real-time radix-4 implementation. Since the TMS320C25 has a limited amount of on-chip data memory space, larger FFTs can require a large transfer of data between program and data memories. As a result, a real-time implementation may not be possible, especially for fast-varying time signals. The addressing mode with bit reversed, discussed in Chapter 2, created for the TMS320C25, can be used in a macro in lieu of the BITR macro shown.

Many other forms of the basic Cooley–Tukey [20] FFT algorithm can also be considered, such as the Winograd [21] or fast Hartley [22–24] transforms. Advantages of transforms with split radices have also been reported [25]. Other sources for the FFT algorithm can be found in [26–29].

```
* PARTIAL 32-POINT FFT PROGRAM,DECIMATION IN FREQUENCY
*     .
*     .
* TWIDDLE CONSTANTS W
COS0      DATA      32767
COS1      DATA      32137
COS2      DATA      30273
COS3      DATA      27245
COS4      DATA      23170
COS5      DATA      18204
COS6      DATA      12539
COS7      DATA      6393
COS8      DATA      0
COS9      DATA      -6393
COS10     DATA      -12539
COS11     DATA      -18204
COS12     DATA      -23170
COS13     DATA      -27245
COS14     DATA      -30273
COS15     DATA      -32137
SIN0      DATA      0
SIN1      DATA      -6393
SIN2      DATA      -12539
SIN3      DATA      -18204
SIN4      DATA      -23170
SIN5      DATA      -27245
SIN6      DATA      -30273
SIN7      DATA      -32137
SIN8      DATA      -32767
SIN9      DATA      -32137
SIN10     DATA      -30273
SIN11     DATA      -27245
SIN12     DATA      -23170
SIN13     DATA      -18204
SIN14     DATA      -12539
SIN15     DATA      -6393
* SAMPLES-REAL PART
XR0       EQU       512
XR1       EQU       514
XR2       EQU       516
XR3       EQU       518
XR4       EQU       520
XR5       EQU       522
XR6       EQU       524
XR7       EQU       526
XR8       EQU       528
XR9       EQU       530
XR10      EQU       532
XR11      EQU       534
XR12      EQU       536
XR13      EQU       538
XR14      EQU       540
```

FIGURE 9.30. Partial program to implement a 32-point FFT (FFT32)

```
XR15      EQU      542
XR16      EQU      544
XR17      EQU      546
XR18      EQU      548
XR19      EQU      550
XR20      EQU      552
XR21      EQU      554
XR22      EQU      556
XR23      EQU      558
XR24      EQU      560
XR25      EQU      562
XR26      EQU      564
XR27      EQU      566
XR28      EQU      568
XR29      EQU      570
XR30      EQU      572
XR31      EQU      574
* SAMPLES-IMAGINARY PART
XI0       EQU      513
XI1       EQU      515
XI2       EQU      517
XI3       EQU      519
XI4       EQU      521
XI5       EQU      523
XI6       EQU      525
XI7       EQU      527
XI8       EQU      529
XI9       EQU      531
XI10      EQU      533
XI11      EQU      535
XI12      EQU      537
XI13      EQU      539
XI14      EQU      541
XI15      EQU      543
XI16      EQU      545
XI17      EQU      547
XI18      EQU      549
XI19      EQU      551
XI20      EQU      553
XI21      EQU      555
XI22      EQU      557
XI23      EQU      559
XI24      EQU      561
XI25      EQU      563
XI26      EQU      565
XI27      EQU      567
XI28      EQU      569
XI29      EQU      571
XI30      EQU      573
XI31      EQU      575
* TEMPORARY STORAGE LOCATION
TR1       EQU      576
TI1       EQU      577
* STORAGE FOR COSINE AND SINE (W)
C0        EQU      580
C1        EQU      581
C2        EQU      582
C3        EQU      583
C4        EQU      584
C5        EQU      585
C6        EQU      586
C7        EQU      587
C8        EQU      588
C9        EQU      589
C10       EQU      590
C11       EQU      591
C12       EQU      592
C13       EQU      593
C14       EQU      594
C15       EQU      595
S0        EQU      596
S1        EQU      597
S2        EQU      598
S3        EQU      599
S4        EQU      600
S5        EQU      601
S6        EQU      602
S7        EQU      603
S8        EQU      604
S9        EQU      605
S10       EQU      606
S11       EQU      607
S12       EQU      608
S13       EQU      609
S14       EQU      610
```

FIGURE 9.30. (continued)

```
S15      EQU     611
***** MACRO DEFINITIONS
* BIT REVERSAL MACRO
BITR     $MACRO R1,I1,R2,I2
         LAC    :R1.S:
         SACL   TR1
         LAC    :R2.S:
         SACL   :R1.S:
         LAC    TR1
         SACL   :R2.S:
         LAC    :I1.S:
         SACL   TR1
         LAC    :I2.S:
         SACL   :I1.S:
         LAC    TR1
         SACL   :I2.S:
         $END BITR
* BUTTERFLY WITHOUT SINE & COSINE TERMS
FREEBF   $MACRO R1,I1,R2,I2
         LAC    :R1.S:
         SUB    :R2.S:
         SACL   TR1
         LAC    :I1.S:
         SUB    :I2.S:
         SACL   TI1
         LAC    :R1.S:
         ADD    :R2.S:
         SACL   :R1.S:
         LAC    :I1.S:
         ADD    :I2.S:
         SACL   :I1.S:
         LAC    TR1
         SACL   :R2.S:
         LAC    TI1
         SACL   :I2.S:
         $END FREEBF
* BUTTERFLY WITH COSINE (K1) AND SINE (K2) TERMS
CONSBF   $MACRO R1,I1,R2,I2,K1,K2
         LAC    :R1.S:
         SUB    :R2.S:
         SACL   TR1
         LAC    :I1.S:
         SUB    :I2.S:
         SACL   TI1
         LAC    :R1.S:
         ADD    :R2.S:
         SACL   :R1.S:
         LAC    :I1.S:
         ADD    :I2.S:
         SACL   :I1.S:
****
         ZAC
         LT     TR1
         MPY    :K1.S:
         LTA    TI1
         MPY    :K2.S:
         SPAC
         SACH   :R2.S:,1
         ZAC
         LT     TR1
         MPY    :K2.S:
         LTA    TI1
         MPY    :K1.S:
         APAC
         SACH   :I2.S:,1
         $END CONSBF
****** THE BEEF ******
*FIRST STAGE (ITERATION)
FIR FREEBF XR0,XI0,XR16,XI16
    CONSBF XR1,XI1,XR17,XI17,C1,S1
    CONSBF XR2,XI2,XR18,XI18,C2,S2
    CONSBF XR3,XI3,XR19,XI19,C3,S3
    CONSBF XR4,XI4,XR20,XI20,C4,S4
    CONSBF XR5,XI5,XR21,XI21,C5,S5
    CONSBF XR6,XI6,XR22,XI22,C6,S6
    CONSBF XR7,XI7,XR23,XI23,C7,S7
    CONSBF XR8,XI8,XR24,XI24,C8,S8
    CONSBF XR9,XI9,XR25,XI25,C9,S9
    CONSBF XR10,XI10,XR26,XI26,C10,S10
    CONSBF XR11,XI11,XR27,XI27,C11,S11
    CONSBF XR12,XI12,XR28,XI28,C12,S12
    CONSBF XR13,XI13,XR29,XI29,C13,S13
    CONSBF XR14,XI14,XR30,XI30,C14,S14
    CONSBF XR15,XI15,XR31,XI31,C15,S15
*****************************************
*SECOND STAGE (ITERATION)
    FREEBF XR0,XI0,XR8,XI8
```

FIGURE 9.30. (continued)

```
        CONSBF XR1,XI1,XR9,XI9,C2,S2
        CONSBF XR2,XI2,XR10,XI10,C4,S4
        CONSBF XR3,XI3,XR11,XI11,C6,S6
        CONSBF XR4,XI4,XR12,XI12,C8,S8
        CONSBF XR5,XI5,XR13,XI13,C10,S10
        CONSBF XR6,XI6,XR14,XI14,C12,S12
        CONSBF XR7,XI7,XR15,XI15,C14,S14
        FREEBF XR16,XI16,XR24,XI24
        CONSBF XR17,XI17,XR25,XI25,C2,S2
        CONSBF XR18,XI18,XR26,XI26,C4,S4
        CONSBF XR19,XI19,XR27,XI27,C6,S6
        CONSBF XR20,XI20,XR28,XI28,C8,S8
        CONSBF XR21,XI21,XR29,XI29,C10,S10
        CONSBF XR22,XI22,XR30,XI30,C12,S12
        CONSBF XR23,XI23,XR31,XI31,C14,S14
****************************************
*THIRD STAGE (ITERATION)
        FREEBF XR0,XI0,XR4,XI4
        CONSBF XR1,XI1,XR5,XI5,C4,S4
        CONSBF XR2,XI2,XR6,XI6,C8,S8
        CONSBF XR3,XI3,XR7,XI7,C12,S12
        FREEBF XR8,XI8,XR12,XI12
        CONSBF XR9,XI9,XR13,XI13,C4,S4
        CONSBF XR10,XI10,XR14,XI14,C8,S8
        CONSBF XR11,XI11,XR15,XI15,C12,S12
        FREEBF XR16,XI16,XR20,XI20
        CONSBF XR17,XI17,XR21,XI21,C4,S4
        CONSBF XR18,XI18,XR22,XI22,C8,S8
        CONSBF XR19,XI19,XR23,XI23,C12,S12
        FREEBF XR24,XI24,XR28,XI28
        CONSBF XR25,XI25,XR29,XI29,C4,S4
        CONSBF XR26,XI26,XR30,XI30,C8,S8
        CONSBF XR27,XI27,XR31,XI31,C12,S12
****************************************
*FOURTH STAGE (ITERATION)
        FREEBF XR0,XI0,XR2,XI2
        CONSBF XR1,XI1,XR3,XI3,C8,S8
        FREEBF XR4,XI4,XR6,XI6
        CONSBF XR5,XI5,XR7,XI7,C8,S8
        FREEBF XR8,XI8,XR10,XI10
        CONSBF XR9,XI9,XR11,XI11,C8,S8
        FREEBF XR12,XI12,XR14,XI14
        CONSBF XR13,XI13,XR15,XI15,C8,S8
        FREEBF XR16,XI16,XR18,XI18
        CONSBF XR17,XI17,XR19,XI19,C8,S8
        FREEBF XR20,XI20,XR22,XI22
        CONSBF XR21,XI21,XR23,XI23,C8,S8
        FREEBF XR24,XI24,XR26,XI26
        CONSBF XR25,XI25,XR27,XI27,C8,S8
        FREEBF XR28,XI28,XR30,XI30
        CONSBF XR29,XI29,XR31,XI31,C8,S8
****************************************
* FIFTH (LAST) STAGE (ITERATION)
        FREEBF XR0,XI0,XR1,XI1
        FREEBF XR2,XI2,XR3,XI3
        FREEBF XR4,XI4,XR5,XI5
        FREEBF XR6,XI6,XR7,XI7
        FREEBF XR8,XI8,XR9,XI9
        FREEBF XR10,XI10,XR11,XI11
        FREEBF XR12,XI12,XR13,XI13
        FREEBF XR14,XI14,XR15,XI15
        FREEBF XR16,XI16,XR17,XI17
        FREEBF XR18,XI18,XR19,XI19
        FREEBF XR20,XI20,XR21,XI21
        FREEBF XR22,XI22,XR23,XI23
        FREEBF XR24,XI24,XR25,XI25
        FREEBF XR26,XI26,XR27,XI27
        FREEBF XR28,XI28,XR29,XI29
        FREEBF XR30,XI30,XR31,XI31
****************************************
*BIT REVERSAL
        BITR XR1,XI1,XR16,XI16
        BITR XR2,XI2,XR8,XI8
        BITR XR3,XI3,XR24,XI24
        BITR XR5,XI5,XR20,XI20
        BITR XR6,XI6,XR12,XI12
        BITR XR7,XI7,XR28,XI28
        BITR XR9,XI9,XR18,XI18
        BITR XR11,XI11,XR26,XI26
        BITR XR13,XI13,XR22,XI22
        BITR XR15,XI15,XR30,XI30
        BITR XR19,XI19,XR25,XI25
        BITR XR23,XI23,XR29,XI29
*GET POWER SPECTRUM
*          .
*          .
        END
```

FIGURE 9.30. (concluded)

251

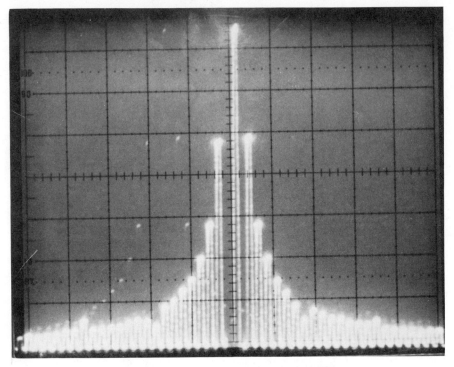

FIGURE 9.31. Power spectrum of 256-point FFT

9.7 FFT Algorithm with a Fixed Geometry

Problem Statement

Implement a 16-point FFT algorithm using a variation of the standard Cooley–Tukey algorithm [20]. Instead of each stage of the signal flow graph following the familiar butterfly pattern, this algorithm produces a signal flow graph that causes each stage to have an identical geometric structure. This makes the indexing easier to implement, but it comes with a price—greater internal storage requirements.

Design Considerations

This algorithm is explained in Stanley [3]. The flow graph is shown in Figure 9.32. The solid lines are weighted by the twiddle factor, the power of which is given by the circle it enters. Sixteen weights are needed to do the calculation. Because of symmetry, only eight need be stored in a lookup table: $W^0 - W^7$.

The input time sequences are entered through the A/D port and the resulting spectrum is displayed on an oscilloscope connected to the D/A port of the AIB. A sample rate of 6400 Hz was chosen. This sample rate gives a range of working

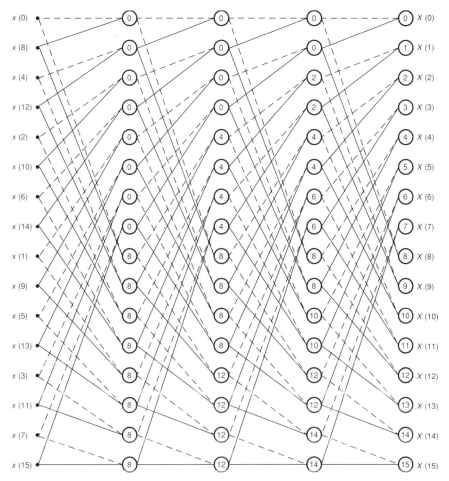

FIGURE 9.32. Signal flow diagram of a 16-point fixed-geometry FFT

frequencies from 400 to 3200 Hz. The 400-Hz minimum is the lowest-frequency sine wave at which at least one cycle will fit into the window and the 3200 Hz is the Nyquist frequency.

A program was written in Basic to check the algorithm and to determine what scaling is necessary to implement it on the TMS320. The results of this simulation showed that if the input is scaled by 16, all the calculations remain in the proper range.

Results and Conclusions

The results for a 400-Hz sine wave input and a 1200-Hz sine wave input are shown in Figures 9.33 and 9.34, respectively. The graphs in each of these figures show a delta function occurring at the frequency of the sinusoidal input and a second delta function occurring at its folded frequency. One of the delta functions

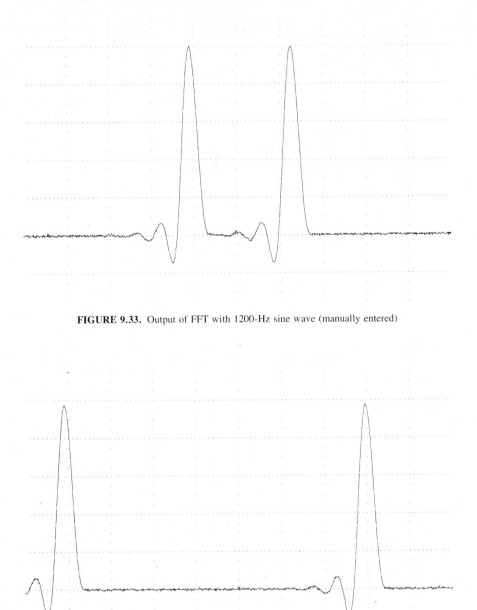

FIGURE 9.33. Output of FFT with 1200-Hz sine wave (manually entered)

FIGURE 9.34. Output of FFT with 400-Hz sine wave (input: function generator)

corresponds to the positive frequency and the other to the negative frequency component. The spacing between the delta functions increases from Figure 9.33 to Figure 9.34 by a factor of approximately three which is the same as the ratio of the input frequencies.

The algorithm was tested with both manually entered data and real-time data entered through the AIB port. The program reliably produced expected results when the inputs were entered by hand into the memory. However, when a real-time signal was the input, occasional spurious results would be obtained. Possible extensions of this project are to window the input data or to average the power spectrum.

9.8 REAL-TIME FFT

Problem Statement

A real-time 128-point FFT can be implemented using the functional block diagram in Figure 9.35. A kit, available from Texas Instruments and other vendors, includes a TMS32020; a combo-codec filter (TCM2916) in lieu of an A/D and D/A; the FFT program, based on the Cooley–Tukey algorithm, stored on two PROMs (TBP38L165); and supporting documents [1,20]. Since the chips and components shown are readily available and reasonably priced through many vendors, an FFT board based on this design can easily be built using Figures 9.36 through 9.38. Although the design work is already done, it is still a good instructional project. In addition to affording an opportunity to study the techniques used to do the FFT, it also results in a real-time spectrum analyzer, with which speech and music can be observed.

Implementation

The real-time spectrum of a sinusoidal waveform is displayed in Figure 9.39. The sinusoidal waveform may be taken from the function generator chip (ICL8038) in Figure 9.36 or from an external function generator. The display shows a delta function occurring at the frequency of the sinusoidal input which is $\simeq 2.4$ kHz,

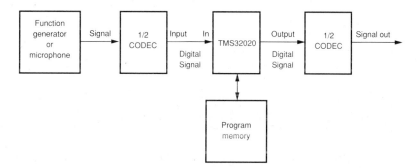

FIGURE 9.35. Functional block diagram of FFT (Courtesy of Texas Instruments Inc.)

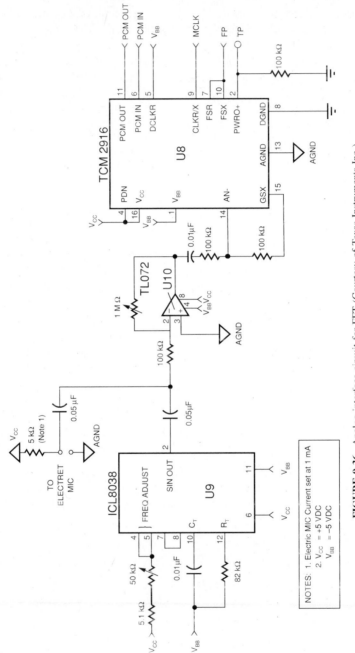

FIGURE 9.36. Analog interface circuit for FFT (Courtesy of Texas Instruments Inc.)

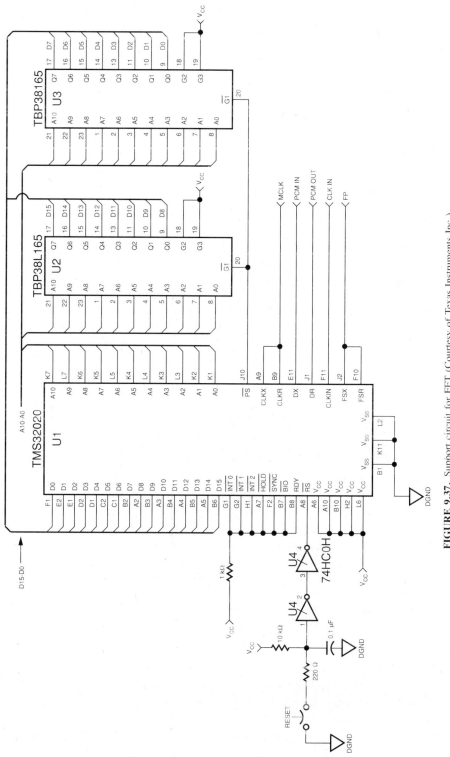

FIGURE 9.37. Support circuit for FFT (Courtesy of Texas Instruments Inc.)

257

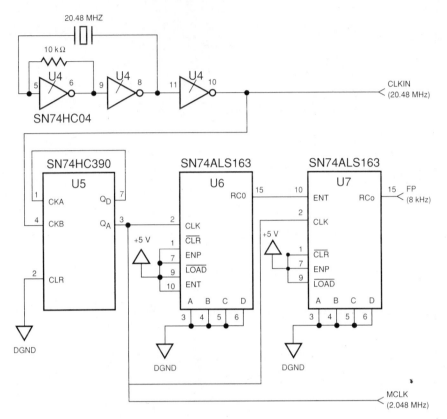

FIGURE 9.38. Clock circuit for FFT (Courtesy of Texas Instruments Inc.)

and a second delta function at its folded frequency. This output was obtained using an 18.375-MHz crystal instead of the 20.48-MHz shown in Figure 9.38. As a result, the signal (FP), in Figure 9.38, is 7.177 kHz and not 8 kHz as shown. The clock, CLKIN, is divided by 10 by a decade counter, then by 16 twice due to the two SN74ALS163 binary counters. FP represents a framing pulse occurring at a frequency of 7.177 kHz and denotes the start of a new frame. A reference or range is provided by having a large negative value at zero frequency of the FFT. The sampling period T_s is $(1/7.177)$ kHz, or 139.3 μs. Since there are 128 points, the distance between the two negative spikes is 128×139.3 μs, or 17.83 ms, as verified in Figure 9.39.

Conclusion

Using the kit described, a crude but inexpensive spectrum analyzer can be built. With music as input, the corresponding output spectrum is displayed on an oscilloscope in real time. Several FFT programs available from Texas Instruments in their utility package, can be used to compute 128-, 256-, and 1024-point FFTs.

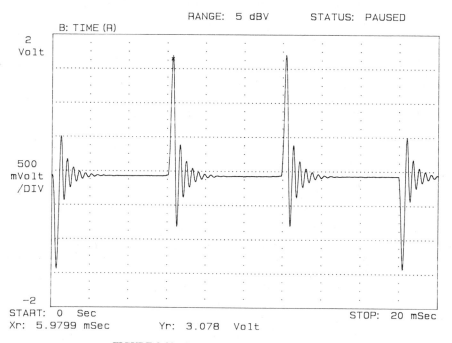

FIGURE 9.39. Output spectrum of sinusoidal input

9.9 IMPLEMENTATION OF A SECOND-ORDER GOERTZEL ALGORITHM

Problem Statement

The purpose of this project is to implement a second-order Goertzel algorithm on the TMS320C25. Goertzel's algorithm performs a DFT using an IIR filter calculation. Compared to a direct N-point DFT calculation, this algorithm uses half the number of real multiplications, the same number of real additions, and requires approximately $1/N$ the number of trigonometric evaluations. The biggest advantage of the Goertzel algorithm over the direct DFT is the reduction of the trigonometric evaluations. Both the direct method and the Goertzel method are more efficient than the FFT when a "small" number of spectrum points is required rather than the entire spectrum. However, for the entire spectrum, the Goertzel algorithm is an N^2 effort, just as is the direct DFT.

Design Considerations

Both the first-order and the second-order Goertzel algorithms are explained in several books [2,4,31]. A discussion of them follows. Since

$$W_N^{-kN} = e^{j2\pi k} = 1$$

both sides of the DFT in (7.2) can be multiplied by it, giving

$$X(k) = W_N^{-kN} \sum_{r=0}^{N-1} x(r) W_N^{+kr} \tag{9.4}$$

which can be written as

$$X(k) = \sum_{r=0}^{N-1} x(r) W_N^{-k(N-r)} \tag{9.5}$$

Define a discrete-time function as

$$y_k(n) = \sum_{r=0}^{N-1} x(r) W_N^{-k(n-r)} \tag{9.6}$$

The discrete transform is then

$$X(k) = y_k(n) \big|_{n=N} \tag{9.7}$$

Equation (9.6) is a discrete convolution of a finite-duration input sequence $x(n), 0 < n < N - 1$, with the infinite sequence W_N^{-kn}. The infinite impulse response is therefore

$$h(n) = W_N^{-kn} \tag{9.8}$$

The Z-transform of $h(n)$ in (9.8) is

$$H(z) = \sum_{n=0}^{\infty} h(n) z^{-n} \tag{9.9}$$

Substituting (9.8) into (9.9) gives

$$H(z) = \sum_{n=0}^{\infty} W_N^{-kn} z^{-n} = 1 + W_N^{-k} z^{-1} + W_N^{-2k} z^{-2} + \cdots \tag{9.10}$$

$$= \frac{1}{1 - W_N^{-2k} z^{-1}}$$

Thus equation (9.10) represents the transfer function of the convolution sum in equation (9.6). Its flow graph represents the first-order Goertzel algorithm and is

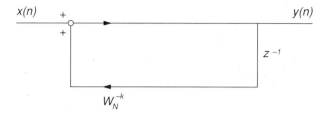

FIGURE 9.40. First-order Goertzel algorithm

shown in Figure 9.40. The DFT of the kth frequency component is calculated by starting with the initial condition $y_k(-1) = 0$ and running through N iterations to obtain the solution $X(k) = y_k(N)$. The $x(n)$'s are processed in time order and processing can start as soon as the first one comes in. This structure needs the same number of real multiplications and additions as the direct DFT but $1/N$ the number of trigonometric evaluations.

The second-order Goertzel algorithm can be obtained by multiplying the numerator and denominator of (9.10) by $1 - W_N^{-k}z^{-1}$ to give

$$H(z) = \frac{1 - W_N^{+k}z^{-1}}{1 - 2\cos(2\pi k/N)z^{-1} + z^{-2}} \qquad (9.11)$$

The flow graph for this equation is shown in Figure 9.41. Notice that the left-half of the graph contains feedback flows and the right half contains only feedforward terms. Therefore, only the left-half of the flow graph must be evaluated each iteration. The feedforward terms need only be calculated once for $y_k(N)$. For real data, there is only one real multiplication in this graph and only one trigonometric evaluation for each frequency. Scaling is a problem for fixed-point arithmetic realizations of this filter structure; therefore, simulation is extremely useful.

Testing and Conclusions

The algorithm was tested on a discrete rectangular pulse and the results agreed with the standard solution found in textbooks. Implementation of this algorithm is made easier by first writing it in a higher-level language and then converting it into TMS320 code. Following this procedure provides a check on both the logic

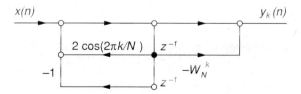

FIGURE 9.41. Second-order Goertzel algorithm

and the scaling of the implementation. The lookup tables are the major difference between the TMS320 code and the higher-level language implementation.

The second-order Goertzel algorithm is more efficient than the first-order Goertzel algorithm. The first-order Goertzel algorithm (assuming a real input function) requires approximately $4N$ real multiplications, $3N$ real additions, and 2 trigonometric evaluations per frequency component as opposed to N real multiplications, $2N$ real additions, and 2 trigonometric evaluations per frequency component for the second-order Goertzel algorithm. The direct DFT requires approximately $2N$ real multiplications, $2N$ real additions, and $2N$ trigonometric evaluations per frequency component.

This Goertzel algorithm is useful in situations where only a few points in the spectrum are necessary, as opposed to the entire spectrum. Detection of several discrete frequency components is a good example. Since the algorithm processes samples in time order, it allows the calculation to begin when the first sample arrives. In contrast, the FFT must have the entire frame in order to start the calculation.

9.10 SIMULATION OF HEARING IMPAIRMENTS

Currently, considerable effort is being directed toward the implementation of hearing aid devices based on the TMS320 digital signal processor [32].

Problem Statement

The objective of this project is to simulate a level of hearing impairment using a representative lowpass filter.

Design Considerations

Hearing impairments can be simulated by passing normal speech through a lowpass filter, which is based on a hearing-impaired person's audiogram; thus demonstrating how certain sounds are "heard" by a hearing-impaired person. An audiogram is basically a frequency response of a person's hearing system. Tones that a person cannot detect appear as low spots on the curve, whereas those that are easily heard appear as peaks on the curve. Figure 9.42 shows three lowpass FIR filters with cutoff and roll-off frequency characteristics based on the audiograms of a person with mild hearing impairment, the person's sister with severe hearing impairment, and the person's mother with deaf hearing.

Results and Conclusions

A tape containing specific test sounds is played using the configuration shown in Figure 9.43 to demonstrate how a hearing impaired person hears. The audio is played through each of the three lowpass filters to simulate the perceived hearing of the three people with different levels of hearing impairments.

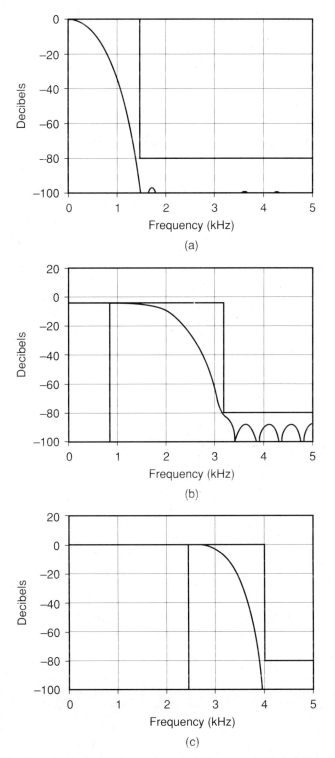

FIGURE 9.42. Filters for hearing impairments (log megnitude response): (*a*) deaf; (*b*) hard of hearing; (*c*) mild hearing

FIGURE 9.43. Configuration for hearing impairments

9.11 DIGITAL TOUCH-TONE DECODER/GENERATOR

Problem Statement

The goal of this project is to design and test a system that can decode and generate standard AT&T dual-tone multiple-frequency (DTMF) Touch-Tone dialing signals. An AT&T DTMF Touch-Tone signal, used in an ordinary American telephone, can send 16 distinct frequency patterns, each representing a key. Each key is represented by two different frequencies, one of which is a column frequency (from the high-frequency group) and the other a row frequency (from the low-frequency group). There are four column and four row frequencies available, which are assigned to keys as shown in Table 9.2. For example, if the key "*" is to be transmitted, the frequency selected is in row 4 = (941 Hz) and column 1 = (1209 Hz). Conversely, if these two frequencies are detected, it means that the * key was received. The AT&T specifications for DTMF signals require that the transmit/receive rate is 10 digits/second or 100 ms per digit. The signal must be present for at least 45 ms and no more than 55 ms. The remainder of the 100-ms interval must be quiet.

Design Considerations

One approach is to use a discrete Fourier transform (DFT) and examine the frequency domain for the eight tones. This approach was taken in a paper by Mock in Lin [1], which is a good source of information on the Touch-Tone scheme. In addition, this design is a good example of the use of the Goertzel algorithm, described in Section 9.9, for the calculation of a small number of frequencies rather than the complete spectrum. However, since this approach was already done, it was not used in this project.

The approach employed uses digital bandpass filters to detect the eight tones. Instead of using eight separate filters running together, one digital bandpass filter can be used and its center frequency changed simply by scaling the sample frequency. By checking each frequency in a sequential scanning fashion, and keeping note of which frequency exceeds a given threshold, the key can be determined.

TABLE 9.2 Touch-Tone Matrix

	Col. 1 (1209 Hz)	Col. 2 (1336 Hz)	Col. 3 (1477 Hz)	Col. 4 (1633 Hz)
Row 1 (697 Hz)	1	2	3	A
Row 2 (770 Hz)	4	5	6	B
Row 3 (852 Hz)	7	8	9	C
Row 4 (941 Hz)	*	0	#	D

This technique has one serious drawback, however, because only one of eight frequencies is being tested at a given time, as opposed to all eight at once. As a result, it will be at least eight times slower than using individual filters of the same type. Some of this time may be reduced if multirate techniques are used.

Touch-Tone Decoder Design

The first problem to be solved is to define the master filter characteristics, such as passband, stopband, and roll-off. To determine the requirements of the filter, all of the Touch-Tone key codes are tabulated with the row–column frequencies and the distance between frequencies (delta frequency) in Table 9.3. As shown in Table 9.3, the most critical filter is needed for the * key with the smallest frequency difference at 268 Hz. This frequency difference is a good starting point in determining the bandwidth for the filter.

An additional piece of information needed to specify the passband of the master filter is the tolerance specified on the Touch-Tone frequencies. The amount of frequency deviation allowable in a given Touch-Tone frequency is ±3%. Taking the 3% into account will bring the most critical frequencies a little closer. Row 4 can be as high as 962.2 Hz and column 1 can be as low as 1172.7 Hz. In the 203.5 Hz between these two frequencies, no Touch-Tone frequencies should exist. Thus the master filter must have an acceptable cutoff frequency of approximately 100 Hz from the passband edge. The passband edge was taken to be ±50 Hz from the center frequency, and the stopband was selected to be ±115 Hz from the center frequency.

When scaling a filter, it is best to design the filter at the highest sampling frequency desired and then scale the sampling rate downward. Since most telephones do not have the fourth column (Table 9.2), it will be omitted from the design. This leaves the 1477 Hz as the highest Touch-Tone frequency. Thus the master filter

TABLE 9.3 Frequency Span between Key Codes

Key Code	Row Freq. (Hz)	Column Freq. (Hz)	Delta Freq. (Hz)
0	941	1336	395
1	697	1209	512
2	697	1336	639
3	697	1477	780
4	770	1209	439
5	770	1336	566
6	770	1477	707
7	852	1209	357
8	852	1336	484
9	852	1477	625
A	697	1633	936
B	770	1633	863
C	852	1633	781
D	941	1633	692
*	941	1209	268
#	941	1477	536

is designed around this frequency. The stopband was made 36 dB down from the passband. A sketch of the filter bands is shown in Figure 9.44.

The master filter was designed using the ASPI Digital Filter Design Package. Since the phase is not important in this application, and since data memory and allowed time delay are limited, a sixth-order elliptic filter was used. The filters for the other bands can be scaled by changing the sample rate accordingly. The sample rates for the other filters are calculated from the ratio

$$F_{cd}/F_{sd} = F_{cm}/F_{sm} = 1477/8000$$

or

$$F_{sd} = 5.416384 F_{cd}$$

where F_{cd} = desired center frequency F_{cm} = master center frequency

F_{sd} = desired sample frequency F_{sm} = master sample frequency

For example, the sample frequency required for Row1 tone (697 Hz) is calculated as follows:

$$F_{sd} = 5.416384(697 \text{ Hz}) = 3775.2 \text{ Hz}$$

This frequency can be obtained on the AIB by loading it with the N counter value, which can be calculated as follows:

$$N = (10 \times 10^6/F_{sd}) - 1 = 1323$$

The number of samples needed to detect a key frequency and the detected threshold voltage were found experimentally. With a 10V peak-to-peak signal at the filter's center frequency, it takes 64 samples to detect a 1.8V peak-to-peak output. These parameters yield a digit rate of 0.278 s/digit, which is about three times slower than specifications that call for 0.100 s/digit. No further effort was spent in trying to meet this specification.

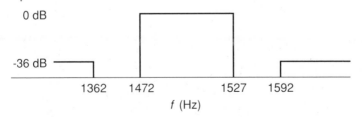

FIGURE 9.44. Master filter bands

Touch-Tone Generator Design

To generate the Touch-Tones, two second-order digital sinusoidal oscillators, similar to the one in Section 4.4, are used. The constants for each frequency are stored in a table and used to initialize the oscillator for a given key. The sinusoidal outputs of the oscillators are summed to create the desired two-tone frequency. The generators are run at a sample frequency of 10 kHz. The number of samples put out per digit is determined from the digit rate and the sample rate. Each tone is followed by a no-tone period of the same length, so that a key release can be detected. The time to transmit S samples is given by

$$\text{transmit time} = 2S/10 \text{ kHz}$$

where 10 kHz is the sample rate and the 2 accounts for the no-tone time. Since the transit time $= 0.3$ s, we can solve for the number of samples, S, as follows

$$S = (0.3\text{s})5000 = 1500 \text{ samples}$$

System Software

The main program consists of a section that initializes the AIB and the master filter and then executes a single loop indefinitely. This main processing loop scans the frequencies and decodes the digits of a 10-digit number and stores it in data memory. Next, the digit stored in data memory is read, and the corresponding two frequencies are found and transmitted for a specified period of time. Then the loop process is repeated.

System Testing

The system was tested using a real AT&T telephone. The connections to the TMS320 system are as shown in Figure 9.45. The audio amplifier shown is designed to drive an 8-Ω load. The output of the TMS320 system is connected to the input of another machine running the same software, and its output was connected to another system, and so on. In the end, five machines were connected in a relay configuration.

To test the system, the user keys-in a 10-digit telephone number, the first machine relays this number to the second, the second to the third, and so on, until the last machine receives the number. At this point the user can interrupt the last machine and examine the data memory location of the received code. The number typed on the first machine should be the same as the number on the last machine. This test was repeated a number of times. It was found that if the first machine (the one connected to the telephone) receives the number correctly, then all of the others will receive and transmit flawlessly.

A long-term test was devised to find how reliable the transmission is between units. To accomplish this, a number was dialed and the output of the last machine was quickly connected in place of the telephone. The dialed number would cycle

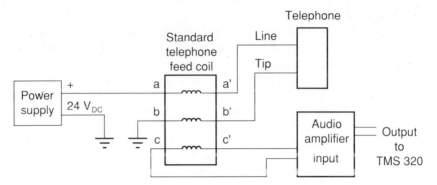

FIGURE 9.45. Telephone connections to the TMS320

through the endless loop until stopped or until one system misses a digit. This test was run for about 1.5 hours. Since it takes about 3 seconds to send and 3 seconds to receive, the total number of transmissions was about 900 ten-digit numbers. This test shows that the system interfacing with itself is extremely reliable.

Conclusion

The software works as intended, except that the decoder does not meet AT&T's specified maximum transmission rate. The system will work if dialing is done without quickly jabbing the keys. The maximum rate specification is not met because of the long time needed for the master filter to reach steady state and the fact that eight scans are made, one for each filter. This problem may be solved by using a filter that reaches steady state faster (has fewer delays), by relaxing the specifications on the existing elliptic filter, or by applying the multirate techniques in Section 9.1 to reduce the number of samples needed for each key test. For example, it should be possible to run at the highest sample rate for the top filter and decimate the rate for excessively lower rates until all the filters are done. Only one scan would be needed for all eight frequencies. The different sample rates would be realized in software, thus considerably reducing the execution time.

9.12 DUAL-TONE MULTIFREQUENCY GENERATION

This project is similar to the preceding project and is documented in [1]. Dialing a number corresponds to generating sinusoids of specific frequencies. Two specific frequencies can be used to encode a digit or a special character as seen in telephone systems. Each digit or character can be uniquely represented from two specific tones.

Problem Statement

Generate a dual-tone multifrequency (DTMF) tone.

Design Considerations

As shown in Table 9.2, a unique pair of frequencies is assigned to each key on a telephone keypad. The column frequencies are the high frequencies and the row frequencies are a group of low frequencies. When a specific number or symbol is chosen, by pressing the associated key, the frequencies corresponding to that key are generated. The digital oscillator program in Section 4.4 can be used to calculate the coefficient values A and C ($B = -1$) to generate specific frequencies.

For example, when the key representing the number 8 is pressed, the third-row frequency (852 Hz) and the second-column frequency (1336 Hz) are to be selected. The composite signal is the addition of both of these signals, and both tones are easily distinguished by listening to the receiver of a telephone while depressing one of the keys.

The tasks needed to produce the dual-tone signals consist of the following:

1. Looking up the dual tones necessary for a given digit
2. Generating these two tones via two sine-wave oscillators and summing them together
3. Putting the composite signal out to the D/A unit, which drives a speaker that is acoustically coupled to a Touch-Tone telephone set

Results and Conclusions

The TMS320 program is successfully tested by generating, or dialing acoustically, a sequence of seven tones (digits). The output speaker is placed close to an off-hook telephone handset. The program is given the desired telephone number, which it dials through the acoustic coupler. An appropriate amount of delay is set in the program to simulate a silent period between each tone. This project can be extended to include the design of an efficient modem.

9.13 DIGITALLY PROGRAMMABLE SPEECH SCRAMBLER

Problem Statement

The objective of this project is to design a programmable audio speech scrambler.

Design Considerations

System Design

This project utilizes the TMS320 processor to create a speech scrambler that will code normal audio into sounds that are unintelligible. The scrambling method used is commonly referred to as *frequency inversion*. This method basically takes an audio range (represented by the band 0.3 to 2.8 kHz) and "folds" it about a carrier signal. The frequency inversion is achieved by multiplying (modulating) the audio input by a carrier signal, causing a shift in the frequency spectrum with upper and lower sidebands. The effect of the multiply operation is shown in Figure 9.46(*a*) and (*b*). On the lower sideband, which represents the audible speech range, the low tones are turned into high tones, and vice versa. The carrier at 3.1 kHz can also be inserted, as shown in Figure 9.46(*b*).

Interception of the speech can be made more difficult by dynamically changing the modulation frequency and including, or omitting, the carrier frequency according to a predefined sequence. This particular design uses a sequence of nine codes, each of which may have any of the following values:

0   No modulation
1   Modulate with frequency FREQ1
2   Modulate with frequency FREQ2

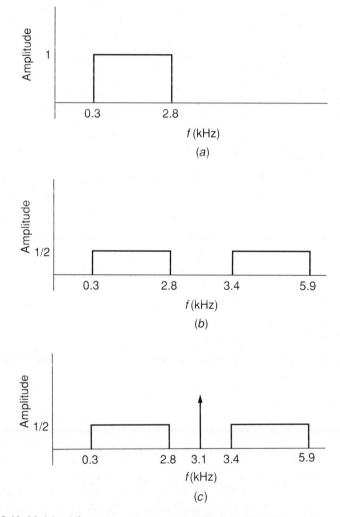

FIGURE 9.46. Modulated frequency spectrum: (*a*) bandlimited audio; (*b*) modulated bandlimited audio; (*c*) modulated bandlimited audio with carrier

The length of time a given code is used is also a parameter that must be predefined and known to both the sender and the receiver.

The system must shift the carrier in frequency according to a security code. The method chosen to generate the carrier frequency is to use a 50% duty cycle square wave which has only odd harmonics, and, after filtering, retains only the fundamental. The carrier must lie around 3100 Hz and the sampling frequency must be an even multiple of the carrier in order to satisfy the 50% duty cycle requirement. The sampling rate is selected at four times the carrier, $\approx$ 12.4 kHz. Since the carrier frequency must change dynamically, the sampling rate must also change. The shift in sampling rate also frequency-scales the filters.

Both sender and receiver require a filter to separate the audio and another filter to separate the carrier frequency band. The frequency range of these filters is chosen appropriately, so that they must pass the desired band at two sampling rates. The two carrier frequencies are chosen to be 3.1 and 3.2 kHz. The carrier frequency bandpass filter must pass frequencies from 3.1 to 3.2 kHz. This filter is designed for a passband from 3.1 to 3.3 kHz at a sampling frequency of 12.8 kHz. With the lower sampling rate of 12.4 kHz, the passband is from 3.0 to 3.1 kHz. Therefore, the frequencies that can possibly pass through this filter are 3.0 to 3.3 kHz, considering both sampling frequencies.

The carrier places another restriction on the low tones of the audio range, since the lowest audio frequency becomes the highest when modulated. This means that to prevent the audio from passing into the carrier band, its lowest frequency must be a safe distance from the limits of the carrier passband. The highest frequency, in the lower sideband, that can be generated is when modulating at 3.2 kHz and is (3.2 kHz −0.3 kHz) = 2.9 kHz. Therefore, the audio band shown in Figure 9.46(a) is within the limits and will be maintained by an input audio filter whose band is 0.3 to 2.8 kHz.

Each transceiver has three filters: one to limit the incoming audio, one to filter the carrier and upper sideband, and one to isolate the carrier. All these filters are realized as elliptic filters in order to keep the processing time within acceptable limits, and they are designed with the ASPI filter design package.

Hardware Setup

To test the scrambler, two TMS320s with AIBs are connected as master and slave, as shown in Figure 9.47. The output of the master is connected to the input of the slave. The master receives audio from its input port, scrambles it, and then transmits it to the slave. The slave receives the scrambled signal, descrambles it, and outputs it to a speaker.

The master transmits a carrier pulse, which is a short burst of the carrier frequency that is used to synchronize the transmission. Upon startup, both units assume that they are the master and both transmit a carrier pulse. Whichever unit receives the pulse first becomes the slave and suspends transmission of the carrier pulse. The other unit then becomes the master and continues to transmit the synchronizing pulse. If the two units are in lockstep, a deadlock can occur, resulting in two slaves. This situation is not critical in a laboratory environment, and can

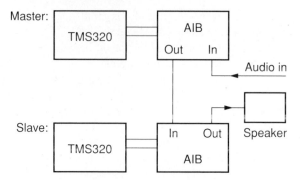

FIGURE 9.47. Experimental setup for speech scrambler

be resolved by having a unit in the slave state become master if a synchronization pulse is not received for a specified period of time. Both the master and the slave continue to listen for the synchronization pulse. If the master receives the carrier pulse, it will become a slave.

When the audio is being modulated, both sidebands are being transmitted, but the receiver's (slave's) audio bandpass filter passes only the lower sideband. For the receiver to demodulate the audio, it must remix the scrambled lower sideband with the correct carrier frequency, in order to reverse the band, and reproduce the original audio. This remixing produces upper and lower audible sidebands, but the one of interest is the lower one. To separate the two, the remixed signal is passed through an additional lowpass filter. This occurs only in the slave, or receiving unit.

Program Design

The program is designed in modules, and then each module (or subroutine) is called as necessary, to facilitate readability and organization. The program revolves around a main loop, which calls the different subroutines, and I/O ports. The subroutine names are listed below with a short description of each.

1. INIT: defines the names of locations in data memory and proceeds to load the initial values to each location. It also calls the initialization routines of the filters designed with the ASPI's Digital Filter Design Package (DFDP). This routine is called once upon initial execution.

2. F1: implements the CARRIER bandpass filter, designed with DFDP software. It is used to detect the presence of the synchronization pulse.

3. F2: implements the AUDIO bandpass filter, designed with the DFDP software. It is used to limit incoming audio to the lower sideband frequency range of 300 to 2800 Hz.

4. LOWPAS: determines the unit's master/slave status, and if master, it returns. If slave, it calls a lowpass filter F3 to remove the upper sideband of the remixed signal.

5. MASTSL: gets the output value of the CARRIER filter and determines if the absolute value is greater than some threshold value. If it is, a carrier has been detected and the unit must enter the slave mode by setting MASTER = 0. Also, the pointers to the code sequence and the code usage time are initialized, since the presence of a carrier indicates that the sending unit is a master and is trying to synchronize. If no carrier is present, nothing further is done and it returns to the main loop.

6. SEQNC: sees if there is a carrier present; and if one exists, it returns to the main loop since the pointers have been adjusted by the MASTSL subroutine. If there is no carrier present, this means that normal operation is being carried out, and the pointers must be adjusted accordingly. The usage time is decremented and checked for zero. This counter determines if the unit should switch to another word of the security code. If time is up, the security pointer is decremented and checked to determine if it is at the end of its sequence. If either pointer is at the end of its sequence, the pointer is loaded with its maximum value and the subroutine then returns.

7. CARMIX: generates the carrier 50% duty cycle square wave by multiplying a register by 1 or -1 every two samples. This creates a square wave at one-fourth of the sampling frequency. The routine also checks the value of the security word. If modulation is necessary, the output of the input audio bandpass filter is multiplied in time by the square wave, which is equivalent to a convolution in frequency. Using a symmetrical carrier with no dc offset produces a frequency spectrum that contains sidebands centered around the carrier with the carrier component being nulled.

8. CARADD: adds a carrier pulse, which will synchronize both systems, to the value to be sent out if the unit is a master and is at the end of the security sequence. A problem that exists here is that there is a group delay associated with the receiver's carrier filter, so that the slave unit will still see a carrier, even though the master has terminated transmission. The master resynchronizes at the end of the sequence, but the slave resynchronizes at the end of the carrier burst. Furthermore, the slave's filter needs some time to pass the carrier, and thus detects it at a later time. To solve this problem, two user-programmable words allow the master to terminate the carrier before it reaches the end of its sequence. In this way, if the carrier is terminated by the master at exactly the group delay of the slave's filter, the slave unit will sense the termination of the carrier at the desired time.

9. SAMPRT: looks up the previous sample rate and compares it to the desired sample rate. If a change in sampling frequency is necessary, a new rate is sent to the sample rate clock. If not, the subroutine returns immediately.

Results and Conclusions

The scrambler is effective in transmitting bandlimited voice, but the clarity was not commendable. The source of the distortion could not be identified. However, with no input, an oscillation is present that consists of two superimposed sine waves. Initially, the system is quiet, but if an input is introduced, then removed, the distortion appears and remains permanently. The program was modified to test each filter separately. The conclusion of this test shows that the noise is due to the nature of the IIR filter, possibly limit cycles in the filters.

Modulated signals have a smaller amplitude than the unmodulated signals. This fact was initially overlooked but was easily adjusted by increasing the gain on the modulated signals. The synchronization scheme needs to be redesigned to make it more robust. Noise in the carrier band may be interpreted as a carrier pulse by the master, resulting in both units being slaves. The current scheme uses simple threshold detection, which can lead to false detection. A more reliable detection method and a better synchronization protocol would improve this project.

9.14 BANDLIMITING BASEBAND COMMUNICATIONS

Problem Statement

Development of a method of reducing the bandwidth of voice modulated signals to relieve congestion in the radio frequency bands allocated for voice communications.

Design Considerations

A narrow band voice modulation system is designed to maintain the important characteristics of speech while reducing the bandwidth by $\simeq$ 34%. It requires compression and expansion of the frequency bandwidth at the transmitting and receiving ends, respectively. From the characteristics of spoken words, a band of frequencies between 600 Hz and 1500 Hz can be filtered out with little degradation in the speech.

Speech from a tape recorder was input to a multiband filter, designed using the DFDP discussed in Chapter 5, with passbands within the range 0 to 600 Hz and 1500 to 2400 Hz. The two passbands of frequencies were selected such that the first two formants in a typical voice would fall within those two bands. The audio output from this filter demonstrated little or no noticeable degradation of the input speech. This result formed the incentive to transmit the audio at a reduced bandwidth, eliminating the frequencies 600–1500 Hz.

Figure 9.48 shows the encoder block diagram at the transmitting end as well as the decoder block diagram at the receiving end. Two systems (implemented with SWDS and AIB) were used; one to run an encoder program and the other for the decoder. The output $Y_i(n)$ from the (AIB) encoder end becomes the input to the (second AIB) decoder end. The output of each of the bandpass filters is multiplied/modulated by a 3.1 kHz signal. Figure 9.49 shows the results from filtering and modulating. The 3.1 kHz signal is obtained using three sample points from a table look-up using a utility program from a simulator package. Figure

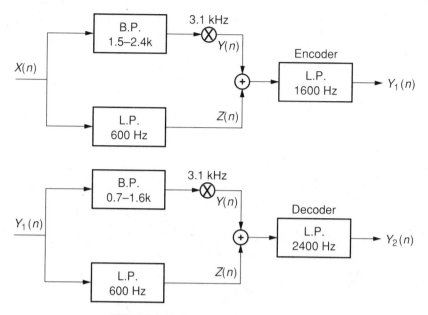

FIGURE 9.48. Encoder/decoder block diagram

9.49(a) shows that, using this scheme, the transmitted signal has a bandwidth of 1600 Hz instead of 2400 Hz, and Figure 9.49(b) shows the procedure to receive the bandlimited signal and reconstruct the two passbands.

Appendix J shows a complete listing of the encoder program which is very similar to the decoder program except for the coefficients of the different filters. For both the encoder and the decoder, all the filters are designed using the DFDP. The bandpass and the first lowpass filters contain 95 coefficients and the lowpass filter at the output contains 59 coefficients. All the filters are FIR-type filters with Kaiser windows. An alternative encoder and decoder system is illustrated in Appendix J.

Results and Conclusions

The original encoding/decoding scheme produced an output speech with a very negligible reduction in quality and noticeable reduction in bandwidth. Figure 9.50 shows the output spectrum of the decoder system. Since the first two dominant formants in a typical voice occur at frequencies of $f_1 \simeq 270$ Hz and $f_2 \simeq 2290$ Hz, more stringent (input) bandpass filters are designed, with the lowpass filters in Figure 9.48 changed to bandpass filters, and the modulating frequency also changed. This resulted in further bandwidth reduction, but only in a slight improvement in speech quality. The overall results show that appropriate encoding and decoding can produce a substantial reduction in bandwidth in many applications in voice communications.

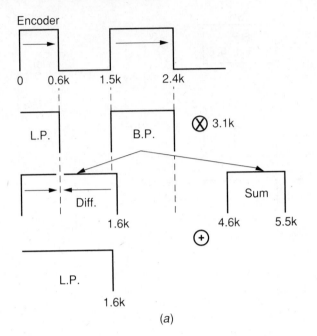

(a)

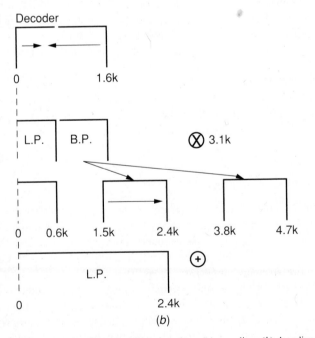

(b)

FIGURE 9.49. Encoding/decoding procedure: (a) encoding; (b) decoding

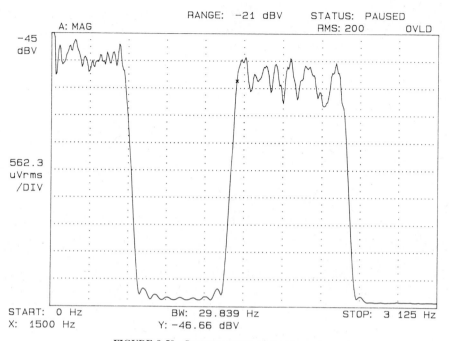

FIGURE 9.50. Output spectrum of decoder system

9.15 ZERO-CROSSING-BASED SPECTRUM ANALYZER

Problem Statement

The purpose of this project is to implement a TMS320 program that calculates the DFT of a signal from its zero crossings and to evaluate its accuracy using a test function.

Design Considerations

The algorithm used is developed and explained in a paper by Kay and Sudhaker [33]. A constant must be added to the signal to ensure that all the zeros are real. This modified signal is then represented by a polynomial of the form

$$P_k(z) = \prod_{i=1}^{k} (z - z_i)$$

$$= z^k + \sum_{i=1}^{k} a_i^k z^{k-i} \tag{9.12}$$

where z_i is a zero of the function. In terms of time,

$$z_i = \cos(2\pi t_i/T) + j\,\sin(2\pi t_i/T) \qquad (9.13)$$

where t_i is the measured time of the zero crossing and T is the length of the time frame. Since the zeros come in sequentially, the polynomial can be represented recursively,

$$P_{k+1}(z) = (z - z_{k+1})P_k(z) \qquad (9.14)$$

By expanding (9.14) in terms of the a_i coefficients in (9.12), one obtains a recursion relationship for the a_i coefficients:

$$a_0{}^k = 1$$

$$a_i{}^{k+1} = a_i{}^k - z_{k+1}a_{i-1}{}^k \qquad i = 1, 2, 3, \ldots, k$$

and,

$$a_i{}^{(k+1)} = -z_{k+1}a_k{}^{(k)} \qquad i = k+1 \qquad (9.15)$$

Thus the coefficients of P_{k+1} can be evaluated from z_{k+1} and the coefficients of P_k.

The Fourier series can also be expressed in terms of a polynomial in z,

$$P(z) = \sum_{m=-M}^{m=M} c_m z^m \qquad (9.16)$$

where $M = WT$, with T being the interval of observation and W the single-sided bandwidth. By changing the summation index on equation (9.12), it can be shown that

$$c_i = a_{M-i} \qquad i = -M, \ldots, 0, \ldots, M \qquad (9.17)$$

The DFT coefficients S_k are related to the Fourier series coefficients by

$$S_k = Nc_k \qquad k = -M, \ldots, 0, \ldots, M \qquad (9.18)$$

where $N = 2M + 1$ is the DFT length. Thus we can get the DFT from the zero crossings. This calculation, based on the recursive algorithm of (9.15), requires N^2 multiplications; therefore, it is not an efficient algorithm compared to the FFT. However, there are applications where it has advantages.

Testing and Results

The algorithm was tested on the following function, which is a sum of different sinusoidal frequencies with precalculated zero crossings:

TABLE 9.4   Actual versus Algorithm-Generated Fourier Coefficients

Harmonic	Actual Value	Zero-Crossing Algorithm (TMS320)
dc	1.0	1.000
1st	2.0	2.065
2nd	0.0	0.039
3rd	4.0	4.060
4th	0.5	0.520
5th	8.0	8.150

$$P(t) = 1 + 2 \cos 2\pi t + 4 \cos 6\pi t + 0.5 \cos 8\pi t + 8 \cos 10\pi t$$

The algorithm was first written in Basic to identify the scaling problems and isolate programming errors and then converted into TMS320 code. The results of the TMS320 program are compared with the known Fourier series coefficients in Table 9.4.

Conclusions

The results in Table 9.4 show a difference of approximately 2% on the average which was better than the results of the single-precision Basic program, which was written to check the scaling. These results were generated from precalculated zeros. The real test of the algorithm is when the zeros come from a real-time signal. Two questions come to mind: (1) How much error in the zero crossings can the algorithm tolerate? and (2) How accurately can the hardware measure the zero crossings?

An easy way to extend this project to real time is to use the processor to read the A/D port continuously, looking for a sign change in order to find the zero crossings. When it detects a sign change, it grabs the time from the on-chip timer of the TMS320C25, calculates the next set of coefficients, then continues to monitor the A/D port. The algorithm running on the TMS320C25 needs about 14 μs to execute using a sine-table lookup. Instead of the lookup table scheme, a digital oscillator could be used to generate the sine and cosine terms needed for z_i. A refinement of this scheme would be to add some hardware that will interrupt upon a change of sign in the A/D output. A more elaborate project would be to design a zero-crossing detector, independent of the A/D, which will interrupt the processor on each zero crossing.

9.16 CORRELATION OF TWO SIGNALS

This project, which can be performed in a couple of weeks, uses a waveform analyzer capable of correlating two signals.

Problem Statement

Design a scheme to find the correlation between two signals.

Design Considerations

This design will use a two-channel waveform analyzer that can accept two signals. Two TMS320 systems are used for this project, as shown in Figure 9.51. Two highly correlated signals are used. Both signals are generated by the same source. One signal is transmitted directly, while the second one is delayed. The signals are received by microphones, transformed into the frequency domain by the TMS320 units, then fed into a two-channel waveform analyzer, which can display the correlation between the two signals.

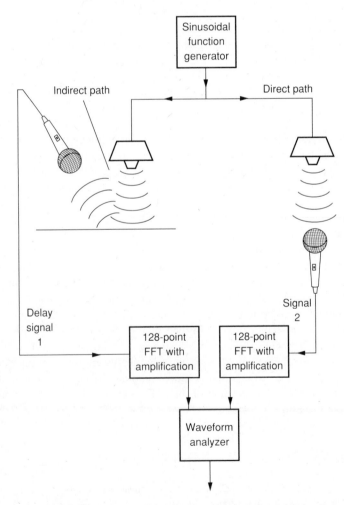

FIGURE 9.51. Block diagram for correlation of two signals

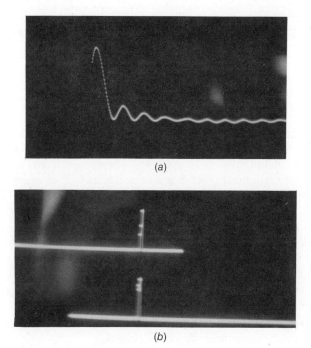

(a)

(b)

FIGURE 9.52. Characteristics of signals for correlation: (a) correlation in time; (b) FFT of each signal

Results and Conclusions

The similarity between the characteristics of the two signals is found by correlating them. The location of the peak of the correlated signals, as observed in the analyzer, corresponds to the amount of skewing in time which exists between the two signals. Changing the amount of the delay associated with one of the signals produces a change in the location of the peak of the correlated signals. A measure of the correlation (in time) between the two signals is shown in Figure 9.52(a). Figure 9.52(b) shows the FFT of each signal. An extension of this project would be to find the correlation integral using the TMS320C25 and display the result on an oscilloscope.

9.17 VOICE RECOGNITION

This project can be performed in approximately a month. Appropriate filter circuits were implemented in hardware.

Problem Statement

To recognize two arbitrary words using a zero-crossing scheme.

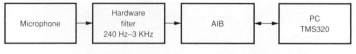

FIGURE 9.53. Speech recognition system

Design Considerations

A zero-crossing speech recognition process described in [34] is used to implement the system, shown in Figure 9.53. A breadboarded circuit, consisting of a differential amplifier and prefilter, a highpass and lowpass filter, and an amplifier, are designed to use the limited on-chip memory efficiently. An amplifier boosts the input voltage from a microphone to a more usable level and the filters reduce hum and noise picked up by the microphone. A 240-Hz highpass filter reduces 60- and 120-Hz hum, and a 3-kHz lowpass filter eliminates frequencies above 3 kHz. A sample frequency of 8 kHz is used, since most of the speech information is contained within a band of frequencies below 3 kHz.

The speech is first amplified, then passed through highpass and lowpass filters to reduce hum and aliasing, and then is amplified again. The amplifier contains the gain control for the speech system and is preset so that at normal speaking amplitudes, the output swing is approximately 1 V peak to peak (limited by a 50-kΩ pot from the hardware amplifier/filter circuit).

The program is set to accept two words and requires approximately 0.1s per word for recognition. An FIR filter is implemented with the TMS320 as a moving-average filter for computation on the sampled speech values in memory.

Results and Conclusions

Using 8 of the 12 bits from the A/D, two samples can be stored in one data memory location. Since only 8 bits of A/D conversion are used, the speech signal is limited to 1 V peak to peak (using 12 bits from the A/D would correspond to ≈ 20 V peak to peak). The samples are stored in 122 on-chip data memory locations, using two samples per location. With an 8-kHz sampling frequency, (122 × 2)/8K, or 0.03s of information can be stored. Using the digital oscillator program section, one tone (with a speaker connected to the audio port on the AIB) was generated to designate that the first word was recognized successfully and two tones to designate that the second word was recognized. This project can be improved by using more available on-chip memory for a longer duration of speech information, as well as by using an additional word recognition algorithm, since different words can have the same number of zero crossings.

9.18 VOICE MAIL

This project requires the use of the TMS32010-based Speech Development System (Part no. 2245198), which includes a board by Texas Instruments, that can be plugged into a slot on an IBM (or compatible) PC, and software that can do text to speech processing.

Spoken instruction can be accomplished with the speech board. This technology is based on phoneme stringing, which breaks down a spoken word into very small units of sound. The phonemes are units that distinguish one sound from another. With the use of the speech board, ASCII text is broken down and converted into natural-sounding English speech. In some cases, a word must be spelled phonetically to return a proper pronunciation.

Text entered into the system is processed in one of two ways by the provided software. First, it is compared to a dictionary; if there is a match, the associated phoneme string and stress is passed directly to the speech processor. If it is not found, the word is given the proper phoneme string and stress as stated by an extensive set of rules, very similar to those used in speaking English aloud.

Problem Statement

Design a scheme using the Speech Development System, so that a user can leave a voice message, which can be played back at a later time.

Design Considerations

Voice storage is the process by which spoken words are sampled and stored in a highly compressed manner on disk. The message can then be played back at a later time when the file is accessed. Four algorithms, available on the board for this storage process, enable the computer to store the voice input in one of four quality levels, ranging from normal quality [2400 bits per second (bps)] to tape-recorder quality (32k bps). The disk space used in the storage of speech is proportional to its quality.

Results and Conclusions

A program written in Basic and shown in Figure 9.54 enables the user to easily use the text-to-speech and voice recording features, and includes the following menu:

```
1. LEAVE MESSAGE
2. READ MESSAGES
3. LEAVE MESSAGE FOR LATER
4. DELETE MESSAGE
5. HELP
6. EXIT
```

For example, selecting the second item in the menu produces a display of available messages from disk and prompts the user to pick one. Selecting the fifth item in the menu, HELP, accesses a routine that gives the user spoken instructions about each menu selection.

```
10  REM ***********************************************************************
30  REM *               PC VOICE MAIL USING THE TI-SPEECH BOARD             *
60  REM ***********************************************************************
90  REM
100 CLS
110 COLOR 7,0
120 GOSUB 6000
130 LOCATE 2,33:PRINT"PC VOICE MAIL";
140 OPEN "TISPCH" FOR OUTPUT AS #1
150 PRINT #1,"RESET"
160 OPEN "TISPCH" FOR INPUT AS #2: INPUT #2,A$
170 PRINT #1,"VOLUME=8"
180 PRINT #1,"^[[14n"
190 PRINT #1,"SPEAKTEXT= THANK YOU FOR USING P C VOICE MAYLE ^[[c"
200 LOCATE 10,27:PRINT"WITH PC VOICE MAIL YOU ARE";
210 LOCATE 11,28:PRINT"ABLE TO LEAVE A MESSAGE";
220 LOCATE 12,26:PRINT"FOR RETRIVAL AT A LATER TIME.";
230 PRINT #1,"SPEAKTEXT= WITH P C VOICE MAYLE, YOU ARE ABLE TO LEAVE A MESSIGE,
240 CLOSE #1:CLOSE #2
250 REM FOR X=1 TO 10000
260 REM NEXT X
270 CLS
280 REM
290 REM *************************** MAIN MENUE ***************************
300 REM
310 CLS
320 COLOR 7,0
330 GOSUB 6000
340 LOCATE 2,34:PRINT"MAIN  MENU";
350 OPEN "TISPCH" FOR OUTPUT AS #1
360 PRINT #1,"RESET"
370 OPEN "TISPCH" FOR INPUT AS #2: INPUT #2,A$
380 PRINT #1,"VOLUME=8"
390 PRINT #1,"^[[14n"
400 PRINT #1,"SPEAKTEXT= MAIN  MEN YOU ^[[c"
410 LOCATE 10,28:PRINT"1. LEAVE MESSAGE";
420 LOCATE 12,28:PRINT"2. READ MESSAGES";
430 LOCATE 14,28:PRINT"3. LEAVE MESSAGE FOR LATER";
440 LOCATE 16,28:PRINT"4. DELETE MESSAGES"
450 LOCATE 18,28:PRINT"5. HELP"
460 LOCATE 20,28:PRINT"6. EXIT"
470 PRINT #1,"SPEAKTEXT= PLEASE SELECT AN OPTION ^[[c"
480 CLOSE #1:CLOSE #2
490 S$=INKEY$
500 V=VAL(S$)
510 IF V>6 OR V<1 GOTO 490
520 IF V=1 THEN GOTO 1000
530 IF V=2 THEN GOTO 2000
540 IF V=3 THEN GOTO 3000
550 IF V=4 GOTO 5000
560 IF V=5 GOTO 4000
570 IF V=6 THEN SYSTEM
580 GOTO 290
1000 REM
1010 REM ********************* LEAVE MESSAGE ROUTINE *********************
1020 REM
1030 CLS
1040 GOSUB 6000
1050 LOCATE 2,32:PRINT"LEAVE A MESSAGE";
1060 OPEN "TISPCH" FOR OUTPUT AS #1
1070 PRINT #1,"RESET"
1080 OPEN "TISPCH" FOR INPUT AS #2: INPUT #2,A$
1090 PRINT #1,"VOLUME=8"
1100 PRINT #1,"^[[14n"
1110 PRINT #1,"SPEAKTEXT= LEAVE A MESSAGE ^[[c"
1120 CLOSE #1:CLOSE #2
1130 LOCATE 10,17:PRINT"PLEASE ENTER THE TITLE OF YOUR MESSAGE"
1140 PRINT
1150 PRINT
1160 PRINT
1170 PRINT
1180 INPUT T$
1190 LOCATE 15,1:PRINT"
1200 LOCATE 10,12:PRINT"   AT THE -BEEP- BEGIN SPEAKING YOUR MESSAGE.
1210 LOCATE 11,20:PRINT"TO END YOUR MESSAGE HIT ANY KEY."
1220 LOCATE 12,23:PRINT"MAKE SURE YOUR MIKE IS ON!"
1230 PRINT
1240 PRINT
1250 PRINT"READY ?"
1260 J$=INKEY$
1270 IF J$="" GOTO 1260
1280 FOR X=1 TO 1000: NEXT X
1290 OPEN "TISPCH" FOR OUTPUT AS #3: PRINT #3, "RESET"
1300 PRINT #3,"RATE=2400"
1310 PRINT #3,"RECORD=";T$;".MSG"
```

FIGURE 9.54. Program listing for voice mail (VOICERUN)

```
1320 CLOSE #3
1330 GOTO 290
2000 REM
2010 REM ********************** READ MESSAGES ROUTINE *********************
2020 REM
2030 CLS
2040 GOSUB 6000
2050 LOCATE 2,33:PRINT"READ MESSAGES";
2060 OPEN "TISPCH" FOR OUTPUT AS #1
2070 PRINT #1,"RESET"
2080 OPEN "TISPCH" FOR INPUT AS #2: INPUT #2,A$
2090 PRINT #1,"VOLUME=8"
2100 PRINT #1,"ˆ[[14n"
2110 PRINT #1,"SPEAKTEXT= READ MESSAGES ˆ[[c"
2120 CLOSE #1:CLOSE #2
2130 LOCATE 10,17:PRINT"THE FOLLOWING MESSAGES HAVE BEEN LEFT."
2140 FOR X=1 TO 12000:NEXT X
2150 CLS
2160 GOSUB 6000
2170 LOCATE 2,30:PRINT"AVAILABLE MESSAGES";
2180 OPEN "TISPCH" FOR OUTPUT AS #1
2190 PRINT #1,"RESET"
2200 OPEN "TISPCH" FOR INPUT AS #2: INPUT #2,A$
2210 PRINT #1,"VOLUME=8"
2220 PRINT #1,"ˆ[[14n"
2230 PRINT #1,"SPEAKTEXT= AVAILABLE MESSAGES ˆ[[c"
2240 CLOSE #1:CLOSE #2
2250 PRINT
2260 PRINT
2270 PRINT
2280 FILES "*.MSG"
2290 PRINT
2300 PRINT
2310 PRINT"WOULD YOU LIKE TO HEAR A MESSAGE? (Y OR N)"
2320 PRINT
2330 PRINT
2340 A$=INKEY$: IF A$="N" OR A$="n" GOTO 290
2350 IF A$="Y" OR A$="y" GOTO 2370
2360 GOTO 2340
2370 PRINT"ENTER MESSAGE NAME INCLUDING EXTENSION"
2380 INPUT M$
2390 OPEN "TISPCH" FOR OUTPUT AS #3: PRINT #3, "RESET"
2400 PRINT #3,"SPEAKFILE=";M$
2410 CLOSE #3
2420 GOTO 2000
3000 REM
3010 REM **************** LEAVE MESSAGES FOR LATER ROUTINE ****************
3020 REM
3030 CLS: GOSUB 6000
3040 LOCATE 2,27:PRINT"LEAVE MESSAGE FOR LATER";
3050 OPEN "TISPCH" FOR OUTPUT AS #1
3060 PRINT #1,"RESET"
3070 OPEN "TISPCH" FOR INPUT AS #2: INPUT #2,A$
3080 PRINT #1,"VOLUME=8"
3090 PRINT #1,"ˆ[[14n"
3100 PRINT #1,"SPEAKTEXT= LEAVE MESSAGE FOR LATER ˆ[[c"
3110 CLOSE #1:CLOSE #2
3120 LOCATE 10,11:PRINT"PLEASE ENTER PRESENT (MILITARY) TIME (HH:MM:SS)"
3130 PRINT
3140 PRINT
3150 INPUT T$
3160 TIME$ = T$
3170 LOCATE 13,1:PRINT"                                                   "
3180 LOCATE 10,11:PRINT"     PLEASE ENTER THE TITLE OF YOUR MESSAGE        "
3190 PRINT
3200 PRINT
3210 INPUT M$
3220 LOCATE 13,1:PRINT"                                                   "
3230 LOCATE 10,11:PRINT" PLEASE ENTER TIME YOU WANT MESSAGE PLAYBACK       "
3240 LOCATE 11,15:PRINT" PLEASE USE MILITARY TIME (HH:MM:SS)    "
3250 PRINT
3260 PRINT
3270 INPUT P$
3280 LOCATE 14,1:PRINT"                                                   "
3290 LOCATE 10,12:PRINT"  AT THE -BEEP- BEGIN SPEAKING YOUR MESSAGE.       "
3300 LOCATE 11,17:PRINT"  TO END YOUR MESSAGE HIT ANY KEY."
3310 LOCATE 12,23:PRINT"MAKE SURE YOUR MIKE IS ON!"
3320 PRINT
3330 PRINT
3340 PRINT"READY ?"
3350 J$=INKEY$
3360 IF J$="" GOTO 3350
3370 FOR X=1 TO 1000: NEXT X
3380 OPEN "TISPCH" FOR OUTPUT AS #3: PRINT #3, "RESET"
```

FIGURE 9.54. (continued)

```
3390 PRINT #3,"RATE=2400"
3400 PRINT #3,"RECORD=";M$;".MSG"
3410 CLOSE #3
3420 CLS
3430 GOSUB 6000
3440 LOCATE 2,32:PRINT"DELAYED MESSAGE";
3450 OPEN "TISPCH" FOR OUTPUT AS #1
3460 PRINT #1,"RESET"
3470 OPEN "TISPCH" FOR INPUT AS #2: INPUT #2,A$
3480 PRINT #1,"VOLUME=8"
3490 PRINT #1,"^[[14n"
3500 PRINT #1,"SPEAKTEXT= DELAYED MESSAGE ^[[c"
3510 CLOSE #1:CLOSE #2
3520 LOCATE 10,15:PRINT"THE MESSAGE ";M$;" WILL BE PLAYED BACK AT ";P$
3530 LOCATE 14,23:PRINT"THE PRESENT TIME IS     ";TIME$
3540 IF P$ = TIME$ THEN 3550 ELSE 3530
3550 OPEN "TISPCH" FOR OUTPUT AS #3: PRINT #3, "RESET"
3560 PRINT #3,"SPEAKFILE=";M$;".MSG"
3570 CLOSE #3
3580 GOTO 280
4000 REM
4010 REM *********************** HELP ROUTINE ************************
4020 REM
4030 LOCATE 10,28:PRINT"                    ";
4040 LOCATE 12,28:PRINT"                    ";
4050 LOCATE 14,28:PRINT"                              ";
4060 LOCATE 16,28:PRINT"                              ";
4070 LOCATE 18,28:PRINT"                              ";
4080 LOCATE 20,28:PRINT"                              ";
4090 OPEN "TISPCH" FOR OUTPUT AS #1
4100 PRINT #1,"RESET"
4110 OPEN "TISPCH" FOR INPUT AS #2: INPUT #2,A$
4120 PRINT #1,"VOLUME=8"
4130 PRINT #1,"^[[14n"
4140 PRINT #1,"SPEAKTEXT= WITH P C VOICE MAYLE YOU HAVE THE FOLLOWING OPTIONS ^[
4150 LOCATE 10,28:PRINT"1. LEAVE MESSAGE";
4160 PRINT #1,"SPEAKTEXT= OPTION 1 IS USED TO LEAVE A MESSAGE ON THE DISK TO BE
4170 LOCATE 12,28:PRINT"2. READ MESSAGES";
4180 PRINT #1,"SPEAKTEXT= OPTION 2 IS USED TO RETREEV ANY MESSAGES LEFT ON THE D
4190 LOCATE 14,28:PRINT"3. LEAVE MESSAGE FOR LATER";
4200 PRINT #1,"SPEAKTEXT= OPTION 3 IS USED TO LEAVE A MESSAGE ON THE DISK TO BE
4210 LOCATE 16,28:PRINT"4. DELETE MESSAGES"
4220 PRINT #1,"SPEAKTEXT= OPTION 4 IS USED TO DELETE 1 OR ALL MESSAGES ^[[c"
4230 LOCATE 18,28:PRINT"5. HELP"
4240 PRINT #1,"SPEAKTEXT= OPTION 5 IS THE HELP ROUTINE, YOU ARE NOW HEARING ^[[c
4250 LOCATE 20,28:PRINT"6. EXIT"
4260 PRINT #1,"SPEAKTEXT= OPTION 6 IS USED TO EXIT THE PROGRAM ^[[c"
4270 PRINT #1,"SPEAKTEXT= PLEASE SELECT AN OPTION ^[[c"
4280 CLOSE #1:CLOSE #2
4290 GOTO 490
5000 REM
5010 REM ********************* DELETE MESSAGE ROUTINE ********************
5020 REM
5030 CLS: GOSUB 6000
5040 LOCATE 2,31:PRINT"DELETE MESSAGES";
5050 OPEN "TISPCH" FOR OUTPUT AS #1
5060 PRINT #1,"RESET"
5070 OPEN "TISPCH" FOR INPUT AS #2: INPUT #2,A$
5080 PRINT #1,"VOLUME=8"
5090 PRINT #1,"^[[14n"
5100 PRINT #1,"SPEAKTEXT= DELETE MESSAGES ^[[c"
5110 CLOSE #1:CLOSE #2
5120 PRINT
5130 PRINT
5140 PRINT
5150 FILES "*.MSG"
5160 PRINT
5170 PRINT
5180 PRINT"WOULD YOU LIKE TO DELETE ANY OR ALL MESSAGES? (Y OR N)"
5190 PRINT
5200 PRINT
5210 A$=INKEY$: IF A$="N" OR A$="n" GOTO 290
5220 IF A$="Y" OR A$="y" GOTO 5240
5230 GOTO 5210
5240 PRINT"INPUT MESSAGE NAME YOU WANT DELETED PLUS EXTENSION"
5250 PRINT"IF YOU WOULD LIKE ALL MESSAGES DELETED TYPE A"
5260 INPUT X$
5270 IF X$="A" OR X$="a" GOTO 5350
5280 PRINT
5290 PRINT"PRESS Y TO DELETE ";X$;" PRESS N TO ABORT"
5300 A$=INKEY$: IF A$="N" OR A$="n" GOTO 290
5310 IF A$="Y" OR A$="y" GOTO 5330
5320 GOTO 5300
5330 KILL X$
5340 GOTO 290
```

FIGURE 9.54. (continued)

```
5350 PRINT
5360 PRINT"PRESS Y TO DELETE ALL MESSAGES, N TO ABORT"
5370 A$=INKEY$: IF A$="N" OR A$="n" GOTO 290
5380 IF A$="Y" OR A$="y" GOTO 5400
5390 GOTO 5370
5400 KILL "*.MSG"
5410 GOTO 290
6000 REM
6010 REM ********************** TOP OF SCREEN ROUTINE **********************
6020 REM
6030 FOR X=2 TO 79
6040 LOCATE 1,X:PRINT CHR$(205);
6050 NEXT X
6060 FOR X=2 TO 79
6070 LOCATE 3,X:PRINT CHR$(205);
6080 NEXT X
6090 RETURN
```

FIGURE 9.54. (concluded)

The program provides an interface between the board and the user. The menu selections are translated into commands sent to the board via the print #1 function. All the functions were tested and worked as expected. Many applications can make use of Texas Instruments' Speech Development System; telephone directories, for instance (see Sections 9.19, 9.20, and 9.21).

9.19 SPEAKER IDENTIFICATION

Problem Statement

The goal of this project is to obtain a system capable of identifying a speaker. This project uses the Texas Instruments speech board as did the preceding project.

Design Considerations

The speech board is programmed so that, upon being identified, a speaker is able to use certain voice commands within a specified vocabulary. This vocabulary is a list of words/phrases available to the speaker. This project allows entry of two users each with a vocabulary. When access is granted, one of the two vocabularies is set up, depending on which user is being acknowledged. The two different vocabularies are chosen to demonstrate two specific applications. One (DIRECTOR) is a voice-controlled telephone directory and the other (SPEAKDOS) is a voice-controlled DOS. A common vocabulary to both speakers is the REQUEST vocabulary, which contains templates that differentiate between the two users and acts based on the result.

Results and Conclusions

To initiate the system, one of the two users says "HELLO," and either the message, "Welcome David, you may use telephone directory", or the message, "Welcome

Peter, you may use DOS", appears. The user is next prompted to state "Request Entry". After this phrase, the decision of who has spoken is made from comparison of stored templates. In this project, only two user-voice patterns are stored. For example, after accessing the SPEAKDOS vocabulary, the user can execute commands such as: TIME, LIST FILES, CHANGE DIR, etc. Both user-specific vocabularies include the phrase "GOOD BYE" to execute a batch file which reinitiates the REQUEST vocabulary. The functionality of the system with two users can be expanded to several users, each with a specific vocabulary.

9.20 VOICE CONTROL OF A RHINO ROBOT

This project, which can be performed in approximately a month, requires special hardware: a Rhino XR-2 Robot, manufactured by Sandhu Machine Design Inc., and a TMS32010-based Speech Development System (part no. 2245198) made by Texas Instruments. The Speech Development System includes a board that can be plugged into an IBM (or compatible) PC and software (see Section 9.18).

Problem Statement

The objective of this project is to control a robot by voice using the Speech Development System and the IBM PC, which interfaces to the robot via a serial line.

Design Considerations

Through an RS-232 serial interface, each individual motor of the Rhino robot can be accessed and controlled. The Speech Development System includes routines for voice recognition. Since the system is voice dependent, "voice training" is first accomplished to store the desired commands.

Results and Conclusions

A total of 12 commands, such as MOVE ARM UP, ROTATE WAIST LEFT, and STOP, are used, each of which has an equivalent code that is transmitted to the Rhino robot. Each code selects a specific motor in the robot and causes it to rotate appropriately. For example, speaking into a microphone connected to the corresponding to the right hand, to rotate. As a motor rotates, a program section is included to display continuously, on the monitor of the PC, a measure of the amount of such rotation. The space bar on the PC can be used to STOP the robot as a safety measure, in case the command STOP is not recognized. Figure 9.55 shows a listing of the program that includes the codes transmitted to the robot, which correspond to each command.

```
120 '  |   -----------------------------------------------
130 '  |   Program to use voice recognition to control
140 '  |   the Rhino XR-2 Robot.
150 '  |   -----------------------------------------------
200 CLS
210 SCREEN 0
220 KEY OFF
230 DIM STACK$(1000),CMD$(30),VOICE$(30)
240 FOR X=1 TO 19:READ CMD$(X),VOICE$(X):NEXT X
250 'OPEN "TISPCH" FOR OUTPUT AS #1
252 'PRINT #1,"RESET"
254 'OPEN "tispch" FOR INPUT AS #2
256 'INPUT #2,A$
258 IF A$<>" " THEN PRINT A$
260 'PRINT #1,"speaktext=OH-KE"
262 ED$="^[[C"
264 'PRINT #1,"^[[14N";ED$
268 'PRINT #1,"^[[#240"
270 OPEN "COM1:9600,E,7,1,RS,CS,DS,CD" AS #3
300 ' ------------------------
310 '      Draw Main Screen
320 ' ------------------------
330 CLS
340 COLOR 7,0
350 FOR X=2 TO 79
360 LOCATE 1,X:PRINT CHR$(205);
370 LOCATE 24,X:PRINT CHR$(205);
380 LOCATE 3,X:PRINT CHR$(196);
390 LOCATE 22,X:PRINT CHR$(196);
400 NEXT X
410 LOCATE 1,1:PRINT CHR$(201):LOCATE 1,80:PRINT CHR$(187);
420 LOCATE 24,1:PRINT CHR$(200)::LOCATE 24,80:PRINT CHR$(188);
430 COLOR 15,0
440 LOCATE 2,26:PRINT "Speech Recognition - Robot Control"
450 COLOR 7,0
460 FOR X=2 TO 23
470 LOCATE X,1:PRINT CHR$(186);
480 LOCATE X,80:PRINT CHR$(186);
490 IF X>3 AND X<22 THEN LOCATE X,39:PRINT CHR$(179);
500 NEXT X
510 LOCATE 3,39:PRINT CHR$(194);
520 LOCATE 22,39:PRINT CHR$(193);
530 COLOR 15,0
540 LOCATE 4,12:PRINT "Motor Positions"
550 COLOR 7,0
560 FOR X=2 TO 38
570 LOCATE 5,X:PRINT CHR$(196);
580 NEXT X
590 COLOR 15,0
600 LOCATE 4,50:PRINT "Speech Commands"
610 COLOR 7,0
620 FOR X=40 TO 79
630 LOCATE 5,X:PRINT CHR$(196);
640 NEXT X
700 ' --------------------
710 ' Display Valid Data
720 ' --------------------
730 FOR X=1 TO 16 STEP 2
740 LOCATE X+5,41:PRINT CMD$(X):LOCATE X+5,61:PRINT CMD$(X+1)
750 NEXT X
800 ' --------------------
810 ' Display Motor Positions
820 ' --------------------
830 COLOR 7,0
840 LOCATE 7,8:PRINT "Wrist "
850 LOCATE 9,8:PRINT "Arm"
860 LOCATE 11,8:PRINT "Shoulder"
870 LOCATE 13,8:PRINT "Grasp"
880 LOCATE 15,8:PRINT "Hand"
890 LOCATE 17,8:PRINT "Waist"
900 COLOR 7,0
910 FOR X=1 TO 6
920 LOCATE X*2+5,20:PRINT MOT(X)
930 NEXT X
1000 ' --------------------
1010 ' Start Recognition
1020 ' --------------------
1025 LOCATE 23,5:COLOR 0,7:PRINT SPACE$(30);"Enter Command";SPACE$(30):COLOR 7,0
1026 'PRINT #1,"speaktext=Please Enter Command";ED$
```

FIGURE 9.55. Program listing for control of Rhino robot (ROBOT)

```
1028 A1$="":A2$=""
1030 A1$=INKEY$
1040 IF A1$="" THEN 1030
1050 A2$=INKEY$
1060 IF A2$="" THEN 1050
1070 V=VAL(A1$)*10+VAL(A2$)
1080 IF V<1 OR V>22 THEN ER=1:GOTO 5000
1090 COUNT=0
1100 IF V<=12 THEN 2000
1110 IF V=13 THEN 4000
1120 GOTO 1000
2000 ' ---------------
2010 ' Move Robot Part
2020 ' ---------------
2030 ON V GOSUB 2100,2110,2120,2130,2140,2150,2160,2170,2180,2190,2200,2210
2035 COLOR 0,7:IF INT(V/2)=V/2 THEN LOCATE V+4,61 ELSE LOCATE V+5,41
2036 PRINT CMD$(V):COLOR 7,0
2037 'PRINT #1,"SPEAKTEXT=";VOICE$(V);ED$
2040 MOT(VV)=MOT(VV)+D:LOCATE VV*2+5,20:PRINT MOT(VV)
2045 'IF MOT(VV)<=MIN OR MOT(VV)=>MAX THEN 2065
2046 COUNT=COUNT+1
2050 PRINT #3,MOT$+STR$(D)
2060 A1$=INKEY$:IF A1$="" THEN 2040
2064 A2$=""
2065 COLOR 7,0:IF INT(V/2)=V/2 THEN LOCATE V+4,61 ELSE LOCATE V+5,41
2066 PRINT CMD$(V):COLOR 7,0
2068 'IF MOT(VV)<=MIN THEN MOT(VV)=MIN
2069 'IF MOT(VV)>=MAX THEN MOT(VV)=MAX
2070 IF VAL(A1$)<1 OR VAL(A1$)>9 THEN 1000 ELSE 1050
2100 MOT$="C+":VV=1:D=5:MIN=-100:MAX=100:RETURN          ' Wrist
2110 MOT$="C-":VV=1:D=-5:MIN=-100:MAX=100:RETURN
2120 MOT$="D+":VV=2:D=5:MIN=-100:MAX=100:RETURN          ' Arm
2130 MOT$="D-":VV=2:D=-5:MIN=-100:MAX=100:RETURN
2140 MOT$="E-":VV=3:D=-5:MIN=-150:MAX=150:RETURN          ' Shoulder
2150 MOT$="E+":VV=3:D=5:MIN=-150:MAX=150:RETURN
2160 MOT$="A+":VV=4:D=-5:MIN=-40:MAX=40:RETURN            ' Grasp
2170 MOT$="A-":VV=4:D=5:MIN=-40:MAX=40:RETURN
2180 MOT$="B+":VV=5:D=-5:MIN=-100:MAX=100:RETURN          ' Hand
2190 MOT$="B-":VV=5:D=-5:MIN=-100:MAX=100:RETURN
2200 MOT$="F+":VV=6:D=5:MIN=-100:MAX=100:RETURN          ' Waist
2210 MOT$="F-":VV=6:D=-5:MIN=-100:MAX=100:RETURN
4000 ' ---------------
4010 ' Stop Motors
4020 ' ---------------
4030 COLOR 0,7:LOCATE 18,41:PRINT CMD$(13):COLOR 7,0
4050 FOR M=65 TO 70
4060 PRINT #3,CHR$(M)+"X"
4070 NEXT M
4080 COLOR 7,0:LOCATE 18,41:PRINT CMD$(13):COLOR 7,0
4090 GOTO 1000
5000 ' ------------------
5010 ' Error Subroutine
5020 ' ------------------
5030 COLOR 0,15:BEEP
5040 IF ER=1 THEN LOCATE 23,5:PRINT SPACE$(25);"Invalid Command Entered";SPACE$(
5050 COLOR 0,7
6000 FOR Z=1 TO 1000:NEXT Z
6040 COLOR 7,0
6050 ON ER GOTO 1000
10000 DATA Move Wrist Up,Moveing Wrist Up
10010 DATA Move Wrist Down,Moveing Wrist Down
10020 DATA Move Arm Up,Moveing Arm Up
10030 DATA Move Arm Down,Moveing Arm Down
10040 DATA Move Shoulder Up,Moveing Shoulder Up
10050 DATA Move Shoulder Down,Moveing Shoulder Down
10060 DATA Open Hand,Opening Hand
10070 DATA Close Hand,Closeing Hand
10080 DATA Rotate Hand Left,Rotateing Hand Left
10090 DATA Rotate Hand Right,Rotateing Hand Right
10100 DATA Rotate Waist Left,Rotateing Waist Left
10110 DATA Rotate Waist Right,Rotateing Waist Right
10120 DATA Stop,Stop
10130 DATA Reset,Reset
10140 DATA Start Programming,Start Programming
10150 DATA Load Program,Load Program
10160 DATA Stop Programming,Stop Programming
10170 DATA Enter Command,Enter Command
10180 DATA Invalid Command Entered,Invalid Command Command Entered
```

FIGURE 9.55. (concluded)

9.21 VOICE CONTROL OF A RHINO ROBOT THROUGH FIBER OPTICS

Problem Statement

The goal of this project is to control the Rhino robot, through a fiber optics medium, using the Texas Instruments speech board. The preceding project in Section 9.20 also makes use of the speech board to control the Rhino robot.

Design Considerations

Fiber optics communication [35,36] offers several advantages over metallic systems. For example,

1. Signals transmitted through fiber optics are not destroyed by outside electronic, magnetic, or radio frequency interference.
2. Optical systems are less susceptible to lightning or high-voltage interference.
3. Optical fibers do not emit radiation.
4. Since optical signals do not require grounding connections, the transmitter and receiver are isolated electrically and are free from ground loop problems.
5. With safety from sparking and shock, fiber optics can be the choice for many processing applications where a safe operation in a hazardous or flammable environment is needed.
6. Digital computing, telephone, and video broadcast systems require new avenues for improved transmission; the high signal bandwidth of optical fibers means increased channel capacity.
7. Longer cable runs require fewer repeaters because fiber optic cables have very low attentuation rates, which ideally suits them for broadcast and telecommunications use.

Fiber Optic Link

The system block diagram, shown in Figure 9.56, consists of an optical transmitter and receiver connected by a length of optical cable in a point-to-point link. The optical transmitter converts an electronic signal voltage into optical power, and into the fiber by a light emitting diode (LED), laser diode (LD), or laser. At the photodetector point, either a positive-intrinsic-negative (PIN), or an avalanche photodiode (APD) captures the lightwave pulses for conversion back to electrical current.

The transmitter circuit is shown in Figure 9.57. The microphone input is approx-

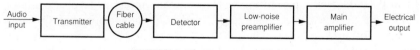

FIGURE 9.56. Fiber optics system

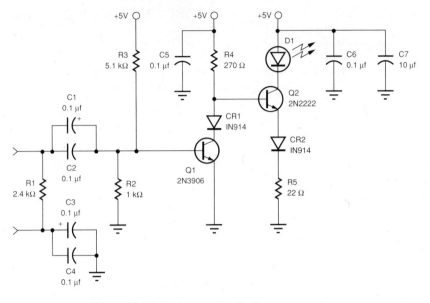

FIGURE 9.57. Audio transmitter for fiber optics system

imately 2 mV peak to peak across R1. C3 (electrolytic) and C4 act as an audio ground, while C1 (electrolytic) and C2 couple the audio signal to the base of Q1, which is reversed-biased at approximately 0.8 Vdc. Q2 is forward-biased at approximately 5 Vdc by the action of R4 and CR1. C5 is used as a bypass capacitor. The input applied to the base of Q1 makes it conduct which causes the emitter voltage to change. This change is felt at the base of Q2, causing it in turn to conduct. The collector current ($\approx$ 20 mA) is applied to the cathode of D1 (LED Transmitter). The LED is forward-biased and always on (lighted). The light is modulated at the rate of change of the collector current of Q2. CR2 and R5 are used to limit the collector current from Q2.

The LED is a Gallium Aluminum Arsenide (GaAlAs) semiconductor, with data rates from dc to 85 MHz. It operates in the 850 nm range, which is infrared towards the red end of the spectrum. The cable attenuation is 3 dB/Km at 850 nm, with bandwidth from dc to 100 MHz at 850 nm.

The photodiode in the receiver circuit (see Figure 9.58) is reversed-biased, or in cutoff. When modulated light is detected at the photodiode, the power output of the photodiode is approximately 10–20 μW at approximately 450 nA. When the photodiode conducts, the base of Q2 goes more positive and conducts through the voltage limiter formed by VR1 and R2, allowing Q1 to conduct also. Q3 in turn goes into conduction through the voltage limiter formed by VR2 and R6. C4 and R7 pass back a percentage of Q3's emitter current as feedback to insure Q2 does not get overdriven. The output of Q3, in conjunction with Q4, forms an emitter–follower circuit, which is the final current amplification stage, with the output current being approximately 1 μA. The modulated signal is capacitively coupled to the potentiometer R11, used to control the gain of operational amplifiers U1,

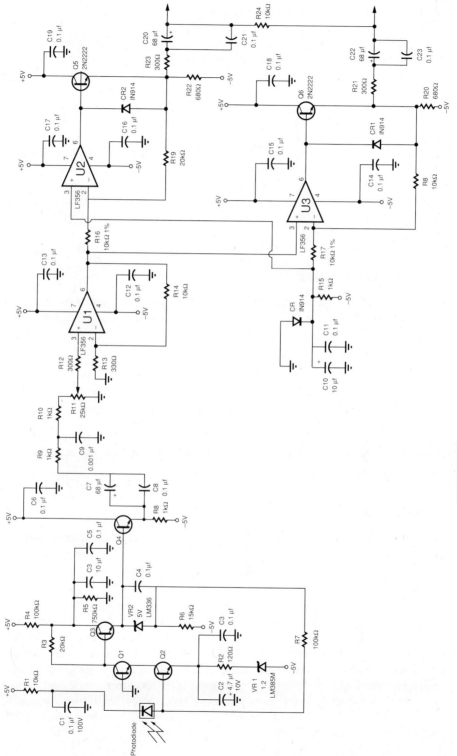

FIGURE 9.58. Audio receiver for fiber optics system

293

U2, and U3. U1, U2, and U3 form a current-to-voltage converter. The cross connects between U2 and U3 form a common noise limiter circuit. Q5 and Q6 have a gain of approximately 20 giving the final output of the circuit a range from 0 to 1 Vdc.

The photodiode at the receiver in Figure 9.58 is a PIN, a semi-conductor detector with an intrinsic region separating the p-doped and n-doped regions. This design gives fast linear response and is widely used in fiber optic receivers. The output of the receiver is interfaced with the speech board. The voice commands are through the microphone via the fiber optic link. The Rhino robot is connected to a PC via an RS232 serial port, through which each of the robot's motors can be driven. The speech development package includes the software necessary to train the robot. Since the system is voice dependent, it is trained and controlled by the same individual.

Results and Conclusions

Commands such as MOVE ARM UP, OPEN HAND, ROTATE WAIST RIGHT, and STOP are used, as in the preceding project. The results indicate that the fiber optic transmitter and receiver are operating successfully within the speech board specifications.

9.22 GRAPHIC EQUALIZER

A graphic equalizer is used in the audio field. Commonly found features such as bass, treble, and multispeaker designs and amplifiers contribute to one's listening pleasure.

Problem Statement

The objective of this project is to design a multiband filter to implement a digital graphic equalizer.

Design Considerations

An equalizer can be used to attenuate the gain throughout the bandwidth. This project uses the TMS320 to implement a graphic equalizer with a single multiband FIR filter using a Kaiser window. The multiband functional block diagram is shown in Figure 9.59 with the range of filtering shown. A program written in Turbo Pascal adjusts the gain of each band filter for either +6, 0, or –6 dB, yielding a total of 27 possible configurations. The program includes a communication package section to select one of 27 object files which correspond to the 27 possible configurations.

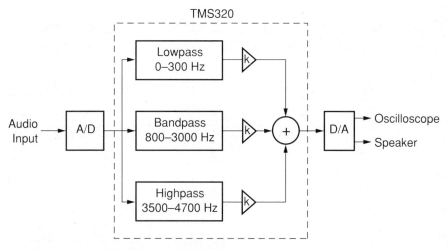

FIGURE 9.59. Multiband functional block diagram

Results and Conclusions

When the program is run, the user selects the gain (+6, 0, or –6 dB) for each of the three filter bands, to accentuate a specific range of frequencies. Downloading and execution of the appropriate (1 out of 27) object file is performed automatically based on the gain selection of the filters. While the object file is being downloaded, a graphic representation of the desired multiband filter is displayed to show the gain levels selected. With music as an input to the AIB, one can hear at the output a specific range of frequencies accentuated, which represent the bass, treble, or midrange frequencies.

9.23 HARMONIC SCANNER

Problem Statement

Design a system that will show which of the four frequencies (1, 2, 3, or 4 kHz) is present in a signal.

Design Considerations and Results

This design problem is solved by a simple harmonic scanner having four bandpass filters executed sequentially. The output of each of the filters is also displayed sequentially. The display of the output is broken into four time slices, one for each filter output. With an input sine wave of one of the frequencies (1, 2, 3, or 4 kHz), only the filter corresponding to that frequency will produce an output. By looking at the displayed regions, one can tell how many of the frequencies

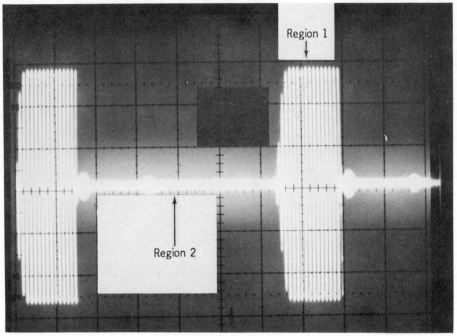

FIGURE 9.60. Harmonic scanner

are present; a casual glance at the relative number of zero crossings will indicate which frequency belongs to which region.

The harmonic scanner consists of a series of four 41-coefficient FIR bandpass filters, with Hamming windows, running sequentially in a continuous loop. An output of either 1, 2, 3, or 4 kHz (each $\pm$ 50 Hz) is obtained. A counter is set so that a set number of samples (160) was taken for each filter. The output, from an oscilloscope, is a series of cycles of the input frequency (region 1), seen for only one-fourth of the time, as shown in Figure 9.60. The "off" time (region 2) is a result of each of the other three filters blocking the original input frequency.

9.24 ADAPTIVE DIFFERENTIAL PULSE-CODE MODULATION

Problem Statement

A kit available from Texas Instruments includes an extensive adaptive differential pulse-code modulation (ADPCM) program on two PROMs, as well as associated

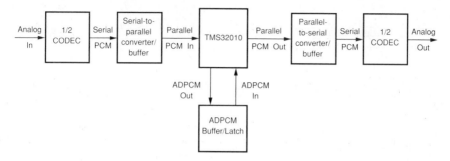

FIGURE 9.61. ADPCM block diagram (Courtesy of Texas Instruments Inc.)

support circuitry and documentation [1]. The 32-kbps ADPCM block diagram shown in Figure 9.61 includes the first-generation digital signal processor, TMS32010, and a combo-codec filter, TCM2916 [17].

Implementation and Testing

The system shown in Figure 9.61 can easily be breadboarded using chips and components readily available. Figures 9.62 through 9.65 show the interface, support, and clock circuits. Many chips and components are the same as for the real-time FFT system discussed in Section 9.8.

The ADPCM scheme reduces by half the bit transmission requirements of pulse-code-modulated (PCM) data while still preserving the quality of the signal. An analog input signal shown in Figure 9.61 on the far left (such as speech) is converted into digital form by the codec and compressed into 8-bit PCM data. The PCM MU-LAW method [1] in the program is used to compress the signal into a logarithmic scale when coding for transmission, which is simulated in this project. These data are linearized into 13-bit data and converted back into 4-bit ADPCM data for transmission. In this specific project, at the receiver end, these data are fed directly back into the TMS32010, decoded to 13-bit data and converted back into 8-bit PCM data. ADPCM transcodes 8-bit PCM into 4-bit ADPCM data by taking advantage of high sample-to-sample correlation of speech waveforms, with the step size of the coder adapted as the signal changes.

Conclusion

The 32-kbps ADPCM scheme provides twice the channel capacity of a 64-kbps PCM method of transmission. A 32-kbps ADPCM Transcoder (TMS320SA32) chip is now available from Texas Instruments.

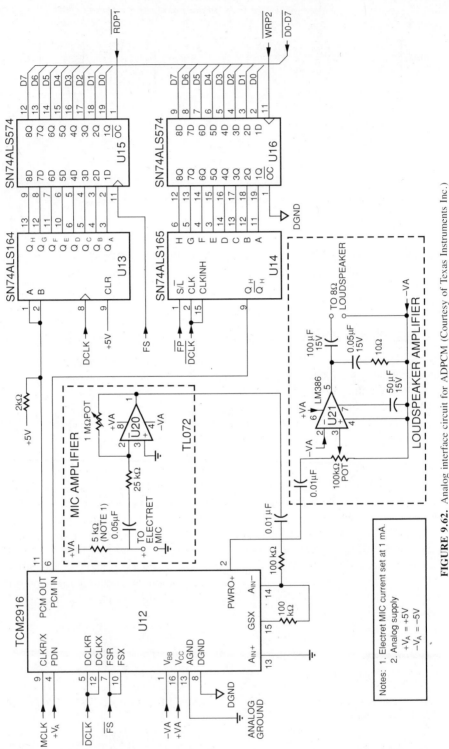

FIGURE 9.62. Analog interface circuit for ADPCM (Courtesy of Texas Instruments Inc.)

298

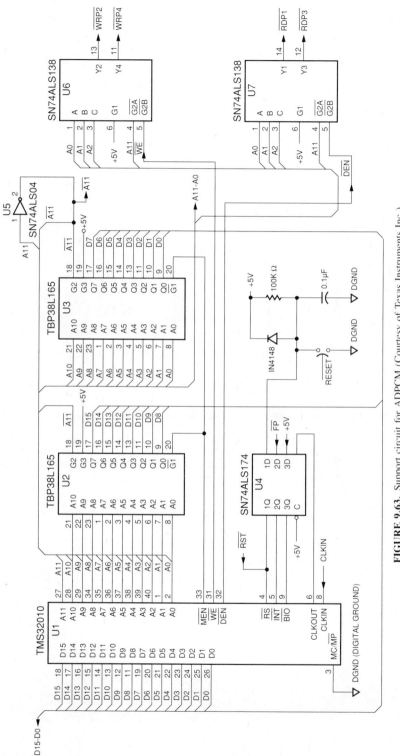

FIGURE 9.63. Support circuit for ADPCM (Courtesy of Texas Instruments Inc.)

299

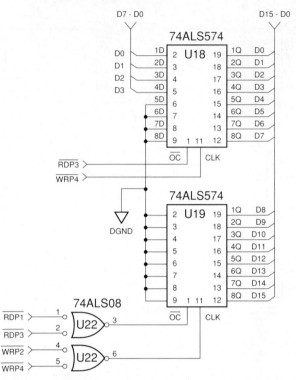

FIGURE 9.64. ADPCM interface circuit (Courtesy of Texas Instruments Inc.)

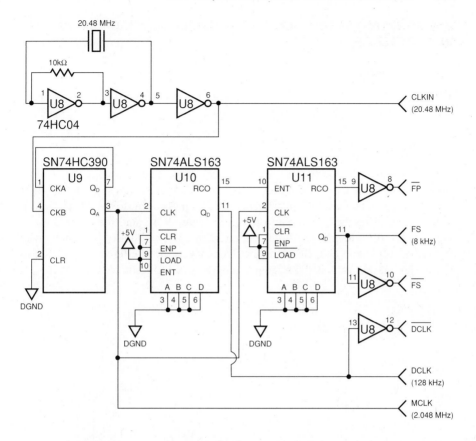

FIGURE 9.65. Clock circuit for ADPCM (Courtesy of Texas Instruments Inc.)

9.25 ADAPTIVE NOTCH FILTER FOR CANCELLATION OF SINUSOIDAL INTERFERENCE

Problem Statement

The goal of this project is to design a real-time adaptive notch filter which cancels a single-frequency sinusoidal interference noise.

Design Consideration

Figure 9.66 shows a block diagram of the two-weight linear combiner for single-frequency interference cancellation. The linear combiner forms the notch filter capable of providing a 37-dB interference reduction. The sinusoidal noise is added to the signal using the conditioning circuit in Figure 9.67, and forms the primary input (IN) to the AIC module discussed in Chapter 3. Two sinusoidal signals, with 90° phase difference, and with the same frequency as the interference noise (added to the signal), are used as the reference input (AUX IN) to the AIC module.

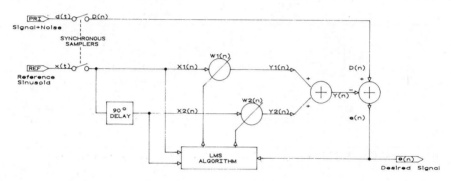

FIGURE 9.66. Two-weight adaptive linear combiner for single-frequency sinusoidal interference cancellation

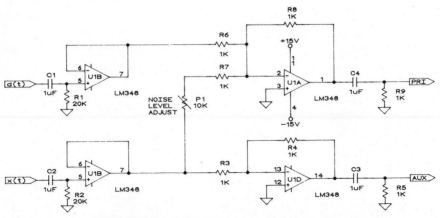

FIGURE 9.67. Signal conditioning circuit for notch filter

The conditioning circuit board is designed to provide isolation of the two signal generators, the elimination of dc offsets, the addition of the signal and noise, and a potentiometer P1 to adjust the level of the noise. A two-point differentiation of the reference input $x_1(n)$ is used to obtain $x_2(n)$, shifted by 90°, and shown in Figure 9.66. Both $x_1(n)$ and $x_2(n)$ become the inputs to the adaptive notch filter. The output of the filter is $y(n)$, where

$$y_k(n) = \sum_{k=1}^{2} w_k(n)x_k(n), \text{ and}$$

the weight w_k is adapted using the LMS criterion,

$$w_k(n + 1) = w_k(n) + \beta e(n)x_k(n)$$

The output signal $e(n)$, which is the difference between the signal plus noise $D(n)$ and the filter's output $Y(n)$ provides the desired signal. The LMS convergence gain β is chosen large enough to produce an optimum convergence time without affecting the stability of the system. Background materials and experiments/projects in adaptive filtering are covered in Chapter 8.

Results and Conclusions

Figure 9.68(a) shows a plot of a sine and cosine using a Basic simulation of the two-point differentiator, and Figure 9.68(b) shows a Basic simulation of the output of the two-weight notch filter. Figure 9.68 shows that Dn (at D(n) in Figure 9.66) contains the signal and sinusoidal noise, and the output displays the convergence toward the desired signal. Figure 9.69 contains a complete listing of the adaptive notch filter program, which includes the two-point differentiator section for producing the 90° phase difference between the two inputs.

The macros (INIT, INN, OUTS) and the subroutine (INITS) for communication between the AIC and the TMS320C25, used in the program, are discussed in Appendix C. Figure 9.70(a) shows a plot of a 1 kHz signal with a 335 Hz sinusoidal interference, and Figure 9.70(b) shows the noise reduced by $\approx$ 37 dB.

Adaptive Notch Variation

This variation of the preceding adaptive notch filter also uses the algorithm described in Figure 9.66. Instead of using an external source to generate the sinusoidal references $x_1(n)$ and $x_2(n)$, they are generated in software. This method is useful for situations where a second input channel, or the sinusoidal source, is not available.

The sine wave and the 90° phase shifted sine wave are generated using a method similar to the oscillator example (see Example 4.4). In the final analysis, the only difference between the cosine wave generator and the sine wave generator is that

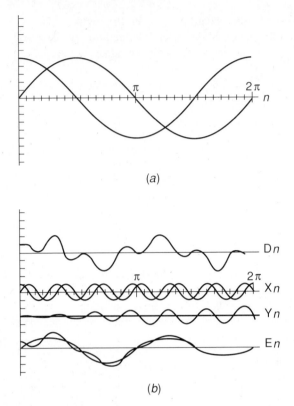

(a)

(b)

FIGURE 9.68. Output from simulation in Basic: (a) two-point differentiator; (b) two-weight adaptive notch filter

```
0001                        IDT        'NOTCH'
0002             * SINGLE-FREQUENCY ADAPTIVE NOISE CANCELLER
0003             * PRI INPUT IS DESIRED SIGNAL + NOISE (Dn)
0004             * AUX INPUT IS REFERENCE SIGNAL (Xn),OUTPUT ERROR SIGNAL (En)
0005 0000                   MLIB       'C:'
0006             *
0007      0073   CSR        EQU        >73         SELECT AUX. INPUT,SYNC MODE
0008      244A   SRATE      EQU        >244A       SAMPLE RATE @ 16kHz
0009      2448   FLTCLK     EQU        >2448       SWITCHED CAPACITOR CLK RATE⁻=288 KHZ
0010      0073   AUX        EQU        >73         SELECT AUX INPUT=X1N
0011      0063   PRI        EQU        >63         SELECT PRIMARY INPUT=DN
0012      0C00   MU         EQU        >0C00       ERROR WEIGHTING FACTOR
0013      2000   FREQ       EQU        >2000       DIFFERENTIATOR CONSTANT @ 350Hz
0014 0000                   DSEG
0015             *
0016             * STORAGE CONSTANTS
0017             *
0018 0000        TEMP       BSS 1                  TEMPORARY REGISTER
0019 0001        WT         BSS 1                  WT = fs/(2 pi fo)
0020 0002        BETA       BSS 1                  CONVERGENCE GAIN
0021 0003        W1N        BSS 1                  CURRENT WEIGHT ONE
0022 0004        W2N        BSS 1                  CURRENT WEIGHT TWO
0023 0005        DN         BSS 1                  DESIRED SIGNAL
0024 0006        YN         BSS 1                  INTERFERENCE SIGNAL
0025 0007        X1N        BSS 1                  CURRENT SAMPLE X(N)
0026 0008        X1NN       BSS 1                  PREVIOUS SAMPLE X(N-1)
0027 0009        X2N        BSS 1                  CURRENT PHASE SHIFTED SAMPLE
0028 000A        X2NN       BSS 1                  PREVIOUS PHSE SHIFTED SAMPLE
0029 000B        EN         BSS 1                  CURRENT ERROR SIGNAL
```

FIGURE 9.69. Program listing for adaptive notch filter (NOTCH.LST)

```
0030                      *
0031 000C                           DEND
0032 0000                           AORG    0              INITIALIZE AIC AND THE
0033 0000 FF80                      B       START          MAIN PROGRAM PARAMETERS
     0001 0000'
0034 0002
0035 0000                           PSEG
0036                      *          INITIALIZE PROCESSOR AND AIC
0037               START    INIT CSR,SRATE,FLTCLK,TABLE
0001                        REF  INITS,DTAB,STATUS
0002          0001 DXR      EQU     >1
0003          0000 DRR      EQU     >0
0004 0000 FF80             B       CODE
     0001 0009'
0005 0002 0000 TABLE       DATA    0,03,FLTCLK,3,SRATE,3,CSR
     0003 0003
     0004 2448
     0005 0003
     0006 244A
     0007 0003
     0008 0073
0006 0009 7800 CODE        SST     STATUS         SAVE DATA PAGE POINTER
0007 000A C800             LDPK    0              POINT TO PAGE 0
0008 000B C000             LARK    AR0,DTAB       SET UP DATA POINTER
0009 000C CB06             RPTK    >06
0010 000D FCA0             BLKP    TABLE,*+
     00DE 0002'
0011 000F FE80             CALL INITS             INIT PROCESSOR AND AIC
     0010 0000
0012 0011 5000             LST     STATUS
0038 0012 C800             LDPK    0              INIT DATA PAGE POINTER TO 0
0039                      *
0040 0013 D001             LALK    FREQ
     0014 2000
0041 0015 6001"            SACL    WT             TRANSFER DIFFERENTIATOR CONSTANT
0042                      *
0043 0016 D001             LALK    MU
     0017 0C00
0044 0018 6002"            SACL    BETA           TRANSFER CONVERGENCE GAIN
0045                      *
0046                      * INITIALIZE VARIABLES
0047                      *
0048 0019 5588             LARP    AR0
0049 001A D000             LRLK    AR0,W1N
     001B 0003"
0050 001C CA00             ZAC
0051 001D CB09             RPTK    >9
0052 001E 60A0             SACL    *+             ZERO FILL >63 TO >6C
0053 001F
0054 001F CE07             SSXM                   SET SIGN EXTENSION MODE
0055                      *
0056                      * GET INPUT SIGNAL X(n)
0057                      *
0058 0020 CE1F NEXT        IDLE                   WAIT FOR XMIT OF DATA TO THE AIC
0059              INN       X1N            INPUT REF SAMPLE FROM AUX. IN
0001              REF STATUS
0002          0000 DRR      EQU     0
0003 0021 7800             SST     STATUS         SAVE DATA PAGE POINTER
0004 0022 C800             LDPK    0              DATA PAGE POINTER TO PAGE 0
0005 0023 2000             LAC     DRR            INPUT A VALUE
0006 0024 6007"            SACL    X1N            STORE IT
0007 0025 5000             LST     STATUS         RESTORE DATA PAGE POINTER
0060              OUTS      EN,PRI  OUTPUT AND SWITCH TO PRI INPUT
0001              REF STATUS
0002          0001 DXR      EQU     1
0003 0026 7800             SST     STATUS         SAVE DATA PAGE POINTER
0004 0027 C800             LDPK    0              POINT TO PAGE 0
0005 0028 200B"            LAC     EN             LOAD OUTPUT VALUE
0006 0029 D005             ORK     >03            SET UP FOR SEC COMM
     002A 0003
0007 002B 6001             SACL    DXR            OUTPUT DATA + SEC CODE (11)
0008 002C CA63             LACK    PRI            LOAD THE SWITCH CODE
0009 002D CE1F             IDLE                   WAIT UNTIL READY
0010 002E 6001             SACL    DXR            OUTPUT IT
0011 002F CB10             RPTK    >10            DELAY 17 CLK CYCLES
0012 0030 5500             NOP
0013 0031 200B"            LAC     EN             LOAD OUTPUT VALUE AGAIN
0014 0032 D004             ANDK    >FFFC          ZERO LOW 2 BITS
     0033 FFFC
0015 0034 6001             SACL    DXR            OUTPUT IT
0016 0035 5000             LST     STATUS         RESTORE THE PAGE POINTER
0061 0036
0062                      *
0063                      * GENERATE 90 DEGREE PHASE SHIFTED REF SIGNAL
0064                      *
```

FIGURE 9.69. (continued)

```
0065 0036 2007"  DIFF    LAC    X1N           PUT X(N) IN ACCUM
0066 0037 1008"          SUB    X1NN          SUBTRACT: X(N)-X(N-1)
0067 0038 6009"          SACL   X2N           TEMPORARY RESULT
0068                 *
0069 0039 3C09"          LT     X2N           SAMPLE FREQ TO T-REG
0070 003A 3801"          MPY    WT            MULTIPLY: WT*[X(N)-X(N-1)]
0071 003B CE14           PAC                  TRANSFER PROD TO ACCUM
0072 003C DE02           ADLK   1,14          ROUND RESULT
     003D 0001
0073 003E 6D09"          SACH   X2N,5         FINAL RESULT TO X2(N)
0074                 *
0075                 *
0076                 * CALCULATE ERROR  E(n) = D(n) - Y(n)
0077                 *
0078 003F 2005"          LAC    DN            GET SIGNAL + "NOISE"
0079 0040 1006"          SUB    YN            SUBTRACT "NOISE"
0080 0041 DE02           ADLK   1,14          ROUND RESULT
     0042 0001
0081 0043 600B"          SACL   EN            EN = DESIRED OUTPUT
0082                 *
0083                 * CALCULATE ERROR x CONVERGENCE GAIN
0084                 *
0085 0044 CA00           ZAC
0086 0045 3C02"          LT     BETA          CONVERGENCE GAIN
0087 0046 380B"          MPY    EN            TEMP=BETA x E(n)
0088 0047 CE14           PAC
0089 0048 DE02           ADLK   1,14          ROUND RESULT
     0049 0001
0090 004A 6800"          SACH   TEMP          STORE IN TEMP
0091                 *
0092                 * GENERATE AND UPDATE WEIGHT COEFFICIENTS
0093                 *
0094 004B 3C00"          LT     TEMP          TREG=BETA x E(n)
0095                 *
0096 004C CA00           ZAC                  WEIGHT ONE
0097 004D 3808"          MPY    X1NN
0098 004E CE14           PAC
0099 004F DE02           ADLK   1,14          ROUND RESULT
     0050 0001
0100 0051 4803"          ADDH   W1N
0101 0052 6803"          SACH   W1N           STORE W1(n+1)
0102                 *
0103 0053 CA00           ZAC                  WEIGHT TWO
0104 0054 380A"          MPY    X2NN
0105 0055 CE14           PAC
0106 0056 DE02           ADLK   1,14          ROUND RESULT
     0057 0001
0107 0058 4804"          ADDH   W2N
0108 0059 6804"          SACH   W2N           STORE W2(n+1)
0109 005A CE1F           IDLE
0110                      INN    DN            INPUT SIGNAL + NOISE
0001                      REF    STATUS
0002      0000  DRR       EQU    0
0003 005B 7800           SST    STATUS        SAVE DATA PAGE POINTER
0004 005C C800           LDPK   0             DATA PAGE POINTER TO PAGE 0
0005 005D 2000           LAC    DRR           INPUT A VALUE
0006 005E 6005"          SACL   DN            STORE IT
0007 005F 5000           LST    STATUS        RESTORE DATA PAGE POINTER
0111                 *
0112                 * OUTPUT DESIRED SIGNAL
0113                 *
0114                      OUTS   EN,AUX
0001                      REF    STATUS
0002      0001  DXR       EQU    1
0003 0060 7800           SST    STATUS        SAVE DATA PAGE POINTER
0004 0061 C800           LDPK   0             POINT TO PAGE 0
0005 0062 200B"          LAC    EN            LOAD OUTPUT VALUE
0006 0063 D005           ORK    >03           SET UP FOR SEC COMM
     0064 0003
0007 0065 6001           SACL   DXR           OUTPUT DATA + SEC CODE (11)
0008 0066 CA73           LACK   AUX           LOAD THE SWITCH CODE
0009 0067 CE1F           IDLE                 WAIT UNTIL READY
0010 0068 6001           SACL   DXR           OUTPUT IT
0011 0069 CB10           RPTK   >10           DELAY 17 CLK CYCLES
0012 006A 5500           NOP
0013 006B 200B"          LAC    EN            LOAD OUTPUT VALUE AGAIN
0014 006C D004           ANDK   >FFFC         ZERO LOW 2 BITS
     006D FFFC
0015 006E 6001           SACL   DXR           OUTPUT IT
0016 006F 5000           LST    STATUS        RESTORE THE PAGE POINTER
0115                 *
0116                 * GENERATE Y(n)=[W1(n) x X1(n)] + [W2(n) x X2(n)]
0117                 *
0118 0070 CA00           ZAC
0119 0071 3C03"          LT     W1N           W1(n)xX1(n)
0120 0072 3807"          MPY    X1N
0121 0073 CE14           PAC
```

FIGURE 9.69. (continued)

306

```
0122 0074 3C04"              LT      W2N        W2(n)xX2(n)
0123 0075 3809"              MPY     X2N
0124 0076 CE15               APAC               ADD THE WEIGHTS
0125 0077 DE02               ADLK    1,14       ROUND RESULT
     0078 0001
0126 0079 6906"              SACH    YN,1
0127             *
0128             * MOVE SAMPLES
0129             *
0130 007A 5607"              DMOV    X1N        X1N -> X1NN
0131 007B 5609"              DMOV    X2N        X2N -> X2NN
0132             *
0133 007C FF80               B       NEXT       RETURN FOR NEXT REF SAMPLE
     007D 0020'
0134             *
0135 007E                    PEND
0136 007E
0137                         END
NO ERRORS, NO WARNINGS
```

FIGURE 9.69. (concluded)

they start with different initial conditions. The Z-transform of the sample cosine wave is given by

$$Z\{\cos n\omega t\} = \frac{1 - \frac{A}{2}z^{-1}}{1 - Az^{-1} - Bz^{-2}} \tag{9.19}$$

where $A = \cos(wT)$ and $B = -1$. This transform defines a transfer function,

$$H(z) = \frac{U(z)}{X(z)} = \frac{1 - \frac{A}{2}z^{-1}}{1 - Az^{-1} - Bz^{-2}} \tag{9.20}$$

The transfer function yields the following difference equation:

$$u(n) = Au(n-1) + Bu(n-2) + x(n) - Ax(n-1) \tag{9.21}$$

If the input to (9.21) is an impulse function, the output is a cosine wave. The constant A determines the frequency, and the amplitude of the impulse determines the amplitude of the cosine. Since equation (9.21) defines an oscillator, the initial energy to start it can come from the input, via the impulse, or from the initial conditions. Because we are starting the oscillations from the initial conditions, we need to know what they must be for both a sine and a cosine signal.

By letting $x(n) = \delta(n)$ in equation (9.21), we get the following two initial conditions:

$$\text{for } n = 0, \qquad u(0) = 1.0$$

$$\text{for } n = 1, \qquad u(1) = A$$

When these conditions are applied to (9.21),

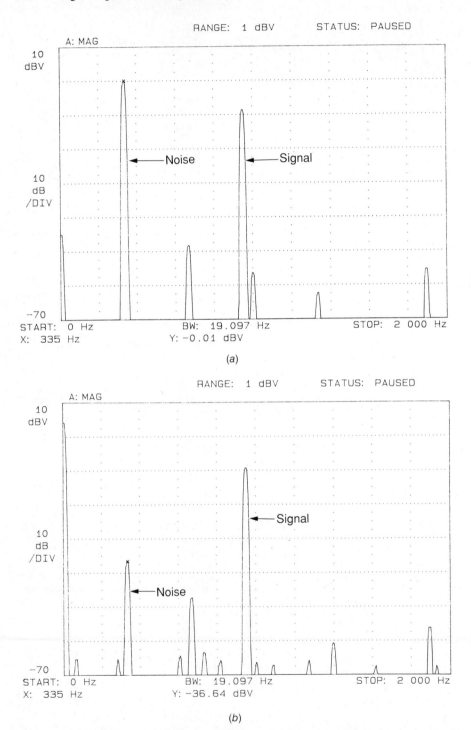

FIGURE 9.70. Frequency spectrum of adaptive notch filter: (*a*) input; (*b*) output

$$u(n) = Au(n - 1) + Bu(n - 2) \qquad (9.22)$$

and the output is a cosine wave. Similarly, the Z-transform of the sine is used to define the difference equation for the sine-wave generator:

$$u(n) = Au(n - 1) + Bu(n - 2) + Cx(n - 1) \qquad (9.23)$$

where $C = \sin(wT)$. Notice that the unforced difference equation is identical to (9.22), the difference equation of the cosine. The initial conditions for the sine wave are generated by solving (9.23), yielding

$$\text{for } n = 0, \qquad u(0) = 0.0$$

$$\text{for } n = 1, \qquad u(1) = C$$

If these initial conditions are applied to equation (9.22), the output $u(n)$ results in a sine wave. Since the only difference between the sine and cosine waves is the phase, it is important that the initial conditions be applied at exactly the same time for each generator. The outputs of these generators are used to simulate the sinusoidal noise in the signal and to provide the sinusoidal references $x_1(n)$ and $x_2(n)$ shown in Figure 9.66. The noise, which is a sinusoidal with an arbitrary phase, is simulated with a weighted sum of the sine and cosine generator outputs. The noise is then added to the sampled input signal to create the signal plus sinusoidal noise. Both the noise mixing and the generation of $x_1(n)$ and $x_2(n)$ take place in software.

Results and Conclusions

This design variation was tested with several different inputs: sine wave, triangle, square wave, and random noise. The noise cancellation was dramatic for the deterministic signals. With a sine wave input, the noise reduction was about 28 dB. On the other hand, the noise reduction for a random input was only about 3 dB. One interesting result observed with the square wave input was that the adaptive notch eliminated noise at the same frequency as the harmonics without eliminating the harmonic itself. Figure 9.71 shows the combined square wave plus 9th harmonic noise and its frequency spectrum before adaptation; Figure 9.72 shows the output of the adaptive filter $e(n)$ after adaptation. Notice the harmonic is left nearly intact. Figures 9.71 and 9.72 demonstrate the discriminating ability of the adaptive notch filter; a fixed frequency domain notch filter would completely eliminate the harmonic.

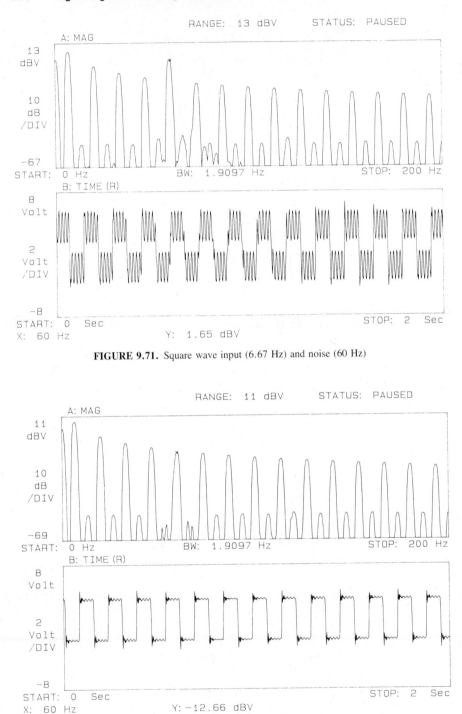

FIGURE 9.71. Square wave input (6.67 Hz) and noise (60 Hz)

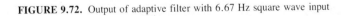

FIGURE 9.72. Output of adaptive filter with 6.67 Hz square wave input

9.26 GUITAR TUNING WITH ADAPTIVE FILTERING FOR NOISE REDUCTION

Problem Statement

The goal of this project is to accurately tune a guitar using digital signal processing techniques.

Design Considerations

Adaptive filtering is used to minimize a bandlimited noise. An interface program written in Pascal is used to configure a Hewlett-Packard spectrum analyzer to communicate with an IBM PC. Figure 9.73 shows the guitar tuning system block diagram. A program written in Turbo Pascal is used to configure a Hewlett-Packard spectrum analyzer for communication over a HPIB (or IEEE 488) bus with an IBM PC. The HPIB card resides in the IBM PC. The interface program is also used to retrieve certain relevant information from the spectrum analyzer such as the marker ON to capture the peak of the guitar signal. Appendix K contains the program listing in Turbo Pascal for communication between an IBM PC and a H-P 70000 spectrum analyzer.

Figure 9.74 shows the analog circuit used to create the desired test signals. Since the guitar signal is weak (≈ 2 mV), an amplifier is used to provide sufficient gain (≈ 255) to drive the next stage. The bandpass filter is designed to pass a guitar input signal between a desired range of frequencies. This signal forms the primary input to the AIC (IN) module. A second bandpass filter is used to bandlimit the random noise which is summed with the guitar input signal, simulating a noisy guitar amplifier. The signal and noise form the reference input to the AIC (AUX IN) module.

The least mean squared (LMS) algorithm is used to adaptively filter out or minimize the bandlimited noise. The AIC module as well as adaptive filtering are discussed in Chapters 3 and 8, respectively. The adaptive filter block diagram in Figure 8.20, and the continuous adaptation filter program, similar to the program listed in Figure 8.22, are used for this project.

The gain of the amplifier is $R_6/R_5 = 255$. Since the desired frequency for tuning the guitar is chosen to be 440 Hz, the "A" tone, the bandpass filter is designed to bandlimit the guitar input around 440 Hz. The transfer function of the bandpass filter is

$$H(s) = \frac{Ks}{s^2 + Bs + w_0^2}$$

where

$$w_o^2 = \left[\frac{1}{R_3C^2}\right]\left|(1/R_1) + (1/R_2)\right| = 6.85 \times 10^6 \text{ rad/s}$$

which gives a center frequency of $f_0 \approx 416$ Hz, and

$$B = \left[\frac{2}{R_3C}\right] = \frac{2}{(28 \times 10^3)(0.15 \times 10^{-6})} = 476.19 \text{ rad/s}$$

resulting in a bandwidth of ≈ 75 Hz.

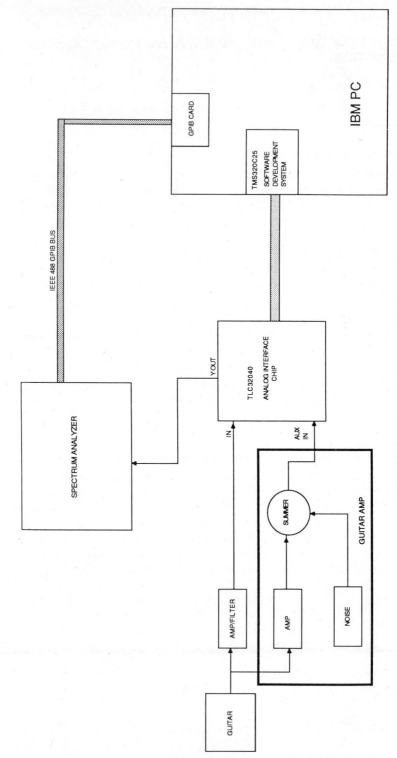

FIGURE 9.73. System block diagram for tuning a guitar

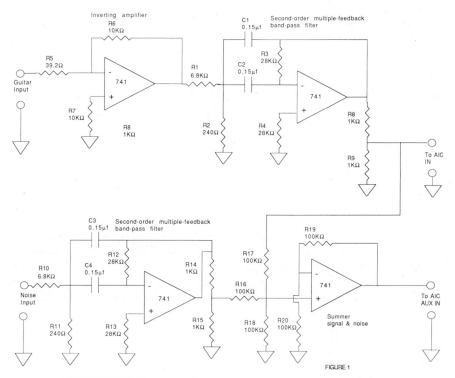

FIGURE 9.74. Hardware support to create the signal for tuning a guitar

Results and Conclusions

Upon running the program, the following menu appears on the IBM PC monitor:

```
Directions for tuning a guitar in the ''A'' note (440 Hz):
The strings are numbered 1-6; 1 being the thickest string.
To tune string 1 ''E'', depress 5th fret and strike.
To tune string 2 ''A'', strike.
To tune string 3 ''D'', depress 7th fret and strike.
To tune string 4 ''G'', depress 2nd fret and strike.
To tune string 5 ''B'', depress 10th fret and strike.
To tune string 6 ''E'', depress 5th fret and strike.
```

A message to either tighten or loosen a particular string is also displayed. The spectrum analyzer shows the spectrum of the guitar signal, with the marker showing the peak frequency.

9.27 MUSICAL TONE OCTAVE GENERATOR

Problem Statement

The goal of this project is to simulate musical tones from software-generated signals.

Design Considerations

One method consists of three oscillators to generate the fundamental frequency and two harmonic frequencies, with the three resulting waveforms summed to produce a complex waveform for each tone. A second method uses a sawtooth generation procedure. The tones in a piano are from A0 = 27.5 Hz to C8 = 4186 Hz, with the fourth octave corresponding to C4 = 261.63 Hz (the middle C) up to B4 = 493.88 Hz. C5 is one octave higher, with twice the frequency of C4, or 523.25 Hz. The musical tone is a complex waveform consisting of a fundamental frequency and harmonic frequencies with reduced amplitudes. The first harmonic frequency is two times the frequency and one-half the amplitude of the fundamental; and the second harmonic frequency is three times the frequency and one-third the amplitude of the fundamental, and so forth, with higher harmonics above the seventh being negligible.

Three types of waveforms can be created in musical synthesizers. The first method uses pure sine-wave summation, incorporating as many harmonics as needed for tone quality. The second method uses sawtooth generation which includes all the harmonics at their relative-reduced amplitudes, and the third method uses square-wave generation with all the odd harmonics at their relative-reduced amplitudes.

The oscillator program example in Figure 4.3 provides some background. In this project, the fundamental and the first two harmonic waveforms are summed. From the difference equation,

$$y(n) = Ay(n-1) + By(n-2) + Cx(n-1)$$

the coefficient A controls the oscillation frequency, B = -1 and C controls the amplitude of the oscillation, or the amplitude envelope. The amplitude envelope, which produces differences between the sounds of each instrument, consists of an attack, decay, sustain, and release section. Sounds from different instruments can be produced by controlling the various sections of the envelope. Some special effects, such as reverberation, can also be produced by feeding back past waveforms through digital delay and phase shifting.

The triangular waveform generation program in Figure 3.9 provides some background for the sawtooth waveform generation method.

Results and Conclusion

For the oscillation method, the frequencies and magnitude levels of each of the complex output signals (fundamental and two harmonics), are verified on a spectrum analyzer. Appendix L lists the frequencies for one octave, the coefficient values, as well as a complete listing of the program. Figure 9.75 shows the frequency spectrum for tone C5.

The oscillator summation method can create more precise frequency accuracy and can be modified to select a desired number of harmonics for different effects. The sawtooth generation method sounds richer due to the presence of all the harmonics. Figure 9.76 shows the frequency spectrum for tone C5 using the sawtooth method.

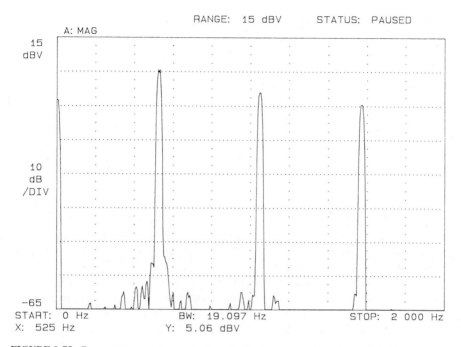

FIGURE 9.75. Frequency spectrum showing tone C5 (fundamental and two harmonics) for oscillator summation method

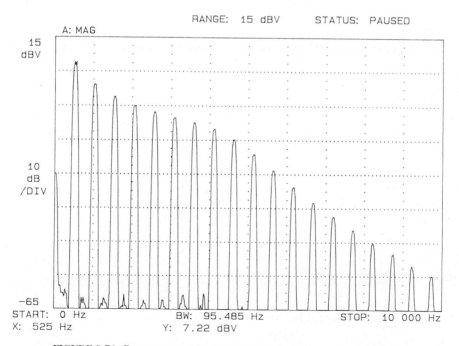

FIGURE 9.76. Frequency spectrum showing all harmonics for sawtooth method

9.28 LABORATORY EXERCISES

This section is devoted to laboratory exercises.

Exercise 9.1   Sine-Wave Generation 1

Purpose:

To design two sinusoidal generators that are 90° out of phase or are in quadrature.

Discussion:

Sinusoidal waveforms in quadrature are required frequently in DSP applications; therefore, this design may be used later in a more elaborate project. Using the direct form transpose structure for an IIR filter, design and implement two sinusoidal generators from the Z-transform for sine and cosine. The cosine and sine generators will be identical filters. The only thing different will be the initial conditions, which can be found using the techniques of Section 4.4. Use a sample rate of 10 kHz and test several frequencies by changing the denominator. Measure the phase difference with an oscilloscope. Note that testing this algorithm directly in real-time requires two analog output channels.

Exercise 9.2   Sine-Wave Generation 2-Backward Difference Equation

Purpose:

Design a sine-wave generator using the backward difference approximation of an analog oscillator.

Discussion:

An analog oscillator is given by the differential equation

$$y'' + K^2 y = 0$$

where k is the frequency of oscillation. The backward difference approximation can be used to obtain a finite difference equation. Using this difference equation, repeat Exercise 9.1. Compare the results of this exercise with Exercise 9.1.

The backward difference approximation for the derivative is given by

$$y'(n) = \frac{y(n) - y(n-1)}{T}$$

where T is the time between samples. This approximation is valid when $\omega <<$

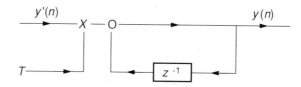

FIGURE 9.77. Backward difference integral

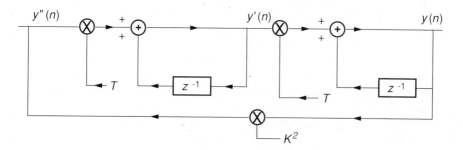

FIGURE 9.78. Backward difference oscillator

$1/T$; therefore, a much higher sample rate may be required to get the same quality output as in Exercise 9.1. An alternative structure which more closely matches the differential equation can be used by observing that the backward difference approximation defines the integral operation, as shown in Figure 9.77. Applying the integral structure to the differential equation gives the structure shown in Figure 9.78.

Purpose:

Generate a sine-wave from a software-generated square wave.

Exercise 9.3   Sine-Wave Generation from a square wave

Discussion:

A sine-wave can be obtained from a square wave by passing it through a lowpass filter which eliminates all the harmonics and leaves the fundamental frequency. A square wave can easily be generated, whose period is a constant fraction of the sample rate; therefore, changing the sample rate will change the frequency of the fundamental. If the lowpass filter is designed to work at the same sample rate, it will automatically be frequency-scaled along with the fundamental of the square wave. The filter design package can be used to design an FIR lowpass filter.

Exercise 9.4   Pseudorandom Noise Generator

Purpose:

Design a pseudorandom noise generator using the remainder of a modulus operation as the random number.

Discussion:

A random sequence of random numbers can be generated by using the recursive relationship

$$I(n + 1) = [J \times I(n) + 1] \bmod M \qquad n = 1, 2, 3, \ldots, M - 1$$

If I is interpreted as a two's-complement fixed-point fraction, the sequence generated is a number between $+1$ and -1. The modulus operation is automatically performed by the finite-length registers which exist on the TMS320C25. M is fixed by the length of the register used and is given by

$$M = 2^L$$

where L is the number of bits in the register used to retain the result for the next calculation.

The process needs a seed, $I(0)$, to start it. $I(0)$ obviously cannot exceed M. The values for $L = 16$ or $L = 32$ are the most convenient moduli for the TMS320C25. If $L = 16$ is chosen, the multiplications can be performed directly by the hardware. If $L = 32$ is chosen, a double-precision multiplication is needed. J is a constant that must be picked to give a good random sequence. If J is picked too small, the sequence becomes more monotonic. For example, if $J = 1$ and $I(0) = 1$, the sequence is exactly monotonic: $I(n) = 1, 2, 3, \cdots, M - 1$. Therefore, J should be picked large, but not too large. A good rule is to choose it one decimal digit less than the modulus. To avoid certain repetitive patterns, J should be chosen so that

$$J = x21$$

where x is an even number and 21 is the last two decimal digits [39]. Accordingly, for $M = 2^{16}$, we can set $J = 6536$ and choose $I(0) = 12,357$. The modulus operation can be performed (for $M = 2^{16}$) by selecting the lower or the higher 16 bits of the accumulator.

REFERENCES

[1] K. S. Lin (editor), *Digital Signal Processing Applications with the TMS320 Family*, Vol. 1, Prentice-Hall, Englewood Cliffs, N.J., 1988.

[2] C. S. Burrus and T. W. Parks, *DFT/FFT and Convolution Algorithms: Theory and Implementation*, Wiley, New York, 1988.

[3] W.D. Stanley, *Digital Signal Processing*, Reston, Reston, Va., 1975.

[4] A. V. Oppenheim and R. Schafer, *Digital Signal Processing*, Prentice-Hall, Englewood Cliffs, N.J., 1975.

[5] B. Widrow and S. D. Stearns, *Adaptive Signal Processing*, Prentice-Hall, Englewood Cliffs, N.J., 1985.

[6] A. V. Oppenheim and R. Schafer, *Discrete-Time Signal Processing*, Prentice-Hall, Englewood Cliffs, N.J., 1989.

[7] T. W. Parks and C. S. Burrus, *Digital Filter Design*, Wiley, New York, 1987.

[8] D. J. DeFatta, J. G. Lucas and W. S. Hodgkiss, *Digital Signal Processing: A System Approach*, Wiley, New York, 1988.

[9] R. W. Schafer and L. R. Rabiner, "A Digital Signal Processing Approach to Interpolation," *Proceedings of the IEEE*, **61**(6), June 1973.

[10] R. E. Crochiere and L. R. Rabiner, "Optimum FIR Digital Filter Implementations for Decimation, Interpolation and Narrow-Band Filtering," *IEEE Transactions on Acoustics, Speech, and Signal Processing*, **ASSP-23**, October 1975.

[11] R. E. Crochiere and L. R. Rabiner, "Further Considerations in the Design of Decimators and Interpolators," *IEEE Transactions on Acoustics, Speech, and Signal Processing, ASSP*, **ASSP-24**, August 1976.

[12] R. E. Crochiere and L. R. Rabiner, *Multirate Digital Signal Processing*, Prentice-Hall, Englewood Cliffs, N.J., 1983.

[13] M. G. Bellanger, J. L. Dagnet and G. P. Lepagnol, "Interpolation, Extrapolation, and Reduction of Computation Speed in Digital Filters," *IEEE Transactions on Acoustics, Speech, and Signal Processing*, **ASSP-22**, August 1974.

[14] L. B. Jackson, *Digital Filters and Signal Processing*, Kluwer Academic, Norwell, Mass., 1986.

[15] J. G. Proakis and D. G. Manolakis, *Introduction to Digital Signal Processing*, MacMillan, New York, 1988.

[16] R. Chassaing, "Digital Broadband Noise Synthesis by Multirate Filtering Using the TMS320C25," *1988 ASEE Annual Conference Proceedings*, Vol. 1.

[17] *Second-Generation TMS320 User's Guide*, Texas Instruments Inc., Dallas, Tex., 1987.

[18] *TLC32040I, TLC32040C, TLC32041I, TLC32041C Analog Interface Circuits, Advanced Information*, Texas Instruments Inc., Dallas, Tex., September 1987.

[19] R. Chassaing, "A Senior Project Course in Digital Signal Processing with the TMS320," *IEEE Transactions on Education*, August 1989.

[20] J. W. Cooley and J. W. Tukey, "An Algorithm for Machine Computation of Complex Fourier Series," *Mathematics of Computation*, **19**, 1965.

[21] H. F. Silverman, "An Introduction to Programming the Winograd Fourier Transform Algorithm (WFTA)," *IEEE Transactions on Acoustics, Speech, and Signal Processing*, **ASSP-25**, April 1977.

[22] R. N. Bracewell, "The Fast Hartley Transform," *Proceedings of the IEEE*, **72**(8), August 1984.

[23] A. Zakhor and A. V. Oppenheim, "Quantization Errors in the Computation of the Discrete Hartley Transform," *IEEE Transactions on Acoustics, Speech, and Signal Processing,* **ASSP-35***(2), October 1987.*

[24] H. V. Sorensen, D. L. Jones, C. S. Burrus, and M. T. Heideman, "On Computing the Discrete Hartley Transform," *IEEE Transactions on Acoustics, Speech, and Signal Processing,* **ASSP-33** (4) October 1985.

[25] M. Vetterli and P. Duhamel, "Split-Radix Algorithms for Length-P^m DFT's," *IEEE Transactions on Acoustics, Speech, and Signal Processing,* **ASSP-37**, January 1989.

[26] L. R. Rabiner and B. Gold, *Theory and Application of Digital Signal Processing,* Prentice-Hall, Englewood Cliffs, N.J., 1975.

[27] G. D. Bergland, "A Guided Tour of the Fast Fourier Transform," *IEEE Spectrum,* **6**, July 1969.

[28] E. O. Brigham, *The Fast Fourier Transform*, Prentice-Hall, Englewood Cliffs, N.J., 1974.

[29] DSP Committee, IEEE ASSP (editors), *Selected Papers in Digital Signal Processing II*, IEEE Press, New York, 1976.

[30] DSP Committee, IEEE ASSP (editors), *Programs for Digital Signal Processing*, IEEE Press, New York, 1979.

[31] G. Goertzel, "An Algorithm for the Evaluation of Finite Trigonometric Series," *American Mathematics Monthly*, **65**, January 1958.

[32] A. M. Engebretson, R. E. Morley, and M. P. O'Connell, "A Wearable, Pocket Sized Processor for Digital Hearing Aid and Other Hearing Prostheses Applications," *International Conference on Acoustics, Speech, and Signal Processing*, Tokyo, 1986.

[33] S. Kay and R. Sudhaker, "A Zero Crossing Spectrum Analyzer," *IEEE Transactions on Acoustics, Speech, and Signal Processing,* **ASSP-34**, February 1986.

[34] J. P. Cater, *Electronically Hearing: Computer Speech Recognition*, Howard W. Sams, Indianapolis, In., 1984.

[35] J. Hecht, *Understanding Fiber Optics*, Sams Understanding Series, H. W. Sams, Indianapolis, In., 1987.

[36] J. Senior, *Optical Fiber Communications Principles and Practice*, Prentice-Hall International Series in Optoelectronics, Prentice-Hall, Englewood Cliffs, N.J., 1985.

[37] C. A. Taylor, *The Physics of Musical Sounds*, English Universities Press, London, 1965.

[38] K. C. Pohlmann, *Principles of Digital Audio*, H. W. Sams, Indianapolis, In., 1989.

[39] D. E. Knuth, *Art of Computer Programming, Seminumerical Algorithms*, Vol. 2, Addison-Wesley, Reading, Mass., 1969.

[40] *Digital Filter Design Package (DFDP)*, Atlanta Signal Processors Inc., Atlanta, Ga., 1977.

[41] A. V. Oppenheim, *Applications of Digital Signal Processing*, Prentice-Hall, Englewood Cliffs, N.J., 1978.

[42] R. Chassaing, "Applications in Digital Signal Processing with the TMS320 Digital Signal Processor in an Undergraduate Laboratory," *1987 ASEE Annual Conference Proceedings*, Vol. 3.

[43]   *TMS320C1x/TMS320C2x Assembly Language Tools User's Guide*, Texas Instruments Inc., Dallas, Tex., 1987.

[44]   *TMS320C2x Software Development System User's Guide*, Texas Instruments Inc., Dallas, Tex., 1988.

[45]   *TMS320C1x/TMS320C2x Source Conversion Reference Guide*, Texas Instruments Inc., Dallas, Tex., 1988.

[46]   *TMS32010 Analog Interface Board User's Guide*, Texas Instruments Inc., Dallas, Tex., 1984.

[47]   S. Waser and M. Flynn, *Introduction to Arithmetic for Digital Systems Designers*, Holt, Rinehart and Winston, New York, 1982.

[48]   H. L. Garner, "Theory of Computer Addition and Overflows," *IEEE Transactions on Computers*, **C-27**(4), April 1978, pp. 297–301.

[49]   F. Harris, "On the Use of Windows for Harmonic Analysis with the Discrete Fourier Transform," *Proceedings of the IEEE*, **66**(1), January 1978.

[50]   B. Widrow, et al., "Adaptive Noise Cancelling: Principles and Applications," *Proceedings of the IEEE*, **63**(12), December 1975.

[51]   J. R. Treichler, C. R. Johnson Jr., and M. G. Larimore, *Theory and Design of Adaptive Filters*, Wiley, New York, 1987.

[52]   S. T. Alexander, *Adaptive Signal Processing Theory and Applications*, Springer-Verlag, New York, 1986.

[53]   S. T. Alexander, "Fast Adaptive Filters: A Geometrical Approach," *IEEE Transactions on Acoustics, Speech, and Signal Processing*, **ASSP-34**, October 1986.

[54]   M. L. Honig and D. G. Messershmidt, *Adaptive Filters: Structures, Algorithms and Applications*, Kluwer Academic, Nowell, Mass., 1984.

[55]   C. F. Cowan and P. F. Grant, *Adaptive Filters*, Prentice-Hall, Englewood Cliffs, N.J., 1985.

[56]   S. Haykin, *Adaptive Filter Theory*, Prentice-Hall, Englewood Cliffs, N.J., 1986.

[57]   R. Chassaing, "A Senior Project Course on Applications In Digital Signal Processing With the TMS320," *1988 ASEE Annual Conference Proceedings*, Vol. 1.

[58]   R. Chassaing, "A Course in Microprocessors Based on the 16/32 Bit 68000 μP," *IEEE Transactions on Education*, August 1987.

[59]   J. F. Kaiser, "Nonrecursive Digital Filter Design Using the I_0-sinh Window Function," *Proceedings of the IEEE International Symposium on Circuits and Systems*, 1974.

[60]   T. W. Parks and J. H. McClellan, "Chebychev Approximation for Nonrecursive Digital Filter with Linear Phase," *IEEE Transactions on Circuit Theory*, **CT-19**, March 1972, pp. 189–194.

[61]   L. B. Jackson, "Roundoff Noise Analysis for Fixed-Point Digital Filters Realized in Cascade or Parallel Form," *IEEE Transactions on Audio and Electroacoustics*, **Au-18**, June 1970, pp. 107–122.

[62] D. W. Horning, "An Undergraduate Digital Signal Processing Laboratory," *1987 ASEE Annual Conference Proceedings*.

[63] N. Ahmed and T. Natarajan, *Discrete-Time Signals and Systems*, Reston, Reston, Va., 1983.

[64] L. C. Ludemen, *Fundamentals of Digital Signal Processing*, Harper & Row, New York, 1986.

[65] C. S. Williams, *Designing Digital Filters*, Prentice-Hall, Englewood Cliffs, N.J., 1986.

[66] M. G. Bellanger, *Digital Filters and Signal Analysis*, Prentice-Hall, Englewood Cliffs, N.J., 1986.

[67] T. Kailath, *Modern Signal Processing*, Hemisphere, New York, 1985.

[68] C. S. Burrus, "Unscrambling for Fast DFT Algorithms," *IEEE Transactions on Acoustics, Speech, and Signal Processing*, **ASSP-36**, July 1988.

[69] D. L. Jones and T. W. Parks, *A Digital Signal Processing Laboratory*, Prentice-Hall, Englewood Cliffs, N.J., 1988.

[70] V. C. Hamacher, Z. G. Vranesic, and S. G. Zaky, *Computer Organization*, McGraw-Hill, New York, 1978.

[71] P. H. Garrett, *Analog I/O Design: Acquisition–Conversion–Recovery*, Reston, Reston, Va., 1981.

[72] R. W. Hamming, *Digital Filters*, Prentice-Hall, Englewood Cliffs, N.J., 1983.

[73] L. B. Jackson, "An Analysis of Limit cycles Due to Multiplicative Rounding in Recursive Digital Filters," *Proceedings of the Seventh Allerton Conference on Circuit System Theory*, 1969, pp. 69–78.

[74] V. B. Lawrence and K. V. Mina, "A New and Interesting Class of Limit Cycles in Recursive Digital Filters," *Proceedings of the IEEE International Symposium on Circuit Systems*," April 1977, pp. 191–194.

[75] J. F. Kaiser, "Some Practical Considerations in the Realization of Linear Digital Filters," *Proceedings of the Third Allerton Conference on Circuit System Theory*, October 1965, pp. 621–633.

A

SWDS Installation
and Conversion to COFF

A.1 SWDS INSTALLATION

Step 1: DOS Modification

Modify (or create) the files AUTOEXEC.BAT and CONFIG.SYS in your MS-DOS.

1. Edit the AUTOEXEC.BAT so that it contains

   ```
   SET SWDS = B:
   PATH = B:
   ```

   B: represents drive B in a dual floppy system. In a hard drive system, B: would typically be replaced by C:\SWDS.

2. Edit the CONFIG.SYS so that it contains

   ```
   FILES = 32
   BUFFERS = 32
   ```

   Boot your PC with the DOS incorporating the procedures above before using the debug monitor.

Step 2: Installation of the Debug Monitor

1. Read the file SWDSINST.DOC on the floppy Software Development disk placed in drive A, and place a formatted disk on drive B.
2. Type the following:

   ```
   SWDSINST B: MACHINE TYPE
   ```

   to copy the necessary software from drive A to drive B. The MACHINE TYPE speicifies the particular machine being used and would be replaced by IBMPCM if using an IBM PC/XT with monochrome monitor.

Again, in a hard drive system, B: would be replaced by C: or C:\SWDS, which would represent a directory SWDS in drive C.

Step 3: Hardware Installation

The SWDS card must be handled with special care during its installation (as with all C-MOS components). The *TMS320C25 Software Development System User's Guide,* pages 1-1 through 1-6, describes the installation of the SWDS based on the MACHINE TYPE.

A.2 CONVERSION TO COFF

In this section we discuss a conversion procedure between the TI-Tag format assembler version 3.1 or earlier, and the common object file format (COFF), version 5.04. We will describe this procedure using the LOOP program of Chapter 1, shown here again in Figure A.1. Other program listings are also included subsequently for comparison between the two formats.

Conversion Utility

To avoid destroying the original LOOP program, rename LOOP.ASM, to LOOPCOFF.ASM. Then type the command

```
DSPCV LOOPCOFF.ASM
```

to create a new source file LOOPCOFF.A00 (the COFF version), shown in Figure A.2. Note the following:

1. The two absolute origin instructions (AORG) are no longer used and are replaced by .sect.
2. The two DATA instructions are replaced by .word.
3. END is replaced by .END.

```
* LOOP PROGRAM (ORIGINAL)-LOOPCOFF.ASM
* INPUT INTO AIB IS OUTPUT - TEST FOR AIB
         AORG     0
         B        START      *INIT. RESET VECTOR
         AORG     >20        *START AT PMA 20
* INITIALIZE AIB
RATE     DATA     999        *N=(10 MHZ/fs)-1 WITH fs=10 KHZ
MODE     DATA     >FA        *MODE FOR AIB
*
START    LDPK     0          *INIT DATA PAGE POINTER TO 0
         LACK     RATE       *GET SAMPLE RATE INTO ACC
         TBLR     >60        *TRANFER FROM PROG MEM TO DATA MEM 60
         OUT      >60,1      *OUTPUT RATE TO AIB PORT 1
         LACK     MODE       *GET MODE INTO ACC
         TBLR     >60        *TRANSFER FROM PROG TO DATA MEM
         OUT      >60,0      *OUTPUT MODE TO AIB PORT 0
         OUT      >60,3      *DUMMY OUTPUT TO AIB PORT 3
WAIT     BIOZ     ALOOP      *WAIT FOR END OF CONVERSION
         B        WAIT       *LOOP BACK
ALOOP    IN       >60,2      *READ DATA FROM PORT 2
         OUT      >60,2      *WRITE DATA OUT TO PORT 2
         B        WAIT       *BRANCH TO CONTINUE
         END
```

FIGURE A.1. LOOP source program (LOOPCOFF.ASM)

```
* LOOP PROGRAM (ORIGINAL)-LOOPCOFF.ASM
* INPUT INTO AIB IS OUTPUT - TEST FOR AIB

;>>>> WARNING, NO ABSOLUTE CODE
;>>>>          AORG    0
        .sect   "AORG0"
        B       START   *INIT. RESET VECTOR

;>>>> WARNING, NO ABSOLUTE CODE
;>>>>          AORG    >20      *START AT PMA 20
        .sect   "AORG1"
* INITIALIZE AIB
RATE:   .word   999      *N==(10 MHZ/fs)-1 WITH fs==10 KHZ
MODE:   .word   0FAh     *MODE FOR AIB
*
START   LDPK    0        *INIT DATA PAGE POINTER TO 0
        LACK    RATE     *GET SAMPLE RATE INTO ACC
        TBLR    060h     *TRANFER FROM PROG MEM TO DATA MEM 60
        OUT     060h,1   *OUTPUT RATE TO AIB PORT 1
        LACK    MODE     *GET MODE INTO ACC
        TBLR    060h     *TRANSFER FROM PROG TO DATA MEM
        OUT     060h,0   *OUTPUT MODE TO AIB PORT 0
        OUT     060h,3   *DUMMY OUTPUT TO AIB PORT 3
WAIT    BIOZ    ALOOP    *WAIT FOR END OF CONVERSION
        B       WAIT     *LOOP BACK
ALOOP   IN      060h,2   *READ DATA FROM PORT 2
        OUT     060h,2   *WRITE DATA OUT TO PORT 2
        B       WAIT     *BRANCH TO CONTINUE
        .end
```

FIGURE A.2. LOOP program converted to COFF (LOOPCOFF.A00)

Assembling with Version 5.04

To assemble the new source file LOOPCOFF.A00, the following changes should be made:

1. Replace all the "*" which represent the comments after the instructions (not the ones in column 1) by ";". Beware of global replacements, since * will be used later as part of an instruction (indirect addressing).
2. Delete the comments created with LOOPCOFF.A00 (;>>>).
3. Rename this file LOOPCOFF.NEW, shown in Figure A.3. You may wish to incorporate some of the new features available with the COFF version.

Type the command

```
DSPA -L LOOPCOFF.NEW
```

to assemble the source file LOOPCOFF.NEW and create an object file LOOP-COFF.OBJ. Use the option -L to obtain the listing file LOOPCOFF.LST, shown in Figure A.4.

Linking to Create an Object File

In order to link, a COMMAND file must first be created. The command file LOOPCOFF.CMD, shown in Figure A.5, contains the appropriate commands to link LOOPCOFF.OBJ and create a map file as well as an output file. Type

```
DSPLNK      LOOPCOFF.CMD
```

to link and create the output file LOOPCOFF.OUT. This output file can be

```
* LOOP PROGRAM (ORIGINAL)-LOOPCOFF.ASM
* INPUT INTO AIB IS OUTPUT - TEST FOR AIB
          .sect    "INT_VEC"
          B        START    ;INIT. RESET VECTOR
          .data
* INITIALIZE AIB
RATE:     .word    999      ;N==(10 MHZ/fs)-1 WITH fs==10 KHZ
MODE:     .word    0FAh     ;MODE FOR AIB
*
          .text
START     LDPK     0        ;INIT DATA PAGE POINTER TO 0
          LACK     RATE     ;GET SAMPLE RATE INTO ACC
          TBLR     060h     ;TRANFER FROM PROG MEM TO DATA MEM 60
          OUT      060h,1   ;OUTPUT RATE TO AIB PORT 1
          LACK     MODE     ;GET MODE INTO ACC
          TBLR     060h     ;TRANSFER FROM PROG TO DATA MEM
          OUT      060h,0   ;OUTPUT MODE TO AIB PORT 0
          OUT      060h,3   ;DUMMY OUTPUT TO AIB PORT 3
WAIT      BIOZ     ALOOP    ;WAIT FOR END OF CONVERSION
          B        WAIT     ;LOOP BACK
ALOOP     IN       060h,2   ;READ DATA FROM PORT 2
          OUT      060h,2   ;WRITE DATA OUT TO PORT 2
          B        WAIT     ;BRANCH TO CONTINUE
          .end
```

FIGURE A.3. LOOP program in COFF corrected (LOOPCOFF.NEW)

```
0001              * LOOP PROGRAM (ORIGINAL)-LOOPCOFF.ASM
0002              * INPUT INTO AIB IS OUTPUT - TEST FOR AIB
0003 0000                  .sect    "INT_VEC"
0004 0000 FF80             B        START    ;INIT. RESET VECTOR
     0001 0000'
0005 0000                  .data
0006              * INITIALIZE AIB
0007 0000 03E7   RATE:     .word    999      ;N==(10 MHZ/fs)-1 WITH fs==10 KHZ
0008 0001 00FA   MODE:     .word    0FAh     ;MODE FOR AIB
0009              *
0010 0000                  .text
0011 0000 C800   START     LDPK     0        ;INIT DATA PAGE POINTER TO 0
0012 0001 CA00"            LACK     RATE     ;GET SAMPLE RATE INTO ACC
0013 0002 5860             TBLR     060h     ;TRANFER FROM PROG MEM TO DATA MEM 60
0014 0003 E160             OUT      060h,1   ;OUTPUT RATE TO AIB PORT 1
0015 0004 CA01"            LACK     MODE     ;GET MODE INTO ACC
0016 0005 5860             TBLR     060h     ;TRANSFER FROM PROG TO DATA MEM
0017 0006 E060             OUT      060h,0   ;OUTPUT MODE TO AIB PORT 0
0018 0007 E360             OUT      060h,3   ;DUMMY OUTPUT TO AIB PORT 3
0019 0008 FA80   WAIT      BIOZ     ALOOP    ;WAIT FOR END OF CONVERSION
     0009 000C'
0020 000A FF80             B        WAIT     ;LOOP BACK
     000B 0008'
0021 000C 8260   ALOOP     IN       060h,2   ;READ DATA FROM PORT 2
0022 000D E260             OUT      060h,2   ;WRITE DATA OUT TO PORT 2
0023 000E FF80             B        WAIT     ;BRANCH TO CONTINUE
     000F 0008'
0024                       .end

No Errors,  No Warnings
```

FIGURE A.4. LOOP program in COFF listing (LOOPCOFF.LST)

```
/* COMMAND FILE FOR LOOPCOFF PROGRAM                                  */
    LOOPCOFF.OBJ                     /* INPUT OBJECT FILE             */
    -O LOOPCOFF.OUT                  /* EXECUTABLE FILE IN COFF       */
    -M LOOPCOFF.MAP                  /* MAP FILE                      */
MEMORY
{
    PAGE 0 :                                      /* PROGRAM MEMORY */
        INTS      : ORIGIN =     0H , LENGTH =    020H
        EXT_PROG  : ORIGIN =   020H , LENGTH = 0FEE0H
    PAGE 1 :                                      /* DATA MEMORY    */
        BLOCK_B2  : ORIGIN =   060H , LENGTH =    020H
}
SECTIONS
{
    INT_VEC  : {
                LOOPCOFF.OBJ (INT_VEC)
              } > INTS         PAGE 0
    .text    : {
                LOOPCOFF.OBJ (.text)
                *(.TEXT)
              } > EXT_PROG     PAGE 0
    .data    : {  } > EXT_PROG PAGE 0
    .bss     : {  } > BLOCK_B2 PAGE 1
}
```

FIGURE A.5. LOOP program linker command file (LOOPCOFF.CMD)

downloaded and executed only if you have the SWDS that accepts the assembler linker version 5.04.

Downloading into the SWDS

The file LOOPCOFF.OUT (COFF version) cannot yet be downloaded into an SWDS that accepts the assembler linker version 3.1 or earlier, with TI-Tag format files. Type

```
* DIFFERENCE EQUATION PROGRAM - COFF VERSION CONVOCOF.ASM
* CONVOLUTION USING MACD,RPTK
* LOAD DATA VALUES (SAMPLES) STARTING AT DMA 3FC

;>>>> WARNING, NO ABSOLUTE CODE
;>>>>          AORG    0
        .sect   "AORG0"
        B       START

;>>>> WARNING, NO ABSOLUTE CODE
;>>>> TABLE   AORG   >20              *START PROGRAM AT PMA >20
TABLE:  .sect   "AORG1"
H0:     .word   01h              *COEFFICIENTS
H1:     .word   02h
H2:     .word   02h
H3:     .word   01h
START   LARP    AR0              *SELECT AR0
        LRLK    AR0,0200h        *POINT TO BLOCK B0
        RPTK    3                *EXECUTE BLOCK MOVE 4 TIMES
        BLKP    TABLE,*+         *P.M. TO D.M. B0 START 0200h
        CNFP    ;CONFIGURE B0 AS P.M.
*
        ZAC     ;CLEAR ACCUMULATOR
        MPYK    0                *CLEAR PRODUCT REGISTER
        LRLK    AR1,03FFh         *POINT TO BOTTOM OF BLOCK B1
        LARP    AR1              *SELECT AR1 FOR INDIR ADDR
* THE BEEF
        RPTK    3                *EXECUTE MACD 4 TIMES
        MACD    65280,*-         *MULT(PMA>FF00)(DMA>3FF),ETC
        APAC    ;ADD LAST PRODUCT TO ACCUM
        .end
```

FIGURE A.6. Difference equation program converted to COFF (CONVOCOF.A00)

```
* DIFFERENCE EQUATION PROGRAM - COFF VERSION CONVOCOF.ASM
* CONVOLUTION USING MACD,RPTK
* LOAD DATA VALUES (SAMPLES) STARTING AT DMA 3FC
        .sect   "INT_VEC"
        B       START
TABLE:  .data
H0:     .word   01h                 ;COEFFICIENTS
H1:     .word   02h
H2:     .word   02h
H3:     .word   01h
        .text
START   LARP    AR0                 ;SELECT AR0
        LRLK    AR0,0200h           ;POINT TO BLOCK B0
        RPTK    3                   ;EXECUTE BLOCK MOVE 4 TIMES
        BLKP    TABLE,*+            ;P.M. TO D.M. B0 START 0200h
        CNFP                        ;CONFIGURE B0 AS P.M.
*
        ZAC                         ;CLEAR ACCUMULATOR
        MPYK    0                   ;CLEAR PRODUCT REGISTER
        LRLK    AR1,03FFh           ;POINT TO BOTTOM OF BLOCK B1
        LARP    AR1                 ;SELECT AR1 FOR INDIR ADDR
* THE BEEF
        RPTK    3                   ;EXECUTE MACD 4 TIMES
        MACD    65280,*-            ;MULT(PMA>FF00)(DMA>3FF),ETC
        APAC                        ;ADD LAST PRODUCT TO ACCUM
        .end
```

FIGURE A.7. Difference equation program in COFF corrected (CONVOCOF.NEW)

```
/* COMMAND FILE FOR CONVOCOF PROGRAM                          */
    CONVOCOF.OBJ                /* INPUT OBJECT FILE          */
    -O CONVOCOF.OUT             /* EXECUTABLE FILE IN COFF    */
    -M CONVOCOF.MAP             /* MAP FILE                   */
MEMORY
{
    PAGE 0 :                            /* PROGRAM MEMORY */
        INTS      : ORIGIN =    0H , LENGTH =   020H
        EXT_PROG  : ORIGIN = 020H , LENGTH = 0FEE0H
    PAGE 1 :                            /* DATA MEMORY    */
        BLOCK_B2  : ORIGIN = 060H , LENGTH =   020H
}
SECTIONS
{
    INT_VEC  : (
                  CONVOCOF.OBJ (INT_VEC)
               )      > INTS        PAGE 0
    .text    : (
                  CONVOCOF.OBJ (.text)
                  *(.TEXT)
               )      > EXT_PROG    PAGE 0
    .data    : (  )   > EXT_PROG    PAGE 0
    .bss     : (  )   > BLOCK_B2    PAGE 1
}
```

FIGURE A.8. Difference equation linker command file (CONVOCOF.CMD)

DSPROM −T LOOPCOFF.OUT

to convert the output to the TI-Tag format (with option T), and create an executable file LOOP.TAG, which can be downloaded into the SWDS, similar to the procedure in Section 1.6 with LOOP.MPO.

This conversion procedure has been used with the programs in Figures A.7 to A.14 to show the differences between the two formats, TI-Tag and COFF. The programs are associated with:

1. A difference equation example discussed in Chapter 2 (CONVOCOF.ASM)

2. A digital oscillator example discussed in Chapter 4 (OSCICOFF.ASM)

3. A lowpass filter example with 11 coefficients discussed in Chapter 5 (LP11-COFF.ASM)

```
* DIGITAL OSCILLATOR PROGRAM - COFF VERSION OSCICOFF.ASM
* Y(n)=A Y(n-1) + B Y(n-2) + C X(n-1)
* Y0 =A Y1 + B Y2   FOR N>1 , X(n-1)=1 FOR n=1 (= 0 OTHERWISE)
* USES INDIRECT ADDRESSING ,  fs = 10 KHZ
* DATA MUST BE LOADED AT DMA 60-66 BEFORE RUNNING PROGRAM
MODE:    .equ    060h        *TRANSPARENT MODE FOR AIB
RATE:    .equ    061h        *RATE==(10 MHZ/fs)-1==999==03E7
Y0:      .equ    062h        *INITIALLY Y0==0, OUTPUT DMA
Y1:      .equ    063h        *Y(1)==C==SIN(2x3.14xFd/fs)
Y2:      .equ    064h        *Y(2)==0
A:       .equ    065h        *A==2 COS(WT)
B:       .equ    066h        *B==-1
*                            *A,B,C DIV BY 2 (MAX VALUE <1)
*                            *A=678E,B=C000,C=259E  Fd=1KHZ
*                            *A=278E,B=C000,C=3CDE  Fd=2KHZ

;>>>> WARNING, NO ABSOLUTE CODE
;>>>>                AORG   >0
         .sect   "AORG0"
             B       START

;>>>> WARNING, NO ABSOLUTE CODE
;>>>>                AORG   >20     *PM START AT >20
         .sect   "AORG1"
START        OUT     MODE,0  *OUTPUT MODE == 0Ah TO PORT 0
             OUT     RATE,1  *OUTPUT RATE ==03E7 TO PORT 1
             OUT     RATE,3  *DUMMY OUTPUT TO PORT 3
WAIT         BIOZ    MAIN    *WAIT TIL BIO==0(EOC OF A/D==0)
*                            *BRANCH TO MAIN AFTER EOC
             B       WAIT    *A/D PROVIDES TIMING
*MAIN SECTION
MAIN         LARK    0,Y2    *LOAD AUX REG AR0 WITH Y2
             LARK    1,B     *LOAD AUX REG AR1 WITH B
             ZAC             ;ZERO THE ACCUMULATOR
             LARP    0       *SELECT AR0 FOR INDIR ADDR
             LT      *-,1    *TR==Y2,DEC AR0,SELECT AR1
             MPY     *-,0    *PR==BxY2,DEC AR1,SELECT AR0
             LTD     *,1     *TR==Y1,ACC==BxY2,MOVE"DOWN"
             MPY     *,0     *PR==AxY1
             APAC            ;ACC=BxY2 + AxY1
             SACH    Y0,1    *SHIFT 1 DUE TO EXTRA BIT
*                            *STORE UPPER 16 BITS
             LAC     Y0      *SINCE A,B,C WAS DIVIDED BY 2
             ADD     Y0      *DOUBLE RESULT
             SACL    Y0      *STORE LOWER 16 BITS IN DMA 62
             DMOV    Y0      *MOVE"DOWN"IN DM FOR NEXT N
             OUT     Y0,2    *OUTPUT (DM 062h) TO PORT 2
             B       WAIT    *BRANCH BACK FOR NEXT   N
         .end
```

FIGURE A.9. Digital oscillator program converted to COFF (OSCICOFF.A00)

```
* DIGITAL OSCILLATOR PROGRAM - COFF VERSION OSCICOFF.ASM
* Y(n)=A Y(n-1) + B Y(n-2) + C X(n-1)
* Y0 =A Y1 + B Y2   FOR N>1 , X(n-1)=1 FOR n=1 (= 0 OTHERWISE)
* USES INDIRECT ADDRESSING ,  fs = 10 KHZ
* DATA MUST BE LOADED AT DMA 60-66 BEFORE RUNNING PROGRAM
         .bss    MODE,1      ;TRANSPARENT MODE FOR AIB
         .bss    RATE,1      ;RATE==(10 MHZ/fs)-1==999==03E7
         .bss    Y0,1        ;INITIALLY Y0==0, OUTPUT DMA
         .bss    Y1,1        ;Y(1)==C==SIN(2x3.14xFd/fs)
         .bss    Y2,1        ;Y(2)==0
         .bss    A,1         ;A==2 COS(WT)
         .bss    B,1         ;B==-1
*                            A,B,C DIV BY 2 (MAX VALUE <1)
*                            A=678E,B=C000,C=259E  Fd=1KHZ
*                            A=278E,B=C000,C=3CDE  Fd=2KHZ
         .sect   "INT_VEC"
             B       START
         .text
START        OUT     MODE,0  ;OUTPUT MODE == 0Ah TO PORT 0
             OUT     RATE,1  ;OUTPUT RATE ==03E7 TO PORT 1
             OUT     RATE,3  ;DUMMY OUTPUT TO PORT 3
WAIT         BIOZ    MAIN    ;WAIT TIL BIO==0(EOC OF A/D==0)
*                            ;BRANCH TO MAIN AFTER EOC
             B       WAIT    ;A/D PROVIDES TIMING
*MAIN SECTION
MAIN         LARK    0,Y2    ;LOAD AUX REG AR0 WITH Y2
```

FIGURE A.10. Digital oscillator program in COFF corrected (OSCICOFF.NEW)

```
          LARK    1,B        ;LOAD AUX REG AR1 WITH B
          ZAC                ;ZERO THE ACCUMULATOR
          LARP    0          ;SELECT AR0 FOR INDIR ADDR
          LT      *-,1       ;TR==Y2,DEC AR0,SELECT AR1
          MPY     *-,0       ;PR==BxY2,DEC AR1,SELECT AR0
          LTD     *,1        ;TR==Y1,ACC==BxY2,MOVE"DOWN"
          MPY     *,0        ;PR==AxY1
          APAC               ;ACC=BxY2 + AxY1
          SACH    Y0,1       ;SHIFT 1 DUE TO EXTRA BIT
                             ;STORE UPPER 16 BITS
          LAC     Y0         ;SINCE A,B,C WAS DIVIDED BY 2
          ADD     Y0         ;DOUBLE RESULT
          SACL    Y0         ;STORE LOWER 16 BITS IN DMA 62
          DMOV    Y0         ;MOVE"DOWN"IN DM FOR NEXT N
          OUT     Y0,2       ;OUTPUT (DM 062h) TO PORT 2
          B       WAIT       ;BRANCH BACK FOR NEXT  N
     .end
```

FIGURE A.10. (concluded)

```
/* COMMAND FILE FOR OSCICOFF PROGRAM                            */
   OSCICOFF.OBJ                 /* INPUT OBJECT FILE            */
   -O OSCICOFF.OUT              /* EXECUTABLE FILE IN COFF      */
   -M OSCICOFF.MAP              /* MAP FILE                     */
MEMORY
{
    PAGE 0 :                              /* PROGRAM MEMORY */
        INTS      : ORIGIN =    0H , LENGTH =   020H
        EXT_PROG  : ORIGIN = 020H , LENGTH = 0FEE0H
    PAGE 1 :                              /* DATA MEMORY    */
        BLOCK_B2  : ORIGIN = 060H , LENGTH =   020H
}
SECTIONS
{
    INT_VEC  : {
                 OSCICOFF.OBJ (INT_VEC)
               }     > INTS        PAGE 0
    .text    : {
                 OSCICOFF.OBJ (.text)
                 *(.TEXT)
               }     > EXT_PROG    PAGE 0
    .data    : {  }  > EXT_PROG    PAGE 0
    .bss     : {  }  > BLOCK_B2    PAGE 1
}
```

FIGURE A.11. Digital oscillator linker command file (OSCICOFF.CMD).

```
* LOWWPASS FILTER ,11 COEFFICIENTS-COFF VERSION-LP11COFF.ASM
* FIR FILTER, fs = 10 KHZ

;>>>> WARNING, NO ABSOLUTE CODE
;>>>>           AORG    0
        .sect   "AORG0"
        B       START

;>>>> WARNING, NO ABSOLUTE CODE
;>>>> TABLE   AORG    >20              *P.M. START AT >20
TABLE:  .sect   "AORG1"
*
* Place filter coefficients here
H0:     .word   00h               *FIRST COEFFICIENT   H0==H10
H1:     .word   05FEh             *SECOND COEFFICIENT  H1==H9
H2:     .word   0CEDh             *THIRD COEFFICIENT   H2==H8
H3:     .word   01363h            *        .
H4:     .word   017F6h            *        .
H5:     .word   0199Dh            *        .
H6:     .word   017F6h            *
H7:     .word   01363h            *
H8:     .word   0CEDh             *
H9:     .word   05FEh             *
H10:    .word   00h               *H10==H0
TABEND
*
*INITIALIZE AIB
MD:     .word   0Ah               *MODE FOR AIB
RATE:   .word   03E7h             *SAMPLING FREQUENCY fs==10 KHZ
*
MODE:   .equ    00h               *MODE FOR AIB(NO OFFSET)
CLOCK:  .equ    01h               *Fs IN DMA+1 (OFFSET BY 1)
YN:     .equ    02h               *OUTPUT IN DMA+2(OFF 2)
```

FIGURE A.12. Lowpass filter converted to COFF (LP11COFF.A00).

```
XN:      .equ    03h                    *NEW SAMPLE IN DMA+3 (OFF 3)
COUNT:   .equ    TABLE-TABEND-1         *COUNT== NO. OF COEFF. - 1
START    LDPK    6                      *SELECT PAGE 6,TOP OF B1 300
         LACK    MD                     *LOAD ACC WITH (MD)
         TBLR    MODE                   *TRANSFER INTO DMA 300
         OUT     MODE,0                 *OUT VALUE MODE TO PORT 0
         LACK    RATE                   *LOAD ACC WITH (RATE)
         TBLR    CLOCK                  *TRANSFER INTO DMA 301
         OUT     CLOCK,1                *OUT VALUE CLOCK TO PORT 1
         OUT     CLOCK,3                *DUMMY OUTPUT TO PORT 3
* LOAD FILTER COEFFICIENTS
         LARP    AR0                    *SELECT AR0 FOR INDIR ADDR
         LRLK    AR0,0200h              *AR0 POINT TO TOP OF BLOCK B0
         RPTK    COUNT                  *REPEAT FOR COEFFICIENTS
         BLKP    TABLE,*+               *P.M. 20-48 ==00h D.M. 200-228
*
         CNFP    ;CONFIGURE BLOCK B0 AS P.M.
WAIT     BIOZ    MAIN                   *NEW SAMPLE WHEN BIO PIN LOW
         B       WAIT                   *CONTINUE UNTIL EOC IS DONE
*
MAIN     IN      XN,2                   *INPUT NEW SAMPLE IN DMA 303
         LRLK    AR1,0303h+COUNT        *AR1 == LAST SAMPLE
         LARP    AR1                    *SELECT AR1 FOR INDIR ADDR
         MPYK    0                      *SET PRODUCT REGISTER TO 0
         ZAC     ;SET ACC TO 0
* THE BEEF
         RPTK    COUNT                  *MULTIPLY THE COEFFICIENTS
         MACD    0FF00h,*-              *H0*X(n-40) + H1*X(n-39) +...
*                                       *FIRST PMA FF00,LAST DMA 32B
         APAC    ;LAST ACCUMULATE
         SACH    YN,1                   *SHIFT LEFT 1, UPPER 16 BITS
         OUT     YN,2                   *OUT RESULT Y(n) FROM DMA 302
         B       WAIT                   *BRANCH FOR NEXT SAMPLE

         .end
```

FIGURE A.12. (concluded)

```
* LOWWPASS FILTER ,11 COEFFICIENTS-COFF VERSION-LP11COFF.ASM
* FIR FILTER, fs = 10 KHZ
         .sect   "INT_VEC"
         B       START
TABLE:   .data
*
* Place filter coefficients here
H0:      .word   00h                    ;FIRST COEFFICIENT   H0==H10
H1:      .word   05FEh                  ;SECOND COEFFICIENT H1==H9
H2:      .word   0CEDh                  ;THIRD COEFFICIENT  H2==H8
H3:      .word   01363h                 ;          .
H4:      .word   017F6h                 ;          .
H5:      .word   0199Dh                 ;          .
H6:      .word   017F6h                 ;
H7:      .word   01363h                 ;
H8:      .word   0CEDh                  ;
H9:      .word   05FEh                  ;
H10:     .word   00h                    ;H10==H0
TABEND
*
*INITIALIZE AIB
MD:      .word   0Ah                    ;MODE FOR AIB
RATE:    .word   03E7h                  ;SAMPLING FREQUENCY fs==10 KHZ
*
         .bss    MODE,1                 ;MODE FOR AIB(NO OFFSET)
         .bss    CLOCK,1                ;Fs IN DMA+1 (OFFSET BY 1)
         .bss    YN,1                   ;OUTPUT IN DMA+2(OFF 2)
         .bss    XN,1                   ;NEW SAMPLE IN DMA+3 (OFF 3)
COUNT:   .equ    TABEND-TABLE-1         ;COUNT== NO. OF COEFF. - 1
         .text
START    LDPK    6                      ;SELECT PAGE 6,TOP OF B1 300
         LACK    MD                     ;LOAD ACC WITH (MD)
         TBLR    MODE                   ;TRANSFER INTO DMA 300
```

FIGURE A.13. Lowpass filter in COFF corrected (LP11COFF.NEW)

```
            OUT      MODE,0           ;OUT VALUE MODE TO PORT 0
            LACK     RATE             ;LOAD ACC WITH (RATE)
            TBLR     CLOCK            ;TRANSFER INTO DMA 301
            OUT      CLOCK,1          ;OUT VALUE CLOCK TO PORT 1
            OUT      CLOCK,3          ;DUMMY OUTPUT TO PORT 3
    * LOAD FILTER COEFFICIENTS
            LARP     AR0              ;SELECT AR0 FOR INDIR ADDR
            LRLK     AR0,0200h        ;AR0 POINT TO TOP OF BLOCK B0
            RPTK     COUNT            ;REPEAT FOR COEFFICIENTS
            BLKP     TABLE,*+         ;P.M. 20-48 ==00h D.M. 200-228
    *
            CNFP                      ;CONFIGURE BLOCK B0 AS P.M.
    WAIT    BIOZ     MAIN             ;NEW SAMPLE WHEN BIO PIN LOW
            B        WAIT             ;CONTINUE UNTIL EOC IS DONE
    *
    MAIN    IN       XN,2             ;INPUT NEW SAMPLE IN DMA 303
            LRLK     AR1,0303h+COUNT  ; AR1 == LAST SAMPLE
            LARP     AR1              ;SELECT AR1 FOR INDIR ADDR
            MPYK     0                ;SET PRODUCT REGISTER TO 0
            ZAC                       ;SET ACC TO 0
    * THE BEEF
            RPTK     COUNT            ;MULTIPLY THE COEFFICIENTS
            MACD     0FF00h,*-        ;H0*X(n-40) + H1*X(n-39) +...
    *                                 ;FIRST PMA FF00,LAST DMA 32B
            APAC                      ;LAST ACCUMULATE
            SACH     YN,1             ;SHIFT LEFT 1, UPPER 16 BITS
            OUT      YN,2             ;OUT RESULT Y(n) FROM DMA 302
            B        WAIT             ;BRANCH FOR NEXT SAMPLE
            .end
```

FIGURE A.13. (concluded)

```
/* COMMAND FILE FOR LP11COFF PROGRAM                           */
    LP11COFF.OBJ                /* INPUT OBJECT FILE            */
    -O LP11COFF.OUT             /* EXECUTABLE FILE IN COFF      */
    -M LP11COFF.MAP             /* MAP FILE                     */
MEMORY
{
    PAGE 0 :                              /* PROGRAM MEMORY */
        INTS      : ORIGIN =   0H , LENGTH =   020H
        EXT_PROG  : ORIGIN = 020H , LENGTH = 0FEE0H
    PAGE 1 :                              /* DATA MEMORY     */
        BLOCK_B1  : ORIGIN = 0300H , LENGTH =  0100H

}
SECTIONS
{
    INT_VEC  : {
                 LP11COFF.OBJ (INT_VEC)
               }        > INTS         PAGE 0
    .text    : {
                 LP11COFF.OBJ (.text)
               *(.TEXT)
               }        > EXT_PROG     PAGE 0
    .data    : {  }     > EXT_PROG     PAGE 0
    .bss     : {  }     > BLOCK_B1     PAGE 1
}
```

FIGURE A.14. Lowpass filter linker command file (LP11COFF.CMD)

B

Analog Interface Board

B.1 INTRODUCTION

Figure B.1 shows the layout of the analog interface board (AIB). The analog input and output jacks are on the right side and the power connectors are on the top left. The AIB has several removable jumpers which are described in Table B.1. The

Power connectors

Audio output

Analog output

Analog input

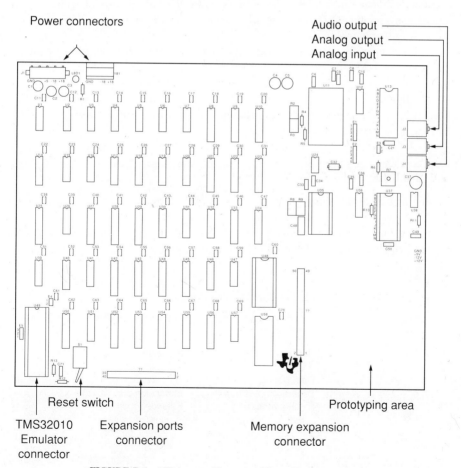

Reset switch

Prototyping area

TMS32010
Emulator
connector

Expansion ports
connector

Memory expansion
connector

FIGURE B.1. AIB layout. (Courtesy of Texas Instruments, Inc.)

TABLE B.1   Factory Jumper Settings

Jumper	Setting[a]	Description
E1	1-2 3-4	Connects J2 input jack to the A/D converter through the input filter
E2	1-2 3-4	Connects the D/A converter to J3 output jack through the output filter
E3	NC	Disables audio amplifier
E4	NC	Leaves V_{cc} pin on target socket open
E5	1-2	Routes A/D end-of-conversion signal to the BIO/pin on target socket
E6	1-2	Connects the sample and hold to the A/D

Source: Texas Instruments, Inc.

[a] NC, no connection.

important jumpers are E5, which connects either the BIO or the interrupt to the end-of-conversion (EOC) signal, and E2, which provides an observation point for the D/A output signal.

The AIB uses six ports to carry out its business. Three of these ports are used for expansion ports and extended memory; the other three (0–2) are used to support the one channel of analog input and one channel of analog output provided by the board. Table B.2 shows the address and function of each port. On most of the ports it is not possible to read back what has just been written because the output and input go to different registers.

B.2 SAMPLE RATE CLOCK

The frequency of the sample rate clock is controlled by the value written to port 1. The binary output of the clock can be programmed to provide the SOC for the A/D converter. This frequency is set by loading a constant, N, into the sample rate counter which is calculated according to the relationship

TABLE B.2   AIB I/O Port Descriptions

Port Address	Input Function	Output Function
0	A/D status register (read)	AIB control register
1	Not used	Sample rate for clock
2	A/D Converter (read) Data (read)	D/A converter Data (write)
3	Expansion port (read)	Expansion port (write)
4	Extended memory address (read)	Extended memory address (write)
5	Extended memory data (read)	Extended memory data (write)

Source: Texas Instruments, Inc.

$$f_s = \frac{f_{clk}}{N + 1}$$

or
$$N = \frac{f_{clk}}{f_s} - 1 \qquad\qquad\qquad \text{(B.1)}$$

where f_{clk} is the system clock frequency, f_s is the sampling frequency and N is a positive constant that is loaded into the sample rate counter. System clock frequency values are

$$f_{clk} = \begin{bmatrix} 5\text{ MHz} \rightarrow \text{TMS32010 board} \\ 5\text{ MHz} \rightarrow \text{TMS320C25 board with slower clock} \\ 10\text{ MHz} \rightarrow \text{TMS320C25 board as shipped} \end{bmatrix}$$

The AIB was not originally designed for the 10-MHz clock rate; consequently, it drops words depending on the orientation of the connecting cable. TI's solution at this writing is to go to the slower clock.

The clock frequency can be continuously programmable in the ranges

$$152.6\text{ Hz to }10\text{ MHz} \qquad (f_{clk} = 10\text{ MHz})$$
$$76.3\text{ Hz to }5\text{ MHz} \qquad (f_{clk} = 5\text{ MHz})$$

The sample rate constant must be loaded and the bit 0 (clock inhibit) of the control register cleared before the clock will run.

B.3 AIB CONTROL REGISTER

The control register (CR) is used to control the modes of operation of the A/D and D/A units along with the expansion and extended memory ports. A specific mode is selected by loading the control register with the proper bit pattern using the OUT DMA, 0 instruction. The control bits are described in Figure B.2.

B.4 D/A MODES OF OPERATION

Bit 1 (DA1) controls the operation mode of the D/A converter and provides two modes:

1. transparent mode and
2. sample delay mode.

The transparent mode is enabled by setting DA1 (bit 1) of the control register. The D/A converter has a double buffered input. The primary buffer is connected to the TMS320C25 and the secondary buffer is connected directly to the D/A converter,

15 · · ·	8	7	6	5	4	3	2	1	0
		DCW	DCR	U/D	AD2	AD1	DA2	DA1	INH

Bit 0   INH   Clock Inhibit
— 1 disable sample rate clock
— 0 enable sample rate clock

Bit 1   DA1   D/A Mode
— 1 Transparent mode for D/A
— 0 Sample delay mode for D/A

Bit 2   DA2   (E) D/A Mode
— 1 Transparent mode for expansion D/A
— 0 Sample delay mode for expansion D/A

Bit 3   AD1   A/D Mode
— 1 Automatic receive mode for A/D
— 0 Asynchonous recieve mode for A/D

Bit 4   AD2   (E) A/D Mode
— 1 Automatic receive mode for expansion A/D
— 0 Asynchonous recieve mode for expansion A/D

Bit 5   U/D   Up/Down Address Counter
— 1 Count up (extended memory address counter)
— 0 Count down (extended memory address counter)

Bit 6   DCR   Disable Count on Read
— 1 Disable counting on data reads (extended memory address counter)
— 0 Enable counting on data reads (extended memory address counter)

Bit 7   DCW   Disable Count On Write
— 1 Disable counting on data writes (extended memory address counter)
— 0 Enable counting on data writes (extended memory address counter)

FIGURE B.2. AIB control register

which means that the converter output will correspond to the digital word in the secondary buffer (allowing for the converter settling time). In the transparent mode, data written to the primary buffer are immediately transferred to the secondary. Under this mode, changes in the D/A output are governed by when the software chooses to write the data to the buffer.

The sample delayed mode is used to ensure that the output of the (D/A) is synchronized with the input (A/D), a requirement of many DSP applications. Typically, a sample is input via the A/D unit; then, after the processor modifies it, the modified sample is sent to the D/A unit. If the software has several different paths, the time between the input sample and the output sample will be variable. The sample delay mode solves this problem. On the start-of-conversion (SOC) pulse for the A/D, data in the primary buffer are moved to the secondary buffer. The effect of the current sample can be loaded immediately into the primary buffer, but it will not appear at the output until the next input sampling starts (the SOC pulse). Therefore, the output sample is delayed by one sample time before it is output. This mode is enabled by setting bit 3 (AD1) to start conversion on the sample rate clock, and clearing bit 1 (DA1) to synchronize the loading of the secondary buffer of the D/A.

B.5 A/D MODES OF OPERATION

The A/D converter has two modes of operation:

1. automatic receive mode and

2. asynchronous receive mode.

The automatic mode causes the input sample to be synchronized with the sample rate clock. In this mode, the sample rate clock drives the SOC pulse of the A/D. The SOC starts the A/D conversion, which takes about 29 μs. Upon receipt of the SOC

JUMPER	SETTING	DESCRIPTION
E1	1-2	Connects J2 input jack to the A/D
	3-4	converter through the input filter.
E2	1-2	Connects the D/A converter to J3
	3-4	output jack through the output filter.
E3	NC	Disables audio amplifier.
E4	NC	Leaves Vcc pin on target socket open.
E5	1-2	Routes A/D end-of-convert signal to
		BIO/ pin on target socket.
E6	1-2	Connects the sample and hold to the A/D.
		NC = No Connection

FIGURE B.3. Automatic receive timing signals

pulse, the A/D status flag is set indicating that conversion is in progress. When the conversion is complete, the status flag resets causing the end-of-conversion (EOC) pulse as seen in Figure B.3. The EOC pulse can be jumpered to drive either the BIO or the INT signals which are used to pass A/D status information to the processor. Reading or writing to port 2 or port 3 resets EOC to its inactive high state. Setting bit 3 (AD1) enables the automatic mode on the primary A/D converter.

The asynchronous mode allows the A/D conversion process to be started by external event rather than the on-board sample rate clock. The way this mode works is that the A/D conversion is initiated by a read to port 2 (the read generates the SOC pulse). The first access will read garbage, but it will start the conversion of the first sample. All subsequent reads will bring in legitimate samples and each will initiate the next conversion. Upon completion of a conversion, the digitized sample is placed in a buffer, where it remains undisturbed until it is read. Clearing bit 3 (AD1) initiates this mode.

B.6 AIB STATUS REGISTER

The status register is needed when the expansion A/D has been added and two A/Ds are being used. The board is built so that both can cause an INT or BIO signal. By reading the status register, the program can tell which A/D caused the BIO or INT signal. Executing any I/O instruction to port 2 or port 3 will clear the respective status register bit. Figure B.4 describes the status register.

B.7 TESTING THE AIB INTERFACE

The error rate on the transmission between the SWDS and the AIB is unacceptably high when using the cable supplied by TI and operating at the 10 MHz clock rate. Folding the cable significantly reduces the error rate. On the other hand, operating with an unfolded cable produces errors that are visible to the naked eye. This test setup is created to give a reliable way to test the interface. The test involves only

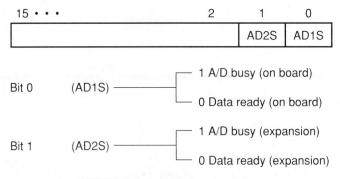

FIGURE B.4. Status register description

the digital transmission paths. Perform the following operations to the AIB prior to running the test:

1. Remove both the A/D unit (U11) and the D/A unit (U35) from the board.

2. Connect the input of the D/A socket (data lines D0-D11) directly to the output of the A/D socket (data lines D0-D11), effectively bypassing both the A/D and D/A.

3. Connect jumper pin 5 of U4 to P1-39 (connecter P1, pin 39) to allow the A/D latch to operate without the EOC signal.

The program shown in Figure B.5 is run to accomplish transmission over the cable. This program simply writes a pattern to the D/A port and reads it from the A/D port. If an error occurs in the transmission, the program traps to the label ERROR, or runs forever if no error occurs.

Test Results

Three different cable configurations were tested as follows:

1. *Standard Length Cable Supplied by TI.*   The program was unable to operate more than a few seconds without an error with the cable unfolded. When

```
*******************************************************************************
*                                                                             *
* This is a loop test to test the A/D-D/A latches & related circuitry on an   *
* AIB connected to a TMS320C25 board. The AIB has to be set up as follows:    *
* 1. Remove U11 & U35                                                         *
* 2. Jumper D0-D15 lines from U23/U34 to U9/U10                               *
* 3. Jumper P1-39 to U4-5 (lets signal Y6 latch data to U9 & U10)             *
* Also, set a breakpoint to label "ERROR"                                     *
*                                                                             *
*******************************************************************************
*
MODE     EQU    >60          used to send mode conditions to AIB
DATOUT   EQU    >61          used for output test data
DATIN    EQU    >62          used for input test data
*
         AORG   0            initialize program memory to location 0
         B      START        start program execution
         AORG   32           initialize rest of code to start here
*
PMODE    DATA   >02          AIB mode - make U23 & U34 transparent
PINIT    DATA   >5555        test data
*
START    LACK   PMODE        send mode setup to AIB
         TBLR   MODE
         OUT    MODE,0

         LACK   PINIT        send test data
         TBLR   DATOUT
LOOP     OUT    DATOUT,2
         OUT    DATOUT,6     latch data to U9 & U10
         IN     DATIN,2      data received is inverted by U9 & U10
         LAC    DATOUT       invert test data for comparison to received data
         CMPL
         ANDK   >FFFF
         SUB    DATIN        ACC=0 if data match
         BNZ    ERROR
         B      LOOP         loop forever if no error
ERROR    NOP
*
         END
```

FIGURE B.5. AIB interface test (AIBTEST.ASM)

the cable was folded the error-free operation time increased but was still unacceptable.

2. *Cable Shielded with Aluminum Foil.* The cable was shielded with foil (with several inches exposed on each end) which was connected to the chassis ground of the PC. The program would not fail for several hours, but would fail on an overnight test.

3. *Short Length Cable.* The cable was shortened to approximately 10 inches. With this cable the program ran overnight without an error.

Conclusions

When using the AIB with the 10 MHz clock, the cable should be shortened to ensure error-free operation. A well-shielded ribbon cable may be a possible solution since the quick-and-dirty shielding significantly reduced the error rate.

C

Analog Interface Chip and Macros

C.1 INTRODUCTION

The analog interface chip (AIC) affords a more economical alternative to the AIB. It provides two input channels and one output channel with several different programmable sample rates (five values ranging from 19.2 to 7.2 KHz). Both the A/D and D/A have 14 bits of resolution with 10-bit linearity over any 10-bit range. In addition, the chip contains antialiasing and reconstruction switched capacitor filters.

Since the AIC connects to the processor serial port, two of the processor registers that are used to communicate with the serial port must be set up as part of the AIC initialization procedure. These two registers are the interrupt mask register and the processor status register 1 (ST1), shown in Figures C.1 and C.2. The serial port interrupts, receive and send, can be masked by writing a 0 to bit positions 5 and 6, respectively. Three bits of the processor status register ST1 help control the serial port. These are the FSM (bit 5), FO (bit 3), and TXM (bit 2).

Figure C.3 shows the internal timing of the AIC and its associated timing registers [1,2]. As the diagram shows, the input and output frequencies are controlled independently, so that the input and output can be operated at two different sampling frequencies. The chip uses the TA and RA registers to divide the system clock down to the switched capacitor filter frequency, and the TB and RB registers to divide the filter frequency down to the sample rate frequency.

The AIC is controlled by four internal registers which are accessed using the secondary communications mode. The bit definitions of each of these registers are shown in Figure C.4. The first three registers control the sample frequency and the switched capacitor filter frequency. The last register is the control register, which selects the input, the gain option, and mode of operation as well as enables the loopback test and the bandpass filter.

C.2 RUNNING THE AIC LOOP PROGRAM USING MACROS

A step-by-step procedure for developing and executing an AIC-based program is given with the AIC LOOP program as an example. This program should be considered as a template for creating programs with the AIC macros. Before running the AIC LOOP program, it is necessary to perform the following steps:

1. Connect the AIC module to the SWDS as shown in Figure 3.15.

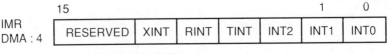

IMR
DMA : 4

15					1	0
RESERVED	XINT	RINT	TINT	INT2	INT1	INT0

Interrupt enable/disable (1/0)

XINT--Serial Port transmit interrupt enable
RINT--Serial Port receive interrupt enable
TINT--Internal Timer interrupt enable
INT2⌉
INT1⊣--External Interrupt enable
INT0⌋

FIGURE C.1. Interrupt mask register

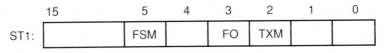

ST1:

15	5	4	3	2	1	0
	FSM		FO	TXM		

FSM--Frame Synchronization Mode
   1 operates with a frame sync pulse
   0 operates without a frame sync pulse

FO ---Format Bit
   1 configures serial port as 8 bit registers
   0 configures serial port as 16 bit registers

TXM--Transmit Mode
   1 configures serial port pin FSX to be an output
   0 configures serial port pin FSX to be an input

FIGURE C.2. ST1 bits that control the serial port

2. Connect a sine-wave generator to the primary input of the AIC. Set the amplitude to less than 3 V and the frequency to approximately 1 kHz. Monitor the input and output with an oscilloscope.

3. Create an executable file from the source file in Figure 3.18 renamed as AICL.ASM. Before starting this step, the following files should be in your default directory: OUTT, INN, INIT, INITS.MPO, and AICL.CTL.

   a. Assemble the source file to obtain an object file such as AICL.MPO.

   b. Link the object files by executing the following commands:

   LINKER

   if outside the SWDS, or

   LINK

   if within the SWDS. Answer the LINKER prompts as shown below:
   Control file [NUL.CTL]: AICL<ENTER>

INTERNAL TIMING CONFIGURATION

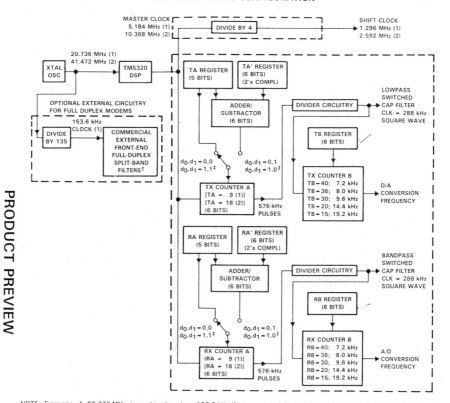

NOTE: Frequency 1, 20.736 MHz, is used to show how 153.6 kHz (for a commercially available modem split-band filter clock), popular speech and modem sampling signal frequencies, and an internal 288-kHz switched-capacitor filter clock can be derived synchronously and as submultiples of the crystal oscillator frequency. Since these derived frequencies are synchronous submultiples of the crystal frequency, aliasing does not occur as the sampled analog signal passes between the analog converter and switched-capacitor filter stages. Frequency 2, 41.472 MHz, is used to show that the AIC can work with high-frequency signals, which are used by high-speed digital signal processors.

†Split-band filtering can alternatively be performed after the analog input function via software in the TMS320.
‡These control bits are described in the AIC DX Data Word Format section.

FIGURE C.3. Internal timing configuration. (Courtesy of Texas Instruments Inc.)

Map file [AICL.MAP]: <ENTER>

Load file [AICL.LOD]: <ENTER>

The .LOD is the executable file, not the .MPO file.

4. Load the AICL.LOD into the TMS320C25 as described in Chapter 1.

5. If your program is successful, the output should be a sine wave that is delayed from the input. Vary the frequency and amplitude and observe the output tracking the input.

The procedure to create an executable run for any AIC-based macro

```
xx|<--TA reg-->|xx|<--RA reg-->|00   Loads the A xmit and
                                     receive reg

x|<---TA'reg-->|x|<---RA'reg-->|01   Loads the A' xmit and
                                     receive inc/dec reg

x|<---TB reg-->|x|<---RB reg-->|10   Loads the B xmit and
                                     receive reg

xxxxxxxxxxxxxxxxx|<--CTL reg-->|11   Loads the control reg
                 b7 ...       b2
    bit
    2 = 0/1 deletes/inserts the bandpass filter
    3 = 0/1 disables/enables the loopback function
    4 = 0/1 disables/enables the AUX port
    5 = 0/1 async/sync operation
    6 = 0/1 gain control*
    7 = 0/1 gain control*
    * b7=0, b6=1 gives single ended = ± 3 V full scale
```

FIGURE C.4. Secondary communication control words

program is essentially the same as outlined for the AIC LOOP program with a new source file taking the place of AICL.ASM. The only other change is that the linker control file in Figure 3.18 must be edited replacing AICL.MPO with the name of the new object file and renaming the control file AICL.CTL to "new name .CTL."

MACRO Descriptions

A brief functional description of each of the AIC macros follows:

Macro Call: INIT CSR, SRATE, FLTCLK, PTAB

Function: Sets up the interrupt vector and initializes the TMS320C25 and the AIC.

Arguments: **CSR:** Selects the configuration, scale option, mode, input port, and gain control (see Figure C.4 for bit settings). Note that the two LSBs must be equal to 1 (b0 = 1, b1 = 1).

   SRATE: Sets the A/D and D/A conversion rates (b0 = 0, b1 = 1).

   FLTCLK: Sets the switched capacitor filter frequency (b0 = 0, b1 = 0).

   PTAB: Gives an internal table a label. This name is arbitrary but should not be the same as another label in your program.

Calls: **INIT**: Initializes the processor and the AIC.

Usage: Can be called more than once. Works only with the synchronous mode (see Figures C.5 and C.6).

Macro Call: INN X

Function: Reads the AIC data receive register (DRR).

Arguments: **X**: User-specified data memory location.

Calls: None.

Usage: An IDLE instruction must precede it to ensure that the data are valid. It may share an idle with an OUTT or OUTS as illustrated:

```
IDLE           IDLE                    IDLE
INN    (or)    INN        (or)    OUTT
OUTS           OUTT                    INN
```

It cannot be used to read a switched input immediately following an OUTS. The sequence

```
IDLE
OUTS
INN
```

will not read the switched input since the switched input data are not placed in the DRR until the next clock pulse. A program listing is shown in Figure C.7.

Macro Call: OUTT Y

Function: Zeros the two LSBs of the content of Y and writes the result to the data transmit register (DXR).

Arguments: **Y**: User-specified data memory location.

Calls: None.

Usage: Must be synchronized with an IDLE instruction. See INN macro for details on the usage and also Figure C.8 for the program listing.

Macro Call: OUTS Y, CTL

Function: Performs an output and then switches the input to the input specified in the CTL argument.

Method: Appends 1s to the two LSBs of Y and writes the result to the DXR, thus placing the AIC in the secondary communication mode. After waiting one sample time, it sends the control information specified in the CTL argument. The DXR is refreshed immediately following the transmission of the control data so that the control data will not be transmitted to the analog output.

Arguments: **Y**: User-specified data memory location.
   CTL: Secondary communication control code.

Calls: None.

Usage: Mainly used to switch between the primary and auxiliary inputs. It requires one sample time to perform. It may also be used to change any of the operating modes that can be set using secondary communication; see Figure C.9 for the program listing.

```
****************************************************
*                                                  *
* INIT--SETS UP THE INTERRUPT VECTOR, INIT         *
*        THE PROCESSOR AND THE AIC                 *
****************************************************
*
INIT      $MACRO   CSR,SRATE,FLTCLK,PTAB
*
*     CSR--DATA FOR THE AIC CONTROL REGISTER
*          [CSR]=XXXXXXX(5 BITS)11
*
*   SRATE--DATA FOR THE TB & TA REGISTERS
*          [SRATE]=[X(TB 6 BITS)X(RA 6 BITS)]
*          SAMPLE RATE = SCFREQ/(B-VALUE)
*
*   FLTCLK--DATA FOR THE TA $ RA REGISTERS
*          [FLTCLK]=XX(TA 5-BIT)X(RA 5-BIT)00
*          DIVIDES THE SYSTEM CLOCK FREQUENCY
*          TO GIVE THE CLOCK RATE FOR THE BP AND
*          LP FILTERS. SCFREQ=SYSTEM FREQ/(A-VALUE)
*
*   PTAB--  ARBITRARY NAME FOR PROGRAM MEMORY DATA TABLE
*
          REF  INITS,DTAB,STATUS
DXR       EQU      >1
DRR       EQU      >0
          B        CODE
:PTAB.S:  DATA     0,03,:FLTCLK.S:,3,:SRATE.S:,3,:CSR.S:
CODE      SST      STATUS          SAVE DATA PAGE POINTER
          LDPK     0               POINT TO PAGE 0
****** TRANSFER PROGRAM TABLE TO DATA MEMORY
          LARK     AR0,DTAB        SET UP DATA POINTER
          RPTK     >06
          BLKP     :PTAB.S:,*+
          CALL INITS               INIT PROCESSOR AND AIC
          LST      STATUS
          $END
```

FIGURE C.5. INIT macro (INIT)

```
****************************************************************
*                                                              *
*INITS.ASM--INITIALIZES THE PROCESSOR AND THE AIC              *
*          CONTAINS THE INITERRUPT  ROUTINES                   *
*                                                              *
****************************************************************
          IDT      'INIT'
          DEF  INITS,DTAB,STATUS
**** REGISTER DEFINITIONS
DXR       EQU  1
IMR       EQU  4              INTERRUPT REGISTER
DRR       EQU  0
IM        EQU >20             ENABLE TRANSMIT INTERRUPT CODE
**** INITIALIZATION TABLE AND TEMP STORAGE
          DSEG
DTAB      BSS  7              INITIAL SET UP DATA
STATUS    BSS  1
TEMP      BSS  1
          DEND
          AORG     28
          B        XINT
          PSEG
************************************************
*                                              *
* XMIT--INTERRUPT ROUTINE                      *
*                                              *
************************************************
XINT
          RET              INT IS USED TO SYNC ONLY

*
INITS
          LALK     >03E0   INITIALIZE STATUS REGISTER NO. 1.
          SACL     TEMP    TXM = 0 - FRAME SYNC PULSES ARE INPUT.
          LST1     TEMP    XF =0 - XF SIGNAL ( LOW ) IS CONNECTED TO THE
*                                  AIC'S RESET PIN #2.
```

FIGURE C.6. Subroutine INITS and interrupt routine XINT (INITS.ASM)

```
*                            FSM = 1 - FRAME SYNC PULSES WILL NOT BE IGNORED.
*
        ZAC
        SACL    DXR      CLEAR DXR ( DATA XMIT REGISTER ) DMA 01.
*
        LACK    IM       LOAD IMR, ( INTERRUPT MASK REGISTER ) DMA 04, TO
        SACL    IMR      MASK ALL INTERRUPT EXCEPT 'XINIT'.
*
*
* INITIALIZE AIC REGISTERS   ( SECONDARY COMMUNICATION ).
*
        LARK    AR0,06   PUT LOOP COUNTER IN AR0 FOR 7 CYCLES.
        LARK    AR1,DTAB PUT DMA STARTING LOCATION OF XMIT DATA IN AR1.
        SXF              DISABLE THE AIC'S RESET BY SETTING THE XF SIGNAL.
SCND    LARP    1        SET ARP TO AR1.
        LAC     *+       LOAD A VALUE
        IDLE             WAIT UNTIL READY
        SACL    DXR      TRANSMIT
        LARP    0        SET ARP TO AR0.
        BANZ    SCND     BRANCH IF AR0 IS NOT 0 AND SUBTRACT 1 FROM AR0.
        RET
        PEND
        END
```

FIGURE C.6. (concluded)

```
*************************************************
*                                               *
* INN--READS THE SERIAL OR AIC INPUT REGISTER *
*************************************************
INN     $MACRO  X
*
*   X--USER INPUT BUFFER
*
        REF     STATUS
DRR     EQU     0
        SST     STATUS          SAVE DATA PAGE POINTER
        LDPK    0               DATA PAGE POINTER TO PAGE 0
        LAC     DRR             INPUT A VALUE
        SACL    :X.S:           STORE IT
        LST     STATUS          RESTORE DATA PAGE POINTER
        $END
```

FIGURE C.7. Input macro (INN)

```
*****************************************
*    OUTT--LOADS THE TRANSMIT REGISTER *
*         OF SERIAL PORT (AIC)         *
*****************************************
OUTT    $MACRO  Y
*
*      Y--USER OUTPUT BUFFER
*
        REF     STATUS
DXR     EQU     1
        SST     STATUS          SAVE DATA PAGE POINTER
        LDPK    0               POINT TO PAGE 0
        LAC     :Y.S:           LOAD OUTPUT VALUE
        ANDK    >FFFC           ZERO 2 LSB'S
        SACL    DXR             OUTPUT IT
        LST     STATUS          RESTORE THE STATUS
        $END
```

FIGURE C.8. Output macro (OUTT)

```
*********************************************
*        OUTS--OUTPUTS DATA, THEN           *
*        SWITCHES TO SPECIFIED INPUT PORT   *
*        TAKES MORE THAN ONE SAMPLE TIME    *
*        TO SWITCH INPUTS.  CAN ALSO BE     *
*        USED TO CHANGE ANY OF THE MODES    *
*        OR CONTROLED BY SECONDARY COMM     *
*********************************************
OUTS     $MACRO   Y,CTL
*        Y--USER OUTPUT BUFFER
*        CTL--SPECIFIES THE PORT
*              USUALLY 63 (PRI) OR 73 (AUX)
         REF      STATUS
DXR      EQU      1
         SST      STATUS          SAVE DATA PAGE POINTER
         LDPK     0               POINT TO PAGE 0
****OUTPUT DATA WITH 1'S IN THE 2 LSB'S
*    TO PUT AIC IN THE SEC. COMMUNICATIONS MODE
*
         LAC      :Y.S:           LOAD OUTPUT VALUE
         ORK      >03             SET UP FOR SEC COMM
         SACL     DXR             OUTPUT DATA + SEC CODE (11)
**** OUTPUT THE SECONDARY CONTROL WORD
         LACK     :CTL.S:         LOAD THE SWITCH CODE
         IDLE                     WAIT UNTIL READY
         SACL     DXR             OUTPUT IT
*** RESTORE VALID OUTPUT TO TRANSMIT REGISTER
         RPTK     >10             DELAY 17 CLK CYCLES
         NOP
         LAC      :Y.S:           LOAD OUTPUT VALUE AGAIN
         ANDK     >FFFC           ZERO LOW 2 BITS
         SACL     DXR             OUTPUT IT
         LST      STATUS          RESTORE THE PAGE POINTER
         $END
```

FIGURE C.9. Macro OUTS to use both sets of AIC inputs (OUTS).

REFERENCES

[1] *TLC32040I, TLC32040C, TLC32041I, TLC32041C Analog Interface Circuits, Advanced Information,* Texas Instruments Inc., Dallas, Tex., September 1987.

[2] *Second-Generation TMS320 User's Guide,* Texas Instruments Inc., Dallas, Tex., 1987.

D

Filter Development Package[†]

D.1 FIR FILTER DEVELOPMENT PACKAGE

The Filter Development Package (FDP) includes an FIR Development Package (FIRDP) to calculate the coefficients of an FIR filter: lowpass, highpass, bandpass, and bandstop, using either the rectangular, Hanning, Hamming, Blackman, or Kaiser windows. The instructions for running FIRDP are given in Figure D.1. Print Screens D.1 to D.3 show a walk-through design of an FIR bandpass filter. Figure D.2 shows the coefficients file, BP41R.COF, generated in TMS320C25 format. The coefficients represent a 41-coefficient bandpass filter using a rectangular window. A second example using a 41-coefficient bandpass filter with a Kaiser window is shown in Screen D.4. Part B of Figure D.1 for running FIRDP can now be used to incorporate the coefficients into a filter program such as RNDBLNK.ASM OR MACBLNK.ASM, shown in Figures D.3 and D.4, respectively. The following program listings, for the FIR Development Package, are shown :

1. FIRDP: FIR control module
2. FIRMM: FIR Development Package main menu, which includes a Kaiser window program module
3. EDITOR: utility program to convert the coefficients file into a format that can be used with MAGPHSE
4. MAGPHSE: to calculate the magnitude and phase of an FIR filter

[†]Contributed by T. Lima.

```
                         FIR DEVELOPMENT PACKAGE



                              Main Menu
                              -_-------

                         1....RECTANGULAR

                         2....HANNING

                         3....HAMMING

                         4....BLACKMAN

                         5....KAISER

                         6....Exit to DOS

          Enter window desired (number only) --> 1




     ***  FIR COEFFICIENT GENERATION USING THE RECTANGULAR WINDOW ***



                    Selections:

                              1....LOWPASS

                              2....HIGHPASS

                              3....BANDPASS

                              4 ...BANDSTOP

                              5....Exit back to Main Menu

          Enter desired filter type (number only) --> 3




     Specifications:
                    BANDPASS
                    Lower Cutoff Frequency = 1000 Hz
                    Upper Cutoff Frequency = 1500 Hz
                    Sampling Frequency (Fs) = 10000 Hz
                    Impulse Duration = 4 msec




     The calculated # of coefficients for the filter is: 41

     Enter # of coefficients desired ONLY if greater than 41
     otherwise, press <Enter> to continue -->
```

SCREEN D.1

```
Send coefficients to:
                    (S)creen
                    (P)rinter
                    (F)ile: contains TMS320 data format
                    (R)eturn to Filter Type Menu
                    (E)xit to DOS

Enter desired path --> P




                *** FIR DEVELOPMENT PACKAGE ***

        41 Coefficient BANDPASS Filter Using the RECTANGULAR Window

Lower Cutoff Frequency = 1000 Hz
Upper Cutoff Frequency = 1500 Hz
Sampling Frequency (Fs) = 10000 Hz
Impulse Duration (D) = 4 msec


            C'(n)                              H(n)
            -----                              ----

                                        DECIMAL              HEX

C'(0)  = +0.10000 = C'(40)    H(0)  = -0.00000 = H(40)  = 0
C'(1)  = +0.07042 = C'(39)    H(1)  = -0.00371 = H(39)  = FF87
C'(2)  = -0.00000 = C'(38)    H(2)  = -0.00000 = H(38)  = 0
C'(3)  = -0.06812 = C'(37)    H(3)  = +0.01202 = H(37)  = 18A
C'(4)  = -0.09355 = C'(36)    H(4)  = +0.02339 = H(36)  = 2FE
C'(5)  = -0.06366 = C'(35)    H(5)  = +0.02122 = H(35)  = 2B7
C'(6)  = +0.00000 = C'(34)    H(6)  = +0.00000 = H(34)  = 0
C'(7)  = +0.05730 = C'(33)    H(7)  = -0.03085 = H(33)  = FC0D
C'(8)  = +0.07568 = C'(32)    H(8)  = -0.05046 = H(32)  = F98B
C'(9)  = +0.04940 = C'(31)    H(9)  = -0.04042 = H(31)  = FAD4
C'(10) = -0.00000 = C'(30)    H(10) = -0.00000 = H(30)  = 0
C'(11) = -0.04042 = C'(29)    H(11) = +0.04940 = H(29)  = 653
C'(12) = -0.05046 = C'(28)    H(12) = +0.07568 = H(28)  = 9B0
C'(13) = -0.03085 = C'(27)    H(13) = +0.05730 = H(27)  = 756
C'(14) = +0.00000 = C'(26)    H(14) = +0.00000 = H(26)  = 0
C'(15) = +0.02122 = C'(25)    H(15) = -0.06366 = H(25)  = F7DA
C'(16) = +0.02339 = C'(24)    H(16) = -0.09355 = H(24)  = F407
C'(17) = +0.01202 = C'(23)    H(17) = -0.06812 = H(23)  = F748
C'(18) = -0.00000 = C'(22)    H(18) = -0.00000 = H(22)  = 0
C'(19) = -0.00371 = C'(21)    H(19) = +0.07042 = H(21)  = 904
C'(20) = -0.00000 = C'(20)    H(20) = +0.10000 = H(20)  = CCD
```

SCREEN D.2

D.2 MAGNITUDE AND PHASE RESPONSE

The program MAGPHSE calculates the magnitude and phase of a filter's response. This program can be used for either FIR or IIR filters where the coefficients can be obtained from a file or entered manually. A resulting file can then be used to plot the filter's response using a PC plot program. A walk-through example, shown in Screens D.5 and D.6, uses BP41R.COF, generated in Section D.1. Such a file must first be edited (using EDITOR.BAS) to create a file BP41R.MP, which can then be used to find the magnitude or phase response. The resulting tabulated magnitude response in Figure D.5 was used to obtain the plot in Figure D.6 using a PC plot program.

```
          Send coefficients to:
                              (S)creen
                              (P)rinter
                              (F)ile: contains TMS320 data format
                              (R)eturn to Filter Type Menu
                              (E)xit to DOS

          Enter desired path --> F




          Enter Filename : BP41R.COF




                              ...saving BP41R.COF
          Send coefficients to:
                              (S)creen
                              (P)rinter
                              (F)ile: contains TMS320 data format
                              (R)eturn to Filter Type Menu
                              (E)xit to DOS
```

SCREEN D.3

```
          FIR DEVELOPMENT PACKAGE



                    Main Menu
                    ---------

                    1....RECTANGULAR

                    2....HANNING

                    3....HAMMING

                    4....BLACKMAN

                    5....KAISER

                    6....Exit to DOS
     Enter window desired (number only) --> 5



     *** FIR COEFFICIENT GENERATION USING THE KAISER WINDOW ***


               Selections:
                              1....LOWPASS

                              2....HIGHPASS

                              3....BANDPASS

                              4....BANDSTOP

                              5....Exit back to Main Menu
     Enter desired filter type (number only) --> 3
```

SCREEN D.4

```
      Specifications:
                      BANDPASS
                      Passband Ripple (AP) = .5 db
                      Stopband Attenuation (AS) = 30 db
                      Lower Passband Frequency = 1000 Hz
                      Upper Passband Frequency = 1500 Hz
                      Lower Stopband Frequency = 600 Hz
                      Upper Stopband Frequency = 1900 Hz
                      Sampling Frequency (Fs) = 10000 Hz


   The calculated # of coefficients required is: 41

   Enter # of coefficients desired ONLY if greater than 41
   otherwise, press <Enter> to continue -->
```

SCREEN D.4 (concluded)

```
       *** Magnitude and Phase Response of a DT System ***


 (F)IR or (I)IR ? F

 Enter coefficients via (T)yping or (F)ile (E = Exit to BASIC) --> E

 (Note: If file then FIRDP file must be edited
        using the FIRDP File Editor.  To run:
        enter <E> above, then load and run
        EDITOR.BAS)




LOAD"EDITOR




                 *** UTILITY - FIRDP File Editor ***




     Enter FIRDP Data File to be edited --> BP41R.COF




          *** Magnitude and Phase Response of a DT System ***

 (F)IR or (I)IR ? F

 Enter coefficients via (T)yping or (F)ile (E = Exit to BASIC) --> F

 (Note: If file then FIRDP file must be edited
        using the FIRDP File Editor.  To run:
        enter <E> above, then load and run
        EDITOR.BAS)
```

SCREEN D.5

```
            *** Magnitude and Phase Response of a DT System ***

    Enter edited filename --> BP41R.MP

    Enter the Sampling Frequency in Hz --> 10000

    Enter the number of steps desired --> 40

    Tabulate (M)agnitude or (P)hase response --> M

    Do you want to normalize the magnitude (Y/N) --> Y

    Do you want (P)ower or (A)mplitude form  --> A

    Are the above entries correct (Y/N) Y
    Send output to:
                    (S)creen
                    (P)rinter
                    (F)ile: (FREQ,AMPLITUDE) Format
                    (RU)n program with same coefficients
                    (R)un program with different coefficients
                    (E)xit to DOS

    Enter desired path --> F



    Enter Filename : BP41R.MAG

    Send output to:
                    (S)creen
                    (P)rinter
                    (F)ile: (FREQ,AMPLITUDE) Format
                    (RU)n program with same coefficients
                    (R)un program with different coefficients
                    (E)xit to DOS

    Enter desired path --> P
```

SCREEN D.6

INSTRUCTIONS FOR RUNNING THE FIR DEVELOPMENT PACKAGE (FIRDP)

A.   1. Run BASIC, for example, type GWBASIC, hit <Enter>

     2. load"FIRDP.BAS",R, <Enter> to load and run

     3. Enter Menu choice and follow menu driven options

     4. If you created and saved a data file containing the
        coefficients, then proceed with step B to create a TMS320
        source code. FIRST, exit FIRDP by following menu options

B.   1. Run your editor, for example, with PC-WRITE,
        Type "ED filename.asm. This filename is a filter program
        with no coefficients, either

                    B>ED RNDBLNK.ASM

                         or

                    B>ED MACBLNK.ASM

        RNDBLNK.ASM includes the noise generator with LTD/MPY, and
        MACBLNK.ASM,the MACD implementation, requires an external source
        (Above assumes that editor and asm file are on same drive)

     2. Press escape <ESC> for no back-up

     3. Scroll down using the cursor key. Place the cursor on the
        line following the statement "Place coefficients here"

FIGURE D.1. Instructions for running the FIR development package (FIRDP)

4. Press <CNTRL-F3>, you will be prompted at the top of the
   screen for a filename. Type in the name of saved file,
   containing the coefficients, generated using FIRDP in part A

5. The coefficient file will appear, Highlighted. Use Backspace
   to remove EOF marker. Press <F5> to remove highlighting

6. Press <F1>, then <F5>, for a prompt to change the
   name of the file. Remember!!! file is still called
   RNDBLNK.ASM or MACBLNK.ASM. Type in new filename

7. Press <F1>, <F3>, to save your filter program under
   the new name. Then <F1>, <F2> to exit PC-WRITE

FIGURE D.1. (concluded)

```
H0      DATA    >0
H1      DATA    >FF87
H2      DATA    >0
H3      DATA    >18A
H4      DATA    >2FE
H5      DATA    >2B7
H6      DATA    >0
H7      DATA    >FC0D
H8      DATA    >F98B
H9      DATA    >FAD4
H10     DATA    >0
H11     DATA    >653
H12     DATA    >9B0
H13     DATA    >756
H14     DATA    >0
H15     DATA    >F7DA
H16     DATA    >F407
H17     DATA    >F748
H18     DATA    >0
H19     DATA    >904
H20     DATA    >CCD
H21     DATA    >904
H22     DATA    >0
H23     DATA    >F748
H24     DATA    >F407
H25     DATA    >F7DA
H26     DATA    >0
H27     DATA    >756
H28     DATA    >9B0
H29     DATA    >653
H30     DATA    >0
H31     DATA    >FAD4
H32     DATA    >F98B
H33     DATA    >FC0D
H34     DATA    >0
H35     DATA    >2B7
H36     DATA    >2FE
H37     DATA    >18A
H38     DATA    >0
H39     DATA    >FF87
H40     DATA    >0
```

FIGURE D.2. Coefficients for bandpass filter with rectangular window

```
* NO EXTERNAL INPUT REQUIRED FILTER SET FOR 41 COEFFICIENTS
* ADJUST ACCORDINGLY FOR DIFFERENT # OF COEFFICIENTS
        IDT     'FILT'
*       DATA MEMORY ALLOCATION/DEFINITION
STAT0   EQU     >60             STATUS REG STORAGE FOR ISR
LH1000  EQU     >61             NOISE GEN POSITIVE SCALER
LHF000  EQU     >62             NOISE GEN NEGATIVE SCALER
RNUMH   EQU     >63             RANDOM NUM UPPER 16-BITS
RNUML   EQU     >64             RANDOM NUM LOWER 16-BITS
MASK15  EQU     >65             BIT-15 MASK
TEMP    EQU     >66             SCRATCH PAD MEMORY AREA
SWDS1   EQU     >7E             RESERVED FOR SWDS USAGE
SWDS2   EQU     >7F             RESERVED FOR SWDS USAGE
CH0     EQU     >0200           FILTER COEFF STORAGE
X0      EQU     >0229           FILTER INPUT SAMPLES
*       MEMORY MAPPED REGISTER LOCATIONS
TIMER   EQU     >0002           INTERVAL TIMER REGISTER
PERIOD  EQU     >0003           PERIOD REGISTER FOR TIMER
INTMSK  EQU     >0004           INTERRUPT MASK REGISTER
        AORG    0
        B       INIT
        AORG    >18
        B       START           TIMER INTERRUPT VECTOR
        AORG    >20     *****PROGRAM MEMORY DATA STORAGE*****
SEEDH   DATA    >7E52           RANDOM NUM SEED,UPPER 16-BITS
SEEDL   DATA    >1603           RANDOM NUM SEED,LOWER 16-BITS
RATE    DATA    >03E7           OUTPUT DATA RATE (10000 Hz)
IMASK   DATA    >0008           ENABLE TIMER INTERRUPT ONLY
*
* Place Coefficients here.
*
INIT    CNFD                    CONFIGURE B0 AS DATA MEMORY
        LDPK    0               POINT TO DATA PAGE 0(forever)
        LALK    SEEDH           POINT TO P.M. STORAGE
        TBLR    RNUMH           XFER SEEDH TO RNUMH IN D.M.
        ADDK    1               INCREMENT READ ADDRESS IN ACC
        TBLR    RNUML           XFER SEEDL TO RNUML IN D.M.
        ADDK    1               INCREMENT READ ADDRESS IN ACC
        TBLR    PERIOD          XFER RATE TO PERIOD REGISTER
        ADDK    1               INCREMENT READ ADDRESS IN ACC
        TBLR    INTMSK          XFER IMASK TO MASK REGISTER
        LALK    >1000
        SACL    LH1000          INITIALIZE LH1000 TO >1000
        NEG
        SACL    LHF000          INITIALIZE LHF000 TO >F000
        LALK    >8000
        SACL    MASK15          INITIALIZE MASK15 TO >8000
* COPY BANDPASS FILTER COEFFS FROM PGMEM TO DMEM
        LRLK    0,CH0           LOAD AR0 WITH WRITE ADX
        LARP    0               POINT TO AR0
        RPTK    >28             LOOP COUNT (41 values)
        BLKP    H0,*+           COPY H0-H40 TO CH0-CH40
        EINT                    ENABLE INTERRUPT
*
WAIT    B       WAIT            WAIT HERE UNTIL AN INTERRUPT
*Modulo 2 sum of the 32-bit noise word bits 31,30,28,and 17
*yields a random bit (1 or 0) shifted into the LSB of the
*noise word, scaled and output to a D/A converter. Since
*only upper 16-bits are involved in determining feedback,
*summation is done in lower accumulator for convenience.
START   LAC     RNUMH           GET RANDOM NUM UPPER 16-BITS
ADD     RNUMH,1         ADD BIT-14 TO BIT-15
AND     MASK15          MASK BIT-15
ADD     RNUMH,3         ADD BIT-12 TO THE SUMMATION
AND     MASK15          MASK BIT-15
ADD     RNUMH,14        ADD BIT-1 TO THE SUMMATION
AND     MASK15          MASK BIT-15
SACH    TEMP,1          SHIFT R 15 (LEFT 1,RIGHT 16)
ZALH    RNUMH           RESTORE RNUMH TO UPPER ACCUM
ADD     RNUML           RESTORE RNUML TO LOWER ACCUM
SFL                     SHIFT 32-BIT ACCUM LEFT 1
ADD     TEMP            ADD FEEDBACK BIT
SACH    RNUMH           STORE UPPER 16-BIT NOISE WORD
SACL    RNUML           STORE LOWER 16-BIT NOISE WORD
LAC     TEMP            RESTORE RANDOM BIT INTO ACCUM
BZ      MINUS
```

FIGURE D.3. Program with PRN as input (RNDBLNK.ASM)

```
              LAC     LH1000            SET OUTPUT POSITIVE IF BIT=1
              B       FILT
MINUS         LAC     LHF000            SET OUTPUT NEGATIVE IF BIT=0
* THE BANDPASS FILTER STARTS HERE
FILT          LRLK    0,X0              LOAD POINTER TO NEWEST SAMPLE
              LARP    0                 POINT TO NEWEST SAMPLE STORAGE
              SACL    *                 STORE SCALED OUTPUT VALUE
              ZAC
              LRLK    0,CH0+>28         LOAD POINTER TO COEFF H40
              LRLK    1,X0+>28          LOAD POINTER TO SAMPLE X40
              LARP    1                 POINT TO X40
              LTD     *-,0              LOAD X40
              MPY     *-,1              H40*X40
              LTD     *-,0              *
              MPY     *-,1              * H39*X39
              LTD     *-,0              *
              MPY     *-,1              * H38*X38
              LTD     *-,0              *
              MPY     *-,1              * H37*X37
              LTD     *-,0              *
              MPY     *-,1              * H36*X36
              LTD     *-,0              *
              MPY     *-,1              * H35*X35
              LTD     *-,0              *
              MPY     *-,1              * H34*X34
              LTD     *-,0              *
              MPY     *-,1              * H33*X33
              LTD     *-,0              *
              MPY     *-,1              * H32*X32
              LTD     *-,0              *
              MPY     *-,1              * H31*X31
              LTD     *-,0              *
              MPY     *-,1              * H30*X30
              LTD     *-,0              *
              MPY     *-,1              * H29*X29
              LTD     *-,0              *
              MPY     *-,1              * H28*X28
              LTD     *-,0              *
              MPY     *-,1              * H27*X27
              LTD     *-,0              *
              MPY     *-,1              * H26*X26
              LTD     *-,0              *
              MPY     *-,1              * H25*X25
              LTD     *-,0              *
              MPY     *-,1              * H24*X24
              LTD     *-,0              *
              MPY     *-,1              * H23*X23
              LTD     *-,0              *
              MPY     *-,1              * H22*X22
              LTD     *-,0              *
              MPY     *-,1              * H21*X21
              LTD     *-,0              *
              MPY     *-,1              * H20*X20
              LTD     *-,0              *
              MPY     *-,1              * H19*X19
              LTD     *-,0              *
              MPY     *-,1              * H18*X18
              LTD     *-,0              *
              MPY     *-,1              * H17*X17
              LTD     *-,0              *
              MPY     *-,1              * H16*X16
              LTD     *-,0              *
              MPY     *-,1              * H15*X15
              LTD     *-,0              *
              MPY     *-,1              * H14*X14
              LTD     *-,0              *
              MPY     *-,1              * H13*X13
              LTD     *-,0              *
              MPY     *-,1              * H12*X12
              LTD     *-,0              *
              MPY     *-,1              * H11*X11
              LTD     *-,0              *
              MPY     *-,1              * H10*X10
              LTD     *-,0              *
              MPY     *-,1              * H9*X9
              LTD     *-,0              *
              MPY     *-,1              * H8*X8
```

FIGURE D.3. (continued)

```
          LTD      *-,0            *
          MPY      *-,1            *  H7*A7
          LTD      *-,0            *
          MPY      *-,1            *  H6*X6
          LTD      *-,0            *
          MPY      *-,1            *  H5*X5
          LTD      *-,0            *
          MPY      *-,1            *  H4*X4
          LTD      *-,0            *
          MPY      *-,1            *  H3*X3
          LTD      *-,0            *
          MPY      *-,1            *  H2*X2
          LTD      *-,0            *
          MPY      *-,1            *  H1*X1
          LTD      *-,0            *
          MPY      *-,1            *  H0*X0
          APAC                     ACC LAST PRODUCT
          SACH     TEMP,1          STORE FILTER OUTPUT
          OUT      TEMP,2          OUTPUT TEMP TO D/A
          EINT
          RET
```

FIGURE D.3. (concluded)

```
* MACBLNK.ASM PROGRAM
* LOOP SET UP TO ACCEPT ANY NUMBER OF COEFFICIENTS (UP TO 256)
* NOTE: fs = 10 KHZ,CHANGE RATE FOR A DIFFERENT SAMPLING FREQ.
          AORG     0
          B        START
TABLE     AORG     >20             *P.M. START AT >20
*
* Place filter coefficients here
TABEND
*
*INITIALIZE AIB
MD        DATA     >000A           *MODE FOR AIB
RATE      DATA     >03E7           *SAMPLING FREQUENCY fs=10 KHZ
*
MODE      EQU      >0              *MODE FOR AIB(NO OFFSET)
CLOCK     EQU      >1              *Fs IN DMA+1 (OFFSET BY 1)
YN        EQU      >2              *OUTPUT IN DMA+2(OFF 2)
XN        EQU      >3              *NEW SAMPLE IN DMA+3 (OFF 3)
COUNT     EQU      TABEND-TABLE-1  *COUNT= NO. OF COEFF. - 1
START     LDPK     6               *SELECT PAGE 6,TOP OF B1 300
          LACK     MD              *LOAD ACC WITH (MD)
          TBLR     MODE            *TRANSFER INTO DMA 300
          OUT      MODE,0          *OUT VALUE MODE TO PORT 0
          LACK     RATE            *LOAD ACC WITH (RATE)
          TBLR     CLOCK           *TRANSFER INTO DMA 301
          OUT      CLOCK,1         *OUT VALUE CLOCK TO PORT 1
          OUT      CLOCK,3         *DUMMY OUTPUT TO PORT 3
* LOAD FILTER COEFFICIENTS
          LARP     AR0             *SELECT AR0 FOR INDIR ADDR
          LRLK     AR0,>200        *AR0 POINT TO TOP OF BLOCK B0
          RPTK     COUNT           *REPEAT FOR COEFFICIENTS
          BLKP     TABLE,*+        *P.M. 20-48 => D.M. 200-228
*
          CNFP                     *CONFIGURE BLOCK B0 AS P.M.
WAIT      BIOZ     MAIN            *NEW SAMPLE WHEN BIO PIN LOW
          B        WAIT            *CONTINUE UNTIL EOC IS DONE
*
MAIN      IN       XN,2            *INPUT NEW SAMPLE IN DMA 303
          LRLK     AR1,>303+COUNT  *AR1 = LAST SAMPLE
          LARP     AR1             *SELECT AR1 FOR INDIR ADDR
          MPYK     0               *SET PRODUCT REGISTER TO 0
          ZAC                      *SET ACC TO 0
* THE BEEF
          RPTK     COUNT           *MULTIPLY THE COEFFICIENTS
          MACD     >FF00,*-        *H0*X(n-40) + H1*X(n-39) +...
*                                  *FIRST PMA FF00,LAST DMA 32B
          APAC                     *LAST ACCUMULATE
          SACH     YN,1            *SHIFT LEFT 1, UPPER 16 BITS
          OUT      YN,2            *OUT RESULT Y(n) FROM DMA 302
          B        WAIT            *BRANCH FOR NEXT SAMPLE
          END
```

FIGURE D.4. Program using MACD implementation (MACBLNK.ASM).

```
UTILITY - MAGNITUDE AND PHASE RESPONSE PROGRAM

    *** AMPLITUDE Response of a DT System ***

Coefficients: BP4IR.MAG


              FREQUENCY              AMPLITUDE
              ---------------------------------
                  0.00                0.01508
                128.21                0.00111
                256.41                0.01673
                384.62                0.00488
                512.82                0.02430
                641.03                0.01555
                769.23                0.05022
                897.44                0.09519
               1025.64                0.48230
               1153.85                0.89406
               1282.05                1.00000
               1410.26                0.70684
               1538.46                0.26505
               1666.67                0.00884
               1794.87                0.04256
               1923.08                0.01023
               2051.28                0.01781
               2179.49                0.00752
               2307.69                0.00978
               2435.90                0.00584
               2564.10                0.00619
               2692.31                0.00469
               2820.51                0.00414
               2948.72                0.00407
               3076.92                0.00298
               3205.13                0.00358
               3333.33                0.00221
               3461.54                0.00321
               3589.74                0.00163
               3717.95                0.00300
               3846.15                0.00127
               3974.36                0.00273
               4102.56                0.00085
               4230.77                0.00264
               4358.97                0.00057
               4487.18                0.00257
               4615.38                0.00034
               4743.59                0.00254
               4871.79                0.00016
               5000.00                0.00245
```

FIGURE D.5. Magnitude response of bandpass filter with rectangular window

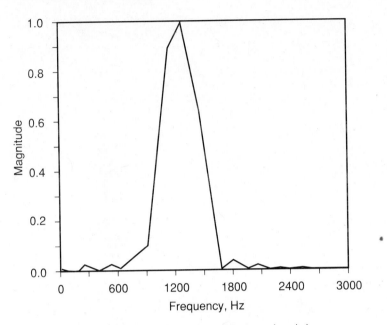

FIGURE D.6. Plot of bandpass filter with rectangular window.

D.3 FIRDP LISTING AND SUPPORT PROGRAM

```
10 REM                              *** FIRDP Control module ***
20 REM
30 REM  This module provides a self-startup of the FIRDP program.
40 REM
50 CLS
60 KEY OFF
70 REM                          *** Display Opening Screen ***
80 REM
90 LOCATE 12
100 PRINT TAB(24);"*** FIR DEVELOPMENT PACKAGE ***"
110 FOR DELAY=1 TO 2000:NEXT DELAY
120 REM
130 REM                          *** Load in Main Menu ***
140 REM
150 LOAD"firmm.bas",R
160 REM
170 END
```

                         FIR DEVELOPMENT PACKAGE

```
20 REM
30 REM   This program module will generate the necessary FIR coefficients
40 REM   using a Rectangular, Hanning, Hamming, and Blackman window
50 REM   sequence.  The user can then define the output path to either
60 REM   the screen, line printer, or an external file which can then
70 REM   be merged with a FIR program to implement the filter.
80 REM
90 REM   NOTE: The Kaiser window sequence is located in a separate module.
```

FIRDP Listing

```
100 REM
110 DIM C(256),H(256),CARY(256),CHEX$(256)
120 PI=3.1415927£
130 CLS
140 KEY OFF
150 REM                        *** Generate Main Menu ***
160 REM
170 LOCATE 3:PRINT TAB(27);"FIR DEVELOPMENT PACKAGE"
180 LOCATE 8
190 PRINT TAB(33);"Main Menu"
200 PRINT TAB(33);"---------"
210 PRINT
220 PRINT TAB(33);"1....RECTANGULAR":PRINT
230 PRINT TAB(33);"2....HANNING":PRINT
240 PRINT TAB(33);"3....HAMMING":PRINT
250 PRINT TAB(33);"4....BLACKMAN":PRINT
260 PRINT TAB(33);"5....KAISER":PRINT
270 PRINT TAB(33);"6....Exit to DOS"
280 PRINT
290 INPUT "          Enter window desired (number only) --> ",WIN
300 XPOS=10
310 IF WIN = 6 THEN CLS:SYSTEM
320 IF WIN = 1 THEN WIN$="RECTANGULAR":XPOS=8
330 IF WIN = 2 THEN WIN$="HANNING"
340 IF WIN = 3 THEN WIN$="HAMMING"
350 IF WIN = 4 THEN WIN$="BLACKMAN"
360 IF WIN = 5 THEN LOAD"Kaiserdp.bas",R
370 IF WIN < 1 OR WIN > 6 THEN 130
380 REM                        *** Generate filter type menu ***
390 CLS
400 LOCATE 4
410 PRINT TAB(XPOS);"***  FIR COEFFICIENT GENERATION USING THE ";WIN$;" WINDOW *
420 LOCATE 8
430 PRINT TAB(22);"Selections:"
440 PRINT
450 PRINT TAB(33);"1....LOWPASS"
460 PRINT
470 PRINT TAB(33);"2....HIGHPASS"
480 PRINT
490 PRINT TAB(33);"3....BANDPASS"
500 PRINT
510 PRINT TAB(33);"4 ...BANDSTOP"
520 PRINT
530 PRINT TAB(33);"5....Exit back to Main Menu"
540 PRINT
550 INPUT "          Enter desired filter type (number only) --> ",TYPE
560 IF TYPE = 5 THEN 130
570 IF TYPE = 1 THEN GOSUB 2000:GOTO 620         'Lowpass Prompts Routine
580 IF TYPE = 2 THEN GOSUB 2120:GOTO 620         'Highpass Prompts Routine
590 IF TYPE=3 OR TYPE=4 THEN GOSUB 2240:GOTO 620 'Bandpass/stop Prompts Routine
600 IF TYPE < 1 OR TYPE > 5 THEN 390
610 GOTO 130
620 REM      *** Prompt for general information and output specifications ***
630 LOCATE 12
640 PRINT "                                                             "
650 MES=0
660 LOCATE 12
670 INPUT "          Enter the sampling frequency (Fs) in Hz --> ",FSAM
680 IF TYPE=1 THEN LCUT=0
690 IF TYPE=2 THEN HCUT=0
700 IF FSAM/2<LCUT OR FSAM/2<HCUT THEN MES=1:GOSUB 4140:GOTO 650   'display error
710 LOCATE 12
720 PRINT "                                                             "
730 IF TYPE=1 OR TYPE=2 THEN YPOS=4 ELSE YPOS=5
740 LOCATE (YPOS):PRINT TAB(24);"Sampling Frequency (Fs) =";FSAM;"Hz"
750 LOCATE 14
760 PRINT "          Number of Coefficients = (D*Fs)+1"
770 PRINT
780 LOCATE 12
790 INPUT "          Enter the duration of the impulse response (D) in msec --> ",
800 LOCATE (YPOS+1):PRINT TAB(24);"Impulse Duration =";D;"msec"
810 LOCATE 12
820 FOR I = 1 TO 3
830 PRINT "
840 NEXT I
850 LOCATE 12:INPUT "          Are the above specifications correct (y/n) ? ",RES$
860 IF RES$="n" OR RES$="N" THEN 390
870 REM
880 REM                        *** calculate number of coefficients required ***
890 D = D/1000
900 NYQST=FSAM/2
```

```
910 Q=CINT((D*FSAM)/2)
920 COEFF=2*Q+1
930 LOW=LCUT/NYQST          'Nu 1
940 IF TYPE=2 THEN GOTO 960      'if highpass then high=1
950 HIGH=HCUT/NYQST         'Nu 2
960 LOCATE 12
970 PRINT "                                                                  "
980 LOCATE 12
990 PRINT "          The calculated f of coefficients for the filter is:";COEFF
1000 PRINT
1010 PRINT "          Enter f of coefficients desired ONLY if greater than";COEFF
1020 INPUT "          otherwise, press <Enter> to continue --> ",TEMP
1030 IF TEMP = 0 THEN 1160
1040 IF TEMP < COEFF THEN 1080
1050 COEFF=TEMP
1060 Q=(COEFF-1)/2
1070 GOTO 1160
1080 FOR BLINK=1 TO 10
1090 LOCATE 20
1100 PRINT TAB(12);"ERROR! : Order will not satisfy specifications - reenter"
1110 FOR DELAY=1 TO 100:NEXT DELAY
1120 LOCATE 20
1130 PRINT "                                                                  "
1140 NEXT BLINK
1150 GOTO 980
1160 CLS
1170 REM
1180 REM
1190 LOCATE 12:PRINT TAB(28)"Please wait ...working"
1200 REM
1210 GOSUB 1420          'Routine to calculate FS coefficients, C'(n)
1220 REM
1230 IF WIN = 2 THEN GOSUB 1640     'Hanning
1240 IF WIN = 3 THEN GOSUB 1760     'Hamming
1250 IF WIN = 4 THEN GOSUB 1880     'Blackman
1260 REM
1270 REM                 *** Rearrange the coefficients ***
1280 FOR N=0 TO Q
1290 H(N)=C(Q-N)
1300 NEXT N
1310 REM                 *** Generate the symmetry about Q ***
1320 FOR N=1 TO Q
1330 H(Q+N)=H(Q-N)     '(i.e., H°i§ = C°q§-i)
1340 NEXT N
1350 REM                 *** Convert coefficients to Hex ***
1360 GOSUB 3990    'call hex conversion routine
1370 PRINT
1380 GOSUB 2430    'call output menu routine
1390 REM
1400 END
1410 REM ===================== FS Calculation Routine =====================
1420 REM
1430 C(0)=HIGH-LOW
1440 IF TYPE = 4 THEN C(0) = 1-C(0)    'for bandstop
1450 FOR I=1 TO Q
1460 C(I)=(SIN(HIGH*I*PI)/(I*PI))-(SIN(LOW*I*PI)/(I*PI))    'Fourier Series
1470 IF TYPE = 4 THEN C(I)=-C(I)       'for bandstop
1480 NEXT I
1490 RETURN
1500 REM =============================================================
1510 REM
1520 REM ================= Rectangular Window Routine =================
1530 REM
1540 REM       This trivial routine is placed here for documentation
1550 REM       purposes only, it is not called from anywhere within the
1560 REM       main program.
1570 REM
1580 REM       The Rectangular window sequence is given by:
1590 REM
1600 REM          W(n) = 1, ùnù <= Q; 0, elsewhere
1610 REM
1620 REM =============================================================
1630 REM
1640 REM ================= Hanning Window Routine =================
1650 REM
1660 REM       The Hanning window sequence is given by:
1670 REM
1680 REM          W(n) = 0.5 + 0.5cos(nPI/Q), ùnù <= Q; 0, elsewhere
1690 REM
1700 FOR I=0 TO Q
```

FIRDP Listing (continued)

```
1710 C(I)=C(I)*(.5+.5*COS(I*PI/Q))        ' C'(n) = C(n)*W(n)
1720 NEXT I
1730 RETURN
1740 REM ========================================================================
1750 REM
1760 REM =================== Hamming Window Routine ==========================
1770 REM
1780 REM          The Hamming window sequence is given by:
1790 REM
1800 REM          W(n) = 0.54 + 0.46cos(nPI/Q), ùnù <= Q; 0, elsewhere
1810 REM
1820 FOR I=0 TO Q
1830 C(I)=C(I)*(.54+.46*COS(I*PI/Q))        ' C'(n) = C(n)*W(n)
1840 NEXT I
1850 RETURN
1860 REM ========================================================================
1870 REM
1880 REM =================== Blackman Window Routine =========================
1890 REM
1900 REM          The Blackman window sequence is given by:
1910 REM
1920 REM       .  W(n) = 0.42 + 0.5cos(2nPI/2Q) + 0.08cos(4nPI/2Q)
1930 REM
1940 FOR I=0 TO Q
1950 C(I)=C(I)*(.42+.5*COS((2*I*PI)/(2*Q))+.08*COS((4*I*PI)/(2*Q)))
1960 NEXT I
1970 RETURN
1980 REM ========================================================================
1990 REM
2000 REM =================== Lowpass Prompts Routine =========================
2010 TYPE$="LOWPASS"
2020 CLS
2030 LOCATE 1
2040 PRINT "          Specifications:"
2050 PRINT TAB(24);TYPE$
2060 LCUT=0
2070 LOCATE 12:INPUT "          Enter the 3-db cutoff frequency in Hz --> ",HCUT
2080 LOCATE 3:PRINT TAB(24);"Cutoff Frequency =";HCUT;"Hz"
2090 RETURN
2100 REM ========================================================================
2110 REM
2120 REM =================== Highpass Prompts Routine ========================
2130 TYPE$="HIGHPASS"
2140 CLS
2150 LOCATE 1
2160 PRINT "          Specifications:"
2170 PRINT TAB(24);TYPE$
2180 HIGH=1      'for highpass normalized upper cutoff = 1
2190 LOCATE 12:INPUT "          Enter the 3-db cutoff frequency in Hz --> ",LCUT
2200 LOCATE 3:PRINT TAB(24);"Cutoff Frequency =";LCUT;"Hz"
2210 RETURN
2220 REM ========================================================================
2230 REM
2240 REM =================== Bandpass/stop Prompts Routine ====================
2250 REM
2260 TYPE$="BANDPASS":IF TYPE = 4 THEN TYPE$="BANDSTOP"
2270 CLS
2280 LOCATE 1
2290 PRINT "          Specifications:"
2300 PRINT TAB(24);TYPE$
2310 LOCATE 12
2320 INPUT "          Enter the 3-db lower cutoff frequency in Hz --> ",LCUT
2330 LOCATE 12
2340 PRINT "                                                                    "
2350 LOCATE 3:PRINT TAB(24);"Lower Cutoff Frequency =";LCUT;"Hz"
2360 LOCATE 12
2370 INPUT "          Enter the 3-db upper cutoff frequency in Hz --> ",HCUT
2380 IF HCUT <= LCUT THEN GOSUB 4140:GOTO 2360
2390 LOCATE 4:PRINT TAB(24);"Upper Cutoff Frequency =";HCUT;"Hz"
2400 RETURN
2410 REM ========================================================================
2420 REM
2430 REM ========================= Output Routine ============================
2440 REM
2450 REM          This routine allows the user to define the output
2460 REM          path.  These include the terminal, line printer, or
2470 REM          data file, which includes TMS320 data format for simple
2480 REM          merging into a 'blank' filter program.
2490 REM
2500 CLS
2510 LOCATE 8
2520 PRINT "          Send coefficients to:"
2530 PRINT TAB(29);"(S)creen"
2540 PRINT TAB(29);"(P)rinter"
```

```
2550 PRINT TAB(29);"(F)ile: contains TMS320 data format"
2560 PRINT TAB(29);"(R)eturn to Filter Type Menu"
2570 PRINT TAB(29);"(E)xit to DOS"
2580 PRINT
2590 INPUT "         Enter desired path --> ",PATH$
2600 IF PATH$ = "S" OR PATH$ = "s" THEN GOSUB 2670
2610 IF PATH$ = "P" OR PATH$ = "p" THEN GOSUB 2990
2620 IF PATH$ = "F" OR PATH$ = "f" THEN GOSUB 3400
2630 IF PATH$ = "R" OR PATH$ = "r" THEN GOTO 390
2640 IF PATH$ = "E" OR PATH$ = "e" THEN CLS:SYSTEM
2650 GOTO 2500
2660 REM
2670 REM ------------------- Output Coefficients to Terminal -----------------
2680 REM
2690 CLS
2700 PRINT TAB(13);COEFF;"Coefficient ";TYPE$;" Filter Using the ";WIN$;" Window
2710 PRINT
2720 PRINT "        Filter Coefficients:"
2730 PRINT
2740 PRINT TAB(20);"C'(n)";TAB(53);"H(n)"
2750 PRINT TAB(20);"-----";TAB(53);"----"
2760 PRINT
2770 PRINT TAB(51);"DECIMAL";TAB(70);"HEX"
2780 PRINT
2790 FOR I=0 TO Q
2800 IF INT(I/11) <> I/11 OR I = 0 THEN 2860
2810 LOCATE 23:PRINT "        Press <Enter> to view";Q-I+1;"remaining coefficient
2820 LOCATE 23                                                                 "
2830 PRINT "
2840 GOSUB 3690     'erase coefficients from screen
2850 REM
2860 GOSUB 3790     'convert coeff. index to char, removing spaces in output
2870 PRINT TAB(8);
2880 PRINT "C'(";C1$;")";
2890 PRINT TAB(15);" = ";
2900 PRINT USING "+£.£££££";C(I);:PRINT " = C'(";C2$;")";
2910 PRINT TAB(41);
2920 PRINT "H(";C1$;")";
2930 PRINT TAB(47);" = ";
2940 PRINT USING "+£.£££££";H(I);: PRINT " = H(";C2$;")";
2950 PRINT TAB(67);" = ";CHEX$(I)
2960 NEXT I
2970 LOCATE 23:INPUT "        Press <Enter> to return to Output Path Menu ",RET
2980 RETURN
2990 REM ------------------- Output Coefficients to Printer -----------------
3000 REM
3010 REM        This nested routine allows the user to produce hardcopy of the
3020 REM        coefficients.  NOTE: If this option is specified from the
3030 REM        output path menu and the printer is NOT READY, then a
3040 REM        "device timeout" will occur via GWBASIC.  This will cause
3050 REM        exiting to GWBASIC.  To restart the FILTER DEVELOPMENT
3060 REM        PACKAGE type <F2> <Enter>.  This will restart whichever
3070 REM        window module selected previously from Main Menu.  Program
3080 REM        must be re-run to re-develop filter to obtain the hardcopy
3090 REM        initially desired.
3100 REM
3110 LPRINT TAB(26);"*** FIR DEVELOPMENT PACKAGE ***"
3120 LPRINT
3130 LPRINT TAB(14);COEFF;"Coefficient ";TYPE$;" Filter Using the ";WIN$;" Windo
3140 LPRINT
3150 IF TYPE=1 THEN LPRINT TAB(8);"Cutoff frequency =";HCUT;"Hz":GOTO 3190
3160 IF TYPE=2 THEN LPRINT TAB(8);"Cutoff frequency =";LCUT;"Hz":GOTO 3190
3170 LPRINT TAB(8);"Lower Cutoff Frequency =";LCUT;"Hz"
3180 LPRINT TAB(8);"Upper Cutoff Frequency =";HCUT;"Hz"
3190 LPRINT TAB(8);"Sampling Frequency (Fs) =";FSAM;"Hz"
3200 LPRINT TAB(8);"Impulse Duration (D) =";D*1000;"msec"
3210 LPRINT
3220 LPRINT TAB(20);"C'(n)";TAB(53);"H(n)"
3230 LPRINT TAB(20);"-----";TAB(53);"----"
3240 LPRINT
3250 LPRINT TAB(51);"DECIMAL";TAB(70);"HEX"
3260 LPRINT
3270 FOR I=0 TO Q
3280 GOSUB 3790     'convert coeff. index to char, removing spaces in output
3290 LPRINT TAB(8);
3300 LPRINT "C'(";C1$;")";
3310 LPRINT TAB(15);" = ";
3320 LPRINT USING "+£.£££££";C(I);:LPRINT " = C'(";C2$;")";
3330 LPRINT TAB(41);
3340 LPRINT "H(";C1$;")";
3350 LPRINT TAB(47);" = ";
3360 LPRINT USING "+£.£££££";H(I);:LPRINT " = H(";C2$;")";
3370 LPRINT TAB(67);" = ";CHEX$(I)
```

FIRDP Listing (continued)

```
3380 NEXT I
3390 RETURN
3400 REM ------------------ Output Coefficients to File ---------------------
3410 REM
3420 REM          This routine writes the coefficients to a file (named by the
3430 REM          user).  The file contains TMS320 data format or,
3440 REM
3450 REM                      H0        DATA     >°coeff in HEX§
3460 REM                      H1        DATA     >°coeff in HEX§
3470 REM
3480 REM          This file can then be merged into a 'blank' filter program
3490 REM          via a word processor like PC-WRITE.  NOTE: A directory or
3500 REM          drive specification can be given along with the filename.
3510 CLS
3520 LOCATE 12
3530 INPUT "          Enter Filename : ",NAM$
3540 LOCATE 12
3550 PRINT "                                                                "
3560 LOCATE 12:PRINT TAB(30);"...saving ";NAM$
3570 OPEN NAM$ FOR OUTPUT AS 1
3580 FOR I=0 TO COEFF-1
3590 S$=STR$(I)
3600 IF I>9 AND I<100 THEN PRINT£1,"H";RIGHT$(S$,2);"      DATA      >";CHEX$(I) :
3610 IF I>99 THEN PRINT£1,"H";RIGHT$(S$,3);"     DATA      >";CHEX$(I):GOTO 3630
3620 PRINT£1,"H";RIGHT$(S$,1);"       DATA     >";CHEX$(I)
3630 NEXT I
3640 CLOSE 1
3650 RETURN
3660 REM ----------------------------------------------------------------------
3670 REM ======================================================================
3680 REM
3690 REM ===================== Clear Coefficients ============================
3700 REM
3710 LOCATE 10
3720 FOR N=1 TO 12
3730 PRINT "
3740 NEXT N
3750 LOCATE 10
3760 RETURN
3770 REM ======================================================================
3780 REM
3790 REM ===================== Integer to Character ==========================
3800 REM
3810 REM          This routine eliminates the spaces produced in the output
3820 REM          when an Integer TYPE is printed.  The spaces before and after
3830 REM          the integer value are set aside in the event that the integer
3840 REM          value is negative.  Since the Coefficient index is always
3850 REM          non-negative, this value is converter to a character type,
3860 REM          and output from this routine as c1$ and c2$.
3870 REM
3880 S1$=STR$(I)
3890 S2$=STR$(COEFF-I-1)        'COEFF-I-1 is the symmetrical alternate of index
3900 C1$=RIGHT$(S1$,1)
3910 C2$=RIGHT$(S2$,1)
3920 IF I > 9 THEN C1$=RIGHT$(S1$,2)
3930 IF I > 99 THEN C1$=RIGHT$(S1$,3)
3940 IF COEFF-I-1 > 9 THEN C2$=RIGHT$(S2$,2)
3950 IF COEFF-I-1 > 99 THEN C2$=RIGHT$(S2$,3)
3960 RETURN
3970 REM ======================================================================
3980 REM
3990 REM =================== Convert Coefficients to Hex =====================
4000 REM
4010 REM          This routine converts the coefficients to Hexidecimal
4020 REM          via a routine in GWBASIC called HEX$().  The HEX values
4030 REM          are accurate to + or - 1.
4040 REM
4050 FOR M=0 TO COEFF-1
4060 CARY(M)=CINT(H(M)*32768!)      'scale by 2^15
4070 IF CARY(M) >= 0 GOTO 4090
4080 CARY(M)=65536!+CARY(M)
4090 CHEX$(M)=HEX$(CARY(M))
4100 NEXT M
4110 RETURN
4120 REM ======================================================================
4130 REM
4140 REM ===================== Error Message Routine =========================
4150 REM
4160 FOR BLINK=1 TO 10
4170 IF MES<>1 THEN 4210
4180 LOCATE 20
4190 PRINT TAB(14);"ERROR! : Sampling Frequency (Fs) >= 2*Nyquist - reenter"
4200 GOTO 4230
4210 LOCATE 20
4220 PRINT TAB(15);"ERROR! : Frequency value is inconsistant - reenter"
```

```
4230 FOR DELAY=1 TO 100:NEXT DELAY
4240 LOCATE 20
4250 PRINT"
4260 NEXT BLINK
4270 RETURN
4280 REM =======================================================================
10 REM                    *** Kaiser Window Program Module ***
20 REM
30 REM    This program module will generate the necessary FIR coefficients
40 REM    using the Kaiser window sequence.  The user can then define the
50 REM    output path to either the screen, printer, or an external file
60 REM    which can be merged with a FIR program to implement the filter.
70 REM
80 DIM C(256),H(256),FACT(256),WK(256),T(256),CHEX$(256),CARY(256)
90 PI = 3.1415927£
100 CLS
110 REM
120 REM                     *** Generate filter type menu ***
130 REM
140 LOCATE 4
150 PRINT TAB(11);"*** FIR COEFFICIENT GENERATION USING THE KAISER WINDOW ***"
160 LOCATE 8
170 PRINT TAB(22);"Selections:":PRINT
180 PRINT TAB(33);"1....LOWPASS":PRINT
190 PRINT TAB(33);"2....HIGHPASS":PRINT
200 PRINT TAB(33);"3....BANDPASS":PRINT
210 PRINT TAB(33);"4....BANDSTOP":PRINT
220 REM
230 PRINT TAB(33);"5....Exit back to Main Menu":PRINT
240 INPUT "       Enter desired filter type (number only) --> ",TYPE
250 IF TYPE = 5 THEN LOAD "firmm.bas",R
260 IF TYPE = 1 THEN GOSUB 1290:GOTO 300        'Lowpass prompts routine
270 IF TYPE = 2 THEN GOSUB 1510:GOTO 300        'Highpass prompts routine
280 IF TYPE=3 OR TYPE=4 THEN GOSUB 1730:GOTO 300 'bandpass/stop prompts routine
290 GOTO 100
300 REM
310 REM      *** Prompt for general information and output specifications ***
320 REM
330 MES=0                     'flag to which error has occured
340 LOCATE 12
350 INPUT "        Enter the Sampling Frequency (Fs) in Hz --> ",FSAM
360 IF TYPE=1 THEN FPASS1=0:FSTOP1=0
370 IF TYPE=2 THEN FPASS2=0:FSTOP2=0
380 IF FSAM/2<FPASS1 OR FSAM/2<FPASS2 OR FSAM/2<FSTOP1 OR FSAM/2<FSTOP2 THEN MES
390 REM
400 IF TYPE=1 OR TYPE=2 THEN LOCATE 7
410 IF TYPE=3 OR TYPE=4 THEN LOCATE 9
420 PRINT TAB(24);"Sampling Frequency (Fs) =";FSAM;"Hz"
430 LOCATE 12
440 PRINT "                                                              "
450 LOCATE 12
460 INPUT "        Are the above specifications correct (y/n)? --> ",RES$
470 IF RES$="n" OR RES$="N" THEN 100
480 LOCATE 12
490 PRINT "                                                              "
500 REM                     *** Calculate the Order required ***
510 FCUT2=ABS(FPASS2-FSTOP2)
520 FCUT1=ABS(FPASS1-FSTOP1)
530 IF TYPE=1 THEN DELTAF=FCUT2
540 IF TYPE=2 THEN DELTAF=FCUT1
550 IF TYPE <> 3 AND TYPE <> 4 THEN 570
560 IF FCUT2 < FCUT1 THEN DELTAF = FCUT2 ELSE DELTAF = FCUT1
570 D2=10^(-.05*ATT)
580 D1=(10^(.05*RIP)-1)/(10^(.05*RIP)+1)
590 IF D1 < D2 THEN NUM = D1 ELSE NUM = D2
600 AHP=-20*LOG(NUM)/LOG(10)
610 IF AHP <= 21 THEN D = .9222 ELSE D = (AHP-7.95)/14.36
620 COEFF=INT(2+D*FSAM/DELTAF)
630 IF COEFF/2 = INT(COEFF/2) THEN COEFF=COEFF+1
640 Q=CINT(COEFF-1)/2
650 LOCATE 12
660 PRINT "        The calculated £ of coefficients required is:";COEFF
670 PRINT
680 PRINT "        Enter £ of coefficients desired ONLY if greater than";COEFF
690 INPUT "        otherwise, press <Enter> to continue --> ",TEMP
700 IF TEMP = 0 THEN 830
710 IF TEMP < COEFF THEN 750
720 COEFF=TEMP
730 Q=(COEFF-1)/2
740 GOTO 830
750 PRINT
760 FOR BLINK=1 TO 10
```

FIRDP Listing (continued)

```
770 LOCATE 20
780 PRINT TAB(12);"ERROR! : Order will not satisfy specifications - reenter"
790 FOR DELAY=1 TO 100:NEXT DELAY
800 LOCATE 20:PRINT "
810 NEXT BLINK
820 GOTO 650
830 CLS
840 LOCATE 12
850 PRINT TAB(28);"Please wait ...working"
860 REM                          *** Compute Coefficients ***
870 REM
880 IF AHP <= 21 THEN ALP = 0 ELSE ALP = (.1102*(AHP-8.7))
890 IF AHP > 21 AND AHP <= 50 THEN ALP=.5842*(AHP-21)^.4+.07886*(AHP-21)
900 FACT(1)=1
910 FOR I=2 TO 30
920 FACT°I§ = FACT(I-1)*I
930 NEXT I
940 FOR I=0 TO Q
950 BESS=ALP*SQR(1-(2*I/(COEFF-1))^2)
960 IOBES=1:IOALP=1
970 FOR N = 1 TO 30
980 IOBES = IOBES+(((BESS/2)^(N))/FACT(N))^2
990 IOALP = IOALP+(((ALP/2)^(N))/FACT(N))^2
1000 NEXT N
1010 WK(I) = IOBES/IOALP
1020 NEXT I
1030 REM
1040 IF TYPE = 1 THEN GOSUB 2100:GOTO 1080        'calculate coeffs for lowpass
1050 IF TYPE = 2 THEN GOSUB 2210:GOTO 1080        'calculate coeffs for highpass
1060 IF TYPE = 3 THEN GOSUB 2320:GOTO 1080        'calculate coeffs for bandpass
1070 IF TYPE = 4 THEN GOSUB 2430                  'calculate coeffs for bandstop
1080 REM
1090 REM                          *** Rearrange Coefficients ***
1100 FOR I = 0 TO COEFF
1110 C(I)=H(I)            ' C'(i)
1120 T(I)=H(I)            ' T(i) is temporary storage for H(i)'s
1130 NEXT I
1140 FOR I=0 TO Q
1150 T(I)=H(Q-I)
1160 NEXT I
1170 FOR I=1 TO Q
1180 T(Q+I)=T(Q-I)
1190 NEXT I
1200 FOR I=0 TO COEFF
1210 H(I)=T(I)
1220 NEXT I
1230 REM                          *** Convert and Output Coefficients ***
1240 REM
1250 GOSUB 4390            'convert coeffs to hex
1260 REM
1270 GOSUB 2760            'call output routine
1280 REM
1290 REM ======================== Lowpass Prompts Routine ========================
1300 TYPE$="LOWPASS"
1310 CLS
1320 REM
1330 GOSUB 2690            'output filter type
1340 REM
1350 GOSUB 2540            'prompt for AS and AP
1360 REM
1370 LOCATE 12
1380 INPUT "          Enter the passband frequency in Hz --> ",FPASS2
1390 LOCATE 12
1400 PRINT "                                                              "
1410 LOCATE 5:PRINT TAB(24);"Passband Frequency =";FPASS2;"Hz"
1420 LOCATE 12
1430 INPUT "          Enter the stopband frequency in Hz --> ",FSTOP2
1440 IF FSTOP2 <= FPASS2 THEN GOSUB 4540:GOTO 1420        'display error message
1450 LOCATE 12
1460 PRINT "                                                              "
1470 LOCATE 6:PRINT TAB(24);"Stopband Frequency =";FSTOP2;"Hz"
1480 RETURN
1490 REM =========================================================================
1500 REM
1510 REM ======================= Highpass Prompts Routine =======================
1520 TYPE$="HIGHPASS"
1530 CLS
1540 REM
1550 GOSUB 2710            'output filter type
1560 REM
1570 GOSUB 2540            'prompt for AS and AP
1580 REM
1590 LOCATE 12
1600 INPUT "          Enter the passband frequency in Hz --> ",FPASS1
1610 LOCATE 12
```

```
1620 PRINT "                                                            "
1630 LOCATE 5:PRINT TAB(24);"Passband Frequency =";FPASS1;"Hz"
1640 LOCATE 12
1650 INPUT "          Enter the stopband frequency in Hz --> ",FSTOP1
1660 IF FSTOP1 >= FPASS1 THEN GOSUB 4540:GOTO 1640      'display error message
1670 LOCATE 12
1680 PRINT "                                                            "
1690 LOCATE 6:PRINT TAB(24);"Stopband Frequency =";FSTOP1;"Hz"
1700 RETURN
1710 REM ======================================================================
1720 REM
1730 REM ================== Bandpass/stop Prompts routine ====================
1740 REM
1750 IF TYPE=3 THEN TYPE$="BANDPASS" ELSE TYPE$="BANDSTOP"
1760 CLS
1770 REM
1780 GOSUB 2710           'output filter type
1790 REM
1800 GOSUB 2540           'prompt for AS and AP
1810 REM
1820 LOCATE 12
1830 INPUT "          Enter the lower passband frequency in Hz --> ",FPASS1
1840 LOCATE 12
1850 PRINT "                                                            "
1860 LOCATE 5:PRINT TAB(24);"Lower Passband Frequency =";FPASS1;"Hz"
1870 LOCATE 12
1880 INPUT "          Enter the upper passband frequency in Hz --> ",FPASS2
1890 IF FPASS1 >= FPASS2 THEN GOSUB 4540:GOTO 1870     'display error message
1900 LOCATE 12
1910 PRINT "                                                            "
1920 LOCATE 6:PRINT TAB(24);"Upper Passband Frequency =";FPASS2;"Hz"
1930 LOCATE 12
1940 INPUT "          Enter the lower stopband frequency in Hz --> ",FSTOP1
1950 IF TYPE=3 THEN IF FSTOP1>=FPASS1 OR FSTOP1>=FPASS2 THEN GOSUB 4540 : GOTO 1
1960 IF TYPE=4 THEN IF FSTOP1<=FPASS1 OR FSTOP1>=FPASS2 THEN GOSUB 4540 : GOTO 1
1970 LOCATE 12
1980 PRINT "                                                            "
1990 LOCATE 7:PRINT TAB(24);"Lower Stopband Frequency =";FSTOP1;"Hz"
2000 LOCATE 12
2010 INPUT "          Enter the upper stopband frequency in Hz --> ",FSTOP2
2020 IF TYPE=3 THEN IF FSTOP2 <= FPASS2 OR FSTOP2 <= FPASS1 OR FSTOP2 <= FSTOP1
2030 IF TYPE=4 THEN IF FSTOP2 >= FPASS2 OR FSTOP2 <= FPASS1 OR FSTOP2 <= FSTOP1
2040 LOCATE 12
2050 PRINT "                                                            "
2060 LOCATE 8:PRINT TAB(24);"Upper Stopband Frequency =";FSTOP2;"Hz"
2070 RETURN
2080 REM ======================================================================
2090 REM
2100 REM ====================== Kaiser Lowpass Routine =======================
2110 REM
2120 LWCUT=(FPASS2+FSTOP2)/2
2130 H0=2*LWCUT/FSAM
2140 H(0)=H0*WK(0)            'H(0) = C(0)*W(0)
2150 FOR I=1 TO Q
2160 H(I)=H0*((SIN(2*LWCUT*I*PI/FSAM))/(2*LWCUT*I*PI/FSAM))*WK(I)
2170 NEXT I
2180 RETURN
2190 REM ======================================================================
2200 REM
2210 REM ===================== Kaiser Highpass Routine =======================
2220 REM
2230 HWCUT=(FPASS1+FSTOP1)/2
2240 H0=-2*HWCUT/FSAM
2250 H(0)=(H0+1)*WK(0)        'H(0) = C(0)*W(0)
2260 FOR I=1 TO Q
2270 H(I)=H0*((SIN(2*HWCUT*I*PI/FSAM))/(2*HWCUT*I*PI/FSAM))*WK(I)
2280 NEXT I
2290 RETURN
2300 REM ======================================================================
2310 REM
2320 REM ===================== Kaiser Bandpass Routine =======================
2330 REM
2340 FCUT1=FPASS1-DELTAF/2
2350 FCUT2=FPASS2+DELTAF/2
2360 H(0)=(2/FSAM)*(FCUT2-FCUT1)*WK(0)
2370 FOR I=1 TO Q
2380 H(I)=(1/(I*PI))*(SIN((2*PI*I*FCUT2)/FSAM)-(SIN((2*PI*I*FCUT1)/FSAM)))*WK(I)
2390 NEXT I
2400 RETURN
2410 REM ======================================================================
2420 REM
```

FIRDP Listing (continued)

```
2430 REM ===================== Kaiser Bandstop Routine =====================
2440 REM
2450 FCUT1=FPASS1+DELTAF/2
2460 FCUT2=FPASS2-DELTAF/2
2470 H(0)=((2*(FCUT1-FCUT2)/FSAM)+1)*WK(0)
2480 FOR I=1 TO Q
2490 H(I)=(1/(I*PI))*(SIN((2*PI*I*FCUT1)/FSAM)-(SIN((2*PI*I*FCUT2)/FSAM)))*WK(I)
2500 NEXT I
2510 RETURN
2520 REM ==========================================================================
2530 REM
2540 REM ================= Prompt for AS and AP Routine =================
2550 REM
2560 LOCATE 12
2570 INPUT "          Enter the passband ripple in db --> ",RIP
2580 LOCATE 12
2590 PRINT "                                                              "
2600 LOCATE 3:PRINT TAB(24);"Passband Ripple (AP) =";RIP;"db"
2610 LOCATE 12
2620 INPUT "          Enter the minimum stopband attenuation in db --> ",ATT
2630 LOCATE 12
2640 PRINT "                                                              "
2650 LOCATE 4:PRINT TAB(24);"Stopband Attenuation (AS) =";ATT;"db"
2660 RETURN
2670 REM ==========================================================================
2680 REM
2690 REM ===================== Output Specs Header =====================
2700 LOCATE 1
2710 PRINT "          Specifications:"
2720 PRINT TAB(24);TYPE$
2730 RETURN
2740 REM ==========================================================================
2750 REM
2760 REM ======================= Output Routine =======================
2770 REM
2780 REM          This routine allows the user to define the output
2790 REM          path. These include the terminal, line printer, or
2800 REM          data file, which includes TMS320 data format for simple
2810 REM          merging into a 'blank' filter program.
2820 REM
2830 CLS
2840 LOCATE 8
2850 PRINT "          Send coefficients to:"
2860 PRINT TAB(29);"(S)creen"
2870 PRINT TAB(29);"(P)rinter"
2880 PRINT TAB(29);"(F)ile: contains TMS320 data format"
2890 PRINT TAB(29);"(R)eturn to Filter Type Menu"
2900 PRINT TAB(29);"(E)xit to DOS"
2910 PRINT
2920 INPUT "          Enter desired path --> ",PATH$
2930 IF PATH$ = "S" OR PATH$ = "s" THEN GOSUB 3000
2940 IF PATH$ = "P" OR PATH$ = "p" THEN GOSUB 3320
2950 IF PATH$ = "F" OR PATH$ = "f" THEN GOSUB 3800
2960 IF PATH$ = "R" OR PATH$ = "r" THEN GOTO 100
2970 IF PATH$ = "E" OR PATH$ = "e" THEN CLS:SYSTEM
2980 GOTO 2830
2990 REM
3000 REM ------------------ Output Coefficients to Terminal -----------------
3010 REM
3020 CLS
3030 PRINT TAB(15);COEFF;"Coefficient ";TYPE$;" Filter Using the KAISER Window"
3040 PRINT
3050 PRINT "          Filter Coefficients:"
3060 PRINT
3070 PRINT TAB(20);"C'(n)";TAB(53);"H(n)"
3080 PRINT TAB(20);"-----";TAB(53);"----"
3090 PRINT
3100 PRINT TAB(51);"DECIMAL";TAB(70);"HEX"
3110 PRINT
3120 FOR I=0 TO Q
3130 IF INT(I/11) <> I/11 OR I = 0 THEN 3190
3140 LOCATE 23:PRINT "          Press <Enter> to view";Q-I+1;"remaining coefficient
3150 LOCATE 23
3160 PRINT "                                                              "
3170 GOSUB 4090          'erase coefficients from screen
3180 REM
3190 GOSUB 4190          'convert coeff. index to char, removing spaces in output
3200 PRINT TAB(8);
3210 PRINT "C'(";C1$;")";
3220 PRINT TAB(15);" = ";
3230 PRINT USING "+£.£££££";C(I);:PRINT " = C'(";C2$;")";
3240 PRINT TAB(41);
3250 PRINT "H(";C1$;")";
3260 PRINT TAB(47);" = ";
```

```
3270 PRINT USING "+£.£££££";H(I);: PRINT " = H(";C2$;")";
3280 PRINT TAB(67);" = ";CHEX$(I)
3290 NEXT I
3300 LOCATE 23:INPUT "          Press <Enter> to return to Output Path Menu ",RET
3310 RETURN
3320 REM -------------------- Output Coefficients to Printer -----------------
3330 REM
3340 REM          This nested routine allows the user to produce hardcopy of the
3350 REM          coefficients.  NOTE: If this option is specified from the
3360 REM          output path menu and the printer is NOT READY, then a
3370 REM          "device timeout" will occur via GWBASIC.  This will cause
3380 REM          exiting to GWBASIC.  To restart the FILTER DEVELOPMENT
3390 REM          PACKAGE type <F2> <Enter>.  This will restart whichever
3400 REM          window module selected previously from Main Menu.  Program
3410 REM          must be re-run to re-develop filter to obtain the hardcopy
3420 REM          initially desired.
3430 REM
3440 LPRINT TAB(26);"*** FIR DEVELOPMENT PACKAGE ***"
3450 LPRINT
3460 LPRINT TAB(15);COEFF;"Coefficient ";TYPE$;" Filter Using the KAISER Window"
3470 LPRINT
3480 LPRINT TAB(8);"Passband Ripple (AP) =";RIP;"db"
3490 LPRINT TAB(8);"Stopband Attenuation (AS) =";ATT;"db"
3500 IF TYPE<>1 THEN 3560
3510 LPRINT TAB(8);"Passband frequency =";FPASS2;"Hz"
3520 LPRINT TAB(8);"Stopband frequency =";FSTOP2;"Hz"
3530 IF TYPE<>2 THEN 3560
3540 LPRINT TAB(8);"Passband frequency =";FPASS1;"Hz"
3550 LPRINT TAB(8);"Stopband frequency =";FSTOP1;"Hz"
3560 LPRINT TAB(8);"Lower Passband Frequency =";FPASS1;"Hz"
3570 LPRINT TAB(8);"Upper Passband Frequency =";FPASS2;"Hz"
3580 LPRINT TAB(8);"Lower Stopband Frequency =";FSTOP1;"Hz"
3590 LPRINT TAB(8);"Upper Stopband Frequency =";FSTOP2;"Hz"
3600 LPRINT TAB(8);"Sampling Frequency (Fs) =";FSAM;"Hz"
3610 LPRINT
3620 LPRINT TAB(20);"C'(n)";TAB(53);"H(n)"
3630 LPRINT TAB(20);"-----";TAB(53);"----"
3640 LPRINT
3650 LPRINT TAB(51);"DECIMAL";TAB(70);"HEX"
3660 LPRINT
3670 FOR I=0 TO Q
3680 GOSUB 4190         'convert coeff. index to char, removing spaces in output
3690 LPRINT TAB(8);
3700 LPRINT "C'(";C1$;")";
3710 LPRINT TAB(15);" = ";
3720 LPRINT USING "+£.£££££";C(I);:LPRINT " = C'(";C2$;")";
3730 LPRINT TAB(41);
3740 LPRINT "H(";C1$;")";
3750 LPRINT TAB(47);" = ";
3760 LPRINT USING "+£.£££££";H(I);:LPRINT " = H(";C2$;")";
3770 LPRINT TAB(67);" = ";CHEX$(I)
3780 NEXT I
3790 RETURN
3800 REM -------------------- Output Coefficients to File --------------------
3810 REM
3820 REM          This routine writes the coefficients to a file (named by the
3830 REM          user).  The file contains TMS320 data format or,
3840 REM
3850 REM                    H0          DATA      >°coeff in HEX§
3860 REM                    H1          DATA      >°coeff in HEX§
3870 REM
3880 REM          This file can then be merged into a 'blank' filter program
3890 REM          via a word processor like PC-WRITE.  NOTE: A directory or
3900 REM          drive specification can be given along with the filename.
3910 CLS
3920 LOCATE 12
3930 INPUT "          Enter Filename : ",NAM$
3940 LOCATE 12                                                                "
3950 PRINT "
3960 LOCATE 12:PRINT TAB(30);"...saving ";NAM$
3970 OPEN NAM$ FOR OUTPUT AS 1
3980 FOR I=0 TO COEFF-1
3990 S$=STR$(I)
4000 IF I>9 AND I<100 THEN PRINT£1,"H";RIGHT$(S$,2);"      DATA      >";CHEX$(I) :
4010 IF I>99 THEN PRINT£1,"H";RIGHT$(S$,3);"      DATA      >";CHEX$(I):GOTO 4030
4020 PRINT£1,"H";RIGHT$(S$,1);"      DATA      >";CHEX$(I)
4030 NEXT I
4040 CLOSE 1
4050 RETURN ------------------------------------------------------------------
4060 REM ------
```

FIRDP Listing (continued)

```
4070 REM ========================================================================
4080 REM
4090 REM ======================= Clear Coefficients =========================
4100 REM
4110 LOCATE 10
4120 FOR N=1 TO 12
4130 PRINT "
4140 NEXT N
4150 LOCATE 10
4160 RETURN
4170 REM ========================================================================
4180 REM
4190 REM ===================== Integer to Character ==========================
4200 REM
4210 REM          This routine eliminates the spaces produced in the output
4220 REM          when an Integer TYPE is printed.  The spaces before and after
4230 REM          the integer value are set aside in the event that the integer
4240 REM          value is negative.  Since the Coefficient index is always
4250 REM          non-negative, this value is converter to a character type,
4260 REM          and output from this routine as c1$ and c2$.
4270 REM
4280 S1$=STR$(I)
4290 S2$=STR$(COEFF-I-1)          'COEFF-I-1 is the symmetrical alternate of index
4300 C1$=RIGHT$(S1$,1)
4310 C2$=RIGHT$(S2$,1)
4320 IF I > 9 THEN C1$=RIGHT$(S1$,2)
4330 IF I > 99 THEN C1$=RIGHT$(S1$,3)
4340 IF COEFF-I-1 > 9 THEN C2$=RIGHT$(S2$,2)
4350 IF COEFF-I-1 > 99 THEN C2$=RIGHT$(S2$,3)
4360 RETURN
4370 REM ========================================================================
4380 REM
4390 REM =================== Convert Coefficients to Hex =====================
4400 REM
4410 REM          This routine converts the coefficients to Hexidecimal
4420 REM          via a routine in GWBASIC called HEX$().  The HEX values
4430 REM          are accurate to + or - 1.
4440 REM
4450 FOR M=0 TO COEFF-1
4460 CARY(M)=CINT(H(M)*32768!)          'scale by 2^15
4470 IF CARY(M) >= 0 GOTO 4490
4480 CARY(M)=65536!+CARY(M)
4490 CHEX$(M)=HEX$(CARY(M))
4500 NEXT M
4510 RETURN
4520 REM ========================================================================
4530 REM
4540 REM ===================== Error Message Routine =========================
4550 REM
4560 FOR BLINK=1 TO 10
4570 IF MES<>1 THEN 4610
4580 LOCATE 20
4590 PRINT TAB(14);"ERROR : Sampling Frequency (Fs) >= 2*Nyquist - reenter"
4600 GOTO 4630
4610 LOCATE 20
4620 PRINT TAB(15);"ERROR! : Frequency value is inconsistant - reenter"
4630 FOR DELAY=1 TO 100:NEXT DELAY
4640 LOCATE 20
4650 PRINT "                                                                 "
4660 NEXT BLINK
4670 RETURN
4680 REM ========================================================================
```

EDITOR.BAS Program Listing

```
10 REM    *** Utility - Convert FIRDP Output File into MAG_PHSE.BAS Format ***
20 REM
30 REM           This utility edits the file produced by the FIRDP, so that,
40 REM           the magnitude and phase of very high order systems may
50 REM           easily be evaluated via the utility MAG_PHSE.BAS.  Otherwise,
60 REM           each coefficient would have to be entered by hand without
70 REM           error into MAG_PHSE.BAS program.
80 REM
90 DIM CARY(256),CARY$(256),FIND$(256)
100 KEY OFF
110 CLS
120 DCVAL=0       'initialize running decimal sum
130 LOCATE 4
140 PRINT TAB(24);"*** UTILITY - FIRDP File Editor ***"
150 LOCATE 12
160 INPUT "        Enter FIRDP Data File to be edited --> ",NAM$
170 LOCATE 12
180 PRINT "                                                          "
190 LOCATE 12
200 INPUT "        Enter # of coefficients in file --> ",COEFF
210 CLS
220 LOCATE 12
230 PRINT TAB(29);"Please wait ...working"
240 REM
250 REM                    *** Read File Into Character Array ***
260 REM
270 OPEN NAM$ FOR INPUT AS 1
280 FOR I=0 TO COEFF-1
290 INPUT#1,FIND$(I)
300 NEXT I
310 CLOSE 1
320 REM
330 REM                    *** Generate HEX Coefficient Array ***
340 REM
350 FOR I=0 TO COEFF-1
360 CARY$(I)=MID$(FIND$(I),18)
370 NEXT I
380 REM
390 REM                    *** Convert HEX to DECIMAL ***
400 REM
410 FOR I=0 TO COEFF-1
420 HXVAL$=CARY$(I)
430 GOSUB 750           'invoke routine given HXVAL$
440 CARY(I)=DCVAL
450 NEXT I
460 REM
470 REM                    *** Prompt for Edited Filename ***
480 REM
490 CLS
500 LOCATE 12
510 INPUT "        Enter new filename for saving edited file --> ",NAM$
520 CLS
530 REM
540 REM           *** Save Decimal Coefficients Under New Filename ***
550 REM
560 LOCATE 12:PRINT TAB(30);"...saving ";NAM$
570 OPEN NAM$ FOR OUTPUT AS 1
580 FOR I=0 TO COEFF-1
590 PRINT#1, CARY(I)
600 NEXT I
610 CLOSE 1
620 CLS
630 LOCATE 12
640 INPUT "        Do you wish to edit another FIRDP file ? (y/n) ",REPLY$
650 IF REPLY$ = "N" OR REPLY$ = "n" THEN 670
660 GOTO 110
670 LOCATE 12
680 PRINT "                                                          "
690 LOCATE 12
700 INPUT "        Exit to (B)asic or to (D)OS ? ",EX$
710 IF EX$ = "B" OR EX$ = "b" THEN CLS:END
720 IF EX$ = "D" OR EX$ = "d" THEN CLS:SYSTEM
730 GOTO 670
```

FIRDP Listing (continued)

```
740 END
750 REM ===================== Convert HEX to DECIMAL =========================
760 REM
770 REM           This routine converts hexidecimal numbers to decimal.
780 REM
790 REM           Input:  Hexidecimal value called HXVAL$.
800 REM           Output: Decimal value call DCVAL.
810 REM
820 REM
830 DIG=LEN(HXVAL$)
840 REM         *** Search Array and Find HEX Coefficients ***
850 FOR N=0 TO DIG-1
860 HXARY$(N)=RIGHT$(HXVAL$,DIG-N)
870 HXARY$(N)=LEFT$(HXARY$(N),1)
880 NEXT N
890 FOR N=0 TO DIG-1
900 IF HXARY$(N)="F" THEN HXARY(N)=15:GOTO 970
910 IF HXARY$(N)="E" THEN HXARY(N)=14:GOTO 970
920 IF HXARY$(N)="D" THEN HXARY(N)=13:GOTO 970
930 IF HXARY$(N)="C" THEN HXARY(N)=12:GOTO 970
940 IF HXARY$(N)="B" THEN HXARY(N)=11:GOTO 970
950 IF HXARY$(N)="A" THEN HXARY(N)=10:GOTO 970
960 HXARY(N)=VAL(HXARY$(N))      'else, 0 - 9
970 NEXT N
980 REM
990 REM                      *** Compute Decimal Value ***
1000 REM
1010 FOR N=DIG-1 TO 0 STEP -1
1020 DCVAL=DCVAL+HXARY(N)*16^((DIG-1)-N)
1030 NEXT N
1040 IF DCVAL > 32768! THEN DCVAL=DCVAL-65536!
1050 DCVAL=DCVAL/32768!
1060 RETURN
1070 REM ================================================================================
```

MAGPHSE Program Listing

```
10 REM  *** Utility to Determine Magnitude and Phase Response of a DT System ***
20 REM
30 REM           Program to evaluate the magnitude and phase response of
40 REM  a discrete-time system.  This program yields the magnitude and
50 REM  phase response of a given DT transfer function H(z) of the form:
60 REM
70 REM               Ao + A1z^-1 + A2z^-2 + ... + Anz^-n
80 REM  H(z) =  --------------------------------------
90 REM               1 + B1z^-1 + B2z^-2 + ... + Bmz^-m
100 REM
110 REM           Note: For a non-recursive (FIR) filter, all of
120 REM                 the b's are zero.
130 REM
140 REM  The program will prompt the user for output path of screen, printer
150 REM  or file.  The file contains either the frequency and magnitude, or
160 REM  frequency and phase, in an (x,y) format to facilitate plotting via
170 REM  a plotter program.
180 REM
190 KEY OFF
200 DIM AS(256),BS(256),MAG(512),PHASE(512),FREQ(512)
210 CLS
220 REM
230 REM                  *** Initialize Running Variables ***
240 REM
250 XX=0:YY=0
260 REALN=0:IMAGN=0
270 REALD=0:IMAGD=0
280 SUM=0:INCRMT=0
290 ANGLE=0
300 REM
310 REM                      *** Initialize Variables ***
320 REM
330 MAXCOEFFS=256
340 MAXITS=512
350 PI=3.1415927£
360 TWOPI=2*PI
370 REM
380 LOCATE 2
390 PRINT TAB(17);"*** Magnitude and Phase Response of a DT System ***"
400 REM
410 PRINT:PRINT
```

```
420 IF PATH$ = "RU" OR PATH$ = "ru" THEN LOCATE 10:GOTO 690      'use same coeffs
430 INPUT "           (F)IR or (I)IR ? ",TYPE$
440 IF TYPE$ = "I" OR TYPE$ = "i" THEN GOSUB 3440:GOSUB 3030:GOTO 650
450 IF TYPE$ <> "F" AND TYPE$ <> "f" THEN CLS:GOTO 380
460 DTSYS$="FIR"
470 PRINT:PRINT:PRINT
480 PRINT "          (Note: If file then FIRDP file must be edited"
490 PRINT "               using the FIRDP File Editor.  To run:"
500 PRINT "               enter <E> above, then load and run"
510 PRINT "               EDITOR.BAS)"
520 LOCATE 7
530 PRINT TAB(9);
540 INPUT"Enter coefficients via (T)yping or (F)ile (E = Exit to BASIC) --> ",M$
550 IF M$ = "T" OR M$ = "t" THEN GOSUB 3440:GOSUB 3270:GOTO 650
560 IF M$ = "E" OR M$ = "e" THEN CLS:END
570 IF M$ <> "F" AND M$ <> "f" THEN LOCATE 10:GOTO 530
580 PRINT
590 GOSUB 3440            'clear portion of screen
600 LOCATE 5:INPUT "          Enter edited filename --> ",NAM$
610 REM
620 REM              *** Load in Transfer Function from File ***
630 REM
640 GOSUB 3550            'generate transfer function from edited file
650 PRINT
660 INPUT "          Enter the Sampling Frequency in Hz --> ",FSAM
670 NYQST=FSAM/2
680 PRINT
690 INPUT "          Enter the number of steps desired --> ",NUMITS
700 IF NUMITS > MAXITS OR NUMITS < 1 THEN GOTO 210
710 PRINT
720 INPUT "          Tabulate (M)agnitude or (P)hase response --> ",RES$
730 TYPE$="MAGNITUDE"
740 IF RES$ = "P" OR RES$ = "p" THEN TYPE$="PHASE"
750 IF TYPE$ = "PHASE" THEN 840
760 PRINT
770 INPUT "          Do you want to normalize the magnitude (Y/N) --> ",X$
780 NORM=0
790 IF X$ = "Y" OR X$ = "y" THEN NORM=1
800 PRINT
810 INPUT "          Do you want (P)ower or (A)mplitude form  --> ",X$
820 POWER=0:TYPE$="AMPLITUDE"
830 IF X$ ="P" OR X$ = "p" THEN POWER=1:TYPE$=" POWER"
840 PRINT
850 INPUT "          Are the above entries correct (Y/N) ",REPLY$
860 IF REPLY$ = "N" OR REPLY$ = "n" THEN CLS:GOTO 380
870 CLS
880 LOCATE 12
890 PRINT TAB(29); "Please wait ...working"
900 REM
910 REM              *** Calculate Step Size ***
920 STEPSZ=PI/(NUMITS-1)
930 REM
940 REM         *** Determine POWER/AMPLITUDE or PHASE Response ***
950 REM
960 FOR N=1 TO NUMITS
970 REM              *** Evaluate numerator ***
980 REALN=AS(0)
990 IMAGN=0
1000 IF NNC < 2 THEN GOTO 1060
1010 FOR I=2 TO NNC
1020 THETA=INCRMT*(I-1)
1030 REALN=REALN+AS(I-1)*COS(THETA)        'sum of real parts of numerator
1040 IMAGN=IMAGN-AS(I-1)*SIN(THETA)        'sum of imaginary parts of numerator
1050 NEXT I
1060 REM              *** Evaluate Denominator ***
1070 REM
1080 IF DTSYS$ = "FIR" THEN 1180       'FIR system, all B's are zero
1090 REALD=BS(0)
1100 IMAGD=0
1110 IF NDC < 2 THEN GOTO 1180
1120 FOR I=2 TO NDC
1130 THETA=INCRMT*(I-1)
1140 REALD=REALD+BS(I-1)*COS(THETA)        'sum of real parts of demoninator
1150 IMAGD=IMAGD-BS(I-1)*SIN(THETA)        'sum of imaginary parts of denominator
1160 NEXT I
1170 REM
1180 REM         *** Evaluate the Power/Amplitude Response ***
1190 REM
1200 IF TYPE$ = "PHASE" THEN 1260
1210 REM
1220 IF DTSYS$="FIR" THEN MAG(N)=REALN^2+IMAGN^2:GOTO 1240    'evaluate mag - FIR
1230 MAG(N)=(REALN^2+IMAGN^2)/(REALD^2+IMAGD^2)    'evaluate mag - IIR
1240 SUM=SUM+MAG(N)
```

FIRDP Listing (continued)

```
1250 REM
1260 REM                    *** Evaluate Phase Response ***
1270 REM
1280 IF TYPE$ <> "PHASE" THEN 1460
1290 XX=(ATN(IMAGN/REALN))
1300 IF DTSYS$ = "FIR" THEN YY=0:GOTO 1320          'angle is zero, denom.is one
1310 YY=(ATN(IMAGD/REALD))
1320 REM
1330 REM Determine proper quadrant.
1340 REM
1350 IF REALN < 0 AND IMAGN >= 0 THEN XX=PI+XX            'to second quadarant
1360 IF REALN < 0 AND IMAGN < 0 THEN XX=TWOPI-(PI-XX)     'to third quadrant
1370 IF REALN > 0 AND IMAGN <= 0 THEN XX=TWOPI+XX         'to fourth quadrant
1380 REM
1390 IF DTSYS$ = "FIR" THEN 1440          'denominator is one
1400 REM
1410 IF REALD < 0 AND IMAGD >= 0 THEN YY=PI+YY            'to second quadrant
1420 IF REALD < 0 AND IMAGD < 0 THEN YY=TWOPI-(PI-YY)     'to third quadrant
1430 IF REALD > 0 AND IMAGD <= 0 THEN YY=TWOPI+YY         'to fourth quadrant
1440 REM
1450 PHASE(N)=XX-YY               'if FIR, YY=0
1460 INCRMT=INCRMT+STEPSZ
1470 NEXT N
1480 REM
1490 IF TYPE$ = "PHASE" THEN 1770
1500 REM
1510 REM
1520 IF POWER <> 1 THEN 1610          ' "1" equals power response
1530 REM
1540 REM
1550 REM           *** Calculate Power Response and Normalize ***
1560 REM
1570 FOR N=1 TO NUMITS
1580 MAG(N)=MAG(N)/SUM
1590 NEXT N
1600 GOTO 1760
1610 REM
1620 REM            *** Calculate the Amplitude Response ***
1630 REM
1640 FOR N=1 TO NUMITS
1650 MAG(N)=SQR(MAG(N))                     'mag = squareroot(real^2 + Imag^2)
1660 IF TMAG < MAG(N) THEN TMAG=MAG(N)  'find smallest mag value for normalizing
1670 NEXT N
1680 REM
1690 IF NORM <> 1 THEN GOTO 1760   ' "1" equalize normalize response
1700 REM
1710 REM            *** Normalize Amplitude Response ***
1720 REM
1730 FOR N=1 TO NUMITS
1740 MAG(N)=MAG(N)/TMAG
1750 NEXT N
1760 REM
1770 REM           *** Output Magnitude and Phase Response ***
1780 REM
1790 GOSUB 1830          'invoke output routine
1800 CLS
1810 GOTO 210
1820 END
1830 REM ======================= Output Menu routine =========================
1840 REM
1850 CLS
1860 LOCATE 8
1870 PRINT "        Send output to:"
1880 PRINT TAB(23);"(S)creen"
1890 PRINT TAB(23);"(P)rinter"
1900 PRINT TAB(23);"(F)ile: (FREQ,";TYPE$;") Format"
1910 PRINT TAB(23);"(RU)n program with same coefficients"
1920 PRINT TAB(23);"(R)un program with different coefficients"
1930 PRINT TAB(23);"(E)xit to DOS"
1940 PRINT
1950 INPUT "        Enter desired path --> ",PATH$
1960 IF PATH$ = "E" OR PATH$ = "e" THEN CLS:SYSTEM
1970 IF PATH$ = "RU" OR PATH$ = "ru" THEN RETURN
1980 IF PATH$ = "R" OR PATH$ = "r" THEN RETURN
1990 IF PATH$ = "S" OR PATH$ = "s" THEN GOSUB 2040
2000 IF PATH$ = "P" OR PATH$ = "p" THEN GOSUB 2270
2010 IF PATH$ = "F" OR PATH$ = "f" THEN GOSUB 2580
2020 GOTO 1850
2030 REM
2040 REM --------------------- Output Response to Screen ----------------------
2050 REM
```

FIRDP Listing (continued)

```
2060 CLS
2070 FINDX=0
2080 PRINT TAB(21);"*** ";TYPE$;" Response of a DT System ***"
2090 PRINT
2100 PRINT TAB(25);"FREQUENCY";:PRINT TAB(47);TYPE$
2110 PRINT TAB(24);"---------------------------------"
2120 FSTEP=NYQST/(NUMITS-1)          'compute frequency stepsize
2130 FOR N=1 TO NUMITS
2140 IF INT((N-1)/17) <> (N-1)/17 OR N=1 THEN 2170
2150 LOCATE 23:INPUT "        Press <Enter> to continue ",REPLY
2160 GOSUB 2930                 'erase data
2170 FREQ(N)=FINDX
2180 PRINT TAB(25);:PRINT USING "££££.££";FREQ(N);
2190 IF TYPE$ <> "PHASE" THEN PRINT TAB(45);:PRINT USING "££££.£££££";MAG(N)
2200 IF TYPE$ = "PHASE" THEN PRINT TAB(45);:PRINT USING "£££.£££££";PHASE(N)
2210 FINDX=FINDX+FSTEP
2220 NEXT N
2230 LOCATE 23:INPUT "        Press <Enter> to return to Output Path Menu ",REPLY
2240 RETURN
2250 REM --------------------------------------------------------------------
2260 REM
2270 REM -------------------- Output Response to Printer ---------------------
2280 REM
2290 FINDX=0
2300 LPRINT TAB(18);"UTILITY - MAGNITUDE AND PHASE RESPONSE PROGRAM"
2310 LPRINT:LPRINT
2320 LPRINT TAB(21);"*** ";TYPE$;" Response of a DT System ***"
2330 LPRINT:LPRINT
2340 IF DTSYS$="FIR" AND (M$="F" OR M$="f") THEN OUTNAM$=NAM$ ELSE OUTNAM$=""
2350 LPRINT TAB(15);"Coefficients: ";OUTNAM$
2360 IF DTSYS$="FIR" AND (M$="F" OR M$="f") THEN 2440
2370 IF DTSYS$="FIR" AND (M$="T" OR M$="t") THEN NDC=NNC
2380 FOR I=0 TO NDC-1
2390 GOSUB 2840             'Convert index to character
2400 IF I > NNC-1 THEN 2420
2410 LPRINT TAB(29);:LPRINT "a(";C$;:LPRINT USING ") = ££.£££££";AS(I);
2420 LPRINT TAB(50);:LPRINT "b(";C$;:LPRINT USING ") = ££.£££££";BS(I)
2430 NEXT I
2440 LPRINT:LPRINT
2450 LPRINT TAB(25);"FREQUENCY";:LPRINT TAB(47);TYPE$
2460 LPRINT TAB(24);"---------------------------------"
2470 FSTEP=NYQST/(NUMITS-1)
2480 FOR N=1 TO NUMITS
2490 FREQ(N)=FINDX
2500 LPRINT TAB(25);:LPRINT USING "££££.££";FREQ(N);
2510 IF TYPE$ <> "PHASE" THEN LPRINT TAB(45);:LPRINT USING "££££.£££££";MAG(N)
2520 IF TYPE$ = "PHASE" THEN LPRINT TAB(45);:LPRINT USING "£££.£££££";PHASE(N)
2530 FINDX=FINDX+FSTEP
2540 NEXT N
2550 RETURN
2560 REM --------------------------------------------------------------------
2570 REM
2580 REM --------------------- Output Response to File ----------------------
2590 REM
2600 REM         This routine creates a file containing the Frequency and
2610 REM         Magnitude values in an (FREQ,RESPONSE) format.  This allows
2620 REM         plotting of the response of the filter via a plotting
2630 REM         program.
2640 REM
2650 CLS
2660 FINDX=0
2670 LOCATE 12:INPUT "        Enter Filename : ",NAM$
2680 LOCATE 12
2690 PRINT "                                                              "
2700 LOCATE 12:PRINT TAB(30);"...saving ";NAM$
2710 OPEN NAM$ FOR OUTPUT AS 1
2720 FSTEP=NYQST/(NUMITS-1)
2730 FOR N=1 TO NUMITS
2740 FREQ(N)=FINDX
2750 IF TYPE$ <> "PHASE" THEN PRINT£1,FREQ(N);MAG(N)
2760 IF TYPE$ = "PHASE"  THEN PRINT£1,FREQ(N),PHASE(N)
2770 FINDX=FINDX+FSTEP
2780 NEXT N
2790 CLOSE 1
2800 RETURN
2810 REM --------------------------------------------------------------------
2820 REM ====================================================================
2830 REM
2840 REM ====================== Integer to Character ========================
2850 REM
2860 S$=STR$(I)
```

FIRDP Listing (continued)

```
2870 C$=RIGHT$(S$,1)
2880 IF I > 9 THEN C$=RIGHT$(S$,2)
2890 IF I > 99 THEN C$=RIGHT$(S$,3)
2900 RETURN
2910 REM =====================================================================
2920 REM
2930 REM ========================= Clear Data ================================
2940 REM
2950 LOCATE 5
2960 FOR C=1 TO 19
2970 PRINT"                                                                   "
2980 NEXT C
2990 LOCATE 5
3000 RETURN
3010 REM =====================================================================
3020 REM
3030 REM ================== Prompt for IIR Transfer Function ==================
3040 REM
3050 DTSYS$="IIR"
3060 REM
3070 LOCATE 5
3080 PRINT TAB(9);
3090 INPUT"Enter the £ of numerator coefficients (256 = Maximum) --> ",NNC
3100 IF NNC > MAXCOEFFS OR NNC < 1 THEN GOTO 210
3110 FOR I=0 TO NNC-1
3120 GOSUB 2840          'convert index to character
3130 PRINT TAB(13);
3140 PRINT "Enter a(";C$;")z^-";C$;TAB(32);"--> ";:INPUT "",AS(I)
3150 NEXT I
3160 PRINT
3170 INPUT "         Enter the £ of denominator coefficients --> ",NDC
3180 IF NDC > MAXCOEFFS OR NDC < 0 THEN GOTO 210
3190 FOR I=0 TO NDC-1
3200 GOSUB 2840          'convert index to character
3210 PRINT TAB(13);
3220 PRINT "Enter b(";C$;")z^-";C$;TAB(32);"--> ";:INPUT "",BS(I)
3230 NEXT I
3240 RETURN
3250 REM =====================================================================
3260 REM
3270 REM ================== Prompt for FIR Transfer Function ==================
3280 REM
3290 DTSYS$="FIR"
3300 IF PATH$ = "RU" OR PATH$ = "ru" THEN LOCATE 10:GOTO 690    'use same coeffs
3310 REM
3320 LOCATE 5
3330 PRINT TAB(9);
3340 INPUT"Enter the £ of coefficients (256 = Maximum) --> ",NNC
3350 IF NNC > MAXCOEFFS OR NNC < 1 THEN GOTO 210
3360 FOR I=0 TO NNC-1
3370 GOSUB 2840          'convert index to character
3380 PRINT TAB(13);
3390 PRINT "Enter a(";C$;")z^-";C$;TAB(32);"--> ";:INPUT "",AS(I)
3400 NEXT I
3410 RETURN
3420 REM =====================================================================
3430 REM
3440 REM ===================== Clear Portion of Screen =======================
3450 REM
3460 LOCATE 5
3470 FOR C=1 TO 8
3480 PRINT TAB(10);
3490 PRINT "                                                              "
3500 NEXT C
3510 LOCATE 5
3520 RETURN
3530 REM =====================================================================
3540 REM
3550 REM ================== Get FIR Transfer Function from File ==============
3560 REM
3570 NNC=0
3580 OPEN NAM$ FOR INPUT AS 1
3590 IF EOF(1) THEN 3650
3600 INPUT£1,INC$
3610 AS(NNC)=VAL(INC$)
3620 IF LEFT$(INC$,1)="H" THEN CLOSE 1:GOSUB 3690
3630 NNC=NNC+1
3640 GOTO 3590
3650 CLOSE 1
3660 RETURN
3670 REM =====================================================================
3680 REM
3690 REM ========================== File Error ===============================
3700 REM
```

```
3710 FOR BLINK=1 TO 10
3720 LOCATE 20
3730 PRINT TAB(23);"ERROR! : FIRDP File must be edited!"
3740 FOR DELAY=1 TO 100:NEXT DELAY
3750 LOCATE 20
3760 PRINT "                                                        "
3770 NEXT BLINK
3780 LOCATE 20
3790 INPUT "          (R)etype filename or (E)xit to BASIC --> ",WARN$
3800 LOCATE 20
3810 PRINT "                                                        "
3820 IF WARN$="E" THEN CLS:SYSTEM
3830 RETURN 600
3840 REM ====================================================================
```

FIRDP Listing (concluded)

E

ASPI Digital Filter
Design Package (FIR Filters)

E.1 INTRODUCTION

The ASPI Digital Filter Design Package (DFDP) is a powerful filter design tool that allows a person to design and implement a digital filter in a matter of minutes. The package is menu driven and relatively easy to use provided that one is familiar with filter specifications. This design tool runs on the IBM (or compatible) PC or AT with the math coprocessor option highly recommended since the speed is enhanced by a factor of 20 with this option. The DFDP provides the following capabilities:

1. Design of FIR filters using either the Kaiser window or the Parks–McClellan algorithm
2. Design of IIR recursive filters using the bilinear transformation of either the Butterworth, Chebychev, or elliptic analog filter design (covered in Appendix G)
3. Generation of TMS320C25 assembly language code

This appendix is designed to get you started. For more detailed information, you should consult Ref. [1].

E.2 FIR FILTER DESIGN

Two different design programs are available: Kaiser window nonrecursive filter design and Parks–McClellan equiripple filter design. Both programs provide for the design of lowpass, highpass, bandpass, bandstop, multiband with exact linear phase as well as differentiators and Hilbert transformers with a phase shift. The Kaiser design is based on a window method and it uses the Kaiser window. The Parks–McClellan method designs for equiripple and uses the Remez exchange algorithm to optimize the coefficients in the Chebychev sense.

379

These two programs and any of the other package programs can be run by executing them directly or by running the filter package batch file, DFDP, and selecting them from the menu. The Kaiser design program resides in a file called KFIR.EXE and the Parks–McClellan program in a file called PMFIR.EXE. To run the Kaiser program directly, for example, type the command

    KFIR <ENTER>

An example of the Kaiser design will be used to demonstrate some of the features and the usage of the package. Perform the following steps to do the Kaiser design:

1. Run the filter design package by typing

    DFDP<ENTER>

The main menu shown in Figure E.1 will appear.
2. Select the Kaiser window FIR design option (2). The Kaiser window main menu will appear (Figure E.2).
3. Select the bandpass option (3). A series of prompts requesting the filter specifications will follow. These are shown in Figure E.3 with the responses underlined. Brief comments on some of the entries follow.

   a. SAMPLING FREQUENCY. This is the sampling frequency of the filter and it in part determines the values of the coefficients. The sample frequency is in kilohertz, as are all the other frequencies.

   b. PASSBAND CUTOFF FREQUENCY and STOPBAND CUTOFF FREQUENCY. The filter transitions are specified by these two frequencies. Figure E.3 shows relative locations of the cutoff frequencies for the FIR passband filter used in this example. No cutoff frequency should exceed one-half the sampling rate.

   c. STOPBAND RIPPLE. Error tolerances are specified for each band. In the case of the Kaiser design, the overshoot is equal on each side of the

```
                    *** Digital Filter Design Package ***
     (C) COPYRIGHT, 1984, 1985, 1986:  ATLANTA SIGNAL PROCESSORS INC, VERSION 2.1
                              SN:  IBM21306

                         PROGRAM SELECTION MENU

        ENTER THE NUMBER CORRESPONDING TO THE DESIGN TECHNIQUE DESIRED

                 1. RECURSIVE (IIR) FILTER DESIGN
                 2. KAISER WINDOW NONRECURSIVE (FIR) FILTER DESIGN
                 3. PARKS-McCLELLAN EQUIRIPPLE (FIR) FILTER DESIGN
                 4. TMS320 CODE GENERATOR
                 5. QUIT

            OPTION DESIRED =  2
```

FIGURE E.1. Program selection menu

```
                    ** Digital Filter Design Package **

                     FIR KAISER-WINDOW DESIGN PROGRAM
   (C) COPYRIGHT,1984,1985,1986: ATLANTA SIGNAL PROCESSORS INC., VERSION 2.12
                            SN: IBM21306

THIS PROGRAM DESIGNS LINEAR-PHASE FIR DIGITAL FILTERS BY
WINDOWING THE UNIT SAMPLE RESPONSE OF AN IDEAL FILTER


  CR TO CONT

      *** KAISER-WINDOW MAIN MENU ***

ENTER THE NUMBER CORRESPONDING TO THE FILTER TYPE DESIRED

         1. LOWPASS
         2. HIGHPASS
         3. BANDPASS
         4. BANDSTOP
         5. MULTIBAND
         6. DIFFERENTIATOR
         7. HILBERT TRANSFORMER
         8. PULSE SHAPING FILTER

 ⌐R TAKE THE FOLLOWING ACTION

         9.  READ SAVED FILE
         10. RETURN TO PROGRAM SELECTION MENU
         11. QUIT (RETURN TO DOS)


  OPTION DESIRED =  _3_
```

FIGURE E.2. Kaiser main menu

```
         ALL FREQUENCIES MUST BE ENTERED IN KILOHERTZ


            ENTER SAMPLING FREQUENCY (KHZ) =    _10_

            ENTER FILTER CUTOFF FREQUENCIES

         LOWER STOPBAND CUTOFF FREQUENCY (KHZ) =   _.8_
         LOWER PASSBAND CUTOFF FREQUENCY (KHZ) =   _1_
         UPPER PASSBAND CUTOFF FREQUENCY (KHZ) =   _1.8_
         UPPER STOPBAND CUTOFF FREQUENCY (KHZ) =   _2_
         STOPBAND RIPPLE = _.05_

            THE ESTIMATED FILTER LENGTH REQUIRED =     66

            ENTER DESIRED LENGTH (<=511) = _71_


              *** CHARACTERISTICS OF DESIRED FILTER ***

                         BANDPASS FILTER

         FILTER LENGTH      =        71
```

FIGURE E.3. Specifications Menu

```
SAMPLING FREQUENCY=   10.000   (KHZ)

                      BAND 1          BAND 2          BAND 3

LOWER BAND EDGE        .0000          1.0000          2.0000
UPPER BAND EDGE        .8000          1.8000          5.0000
NOMINAL GAIN           .0000          1.0000           .0000
NOMINAL RIPPLE         .0500           .0500           .0500


DO YOU WISH TO CHANGE SPECS ? (Y OR N) N  <ENTER>

DO YOU WANT TO QUANTIZE COEFFICIENTS? (Y OR N) N  Y

  WORKING NUMBER OF BITS FOR QUANTIZATION (4-16) =  16
  WORKING
```

FIGURE E.3. (concluded)

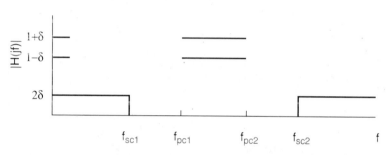

FIGURE E.4 Band specification description

transition; therefore, it is only necessary to specify the ripple in one of the bands. The package uses the stopband and labels it STOPBAND RIPPLE. Thus the smallest allowable error in either band should be entered. The ripple is entered as the absolute ripple; it is not in decibels.

4. After the filter specifications are entered, an estimated filter length will be displayed followed by a prompt for the desired filter length. The question and the response are shown in Figure E.3. The estimated length or any other length may be entered. Notice that we increased the length to 71. If the gain of the highest band is not zero, such as in the case of a highpass, the filter length must be odd in order to meet the linear phase requirement.

5. Next, a summary of the specifications just entered is displayed and is shown in the bottom of Figure E.3. You are also asked if you want to change the specifications. A yes (Y) answer will bring you to the Kaiser main menu, where you will have to reenter all the specifications. Notice that the RIPPLE specification has been converted to decibels using the equation $20 \log \delta$ and $20 \log(1 + \delta)$ for the stopband and passband, respectively.

6. The question

DO YOU WISH TO QUANTIZE THE COEFFICIENTS?

(shown in Figure E.3) should be answered Y if you plan to generate code for the TMS320C25. The prompt

NUMBER OF BITS FOR QUANTIZATION(4-16):

should be answered with 16 to get full resolution on the TMS320C25.

7. The program will find the filter coefficients and then display a summary of the characteristics of the designed filter, shown in Figure E.5, and then ask

DO YOU WANT THE RESULTS SENT TO THE LINE PRINTER? (Y OR N)N

A Y will cause the summary plus a list of all the coefficients to be sent to the line printer (Figure E.6).

8. The FIR filter has now been designed. To display a graph of the filter characteristics, make the selections shown in Figures E.7 and E.8. The resulting graph is shown in Figure E.9.

9. Make the selection shown in Figure E.10 to return to the previous menu shown in Figure E.11. It is very important to save the quantized coefficients before leaving this menu; otherwise, all your work will be lost. Do so by selecting 4 on the menu and answering the subsequent file prompt as indicated in Figure E.11.

The filter design is complete and the filter coefficients have been saved in a file. Figure E.12 shows the selection to return to the top menu, where code generation can be selected. In Section E.3 we explain how to generate both the filter source program and an executable program.

```
             *** CHARACTERISTICS OF DESIGNED FILTER ***
                         BANDPASS FILTER

FILTER LENGTH            = 71
SAMPLING FREQUENCY   = 10.000 (KHZ)

                        BAND 1      BAND 2      BAND 3
LOWER BAND EDGE          .0000      1.0000      2.0000
UPPER BAND EDGE          .8000      1.8000      5.0000
NOMINAL GAIN             .0000      1.0000       .0000
NOMINAL RIPPLE           .0500       .0500       .0500
MAXIMUM RIPPLE           .0401       .0457       .0481
RIPPLE IN DB          -27.9321       .3880    -26.3500

DO YOU WANT RESULTS SENT TO LINE PRINTER ? (Y OR N) N  Y
```

FIGURE E.5. Characteristics of designed filter

```
                   FINITE IMPULSE RESPONSE (FIR)
                   LINEAR-PHASE DIGITAL FILTER DESIGN
                   KAISER-WINDOW ALGORITHM

                      BANDPASS FILTER


           FILTER LENGTH =   71
           SAMPLING FREQUENCY =   10.000   KILOHERTZ
           DESIRED RIPPLE =   26.021   (DB)

                ***** IMPULSE RESPONSE *****

                16-BIT QUANTIZED COEFFICIENTS

               H(  1) =-.724030E-02 = H(  71)
               H(  2) =-.579834E-03 = H(  70)
               H(  3) = .612640E-02 = H(  69)
               H(  4) = .652313E-02 = H(  68)
               H(  5) = .199127E-02 = H(  67)
               H(  6) = .000000E+00 = H(  66)
               H(  7) = .397491E-02 = H(  65)
               H(  8) = .763702E-02 = H(  64)
               H(  9) = .241089E-02 = H(  63)
               H( 10) =-.103149E-01 = H(  62)
               H( 11) =-.182266E-01 = H(  61)
               H( 12) =-.118408E-01 = H(  60)
               H( 13) = .317383E-02 = H(  59)
               H( 14) = .115509E-01 = H(  58)
               H( 15) = .691223E-02 = H(  57)
               H( 16) = .000000E+00 = H(  56)
               H( 17) = .460052E-02 = H(  55)
               H( 18) = .174484E-01 = H(  54)
               H( 19) = .190353E-01 = H(  53)
               H( 20) =-.208282E-02 = H(  52)
               H( 21) =-.306015E-01 = H(  51)
               H( 22) =-.379028E-01 = H(  50)
               H( 23) =-.154800E-01 = H(  49)
               H( 24) = .123444E-01 = H(  48)
               H( 25) = .162964E-01 = H(  47)
               H( 26) = .000000E+00 = H(  46)
               H( 27) =-.131989E-02 = H(  45)
               H( 28) = .330200E-01 = H(  44)
               H( 29) = .712204E-01 = H(  43)
               H( 30) = .531006E-01 = H(  42)
               H( 31) =-.388565E-01 = H(  41)
               H( 32) =-.139610E+00 = H(  40)
               H( 33) =-.149765E+00 = H(  39)
               H( 34) =-.349884E-01 = H(  38)
               H( 35) = .125336E+00 = H(  37)
               H( 36) = .199997E+00 = H(  36)

            *** CHARACTERISTICS OF DESIGNED FILTER ***
```

	BAND 1	BAND 2	BAND 3
LOWER BAND EDGE	.0000	1.0000	2.0000
UPPER BAND EDGE	.8000	1.8000	5.0000
NOMINAL GAIN	.0000	1.0000	.0000
NOMINAL RIPPLE	.0500	.0500	.0500

FIGURE E.6. List of coefficients

```
ENTER CORRESPONDING NUMBER FOR
INSTRUCTION DESIRED

    1. CHANGE FILTER LENGTH OR PARAMETERS
    2. PLOT RESPONSES
    3. DISPLAY FILTER COEFFICIENTS
    4. OUTPUT FILTER COEFFICIENTS
    5. QUANTIZE COEFFICIENTS

    6. RETURN TO KAISER-WINDOW MAIN MENU
    7. RETURN TO PROGRAM SELECTION MENU
    8. QUIT (RETURN TO DOS)

OPTION DESIRED =  2
```

FIGURE E.7. Postdesign menu with plot selection

```
    **   PLOT MENU   **

    1. MAGNITUDE
    2. LOG MAGNITUDE
    3. FREQUENCY ERROR
    4. UNIT SAMPLE RESPONSE
    5. TERMINATE PLOTTING


OPTION DESIRED =  2

HARD COPY ? (Y OR N) N <ENTRY>
```

FIGURE E.8. Plot menu for magnitude selection

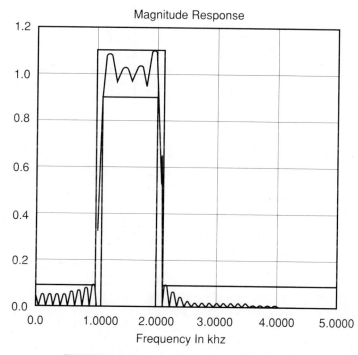

FIGURE E.9. Magnitude plot of bandpass FIR filter

DO YOU WANT AN EXPLODED VIEW ? (Y OR N) N <ENTER>
** PLOT MENU **
1. MAGNITUDE
2. LOG MAGNITUDE
3. FREQUENCY ERROR
4. UNIT SAMPLE RESPONSE
5. TERMINATE PLOTTING
OPTION DESIRED = 5

FIGURE E.10. Plot menu with terminate to next level selection

ENTER CORRESPONDING NUMBER FOR
INSTRUCTION DESIRED

1. CHANGE FILTER LENGTH OR PARAMETERS
2. PLOT RESPONSES
3. DISPLAY FILTER COEFFICIENTS
4. OUTPUT FILTER COEFFICIENTS
5. QUANTIZE COEFFICIENTS

6. RETURN TO KAISER-WINDOW MAIN MENU
7. RETURN TO PROGRAM SELECTION MENU
8. QUIT (RETURN TO DOS)

OPTION DESIRED = 4

FILTER COEFFICIENTS WILL BE WRITTEN TO
YOUR SPECIFIED DISC FILE IN DIRECT FORM
ENTER FILENAME
FILENAME (KFIR.FLT):<ENTER>
   WORKING

ᴊEFFICIENTS WRITTEN TO FILE : KFIR.FLT

CR TO CONT

FIGURE E.11. Save the coefficients screen

ENTER CORRESPONDING NUMBER FOR
INSTRUCTION DESIRED

1. CHANGE FILTER LENGTH OR PARAMETERS
2. PLOT RESPONSES
3. DISPLAY FILTER COEFFICIENTS
4. OUTPUT FILTER COEFFICIENTS
5. QUANTIZE COEFFICIENTS

6. RETURN TO KAISER-WINDOW MAIN MENU
7. RETURN TO PROGRAM SELECTION MENU
8. QUIT (RETURN TO DOS)

OPTION DESIRED = 7

FIGURE E.12. Return to program selection

E.3 TMS320C25 CODE GENERATION

The code generation program, CGEN, will generate a TMS32010, TMS32020, and TMS320C25 source code from the coefficients that were stored in a file during a previous design session. The CGEN (version 2.12) program generates relocatable code instead of absolute code, which earlier versions generated; therefore, the linker must be used to create an executable program. In this description, we will assume that the TMS320C25 is working with the AIB. The following steps outline a procedure to develop an executable filter program.

1. Using the design package, generate the filter coefficients and store them in a file. The responses used for this example are for the FIR design presented in Section E.2.

2. Run CGEN selection (4) of the main menu in Figure E.1. Several questions will be asked, some of which appear innocent but have disastrous effects. Figure E.13 is a print screen of the interaction. A discussion of some of the responses follows:

   a. **Filter Specification Filename? (KFIR.FLT):** <ENTER> will select the default shown, KFIR.FLT. The file we saved in section E.2 was called KFIR.FLT; therefore, we use the <ENTER> to select it.

   b. **Output Code Filename? (KFIR):** FIR <ENTER> is a prompt for the name of the source file. We named it FIR.

   c. **Unique 5 Character Identifier? (FIR):** <ENTER> determines the last five letters used for the labels that will be generated by CGEN. CGEN creates a filter subroutine, init subroutine, and shared variables. These must be named in a predictable way. For example, the filter subroutine will be named FFIR, the init subroutine will be named IFIR, and so on.

   d. **Generate a Fully Executable Filter? (y):N**<ENTER> Answer this NO unless you have an Atlanta Signal Processors I/O board. A NO answer will

```
                    ** Digital Filter Design Package **
                         CODE GENERATION PROGRAM

        (C) Copyright (1987): ATLANTA SIGNAL PROCESSORS, INC.,  version 3.20
                              SN: IBM21306

    Filter Specification Filename? (KFIR.FLT    ) :
              Output Code Filename? (KFIR      ) : FIR
          Unique 5 character Identifier? (FIR  ) :
          Generate a fully executable filter? (Y) : N

    Generating code for FIR - Kaiser Window.  Target processor(s):
        1 - TMS32010; Direct Paged Memory
        2 - TMS32020/c25; Indexed Memory; MACD realization
        3 - TMS32020/c25; Internal Direct Paged Memory
        4 - TMS32020/c25; Indexed Memory; Internal Delay Memory
        5 - TMS32020/c25; Indexed Memory; Internal or External Memory

    Please enter your choice for target processor implementation (1 - 5)? 2

        Code written to file: FIR.ASM
```

FIGURE E.13. FIR code generation screen

generate only the init and the filter subroutines with the proper naming conventions. This is what you want if you have the AIB.

   e. **Target Processor(s):** There are many choices here. Just make sure that you have selected an option for the correct processor. Option 2 was selected for the FIR filter. This option gives us the MACD realization and the addressing is with the auxiliary registers.

3. Source code for the FIR filter has now been generated. Listings of the produced code for the FIR are given in Section E.5. Exit the DFDP and enter the SWDS in order to cross-assemble the source code that was generated by the DFDP. Under the SWDS, the commands look like this:

```
CMD>ASM<ENTER>

   →      Source file [NULL.ASM]: FIR<ENTER>
          Listing file [FIR.LST]:<ENTER>
          Object file [FIR.MPO]:<ENTER>
```

4. A main program must be supplied if you do not have ASPI's board and you elected to generate nonexecutable code. This is what you do if you have the AIB. The program that must be supplied sets up the AIB; calls the filter init program, a subroutine developed by the DFDP; then executes its processing loop, which does the I/O to the AIB; calls the filter subroutine (also developed by the DFDP); and loops back and repeats. Figure E.14 shows a template for the user-supplied main program of an FIR filter. This program must also be assembled to produce a linkable file, which is called FAIB.MPO.

5. Link the CGEN-generated programs to your main program shown in Figure E.14. Here are the commands from within the SWDS:

```
CMD>LINK<ENTER>

   →          Control File [NULL.CTL]: FIR<ENTER>
              Map File [FIR.MAP]: <ENTER>
              Load File [FIR.LOD]: <ENTER>
```

The control file, FIR.CTL, provides the linker with information about what files to combine and where to place certain pieces of data. All the files must be in the same directory as the linker. The following control file was used for the FIR program; each command line has a short explanation to the right of it:

TASK	FIR	Gives an arbitrary name to the program.
PROGRAM	>0	Loads the program at location 0.
COMMON	>300 DFIR	Places the delay variables at location >300.
COMMON	>200 CFIR	Places the coefficients at location >200.
INCLUDE	FAIB.MPO	Main program supplied by the user.

```
*******************************************************************
*MAIN PROGRAM FOR FIR FILTER DESIGNED WITH THE ASPI FILTER    *
*DESIGN PACKAGE                                               *
*ASSUMES THE AIB IS USED.                                     *
*SUBROUTINES CALLED:  IFIR AND FFIR FROM ASPI DESIGN PACKAGE  *
*******************************************************************
        IDT   'AIB'
        REF   FFIR,IFIR              *GENERATED BY CGEN
        DEF      FILTT,VFIR          *REFERENCED BY CGEN
IOPAGE  EQU     0
******** DIRECT PAGE ZERO STARTS AT LOCATION 96
VFIR EQU  96                         *INPUT/OUTPUT SAMPLE FROM FILTER ROUTINE
FILTT     EQU   97                   *TEMPORARY USED BY FILTER SUBROUTINE
DOUT EQU  2                          *D/A OUTPUT
DIN  EQU  2                          *A/D INPUT
TEMP EQU  0                          *TEMPORARY STORAGE FOR AIB INIT
********  AIB INITIALIZATION CONSTANTS
MODE EQU  >FA                        *TRANSPARENT MODE
RATE EQU  999                        *SAMPLE FREQ=10KHZ
     PSEG
        B       START                *JUMP TO MAIN PROGRAM
        BSS     30                   *SKIP 30 ADDRESSES
START
********* INITIALIZE THE TMS320C25
        LDPK    IOPAGE               *SEC UP THE I/O DIRECT PAGE
        SPM     1                    *PRODUCT OUTPUT SHIFT= 1 BIT
        SSXM                         *SET SIGN EXTENT MODE ON
        SOVM                         *SET OVERFLOW ARITH. ON
******** INITIALIZE THE AIB
        LALK    RATE
        SACL    TEMP
        OUT     TEMP,PA1             *SET SAMPLE FREQ
        LALK MODE
        SACL TEMP
        OUT     TEMP,PA0             *SET TRANSPARENT MODE
        OUT     TEMP,PA3             *DUMMY WRITE TO PORT 3
        CALL    IFIR                 *INITIALIZE FILTER
********* START OF MAIN LOOP
IOLOOP
        BIOZ    GET                  *WAIT FOR END OF CONVERSION
        B       IOLOOP
GET  OUT     VFIR,DOUT               *OUTPUT FILTERED SAMPLE
        IN      VFIR,DIN             *AND GET NEW ONE
        CALL    FFIR                 *CALC NEXT OUTPUT
        B       IOLOOP               *REPEAT
        PEND
        END
```

FIGURE E.14. FIR user-supplied main program (FAIB.ASM)

```
INCLUDE   FIR.MPO              DFDP filter routine.
END
```

The executable file produced from the linker is FIR.LOD.

E.4 FREQUENCY RESPONSE OF THE DFDP-GENERATED FIR FILTER

Load the output file from the linker FIR.LOD into the TMS320C25 and run it. The frequency response of the filter developed in Section E.3 is shown in Figure E.15. The frequency response measurement requires a random noise source connected to the input of the AIB and a spectrum analyzer connected to the output.

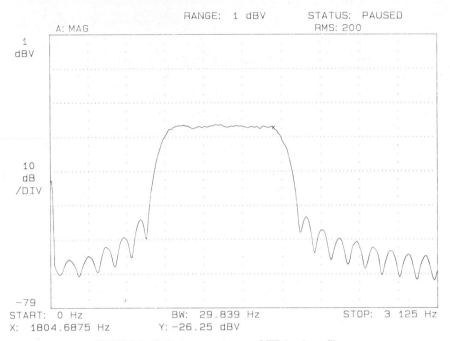

A: MAG

1
dBV

10
dB
/DIV

−79

START: 0 Hz BW: 29.839 Hz STOP: 3 125 Hz
X: 1804.6875 Hz Y: −26.25 dBV

FIGURE E.15. Frequency response of FIR bandpass filter

E.5 ASPI-GENERATED FIR LISTING

```
        IDT     'FIR'           *Unique Identifier
        DEF     FFIR            *Name of filter subroutine
        DEF     IFIR            *Name of filter initialization subroutine
****************************************************
*    ASPI TMS32020 DIGITAL FILTER REALIZATION    *
****************************************************
*          INTERNAL DELAY MEMORY
*          COEF. in PROGRAM MEMORY
*
* NOTE: This realization uses internal memory
*       BANK 0 for program memory.  This allows
*       for very fast execution of the filter
*       by storing the coefficients in BANK 0.
*       However, BANK 0 can not be accessed
*       as data in this mode.  Also, the delay
*       memory must reside in BANK 1 or BANK 2
*
*              KAISER WINDOW
*              71-TAP FIR FILTER
*
*          FILTER GENERATED FROM FILE
*                KFIR.FLT
*          Fri Sep 16 14:01:44 1988
*
*   Filter type: BANDPASS
* Sampling freq: 10000 HZ
*
****************************************************
        PSEG
****************************************************
*          DATA MEMORY DEFINITION              *
****************************************************
*     CONTAINS:
*         STORAGE FOR FILTER INPUT AND OUTPUT
*         STORAGE FOR COEFFICIENTS
*         STORAGE FOR DELAY ELEMENTS
****************************************************
*
* FILTER INPUT and OUTPUT STORAGE
*
        REF     VFIR
*
```

390

```
COEF
*
*   FIR COEFFICIENTS
*
          DATA    -948          *C070
          DATA    -76           *C069
          DATA    803           *C068
          DATA    854           *C067
          DATA    261           *C066
          DATA    0             *C065
          DATA    521           *C064
          DATA    1001          *C063
          DATA    316           *C062
          DATA    -1352         *C061
          DATA    -2389         *C060
          DATA    -1551         *C059
          DATA    416           *C058
          DATA    1514          *C057
          DATA    906           *C056
          DATA    0             *C055
          DATA    603           *C054
          DATA    2286          *C053
          DATA    2494          *C052
          DATA    -273          *C051
          DATA    -4011         *C050
          DATA    -4968         *C049
          DATA    -2029         *C048
          DATA    1618          *C047
          DATA    2136          *C046
          DATA    0             *C045
          DATA    -173          *C044
          DATA    4328          *C043
          DATA    9335          *C042
          DATA    6960          *C041
          DATA    -5093         *C040
          DATA    -18299        *C039
          DATA    -19630        *C038
          DATA    -4586         *C037
          DATA    16428         *C036
          DATA    26214         *C035
          DATA    16428         *C034
          DATA    -4586         *C033
          DATA    -19630        *C032
          DATA    -18299        *C031
          DATA    -5093         *C030
          DATA    6960          *C029
          DATA    9335          *C028
          DATA    4328          *C027
          DATA    -173          *C026
          DATA    0             *C025
          DATA    2136          *C024
          DATA    1618          *C023
          DATA    -2029         *C022
          DATA    -4968         *C021
          DATA    -4011         *C020
          DATA    -273          *C019
          DATA    2494          *C018
          DATA    2286          *C017
          DATA    603           *C016
          DATA    0             *C015
          DATA    906           *C014
          DATA    1514          *C013
          DATA    416           *C012
          DATA    -1551         *C011
          DATA    -2389         *C010
          DATA    -1352         *C009
          DATA    316           *C008
          DATA    1001          *C007
          DATA    521           *C006
          DATA    0             *C005
          DATA    261           *C004
          DATA    854           *C003
          DATA    803           *C002
          DATA    -76           *C001
          DATA    -948          *C000
*
          PEND
          CSEG    'DFIR'
*
*  DELAY STORAGE
*
Z000      BSS     1
          BSS     070-1
ZLAST     BSS     1
```

ASPI-Generated FIR Listing (continued)

```
            BSS     1                     *Extra storage: MACD destroys this
      *                                    Location.
            CEND
      *
      * COEFFICIENT DATA STORAGE
      *
            CSEG    'CFIR'
      FDATA BSS     70+1
      *
            CEND
      *
            PSEG
      **************************************************
      *        FILTER INITIALIZATION SUBROUTINE        *
      **************************************************
      IFIR
            LARP    0
            CNFD
            LRLK    0,FDATA      *POINTER TO COEF. MEMORY
            RPTK    70
            BLKP    COEF,*+      *MOVE COEF.
            CNFP
            LRLK    0,Z000       *POINTER TO DELAY MEMORY
            ZAC                  *CLEAR ACCUMULATOR
            RPTK    070
            SACL    *+           *CLEAR DELAY MEMORY
            RET
      **************************************************
      *              FILTER SUBROUTINE                 *
      **************************************************
      * ASSUMPTIONS:
      *     BANK 0 is set to PROGRAM MEMORY.
      *     COEFFICIENTS are located at 'FDATA'.
      *     DELAY elements are located at 'DELAY'.
      * NOTE: COEF's must be in BANK 0 and DELAY's
      *       must be in BANK 1 or BANK 2.
      *     'DELAR' AR is destroyed by filter routine.
      *     P output shift is set to 1.
      *     SIGN EXTEND mode is ON.
      *     Two's Complement Arithmetic.
      *
      **************************************************
      DELAR EQU     1                     *DELAY AR REGISTER
      *
      FFIR
            LARP    DELAR                 *POINT TO THE DELAY INDEX REGISTER
            LRLK    DELAR,Z000            *INDEX POINTS TO Z-0 (INPUT)
            LAC     VFIR,14 *GET & SCALE INPUT
            SACH    *                     *SAVE SCALED INPUT

            MPYK    0                     *P = 0
            ZAC                           *AC = 0
            LRLK    DELAR,ZLAST           *INDEX POINTS TO Z-N
            RPTK    70
            MACD    FDATA+>FD00,*-  *MULTIPLY, ACCUM. and  DELAY
            APAC                          *FORM RESULT
            SACH    VFIR,0  *SAVE OUTPUT
            RET                           *RETURN
            PEND
            END
```

ASPI-Generated FIR Listing (concluded)

REFERENCE

[1] *Digital Filter Design Package (DFDP)*, Atlanta Signal Processors Inc., Atlanta, Ga.,
     1987.

F

Digital Filter Tools

F.1 CONVERSION FROM *S* TO *Z*

The program bilinear transformation (BLT), in Basic, included at the end of Appendix F, can be used to find a transfer function $H(z)$, given $H(s)$. A walk-through example follows. Given

$$H(s) = \frac{0.03407s^2}{s^4 + 0.26106s^3 + 0.36517s^2 + 0.04322s + 0.0274}$$

discussed in Chapter 6, the BLT is used to find

$$H(z) = \frac{0.02008 - 0.04016z^{-2} + 0.02008z^{-4}}{1 - 2.54948z^{-1} + 3.20244z^{-2} - 2.03596z^{-3} + 0.64136z^{-4}}$$

```
        *** Mapping S to Z Plane Using Bilinear Transformation ***

     Enter the # of numerator coefficients (30 = Max, 0 = Exit) --> 3
          Enter a(0)s^2  --> .03407
          Enter a(1)s^1  --> 0
          Enter a(2)s^0  --> 0

     Enter the # of denominator coefficients --> 5
          Enter b(0)s^4  --> 1
          Enter b(1)s^3  --> .26106
          Enter b(2)s^2  --> .36517
          Enter b(3)s^1  --> .04322
          Enter b(4)s^0  --> .0274

     Are the above coefficients correct ? (y/n) Y


     Send coefficients to:
                    (S)creen
                    (P)rinter
                    (R)un BLT program again
                    (E)xit to DOS
     Enter desired path --> P


            *** BILINEAR TRANSFORMATION PROGRAM ***

     The coefficients in the S - Plane:

               a(0)s^4  =  0.00000      b(0)s^4  =  1.00000
               a(1)s^3  =  0.00000      b(1)s^3  =  0.26106
               a(2)s^2  =  0.03407      b(2)s^2  =  0.36517
               a(3)s^1  =  0.00000      b(3)s^1  =  0.04322
               a(4)s^0  =  0.00000      b(4)s^0  =  0.02740
```

are mapped into the Z – plane using the Bilinear Transformation
and normalized by b0, yielding:

$$
\begin{array}{llllll}
a(0)z^-0 & = & 0.02008 & b(0)z^-0 & = & 1.00000 \\
a(1)z^-1 & = & 0.00000 & b(1)z^-1 & = & -2.54948 \\
a(2)z^-2 & = & -0.04016 & b(2)z^-2 & = & 3.20244 \\
a(3)z^-3 & = & 0.00000 & b(3)z^-3 & = & -2.03596 \\
a(4)z^-4 & = & 0.02008 & b(4)z^-4 & = & 0.64136
\end{array}
$$

FRACTIONAL part of each coefficient is converted to HEX:
(Note: b's are negated for implementation.)

	FRACTION	HEX		FRACTION	HEX
a(0)z^-0 =	0.02008*2^15 =	292	b(0)z^-0 =	0.00000*2^15 =	0
a(1)z^-1 =	0.00000*2^15 =	0	b(1)z^-1 =	0.54948*2^15 =	4655
a(2)z^-2 =	-0.04016*2^15 =	FADC	b(2)z^-2 =	-0.20244*2^15 =	E616
a(3)z^-3 =	0.00000*2^15 =	0	b(3)z^-3 =	0.03596*2^15 =	49A
a(4)z^-4 =	0.02008*2^15 =	292	b(4)z^-4 =	-0.64136*2^15 =	ADE8

F.2 MAGNITUDE AND PHASE CALCULATION

The program MAGPHSE discussed in Appendix D can also be used with IIR filters
to find their magnitude and phase response, as shown in the walk-through example.
The amplitude response of the fourth-order bandpass filter discussed in Chapter 6
is shown in Figure F.1 and plotted in Figure F.2.

```
             *** Magnitude and Phase Response of a DT System ***

    (F)IR or (I)IR ? I
Enter the # of numerator coefficients (256 = Maximum) --> 5
       Enter a(0)z^-0      --> .02008
       Enter a(1)z^-1      --> 0
       Enter a(2)z^-2      --> -.04016
       Enter a(3)z^-3      --> 0
       Enter a(4)z^-4      --> .02008

Enter the # of denominator coefficients --> 5
       Enter b(0)z^-0      --> 1
       Enter b(1)z^-1      --> -2.5494
       Enter b(2)z^-2      --> 3.2024
       Enter b(3)z^-3      --> -2.0359
       Enter b(4)z^-4      --> .6413

Enter the Sampling Frequency in Hz --> 10000

Enter the number of steps desired --> 40

Tabulate (M)agnitude or (P)hase response --> M

Do you want to normalize the magnitude (Y/N) --> Y

Do you want (P)ower or (A)mplitude form  --> A

Are the above entries correct (Y/N) Y
                     Please wait ...working
Send output to:
               (S)creen
               (P)rinter
               (F)ile: (FREQ,AMPLITUDE) Format
               (RU)n program with same coefficients
               (R)un program with different coefficients
               (E)xit to DOS
```

```
Enter desired path --> F
Enter Filename : IIR4.MP

                    ...saving IIR4.MP
Send output to:
                (S)creen
                (P)rinter
                (F)ile: (FREQ,AMPLITUDE) Format
                (RU)n program with same coefficients
                (R)un program with different coefficients
                (E)xit to DOS

Enter desired path -->
```

UTILITY - MAGNITUDE AND PHASE RESPONSE PROGRAM

*** AMPLITUDE Response of a DT System ***

Coefficients:

$a(0)$	=	0.02008	$b(0)$	=	1.00000
$a(1)$	=	0.00000	$b(1)$	=	-2.54940
$a(2)$	=	-0.04010	$b(2)$	=	3.20240
$a(3)$	=	0.00000	$b(3)$	=	-2.03590
$a(4)$	=	0.02008	$b(4)$	=	0.64130

FREQUENCY	AMPLITUDE
0.00	0.00000
128.21	0.00206
256.41	0.00878
384.62	0.02208
512.82	0.04643
641.03	0.09214
769.23	0.18515
897.44	0.39370
1025.64	0.79059
1153.85	0.99473
1282.05	1.00000
1410.26	0.90866
1538.46	0.61398
1666.67	0.37423
1794.87	0.24084
1923.08	0.16563
2051.28	0.11990
2179.49	0.09009
2307.69	0.06955
2435.90	0.05477
2564.10	0.04375
2692.31	0.03530
2820.51	0.02869
2948.72	0.02341
3076.92	0.01913
3205.13	0.01562
3333.33	0.01272
3461.54	0.01030
3589.74	0.00828
3717.95	0.00657
3846.15	0.00514
3974.36	0.00394
4102.56	0.00294
4230.77	0.00211
4358.97	0.00144
4487.18	0.00091
4615.38	0.00050
4743.59	0.00022
4871.79	0.00006
5000.00	0.00000

FIGURE F.1. Amplitude response of IIR bandpass filter

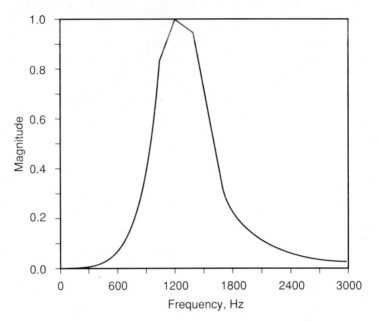

FIGURE F.2. Plot of amplitude response of IIR bandpass filter.

F.3 BILINEAR TRANSFORMATION PROGRAM

```
10 REM                    *** Bilinear Transformation Program ***
20 REM
30 REM             Program to evaluate the BLT given an analog transfer
40 REM      function of a recursive (IIR) system.  The result of the
50 REM      BLT will be of the form:
60 REM
70 REM                   A0 + A1z^-1 + A2z^-2 + ... + Anz^-n
80 REM         H(z)  =  ------------------------------------
90 REM                   1 + B1z^-1 + B2z^-2 + ... + Bmz^-m
100 REM
110 REM               Note: Order of numerator must be less than or equal
120 REM                     to the order of the denominator polynomial.
130 REM
140 REM      The program will prompt the user for output path of screen or
150 REM      or printer.
160 REM
170 DIM AS(32),BS(32),AZ(32),BZ(32),CARY(32,32),FAHEX(32),FBHEX(32),FRACA$(32)
180 DIM FRACB$(32),AZN(32),BZN(32),FRACA(32),FRACB(32)
190 KEY OFF
200 CLS
210 MAXCOEFFS=31
220 LOCATE 3
230 PRINT TAB(13);"*** Mapping S to Z Plane Using Bilinear Transformation ***"
240 REM
250 REM               *** Prompt for Analog Transfer Function ***
260 REM
270 LOCATE 6
280 PRINT TAB(9);
290 INPUT "Enter the # of numerator coefficients (30 = Max, 0 = Exit) --> ",NNC
300 IF NNC = 0 THEN CLS:SYSTEM        'Exit to DOS
310 IF NNC > MAXCOEFFS OR NNC < 1 THEN GOTO 200
320 FOR N=NNC-1 TO 0 STEP -1
330 I=(NNC-1)-N
340 GOSUB 2270             'convert index to character
350 PRINT "              Enter a(";C$;")s^";P$;TAB(29);"--> ";:INPUT "",AS(N)
360 NEXT N
370 PRINT
380 INPUT "        Enter the # of denominator coefficients --> ",NDC
390 IF NDC > MAXCOEFFS OR NDC < 1 THEN GOTO 200
400 FOR N=NDC-1 TO 0 STEP -1
410 I=(NDC-1)-N
420 GOSUB 2270             'convert index to character
430 PRINT "              Enter b(";C$;")s^";P$;TAB(29);"--> ";:INPUT "",BS(N)
440 NEXT N
450 PRINT
460 INPUT "        Are the above coefficients correct ? (y/n) ",C$
```

```
470 IF C$="N" OR C$="n" THEN 200
480 REM
490 REM                     *** Perform Bilinear Transformation ***
500 REM
510 FOR I=0 TO NDC-1
520 CARY(0,I)=1
530 NEXT I
540 NUM=1
550 FOR I=1 TO NDC-1
560 NUM=NUM*((NDC-1)-(I+1)+2)/(I)
570 CARY(I,0)=NUM
580 NEXT I
590 FOR I=1 TO NDC-1
600 FOR J=1 TO NDC-1
610 CARY(I,J)=CARY(I,J-1)-CARY(I-1,J)-CARY(I-1,J-1)
620 NEXT J
630 NEXT I
640 FOR I=0 TO NDC-1
650 SUMOFAS=0
660 SUMOFBS=0
670 FOR J=0 TO NDC-1
680 SUMOFAS=SUMOFAS+CARY(I,J)*AS(J)
690 SUMOFBS=SUMOFBS+CARY(I,J)*BS(J)
700 NEXT J
710 AZ(I)=SUMOFAS
720 BZ(I)=SUMOFBS
730 NEXT I
740 REM
750 REM                     *** Divide all Coefficients by B0 ***
760 REM
770 FOR I=0 TO NDC-1
780 AZN(I)=AZ(I)/BZ(0)
790 BZN(I)=BZ(I)/BZ(0)
800 NEXT I
810 REM
820 REM              *** Generate Fractional Coefficient Arrays ***
830 REM
840 FOR I=0 TO NDC-1
850 IF ABS(AZN(I)) < 1 THEN FRACA(I)=AZN(I):GOTO 880          'already fraction
860 FRACA(I)=ABS(AZN(I))-INT(ABS(AZN(I)))     'FRACA(I)=fractional part of AZN(I)
870 IF AZN(I) < 0 THEN FRACA(I)=-FRACA(I)
880 IF ABS(BZN(I)) < 1 THEN FRACB(I)=BZN(I):GOTO 910          'already fraction
890 FRACB(I)=ABS(BZN(I))-INT(ABS(BZN(I)))     'FRACB(I)=fractional part of BZN(I)
900 IF BZN(I) < 0 THEN FRACB(I)=-FRACB(I)
910 NEXT I
920 REM
930 REM                  *** Negate b's for Filter Implemetation ***
940 REM
950 FOR I=0 TO NDC-1
960 FRACB(I)=-FRACB(I)
970 NEXT I
980 REM
990 REM           *** Convert Fractional Coefficients to Hexidecimal ***
1000 REM
1010 GOSUB 2400           'invoke conversion routine
1020 REM
1030 REM              *** Output Transfer Function Coefficients ***
1040 REM
1050 GOSUB 1090           'invoke output routine
1060 CLS
1070 GOTO 200
1080 END
1090 REM =========================== Output routine ============================
1100 REM
1110 CLS
1120 LOCATE 8
1130 PRINT "       Send coefficients to:"
1140 PRINT TAB(29);"(S)creen"
1150 PRINT TAB(29);"(P)rinter"
1160 PRINT TAB(29);"(R)un BLT program again"
1170 PRINT TAB(29);"(E)xit to DOS"
1180 PRINT
1190 INPUT "       Enter desired path --> ",PATH$
1200 IF PATH$ = "E" OR PATH$ = "e" THEN CLS:SYSTEM
1210 IF PATH$ = "R" OR PATH$ = "r" THEN RETURN
1220 IF PATH$ = "S" OR PATH$ = "s" THEN GOSUB 1260
1230 IF PATH$ = "P" OR PATH$ = "p" THEN GOSUB 1790
1240 GOTO 1110
1250 REM
1260 REM ------------------- Output Coefficients to Screen --------------------
1270 REM
1280 CLS
1290 LOCATE 4
1300 PRINT "       The coefficients in the S - plane:"
```

```
1310 PRINT
1320 FOR N=NDC-1 TO 0 STEP -1
1330 I=(NDC-1)-N
1340 GOSUB 2270           'Convert index to character
1350 PRINT TAB(18);:PRINT"a(";C$;")s^";P$;TAB(27);"= ";
1360 PRINT USING"##.#####";AS(N);
1370 PRINT TAB(43);:PRINT"b(";C$;")s^";P$;TAB(52);"= ";
1380 PRINT USING"##.#####";BS(N)
1390 NEXT N
1400 LOCATE 23
1410 INPUT "         Press <Enter> to view H(z)/b0 ",RPLY
1420 CLS
1430 LOCATE 4
1440 PRINT TAB(8);
1450 PRINT "are mapped into the Z - plane using the Bilinear Transformation"
1460 PRINT TAB(8);
1470 PRINT "and normalized by b0, yielding:"
1480 PRINT
1490 FOR I=0 TO NDC-1
1500 GOSUB 2270           'Convert index to character
1510 PRINT TAB(18);:PRINT"a(";C$;")z^-";C$;TAB(28);"= ";
1520 PRINT USING"##.#####";AZN(I);
1530 PRINT TAB(43);:PRINT"b(";C$;")z^-";C$;TAB(53);"= ";
1540 PRINT USING"##.#####";BZN(I)
1550 NEXT I
1560 LOCATE 23
1570 PRINT TAB(9);
1580 INPUT "Press <Enter> to view fractional parts of the coefficients ",RPLY
1590 CLS
1600 LOCATE 4
1610 PRINT "       FRACTIONAL part of each coefficient is converted to HEX:"
1620 PRINT "       (Note: b's are negated for implementation.)
1630 PRINT
1640 PRINT TAB(19);"FRACTION";TAB(34);"HEX";TAB(56);"FRACTION";TAB(71);"HEX"
1650 PRINT
1660 FOR I=0 TO NDC-1
1670 GOSUB 2270           'Convert index to character
1680 PRINT TAB(6);:PRINT"a(";C$;")z^-";C$;TAB(16);"= ";
1690 PRINT USING"##.#####";FRACA(I);
1700 PRINT TAB(26);"*2^15 = ";FRACA$(I);
1710 PRINT TAB(43);:PRINT"b(";C$;")z^-";C$;TAB(53);"= ";
1720 PRINT USING"##.#####";FRACB(I);
1730 PRINT TAB(63);"*2^15 = ";FRACB$(I)
1740 NEXT I
1750 LOCATE 23:INPUT "         Press <Enter> to return to Output Path Menu ",RPLY
1760 RETURN
1770 REM ---------------------------------------------------------------------
1780 REM
1790 REM ------------------ Output Coefficients to Printer --------------------
1800 REM
1810 LPRINT TAB(20);"*** BILINEAR TRANSFORMATION PROGRAM ***"
1820 LPRINT:LPRINT
1830 LPRINT "       The coefficients in the S - Plane:"
1840 LPRINT
1850 FOR N=NDC-1 TO 0 STEP -1
1860 I=(NDC-1)-N
1870 GOSUB 2270           'Convert index to character
1880 LPRINT TAB(18);
1890 LPRINT"a(";C$;")s^";P$;TAB(27);"= ";:LPRINT USING"##.#####";AS(N);
1900 LPRINT TAB(43);
1910 LPRINT"b(";C$;")s^";P$;TAB(52);"= ";:LPRINT USING"##.#####";BS(N)
1920 NEXT N
1930 LPRINT
1940 LPRINT TAB(8);
1950 LPRINT "are mapped into the Z - plane using the Bilinear Transformation"
1960 LPRINT TAB(8);
1970 LPRINT "and normalized by b0, yielding:"
1980 LPRINT
1990 FOR I=0 TO NDC-1
2000 GOSUB 2270           'Convert index to character
2010 IF I > NDC-1 THEN 2000
2020 LPRINT TAB(18);
2030 LPRINT"a(";C$;")z^-";C$;TAB(28);"= ";:LPRINT USING"##.#####";AZN(I);
2040 LPRINT TAB(43);
2050 LPRINT"b(";C$;")z^-";C$;TAB(53);"= ";:LPRINT USING"##.#####";BZN(I)
2060 NEXT I
2070 LPRINT
2080 LPRINT "       FRACTIONAL part of each coefficient is converted to HEX:"
2090 LPRINT "       (Note: b's are negated for implementation.)"
2100 LPRINT
2110 LPRINT TAB(19);"FRACTION";TAB(34);"HEX";TAB(56);"FRACTION";TAB(71);"HEX"
2120 LPRINT
2130 FOR I=0 TO NDC-1
2140 GOSUB 2270           'Convert index to character
2150 IF I > NDC-1 THEN 2000
2160 LPRINT TAB(6);
```

Bilinear Transformation Program (continued)

```
2170 LPRINT"a(";C$;")z^-";C$;TAB(16);"= ";:LPRINT USING"##.#####";FRACA(I);
2180 LPRINT TAB(26);"*2^15 = ";FRACA$(I);
2190 LPRINT TAB(43);
2200 LPRINT"b(";C$;")z^-";C$;TAB(53);"= ";:LPRINT USING"##.#####";FRACB(I);
2210 LPRINT TAB(63);"*2^15 = ";FRACB$(I)
2220 NEXT I
2230 RETURN
2240 REM ------------------------------------------------------------------------
2250 REM =======================================================================
2260 REM
2270 REM ===================== Integer to Character ========================
2280 REM
2290 S1$=STR$(I)
2300 S2$=STR$(N)
2310 C$=RIGHT$(S1$,1)
2320 P$=RIGHT$(S2$,1)
2330 IF I > 9 THEN C$=RIGHT$(S1$,2)
2340 IF N > 9 THEN P$=RIGHT$(S2$,2)
2350 IF I > 99 THEN C$=RIGHT$(S1$,3)
2360 IF N > 99 THEN P$=RIGHT$(S2$,3)
2370 RETURN
2380 REM =======================================================================
2390 REM
2400 REM ==================== Convert Coefficients to Hex ====================
2410 REM
2420 FOR M=0 TO NDC-1
2430 FAHEX(M)=CINT(FRACA(M)*32767!)
2440 FBHEX(M)=CINT(FRACB(M)*32767!)
2450 IF FAHEX(M) >= 0 THEN 2470
2460 FAHEX(M)=65536!+FAHEX(M)
2470 IF FBHEX(M) >= 0 THEN 2490
2480 FBHEX(M)=65535!+FBHEX(M)
2490 FRACA$(M)=HEX$(FAHEX(M))
2500 FRACB$(M)=HEX$(FBHEX(M))
2510 NEXT M
2520 RETURN
2530 REM =======================================================================
```

Bilinear Transformation Program (concluded)

G

ASPI Digital Filter Design Package (IIR Filters)

G.1 IIR FILTER DESIGN

The IIR filter design component of this package does the analog filter types: Butterworth, Chebychev I and II, and elliptic. The bilinear transformation is used to convert these filters to the z domain. These transfer functions are broken down into cascaded 2nd-order sections that are implemented using the transpose direct form II. While the ordering of the individual sections is mathematically irrelevant, to reduce quantization noise and overflow, the following steps are used for the implementation of the filter:

1. The second-order sections are placed in Q-order with the highest-Q complex pole–pair and the nearest complex zero–pair in the last or output section.
2. Each subsequent complex pole–pair is grouped with its nearest complex zero–pair to make a second-order section.
3. Each section is scaled to a gain of less than 1.

Some general guidelines for setting the specifications are as follows:

1. The sampling frequency should be at least twice the highest-frequency component present.
2. Avoid too-stringent ripple values such as $\delta < 0.001$ (60 dB).
3. Avoid specifying too-narrow passbands, stopbands, and transition regions.

Since the menu interaction for the IIR design is similar to the FIR presented in Section E.2, the screens for it are presented sequentially in Figures G.1 to G.12 without comment. When you have completed the steps indicated in the menus, you will have designed an IIR filter and saved its coefficients. The next step is to generate executable code, and it is covered in the next section.

400

```
                      ** Digital Filter Design Package **

                  IIR BILINEAR TRANSFORM DESIGN PROGRAM


(C) COPYRIGHT,1984,1985,1986: ATLANTA SIGNAL PROCESSORS INC.,VERSION 2.12
                            SN: IBM21306
                     PROGRAM SELECTION MENU

ENTER THE NUMBER CORRESPONDING TO THE DESIGN TECHNIQUE DESIRED

        1.  RECURSIVE (IIR) FILTER DESIGN
        2.  KAISER WINDOW NONRECURSIVE (FIR) FILTER DESIGN
        3.  PARKS-McCLELLAN EQUIRIPPLE (FIR) FILTER DESIGN
        4.  TMS320 CODE GENERATOR
        5.  QUIT


OPTION DESIRED =  1
```

FIGURE G.1. Program selection menu with IIR selection

```
                      ** Digital Filter Design Package **

                  IIR BILINEAR TRANSFORM DESIGN PROGRAM


 (C) COPYRIGHT,1984,1985,1986: ATLANTA SIGNAL PROCESSORS INC.,VERSION 2.12
                            SN: IBM21306
        THIS FILTER DESIGN PROGRAM DESIGNS RECURSIVE DIGITAL
        FILTERS FROM BUTTERWORTH, CHEBYSHEV, AND ELLIPTIC
        ANALOG PROTOTYPES.



    CR TO CONT

        ***  IIR BILINEAR TRANSFORM MAIN MENU  ***

    ENTER THE NUMBER CORRESPONDING TO THE FILTER TYPE DESIRED

            1. LOWPASS
            2. HIGHPASS
            3. BANDPASS
            4. BANDSTOP

    OR TAKE THE FOLLOWING ACTION

            5. READ SAVED FILE
            6. RETURN TO PROGRAM SELECTION MENU
            7. QUIT (RETURN TO DOS)

     OPTION DESIRED:   4
```

FIGURE G.2. Filter-type menu

```
        ALL FREQUENCIES MUST BE ENTERED IN KILOHERTZ

        ENTER SAMPLING FREQUENCY (KHZ) =   10

        ENTER BANDSTOP FILTER CUTOFF FREQUENCIES

        LOWER PASSBAND CUTOFF FREQUENCY (KHZ) =    1
        LOWER STOPBAND CUTOFF FREQUENCY (KHZ) =    1.1
        UPPER STOPBAND CUTOFF FREQUENCY (KHZ) =    1.5
        UPPER PASSBAND CUTOFF FREQUENCY (KHZ) =    1.6
        PASSBAND RIPPLE =  .1
        STOPBAND RIPPLE =  .1
```

FIGURE G.3. Filter specification entries

```
TO MEET YOUR SPECIFICATIONS WOULD REQUIRE FILTERS
OF THE FOLLOWING ORDERS:

          BUTTERWORTH:   16
          CHEBYSHEV I:    8
          CHEBYSHEV II:   8
          ELLIPTIC:       6

IF ONE OF THESE IS SATISFACTORY, ENTER THE
CORRESPONDING NUMBER:

                    1. BUTTERWORTH
                    2. CHEBYSHEV I
                    3. CHEBYSHEV II
                    4. ELLIPTIC
    OTHERWISE
                    5. RETURN TO IIR BILINEAR TRANSFORM MAIN MENU
                    6. RETURN TO PROGRAM SELECTION MENU
                    7. QUIT (RETURN TO DOS)

    OPTION DESIRED:    4

WORKING DO YOU WISH TO QUANTIZE COEFFICIENTS ?  (Y or N)N
```

FIGURE G.4. Design method selection

```
        *** CHARACTERISTICS OF DESIGNED FILTER ***

               ELLIPTIC    BANDSTOP FILTER

         FILTER ORDER =    6
         Sampling frequency =  10.000  KiloHertz

                          BAND  1        BAND  2        BAND  3


    LOWER BAND EDGE        .00000        1.10000        1.60000
    UPPER BAND EDGE       1.00000        1.50000        5.00000
    NOMINAL GAIN          1.00000         .00000        1.00000
    NOMINAL RIPPLE         .10000         .10000         .10000
    MAXIMUM RIPPLE         .06979         .06182         .06926
    RIPPLE IN dB           .58596      -24.17769         .58164


DO YOU WANT RESULTS SENT TO THE LINE PRINTER ? (Y OR N) N Y
```

FIGURE G.5. Designed filter specifications

```
                INFINITE IMPULSE RESPONSE (IIR)
                ELLIPTIC     BANDSTOP FILTER
                16-BIT QUANTIZED COEFFICIENTS


         FILTER ORDER =    6
         SAMPLING FREQUENCY =  10.000  KILOHERTZ

  I      A(I,1)        A(I,2)        B(I,0)        B(I,1)        B(I,2)

  1    -1.039490      .506653       .753357      -1.039551      .753357
  2    -1.039307      .931610       .851074      -1.019775      .851074
  3    -1.557495      .950653      1.160889      -1.772278     1.160889

        *** CHARACTERISTICS OF DESIGNED FILTER ***


                          BAND  1        BAND  2        BAND  3


    LOWER BAND EDGE        .00000        1.10000        1.60000
    UPPER BAND EDGE       1.00000        1.50000        5.00000
    NOMINAL GAIN          1.00000         .00000        1.00000
    NOMINAL RIPPLE         .10000         .10000         .10000
    MAXIMUM RIPPLE         .06979         .06182         .06926
    RIPPLE IN dB           .58596      -24.17769         .58164
```

FIGURE G.6. List of IIR filter coefficients

```
ENTER CORRESPONDING NUMBER FOR
INSTRUCTION DESIRED

   1. AUTOMATICALLY INCREMENT FILTER ORDER
   2. PLOT RESPONSES
   3. DISPLAY FILTER COEFFICIENTS
   4. OUTPUT FILTER COEFFICIENTS
   5. QUANTIZE COEFFICIENTS

   6. RETURN TO IIR BILINEAR TRANSFORM MAIN MENU
   7. RETURN TO PROGRAM SELECTION MENU
   8. QUIT (RETURN TO DOS)

OPTION DESIRED =   2
```

FIGURE G.7. Selection of plotting option

```
** PLOT MENU **

1. MAGNITUDE RESPONSE
2. LOG MAGNITUDE RESPONSE
3. PHASE RESPONSE
4. GROUP DELAY
5. POLE-ZERO PLOT
6. IMPULSE RESPONSE
7. DISPLAY ALL PLOTS
8. TERMINATE PLOTTING

OPTION DESIRED =   1

WORKING
HARD COPY ? (Y OR N) NDO YOU WANT AN EXPLODED VIEW? (Y or N)N  Y
```

FIGURE G.8. Plot menu selections

```
** PLOT MENU **

1. MAGNITUDE RESPONSE
2. LOG MAGNITUDE RESPONSE
3. PHASE RESPONSE
4. GROUP DELAY
5. POLE-ZERO PLOT
6. IMPULSE RESPONSE
7. DISPLAY ALL PLOTS
8. TERMINATE PLOTTING

OPTION DESIRED =    8
```

FIGURE G.9. Terminating the plot option

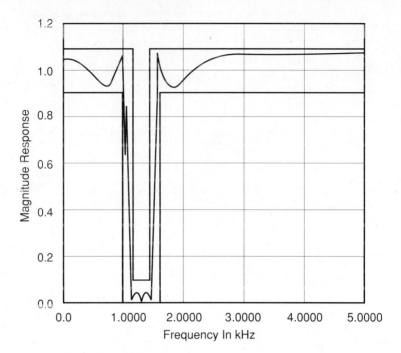

FIGURE G.10. Package-generated frequency response

```
ENTER CORRESPONDING NUMBER FOR
 INSTRUCTION DESIRED

        1. AUTOMATICALLY INCREMENT FILTER ORDER
        2. PLOT RESPONSES
        3. DISPLAY FILTER COEFFICIENTS
        4. OUTPUT FILTER COEFFICIENTS
        5. QUANTIZE COEFFICIENTS
        6. RETURN TO IIR BILINEAR TRANSFORM MAIN MENU
        7. RETURN TO PROGRAM SELECTION MENU
        8. QUIT (RETURN TO DOS)


 OPTION DESIRED = ENTER FILENAME (IIR.FLT) : <ENTER>

 COEFFICIENTS WRITTEN TO FILE : IIR.FLT
  CR TO CONT <ENTER>
```

FIGURE G.11. Saving the coefficients

```
ENTER CORRESPONDING NUMBER FOR
 INSTRUCTION DESIRED

        1. AUTOMATICALLY INCREMENT FILTER ORDER
        2. PLOT RESPONSES
        3. DISPLAY FILTER COEFFICIENTS
        4. OUTPUT FILTER COEFFICIENTS
        5. QUANTIZE COEFFICIENTS

        6. RETURN TO IIR BILINEAR TRANSFORM MAIN MENU
        7. RETURN TO PROGRAM SELECTION MENU
        8. QUIT (RETURN TO DOS)

 OPTION DESIRED =  7
```

FIGURE G.12. Return to program selection menu

G.2 TMS320C25 CODE GENERATION

Source generation for the IIR filter is similar to that for the FIR filter explained in Section E.3. The code generator is evoked by selecting option (4) from the main menu or by running CGEN.EXE directly from MS DOS. In either case you will be prompted to answer the first four questions in Figure G.13. The responses to these questions are explained in Section E.3. The explanations are the same as for the FIR filter, except that now you give the name of a filter coefficients file created from a previous IIR design. Upon seeing that the file given is an IIR filter, the code generator will display the remainder of Figure G.13 and prompt you for a selection. You may select any of the options given but be sure that it is a TMS320C25 option.

When control returns, source code for the IIR filter will have been generated. Exit from the package and assemble the code in the same way as for the FIR filter in Section E.3. From here on, the steps to create an executable file are the same as those for the FIR filter, explained in Section E.3, except replace the FIR.CTL file with an IIR.CTL file that contains the following commands:

TASK	IIR	Task name.
PROGRAM	>0	Start of the program.
COMMON	>300 XIIR	Coefficient storage starts at location >300.
INCLUDE	IAIB.MPO	User-supplied main program.
INCLUDE	IIR.MPO	Filter subroutines generated by the DFDP.
END		

```
           (C) Copyright (1987): ATLANTA SIGNAL PROCESSORS, INC.,  version 3.20
                                  SN: IBM21306

       Filter Specification Filename? (IIR.FLT    ):
              Output Code Filename? (IIR          ):
            Unique 5 character Identifier? (IIR   ):
            Generate a fully executable filter? (Y): n

       Generating code for IIR - 2nd order sections.  Target processor(s):
            1 - TMS32010; Direct Paged Memory
            2 - TMS32010; Direct Paged Memory; Increased Precision
            3 - TMS32020; Direct Paged Memory
            4 - TMS32020; Direct Paged Memory; Increased Precision
   C:       5 - TMS32020; Indexed Memory
            6 - TMS32020; Indexed Memory; Increased Precision
            7 - TMS32020; Index Memory; Looping Control
            8 - TMS32020; Indexed Memory; Looping Control; Increased Precision
            9 - TMS320c25; Direct Paged Memory
           10 - TMS320c25; Direct Paged Memory; Increased Precision
           11 - TMS320c25; Indexed Memory
           12 - TMS320c25; Indexed Memory; Increased Precision
           13 - TMS320c25; Indexed Memory; Looping Control
           14 - TMS320c25; Indexed Memory; Looping Control; Increased Precision

       Please enter your choice for target processor implementation (1 - 14)? 11
```

FIGURE G.13. IIR code generation screen

```
**********************************************
*MAIN PROGRAM FOR IIR FILTER DESIGNED WITH *
*ASPI FILTER DESIGN PACKAGE                 *
*SUBROUTINES CALLED:  FIIR AND IIIR         *
**********************************************
        IDT   'AIB'
        REF   FIIR,IIIR            *GENERATED BY CGEN
        DEF      FILTT,VIIR        *REFERENCED BY CGEN
IOPAGE    EQU   0
****** DIRECT PAGE ZERO STARTS A LOCATION 96
VIIR      EQU   96                 *INPUT/OUTPUT SAMPLE TO/FROM FILTER SUBROUTINE
FILTT     EQU   97                 *TEMPORARY USED BY IIR FILTER SUBROUTINE
DOUT      EQU   2                  *D/A OUTPUT
DIN       EQU   2                  *A/D INPUT
TEMP EQU  0                        *TEMPORARY FOR INITIALIZING THE AIB
********** AIB INITIALIZATION CONSTANTS
MODE      EQU   >FA                *TRANSPARENT MODE
RATE      EQU   999                *SAMPLE FREQ=10KHZ
     PSEG
     B        START                *JUMP TO START OF MAIN PROGRAM
     BSS      30                   *SKIP 30 ADDRESSES
START
****** INITIALIZE THE TMS320C25
        LDPK     IOPAGE            *SET UP THE I/O DIRECT PAGE
        SPM      1                 *PRODUCT OUTPUT SHIFT =1 BIT
        SSXM                       *SET SIGN EXTEND MODE ON
        SOVM                       *SET OVERFLOW ARITH. ON
****** INITIALIZE THE AIB
        LALK     RATE
        SACL     TEMP
        OUT      TEMP,PA1          *INIT SAMPLE CLOCK
        LALK     MODE
        SACL     TEMP
        OUT      TEMP,PA0          *SET TRANSPARENT MODE
        OUT      TEMP,PA3          *DUMMY WRITE TO PORT 3
        CALL     IIIR              *INITIALIZE FILTER
****** START OF MAIN LOOP
IOLOOP
        BIOZ     GET               *WAIT FOR END OF CONVERSION
        B        IOLOOP
GET  OUT      VIIR,DOUT            *OUTPUT FILTERED SAMPLE
        IN       VIIR,DIN          *AND GET NEW ONE
        CALL     FIIR              *CALC THE NEXT OUTPUT VALUE
        B        IOLOOP            *REPEAT
        PEND
        END
```

FIGURE G.14. IIR user-supplied main program (IAIB.ASM)

The IIR.CTL file includes the IIR.MPO instead of the FIR.MPO and a different user-supplied main program, IAIB.MPO (Figure G.14), instead of FAIB.MPO (Figure E.14). Upon completion of the linking, you will have created an executable file with a .LOD extension.

G.3 FREQUENCY RESPONSE OF A DFDP-GENERATED IIR FILTER

Load the output file from the linker, IIR.LOD, created from the preceding sections into the TMS320C25 and run it. The frequency response of the IIR filter is shown in Figure G.15. The frequency response measurement requires a random noise source connected to the input of the AIB and a spectrum analyzer connected to the output.

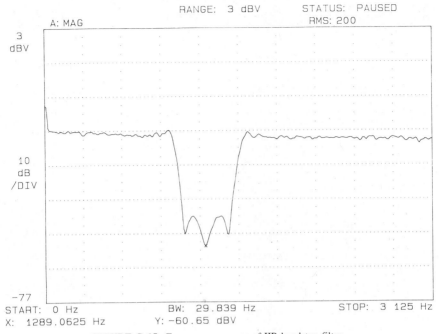

RANGE: 3 dBV   STATUS: PAUSED
A: MAG   RMS: 200

3 dBV

10 dB /DIV

−77

START: 0 Hz   BW: 29.839 Hz   STOP: 3 125 Hz
X: 1289.0625 Hz   Y: −60.65 dBV

FIGURE G.15. Frequency response of IIR bandstop filter

G.4 ASPI-GENERATED IIR FILTER LISTING

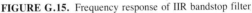

```
        IDT      'IIR'                  *UNIQUE NAME
        DEF      FIIR              *Name of filter subroutine
        DEF      IIIR              *Name of filter initialization subroutine
********************************************************
*    ASPI TMS320c25 DIGITAL FILTER REALIZATION    *
********************************************************
*            3-STAGE RECURSIVE FILTER
*             SECOND ORDER SECTIONS
*           FILTER GENERATED FROM FILE
*                    IIR.FLT
*             Fri Sep 16 16:01:33 1988
*
*        Filter type: BANDSTOP
* Approximation type: ELLIPTIC
*       Sampling freq: 10000 HZ
*
********************************************************
        PSEG
********************************************************
*          DATA MEMORY DEFINITION              *
********************************************************
*    CONTAINS:
*         STORAGE FOR FILTER INPUT AND OUTPUT
*         STORAGE FOR COEFFICIENTS
*         STORAGE FOR DELAY ELEMENTS
********************************************************
*
* FILTER INPUT and OUTPUT STORAGE
*
        REF      FILTT,VIIR
*
COEF
* COEFFICIENT INITIALIZATION STORAGE AREA
*
*
*
*   SECOND-ORDER SECTION # 01
*
        DATA     12343          *B0
        DATA     -17032         *B1
```

```
          DATA     17031          *A1
          DATA     -16602         *A2
          DATA     12343          *B2
*
*
*   SECOND-ORDER SECTION # 02
*
          DATA     13944          *B0
          DATA     -16708         *B1
          DATA     17028          *A1
          DATA     -30527         *A2
          DATA     13944          *B2
*
*
*   SECOND-ORDER SECTION # 03
*
          DATA     19020          *B0
          DATA     -29037         *B1
          DATA     25518          *A1
          DATA     -31151         *A2
          DATA     19020          *B2
*
          PEND
*
          CSEG     'XIIR'
*
*  COEFFICIENT DATA STORAGE AREA
*
FDATA     BSS      14+1
*
*  DELAY STORAGE DATA STORAGE AREA
*
DELAY     BSS      5+1
*
          CEND
*
          PSEG
****************************************************
*         FILTER INITIALIZATION SUBROUTINE        *
****************************************************
IIIR
          LARP     0              *POINT TO AR0
          LRLK     0,FDATA        *POINTER TO DATA MEMORY
          RPTK     14             *COUNT FOR NUMBER OF POINTS
          BLKP     COEF,*+        *BLOCK MOVE OF COEF.
          ZAC                     *CLEAR ACCUMULATOR
          LRLK     0,DELAY
          RPTK     5              *NUMBER OF DELAY POINTS
          SACL     *+             *CLEAR DATA VALUE
          RET                     *INIT RETURN
****************************************************
*              FILTER SUBROUTINE                  *
****************************************************
*    ASSUMPTIONS:
*        SATURATION ARITHMETIC MODE IS ON
*        P REGISTER OUTPUT SHIFT = 1
*        PAGE REGISTER IS SET TO 'IOPAGE'
*        SIGN EXTEND MODE IS ON
*
*    INPUT:    VIIR
*    OUTPUT:   VIIR
*
****************************************************
COEFAR    EQU      1              *Use AR1 to point to COEF's
DELAR     EQU      2              *Use AR2 to point to DELAY's
*
FIIR
          LRLK     COEFAR,FDATA   *Load COEF AR with beginning of COEF's
          LRLK     DELAR,DELAY    *Load DELAY AR with beginning of DELAY's
          LARP     COEFAR         *Select Proper AR for initial use
*
*  SECOND-ORDER FILTER SECTION # 01
*
*Warning: The input will be scaled up which could cause clipping.
          ZALH     VIIR           *GET INPUT
          ADDH     VIIR           *SCALE INPUT
          SACH     FILTT          *SAVE SCALED INPUT
          LT       FILTT          *GET SCALED INPUT
          MPY      *+,DELAR       *P = B0 * INPUT
          ZALH     *+,COEFAR      *AC = Z-1
          MPYA     *+,DELAR       *AC = Z-1 + (B0 * INPUT)
*                                 *P = B1 * INPUT
          SACH     VIIR           *Save in OUTPUT
          LTP      VIIR           *AC = B1 * INPUT
          ADDH     *-,COEFAR      *AC = Z-2 + (B1 * INPUT)
```

```
          MPY     *+                 *P = A1 * OUTPUT
          APAC
          MPYA    *+,DELAR           *AC = Z-2 + (B1 * INPUT) + (A1 * OUTPUT)
*                                    *P = A2 * OUTPUT
          SACH    *+,0,COEFAR        *Save in Z-1
          LTP     FILTT              *AC = A2 * OUTPUT

          MPY     *+,DELAR           *P = B2 * INPUT
          APAC                       *AC = (B2 * INPUT) + (A2 * OUTPUT)
          SACH    *+,0,COEFAR        *Save in Z-2
*
*  SECOND-ORDER FILTER SECTION # 02
*
          ZALH    VIIR               *GET INPUT
          ADDH    VIIR               *SCALE INPUT
          SACH    FILTT              *SAVE SCALED INPUT
          LT      FILTT              *GET SCALED INPUT
          MPY     *+,DELAR           *P = B0 * INPUT
          ZALH    *+,COEFAR          *AC = Z-1
          MPYA    *+,DELAR           *AC = Z-1 + (B0 * INPUT)
*                                    *P = B1 * INPUT
          SACH    VIIR               *Save in OUTPUT
          LTP     VIIR               *AC = B1 * INPUT
          ADDH    *-,COEFAR          *AC = Z-2 + (B1 * INPUT)
          MPY     *+                 *P = A1 * OUTPUT
          APAC
          MPYA    *+,DELAR           *AC = Z-2 + (B1 * INPUT) + (A1 * OUTPUT)
*                                    *P = A2 * OUTPUT
          SACH    *+,0,COEFAR        *Save in Z-1
          LTP     FILTT              *AC = A2 * OUTPUT
          MPY     *+,DELAR           *P = B2 * INPUT
          APAC                       *AC = (B2 * INPUT) + (A2 * OUTPUT)
          SACH    *+,0,COEFAR        *Save in Z-2
*
*  SECOND-ORDER FILTER SECTION # 03
*
          ZALH    VIIR               *GET INPUT
          ADDH    VIIR               *SCALE INPUT
          SACH    FILTT              *SAVE SCALED INPUT
          LT      FILTT              *GET SCALED INPUT
          MPY     *+,DELAR           *P = B0 * INPUT
          ZALH    *+,COEFAR          *AC = Z-1
          MPYA    *+,DELAR           *AC = Z-1 + (B0 * INPUT)
*                                    *P = B1 * INPUT
          SACH    VIIR               *Save in OUTPUT
          LTP     VIIR               *AC = B1 * INPUT
          ADDH    *-,COEFAR          *AC = Z-2 + (B1 * INPUT)
          MPY     *+                 *P = A1 * OUTPUT
          APAC
          MPYA    *+,DELAR           *AC = Z-2 + (B1 * INPUT) + (A1 * OUTPUT)
*                                    *P = A2 * OUTPUT
          SACH    *+,0,COEFAR        *Save in Z-1
          LTP     FILTT              *AC = A2 * OUTPUT
          MPY     *+,DELAR           *P = B2 * INPUT
          APAC                       *AC = (B2 * INPUT) + (A2 * OUTPUT)
          SACH    *+,0,COEFAR        *Save in Z-2
*
          RET                        *RETURN
          PEND
          END
```

ASPI-Generated IIR Filter Listing (concluded)

H

Multirate Filtering

H.1 INTRODUCTION

Multirate processing uses more than one sampling frequency in order to perform a desired signal processing operation. The two basic operations are decimation, which is a sample rate reduction by D, and interpolation, which is a sample rate increase by I. The advantages of multirate processing provide for less computational and storage requirements as well as lower-order filter design. Decimation techniques can be used, for example, in conjunction with filtering. Crochiere and Rabiner have shown that multirate decimators can reduce the computational requirements of the filter. Digital signal processing techniques using decimation and interpolation have also been discussed in Refs. 1 to 8. The interpolation operation, which represents a sampling rate increase by a factor I, consists of "padding" $I - 1$ zeros between pairs of input samples x_i and x_{i+1}.

The operations of decimation by an integer value D and interpolation by an integer value I can be cascaded if a sample rate increase or decrease by a noninteger is desired. For example, if a sampling rate increase of 2.5 is desired, an interpolation operation by I = 5 would pad (add) four zeros between each input sample, and for the decimation operation, the interpolated input samples would be shifted by $D = 2$ before each calculation. Applications for multirate signal processing include the conversion of digital signal codes, transmission of speech, and efficient filtering techniques. It has been shown [1–4] that for large changes in sampling rate, it is usually more efficient to decimate or interpolate with a series of stages.

H.2 RANDOM GENERATOR ALGORITHM

Initially, a 32-bit seed is provided using the 32-bit-wide accumulator. The first PRN output bit is generated as a modulo 2 sum. This output is also shifted into the least significant bit of the 32-bit word, with this process repeated whenever a noise sample output is needed. The PRN sequence, which repeats after $(2^{32} - 1)$ iterations, appears as a series of discrete frequencies spaced at the reciprocal of the rate at which it repeats. With the required 32,754 noise samples per second,

410

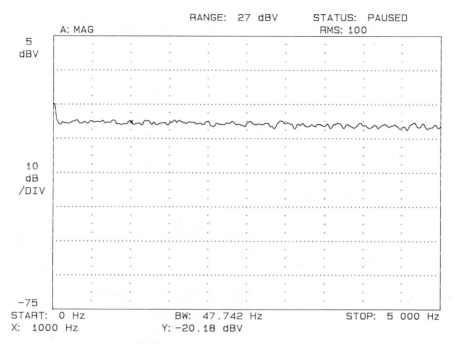

RANGE: 27 dBV   STATUS: PAUSED
A: MAG   RMS: 100

5
dBV

10
dB
/DIV

−75
START: 0 Hz   BW: 47.742 Hz   STOP: 5 000 Hz
X: 1000 Hz   Y: −20.18 dBV

FIGURE H.1. PRN generator output–wide band

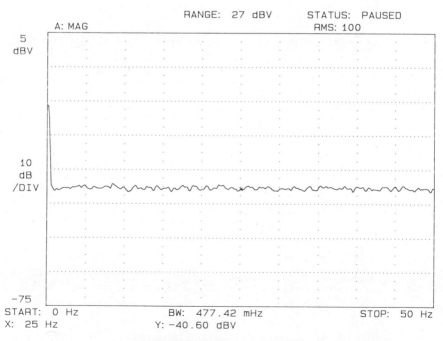

RANGE: 27 dBV   STATUS: PAUSED
A: MAG   RMS: 100

5
dBV

10
dB
/DIV

−75
START: 0 Hz   BW: 477.42 mHz   STOP: 50 Hz
X: 25 Hz   Y: −40.60 dBV

FIGURE H.2. PRN generator output

```
0001                      IDT    'MRFC25'
0002            * THIS PROGRAM IMPLEMENTS A 1/3 OCTAVE MULTIRATE FILTER
0003            * WITH INTERPOLATION USING THE TMS320C25 DSP
0004            *      PERMANENT DATA MEMORY ALLOCATION/DEFINITION
0005    0060    STAT0   EQU    >0060    * status register storage
0006    0061    LH1000  EQU    >0061    * noise gen positive value
0007    0062    LHF000  EQU    >0062    * noise gen negative value
0008    0063    RNUMH   EQU    >0063    * random number upper 16-bits
0009    0064    RNUML   EQU    >0064    * random number lower 16-bits
0010    0065    MASK15  EQU    >0065    * bit 15 mask
0011    0066    TEMP    EQU    >0066    * used by noise gen
0012    0067    TEMP1   EQU    >0067    * used by wait loop
0013    0068    FLAG    EQU    >0068    * flag used by calling routines
0014    0069    I1      EQU    >0069    * interpolator coeffs 0-20
0015    006A    I3      EQU    >006A    *
0016    006B    I5      EQU    >006B    *
0017    006C    I7      EQU    >006C    *
0018    006D    I9      EQU    >006D    *
0019    006E    I11     EQU    >006E    *
0020    006F    I13     EQU    >006F    *
0021    0070    I15     EQU    >0070    *
0022    0071    I17     EQU    >0071    *
0023    0072    I19     EQU    >0072    *
0024    0073    I0      EQU    >0073    *
0025    0074    I2      EQU    >0074    *
0026    0075    I4      EQU    >0075    *
0027    0076    I6      EQU    >0076    *
0028    0077    I8      EQU    >0077    *
0029    0078    I10     EQU    >0078    *
0030    0079    I12     EQU    >0079    *
0031    007A    I14     EQU    >007A    *
0032    007B    I16     EQU    >007B    *
0033    007C    I18     EQU    >007C    *
0034    007D    I20     EQU    >007D    *
0035    007E    TEMP2   EQU    >007E    * USED BY SWDS ISR
0036    007F    TEMP3   EQU    >007F    * USED BY SWDS ISR
0037    0200    X100    EQU    >0200    * past sample table, stage 1
0038    0228    X140    EQU    >0228    *
0039    0229    X200    EQU    >0229    * past sample table, stage 2
0040    0251    X240    EQU    >0251    *
0041    0252    X300    EQU    >0252    * past sample table, stage 3
0042    027A    X340    EQU    >027A    *
0043    027B    X400    EQU    >027B    * past sample table, stage 4
0044    02A3    X440    EQU    >02A3    *
0045    02A4    X500    EQU    >02A4    * past sample table, stage 5
0046    02CC    X540    EQU    >02CC    *
0047    02CD    X600    EQU    >02CD    * past sample table, stage 6
0048    02F5    X640    EQU    >02F5    *
0049    02F6    X700    EQU    >02F6    * past sample table, stage 7
0050    031E    X740    EQU    >031E    *
0051    031F    X800    EQU    >031F    * past sample table, stage 8
0052    0347    X840    EQU    >0347    *
0053    0348    X900    EQU    >0348    * past sample table, stage 9
0054    0370    X940    EQU    >0370    *
0055    0371    IX100   EQU    >0371    * past interp inputs, stage 1
0056    037B    IX110   EQU    >037B    *
0057    037C    IX200   EQU    >037C    * past interp inputs, stage 2
0058    0386    IX210   EQU    >0386    *
0059    0387    IX300   EQU    >0387    * past interp inputs, stage 3
0060    0391    IX310   EQU    >0391    *
0061    0392    IX400   EQU    >0392    * past interp inputs, stage 4
0062    039C    IX410   EQU    >039C    *
0063    039D    IX500   EQU    >039D    * past interp inputs, stage 5
0064    03A7    IX510   EQU    >03A7    *
0065    03A8    IX600   EQU    >03A8    * past interp inputs, stage 6
0066    03B2    IX610   EQU    >03B2    *
0067    03B3    IX700   EQU    >03B3    * past interp inputs, stage 7
0068    03BD    IX710   EQU    >03BD    *
0069    03BE    IX800   EQU    >03BE    * past interp inputs, stage 8
0070    03C8    IX810   EQU    >03C8    *
0071    03C9    A100    EQU    >03C9    * bandpass coeffs, stage 1
0072    03F1    A140    EQU    >03F1    *
0073    03F2    A200    EQU    >03F2    * bandpass coeffs, stage 2
0074    041A    A240    EQU    >041A    *
0075    041B    A300    EQU    >041B    * bandpass coeffs, stage 3
0076    0443    A340    EQU    >0443    *
0077    0444    A400    EQU    >0444    * bandpass coeffs, stage 4
0078    046C    A440    EQU    >046C    *
0079    046D    A500    EQU    >046D    * bandpass coeffs, stage 5
0080    0495    A540    EQU    >0495    *
0081    0496    A600    EQU    >0496    * bandpass coeffs, stage 6
0082    04BE    A640    EQU    >04BE    *
0083    04BF    A700    EQU    >04BF    * bandpass coeffs, stage 7
0084    04E7    A740    EQU    >04E7    *
```

FIGURE H.3. Multirate filter listing program using the AIB (MULTIFLT.LST)

```
0085        04E8   A800      EQU     >04E8       * bandpass coeffs, stage 8
0086        0510   A840      EQU     >0510       *
0087        0511   A900      EQU     >0511       * bandpass coeffs, stage 9
0088        0539   A940      EQU     >0539       *
0089        053A   ABF256    EQU     >053A       * end of working buffer
0090        0639   ABUFF1    EQU     >0639       * start of working buffer
0091        063A   BBF256    EQU     >063A       * end of output buffer
0092        0739   BBUFF1    EQU     >0739       * start of output buffer
0093               *
0094               *         TEMPORARY DATA MEMORY ALLOCATION/DEFINITION
0095               *
0096        0200   TL1       EQU     X100        * 1/3 octave levels 1-27
0097        021B   TAL0      EQU     X100+>1B    * low 1/3 octave bandpass coeff
0098        0244   TAM0      EQU     X100+>44    * mid 1/3 octave bandpass coeff
0099        026D   TAH0      EQU     X100+>6D    * hi  1/3 octave bandpass coeff
0100               *
0101               *         MEMORY MAPPED REGISTER LOCATIONS
0102               *
0103        0002   TIMER     EQU     >0002       * interval timer register
0104        0003   PERIOD    EQU     >0003       * period register for timer
0105        0004   INTMSK    EQU     >0004       * interrupt mask register
0106               *
0107  0000                   AORG    0           * branch around p.m.
0108  0000 FF80              B       INIT        *
      0001 00EF
0109               *
0110               *         DEFINE INTERRUPT SERVICE ROUTINE
0111               *
0112  0018                   AORG    >0018       * ISR begins at address >18
0113  0018 7860   ISR        SST     STAT0       * save current status register
0114  0019 558F              LARP    7           * =>AR7 (output buffer pointer)
0115  001A E290              OUT     *-,2        * out filtered noise sample
0116  001B 5060              LST     STAT0       * retrieve past status register
0117  001C CE00              EINT                * re-enable interrupts
0118  001D CE26              RET                 * return from ISR
0119               *
0120               *         PROGRAM MEMORY DATA TABLE ALLOCATION/DEFINITION
0121               *
0122  0040                   AORG    >0040       * pgrm mem data storage begins
0123  0040 0000   L1         DATA    >0000       * levels,1/3 octave bands 0-27
0124  0041 007F   L2         DATA    >007F       *
0125  0042 0000   L3         DATA    >0000       *
0126  0043 0000   L4         DATA    >0000       *
0127  0044 007F   L5         DATA    >007F       *
0128  0045 0000   L6         DATA    >0000       *
0129  0046 0000   L7         DATA    >0000       *
0130  0047 007F   L8         DATA    >007F       *
0131  0048 0000   L9         DATA    >0000       *
0132  0049 0000   L10        DATA    >0000       *
0133  004A 007F   L11        DATA    >007F       *
0134  004B 0000   L12        DATA    >0000       *
0135  004C 0000   L13        DATA    >0000       *
0136  004D 007F   L14        DATA    >007F       *
0137  004E 0000   L15        DATA    >0000       *
0138  004F 0000   L16        DATA    >0000       *
0139  0050 007F   L17        DATA    >007F       *
0140  0051 0000   L18        DATA    >0000       *
0141  0052 0000   L19        DATA    >0000       *
0142  0053 007F   L20        DATA    >007F       *
0143  0054 0000   L21        DATA    >0000       *
0144  0055 0000   L22        DATA    >0000       *
0145  0056 007F   L23        DATA    >007F       *
0146  0057 0000   L24        DATA    >0000       *
0147  0058 0000   L25        DATA    >0000       *
0148  0059 007F   L26        DATA    >007F       *
0149  005A 0000   L27        DATA    >0000       *
0150  005B 003D   AL0        DATA    >003D       * low 1/3 octave coeffs  0-40
0151  005C 0140   AL1        DATA    >0140       *
0152  005D 0096   AL2        DATA    >0096       *
0153  005E FF0C   AL3        DATA    >FF0C       *
0154  005F FF4C   AL4        DATA    >FF4C       *
0155  0060 FFF7   AL5        DATA    >FFF7       *
0156  0061 FDF7   AL6        DATA    >FDF7       *
0157  0062 FD7E   AL7        DATA    >FD7E       *
0158  0063 04CE   AL8        DATA    >04CE       *
0159  0064 0B70   AL9        DATA    >0B70       *
0160  0065 00A4   AL10       DATA    >00A4       *
0161  0066 EB91   AL11       DATA    >EB91       *
0162  0067 ED9B   AL12       DATA    >ED9B       *
0163  0068 0FED   AL13       DATA    >0FED       *
0164  0069 265E   AL14       DATA    >265E       *
0165  006A 0858   AL15       DATA    >0858       *
0166  006B D6B4   AL16       DATA    >D6B4       *
0167  006C D875   AL17       DATA    >D875       *
```

FIGURE H.3. (continued)

```
0168 006D 1201   AL18   DATA   >1201      *
0169 006E 35F2   AL19   DATA   >35F2      *
0170 006F 1201   AL20   DATA   >1201      *
0171 0070 D875   AL21   DATA   >D875      *
0172 0071 D6B4   AL22   DATA   >D6B4      *
0173 0072 0858   AL23   DATA   >0858      *
0174 0073 265E   AL24   DATA   >265E      *
0175 0074 0FED   AL25   DATA   >0FED      *
0176 0075 ED9B   AL26   DATA   >ED9B      *
0177 0076 EB91   AL27   DATA   >EB91      *
0178 0077 00A4   AL28   DATA   >00A4      *
0179 0078 0B70   AL29   DATA   >0B70      *
0180 0079 04CE   AL30   DATA   >04CE      *
0181 007A FD7E   AL31   DATA   >FD7E      *
0182 007B FDF7   AL32   DATA   >FDF7      *
0183 007C FFF7   AL33   DATA   >FFF7      *
0184 007D FF4C   AL34   DATA   >FF4C      *
0185 007E FF0C   AL35   DATA   >FF0C      *
0186 007F 0096   AL36   DATA   >0096      *
0187 0080 0140   AL37   DATA   >0140      *
0188 0081 003D   AL38   DATA   >003D      *
0189 0082 0000   AL39   DATA   >0000      *
0190 0083 0000   AL40   DATA   >0000      *
0191 0084 FFD4   AM0    DATA   >FFD4      * mid 1/3 octave coeffs 0-40
0192 0085 FFF0   AM1    DATA   >FFF0      *
0193 0086 FF87   AM2    DATA   >FF87      *
0194 0087 0073   AM3    DATA   >0073      *
0195 0088 0130   AM4    DATA   >0130      *
0196 0089 FF6D   AM5    DATA   >FF6D      *
0197 008A FF01   AM6    DATA   >FF01      *
0198 008B 000D   AM7    DATA   >000D      *
0199 008C FE8A   AM8    DATA   >FE8A      *
0200 008D 0140   AM9    DATA   >0140      *
0201 008E 0757   AM10   DATA   >0757      *
0202 008F FCFD   AM11   DATA   >FCFD      *
0203 0090 EF1F   AM12   DATA   >EF1F      *
0204 0091 0483   AM13   DATA   >0483      *
0205 0092 1CE5   AM14   DATA   >1CE5      *
0206 0093 FB0E   AM15   DATA   >FB0E      *
0207 0094 D71B   AM16   DATA   >D71B      *
0208 0095 03DE   AM17   DATA   >03DE      *
0209 0096 31CE   AM18   DATA   >31CE      *
0210 0097 FE89   AM19   DATA   >FE89      *
0211 0098 CAE6   AM20   DATA   >CAE6      *
0212 0099 FE89   AM21   DATA   >FE89      *
0213 009A 31CE   AM22   DATA   >31CE      *
0214 009B 03DE   AM23   DATA   >03DE      *
0215 009C D71B   AM24   DATA   >D71B      *
0216 009D FB0E   AM25   DATA   >FB0E      *
0217 009E 1CE5   AM26   DATA   >1CE5      *
0218 009F 0483   AM27   DATA   >0483      *
0219 00A0 EF1F   AM28   DATA   >EF1F      *
0220 00A1 FCFD   AM29   DATA   >FCFD      *
0221 00A2 0757   AM30   DATA   >0757      *
0222 00A3 0140   AM31   DATA   >0140      *
0223 00A4 FE8A   AM32   DATA   >FE8A      *
0224 00A5 000D   AM33   DATA   >000D      *
0225 00A6 FF01   AM34   DATA   >FF01      *
0226 00A7 FF6D   AM35   DATA   >FF6D      *
0227 00A8 0130   AM36   DATA   >0130      *
0228 00A9 0073   AM37   DATA   >0073      *
0229 00AA FF87   AM38   DATA   >FF87      *
0230 00AB FFF0   AM39   DATA   >FFF0      *
0231 00AC FFD4   AM40   DATA   >FFD4      *
0232 00AD 0000   AH0    DATA   >0000      * hi 1/3 octave coeffs 0-40
0233 00AE 0000   AH1    DATA   >0000      *
0234 00AF 0040   AH2    DATA   >0040      *
0235 00B0 FF94   AH3    DATA   >FF94      *
0236 00B1 FFF3   AH4    DATA   >FFF3      *
0237 00B2 003D   AH5    DATA   >003D      *
0238 00B3 0021   AH6    DATA   >0021      *
0239 00B4 006E   AH7    DATA   >006E      *
0240 00B5 FEA5   AH8    DATA   >FEA5      *
0241 00B6 0050   AH9    DATA   >0050      *
0242 00B7 00E3   AH10   DATA   >00E3      *
0243 00B8 0079   AH11   DATA   >0079      *
0244 00B9 00E2   AH12   DATA   >00E2      *
0245 00BA F7B2   AH13   DATA   >F7B2      *
0246 00BB 072B   AH14   DATA   >072B      *
0247 00BC 0E2B   AH15   DATA   >0E2B      *
0248 00BD E297   AH16   DATA   >E297      *
0249 00BE 02A0   AH17   DATA   >02A0      *
0250 00BF 2AD8   AH18   DATA   >2AD8      *
0251 00C0 D8B8   AH19   DATA   >D8B8      *
0252 00C1 EAB9   AH20   DATA   >EAB9      *
```

FIGURE H.3. (continued)

```
0253  00C2 3B5F    AH21   DATA    >3B5F       *
0254  00C3 EAB9    AH22   DATA    >EAB9       *
0255  00C4 D8B8    AH23   DATA    >D8B8       *
0256  00C5 2AD8    AH24   DATA    >2AD8       *
0257  00C6 02A0    AH25   DATA    >02A0       *
0258  00C7 E297    AH26   DATA    >E297       *
0259  00C8 0E2B    AH27   DATA    >0E2B       *
0260  00C9 072B    AH28   DATA    >072B       *
0261  00CA F7B2    AH29   DATA    >F7B2       *
0262  00CB 00E2    AH30   DATA    >00E2       *
0263  00CC 0079    AH31   DATA    >0079       *
0264  00CD 00E3    AH32   DATA    >00E3       *
0265  00CE 0050    AH33   DATA    >0050       *
0266  00CF FEA5    AH34   DATA    >FEA5       *
0267  00D0 006E    AH35   DATA    >006E       *
0268  00D1 0021    AH36   DATA    >0021       *
0269  00D2 003D    AH37   DATA    >003D       *
0270  00D3 FFF3    AH38   DATA    >FFF3       *
0271  00D4 FF94    AH39   DATA    >FF94       *
0272  00D5 0040    AH40   DATA    >0040       *
0273  00D6 019B    TI1    DATA    >019B       * 1st interpolator coeffs
0274  00D7 FB98    TI3    DATA    >FB98       *
0275  00D8 0A37    TI5    DATA    >0A37       *
0276  00D9 E8ED    TI7    DATA    >E8ED       *
0277  00DA 500E    TI9    DATA    >500E       *
0278  00DB 500E    TI11   DATA    >500E       *
0279  00DC E8ED    TI13   DATA    >E8ED       *
0280  00DD 0A37    TI15   DATA    >0A37       *
0281  00DE FB98    TI17   DATA    >FB98       *
0282  00DF 019B    TI19   DATA    >019B       *
0283  00E0 00EE    TI0    DATA    >00EE       * 2nd interpolator coeffs
0284  00E1 FE76    TI2    DATA    >FE76       *
0285  00E2 026F    TI4    DATA    >026F       *
0286  00E3 FCAE    TI6    DATA    >FCAE       *
0287· 00E4 03EE    TI8    DATA    >03EE       *
0288  00E5 7BD2    TI10   DATA    >7BD2       *
0289  00E6 03EE    TI12   DATA    >03EE       *
0290  00E7 FCAE    TI14   DATA    >FCAE       *
0291  00E8 026F    TI16   DATA    >026F       *
0292  00E9 FE76    TI18   DATA    >FE76       *
0293  00EA 00EE    TI20   DATA    >00EE       *
0294  00EB 7E52    SEEDH  DATA    >7E52       * random no. seed upper 16-bits
0295  00EC 1603    SEEDL  DATA    >1603       * random no. seed lower 16-bits
0296  00ED 0262    RATE   DATA    >0262       * output data rate 16393 Hz
0297  00EE 0008    IMASK  DATA    >0008       * enable timer interrupt only
0298                 *
0299                 *      END OF MEMORY ALLOCATION/DEFINITION
0300                 *
0301                 *      BEGIN INITIALIZATION
0302                 * INITIALIZE DATA MEMORY TO ZERO
0303  00EF CE04    INIT   CNFD                * configure memory to dmem
0304  00F0 C800           LDPK    0           * point to data page 0, forever
0305  00F1 5588           LARP    0           *
0306  00F2 CA00           ZAC                 *
0307  00F3 D000           LRLK    0,>0060     * point to 1st usable data page
      00F4 0060
0308  00F5 CB1F           RPTK    >1F         * zero out data page 0&1
0309  00F6 60A0           SACL    *+          *
0310  00F7 D000           LRLK    0,>0200     * zero data memory >200 to >7FF
      00F8 0200
0311  00F9 D100           LRLK    1,>5FF      *
      00FA 05FF
0312  00FB 5588    CINIT  LARP    0           *
0313  00FC 60A9           SACL    *+,0,1      *
0314  00FD FB90           BANZ    CINIT       *
      00FE 00FB
0315  00FF D000           LRLK    0,TL1       * 1/3 oct coeffs & lev. from pm
      0100 0200
0316  0101 5588           LARP    0           *
0317  0102 CB95           RPTK    >95         * ((41*3)+27-1)
0318  0103 FCA0           BLKP    L1,*+       *
      0104 0040
0319                 *      CALCULATE COEFFICIENTS FOR 9 OCTAVE BANDS
0320                 *      for each of 9 octave bands
0321                 *        for each of 41 coeffs
0322                 *          for each of 3 subbands (1/3 octave bands)
0323                 *            scale the coeff by the proper level
0324                 *            control value and accumulate the sum
0325                 *          next  subband
0326                 *          store the octave band coeff in data memory
0327                 *        next coeff
0328                 *      next octave band
0329  0105 D300           LRLK    3,A100      * 1st octave coeff adx
      0106 03C9
0330  0107 D400           LRLK    4,TL1       * 1st 1/3 oct level adx
```

FIGURE H.3. (continued)

```
        0108 0200                             * loop counter (9 octave bands)
0331    0109 C608          LARK   6,8         * 1st low 1/3 oct coeff adx
0332    010A D000   CONT   LRLK   0,TAL0      * 1st low 1/3 oct coeff adx
        010B 021B
0333    010C D100          LRLK   1,TAM0      * 1st mid 1/3 oct coeff adx
        010D 0244
0334    010E D200          LRLK   2,TAH0      * 1st hi  1/3 oct coeff adx
        010F 026D
0335    0110 C528          LARK   5,40        * loop counter (41 coeffs)
0336    0111 5588   CONT1  LARP   0           *
0337    0112 CA00          ZAC                * clear acc before summation
0338    0113 3CAC          LT     *+,4        * load low 1/3 oct coeff
0339    0114 38A9          MPY    *+,1        * scale it
0340    0115 3DAC          LTA    *+,4        * load med 1/3 oct coeff
0341    0116 38AA          MPY    *+,2        * scale it
0342    0117 3DAC          LTA    *+,4        * load hi 1/3 oct coeff
0343    0118 38AB          MPY    *+,3        * scale it
0344    0119 CE15          APAC               * accumulate final result
0345    011A 69AC          SACH   *+,1,4      * store octave band coeff to dm
0346    011B 7F03          SBRK   3           * =>low 1/3 oct level again
0347    011C 558D          LARP   5           * check coeff loop counter
0348    011D FB90          BANZ   CONT1       * end of coeff loop
        011E 0111
0349    011F 558C          LARP   4           * select 1/3 oct level pointer
0350    0120 7E03          ADRK   3           * =>next band low 1/3 oct level
0351    0121 558E          LARP   6           * check band loop counter
0352    0122 FB90          BANZ   CONT        * end of band 1 to 9 loop
        0123 010A
0353               *
0354               *       RE-ZERO TEMPORARY DATA MEMORY
0355               *
0356               * used for calculating the 9 octave band coeff sets
0357               * and needs to be re-zeroed
0358               *
0359    0124 5588          LARP   0           *
0360    0125 CA00          ZAC                *
0361    0126 D000          LRLK   0,TL1       *
        0127 0200
0362    0128 CBFF          RPTK   >FF         *
0363    0129 60A0          SACL   *+          *
0364               *
0365               * INITIALIZE DATA MEMORY CONSTANTS
0366               *
0367    012A D001          LALK   >8000       *
        012B 8000
0368    012C 6065          SACL   MASK15      * bit 15 mask for noise gen
0369    012D D001          LALK   >1000       *
        012E 1000
0370    012F 6061          SACL   LH1000      * noise gen positive out level
0371    0130 D001          LALK   >FC00       *
        0131 FC00
0372    0132 6062          SACL   LHF000      * noise gen negative out level
0373    0133 D001          LALK   SEEDH       * =>random number seed storage
        0134 00EB
0374    0135 5863          TBLR   RNUMH       * get random no. upper 16-bits
0375    0136 CC01          ADDK   1           *
0376    0137 5864          TBLR   RNUML       * get random no. lower 16-bits
0377    0138 CC01          ADDK   1           * => timer interval value
0378    0139 5803          TBLR   PERIOD      * timer interval (16393 Hz)
0379    013A CC01          ADDK   1           * point to interrupt mask
0380    013B 5804          TBLR   INTMSK      * unmasks only timer interrupt
0381    013C D001          LALK   TI1         * =>interp coeff pgmem storage
        013D 00D6
0382    013E D000          LRLK   0,I1        * =>interp coeff dmem storage
        013F 0069
0383    0140 CB14          RPTK   20          * there are 21 coeffs to copy
0384    0141 58A0          TBLR   *+          * interp coeffs from pm to dm
0385               *
0386               * END OF INITIALIZATION
0387               *
0389               *
0390    0142 D700   START  LRLK   7,BBUFF1    * init out buffer pointer
        0143 0739
0391    0144 CE00          EINT               * enable interrupts
0392               *
0393               * CALLING ROUTINE FOR EACH FILTER STAGE (1 to 9)
0394               *
0395               * these calling routines set auxilliary pointers for addressing
0396               * and loop counters, and then call subroutine 'FILT' which
0397               * produces noise samples, and executes the correct number of
0398               * bandpass filters and interpolations for stage being called
0399               * auxilliary register usage is as follows:
0400               *      AR0=number of filter passes for the stage being called
0401               *      AR1=pointer to interpolator coeffs
0402               *      AR2=pointer to interpolator past sample table
```

FIGURE H.3. (continued)

```
0403               *         AR3=pointer to bandpass filter past sample table
0404               *         AR4=pointer to bandpass filter coeffs
0405               *         AR5=pointer to working buffer write address
0406               *         AR6=pointer to working buffer read address
0407               *         AR7=RESERVED FOR ISR OUTPUT BUFFER ADDRESS
0408               *
0409               *   FLAG=variable to tell FILT when the first stage is being
0410               *   processed (no previous stage outputs in the working
0411               *   buffer)(FLAG=-1), and when the last stage is being
0412               *   processed (no interpolation required)(FLAG=+1)
0413               *
0414 0145 D600   STAGE1  LRLK    6,ABF256
     0146 053A
0415 0147 D500           LRLK    5,ABF256+>1
     0148 053B
0416 0149 D400           LRLK    4,A140
     014A 03F1
0417 014B D300           LRLK    3,X100
     014C 0200
0418 014D D200           LRLK    2,IX100
     014E 0371
0419 014F C17D           LARK    1,I20
0420 0150 C000           LARK    0,0
0421 0151 CA01           LACK    1
0422 0152 CE23           NEG
0423 0153 6068           SACL    FLAG
0424 0154 FE80           CALL    FILT
     0155 01D4
0425 0156 D600   STAGE2  LRLK    6,ABF256+>1
     0157 053B
0426 0158 D500           LRLK    5,ABF256+>3
     0159 053D
0427 015A D400           LRLK    4,A240
     015B 041A
0428 015C D300           LRLK    3,X200
     015D 0229
0429 015E D200           LRLK    2,IX200
     015F 037C
0430 0160 C17D           LARK    1,I20
0431 0161 C001           LARK    0,1
0432 0162 CA00           ZAC
0433 0163 6068           SACL    FLAG
0434 0164 FE80           CALL    FILT
     0165 01D4
0435 0166 D600   STAGE3  LRLK    6,ABF256+>3
     0167 053D
0436 0168 D500           LRLK    5,ABF256+>7
     0169 0541
0437 016A D400           LRLK    4,A340
     016B 0443
0438 016C D300           LRLK    3,X300
     016D 0252
0439 016E D200           LRLK    2,IX300
     016F 0387
0440 0170 C17D           LARK    1,I20
0441 0171 C003           LARK    0,3
0442 0172 FE80           CALL    FILT
     0173 01D4
0443 0174 D600   STAGE4  LRLK    6,ABF256+>7
     0175 0541
0444 0176 D500           LRLK    5,ABF256+>F
     0177 0549
0445 0178 D400           LRLK    4,A440
     0179 046C
0446 017A D300           LRLK    3,X400
     017B 027B
0447 017C D200           LRLK    2,IX400
     017D 0392
0448 017E C17D           LARK    1,I20
0449 017F C007           LARK    0,7
0450 0180 FE80           CALL    FILT
     0181 01D4
0451 0182 D600   STAGE5  LRLK    6,ABF256+>F
     0183 0549
0452 0184 D500           LRLK    5,ABF256+>1F
     0185 0559
0453 0186 D400           LRLK    4,A540
     0187 0495
0454 0188 D300           LRLK    3,X500
     0189 02A4
0455 018A D200           LRLK    2,IX500
     018B 039D
0456 018C C17D           LARK    1,I20
0457 018D C00F           LARK    0,>F
0458 018E FE80           CALL    FILT
```

FIGURE H.3. (continued)

```
        018F 01D4
0459    0190 D600    STAGE6   LRLK     6,ABF256+>1F
        0191 0559
0460    0192 D500             LRLK     5,ABF256+>3F
        0193 0579
0461    0194 D400             LRLK     4,A640
        0195 04BE
0462    0196 D300             LRLK     3,X600
        0197 02CD
0463    0198 D200             LRLK     2,IX600
        0199 03A8
0464    019A C17D             LARK     1,I20
0465    019B C01F             LARK     0,>1F
0466    019C FE80             CALL     FILT
        019D 01D4
0467    019E D600    STAGE7   LRLK     6,ABF256+>3F
        019F 0579
0468    01A0 D500             LRLK     5,ABF256+>7F
        01A1 05B9
0469    01A2 D400             LRLK     4,A740
        01A3 04E7
0470    01A4 D300             LRLK     3,X700
        01A5 02F6
0471    01A6 D200             LRLK     2,IX700
        01A7 03B3
0472    01A8 C17D             LARK     1,I20
0473    01A9 C03F             LARK     0,>3F
0474    01AA FE80             CALL     FILT
        01AB 01D4
0475    01AC D600    STAGE8   LRLK     6,ABF256+>7F
        01AD 05B9
0476    01AE D500             LRLK     5,ABF256+>FF
        01AF 0639
0477    01B0 D400             LRLK     4,A840
        01B1 0510
0478    01B2 D300             LRLK     3,X800
        01B3 031F
0479    01B4 D200             LRLK     2,IX800
        01B5 03BE
0480    01B6 C17D             LARK     1,I20
0481    01B7 C07F             LARK     0,>7F
0482    01B8 FE80             CALL     FILT
        01B9 01D4
0483    01BA D600    STAGE9   LRLK     6,ABF256+>FF
        01BB 0639
0484    01BC D500             LRLK     5,ABF256+>FF
        01BD 0639
0485    01BE D400             LRLK     4,A940
        01BF 0539
0486    01C0 D300             LRLK     3,X900
        01C1 0348
0487    01C2 C0FF             LARK     0,>FF
0488    01C3 CA01             LACK     1
0489    01C4 6068             SACL     FLAG
0490    01C5 FE80             CALL     FILT
        01C6 01D4
0491            *
0492            * END OF CALLING ROUTINES
0493            *
0494            *
0496            *
0497            * WAIT HERE UNTIL OUTPUT BUFFER IS EMPTY
0498            *
0499    01C7 7767    WAIT     SAR      7,TEMP1
0500    01C8 D001             LALK     BBF256
        01C9 063A
0501    01CA 1067             SUB      TEMP1
0502    01CB F280             BLEZ     WAIT
        01CC 01C7
0503            *
0504            * NOW THAT OUTPUT BUFFER IS EMPTY, REFILL IT
0505            *   copy working buffer to output buffer
0506            *
0507    01CD 558F             LARP     7
0508    01CE 7E01             ADRK     1
0509    01CF CBFF             RPTK     >FF
0510    01D0 FDA0             BLKD     ABF256,*+
        01D1 053A
0511    01D2 FF80             B        START        * restart the entire process
        01D3 0142
0513            *
0514            * SUBROUTINE "FILT" CONTAINS 32-BIT NOISE GENERATOR,9 OCTAVE
0515            * BAND FILTERS, AND INTERPOLATORS
0516            *
0517    01D4 2063    FILT     LAC      RNUMH        * get random no. upper 16-bits
```

FIGURE H.3. (continued)

```
0518 01D5 4E65              AND     MASK15          * mask bit 15
0519 01D6 0163              ADD     RNUMH,1         * add bit 14 to bit 15
0520 01D7 4E65              AND     MASK15          * mask bit 15
0521 01D8 0363              ADD     RNUMH,3         * add bit 12 to bit 15
0522 01D9 4E65              AND     MASK15          * mask bit 15
0523 01DA 0E63              ADD     RNUMH,14        * add bit 1 to bit 15
0524 01DB 4E65              AND     MASK15          * mask bit 15
0525 01DC 6966              SACH    TEMP,1          * right 15 (left 1,right 16)
0526 01DD 4063              ZALH    RNUMH           * load no. random upper 16-bits
0527 01DE 0064              ADD     RNUML           * add random no.lower 16-bits
0528 01DF CE18              SFL                     * shift the 32-bit accum left 1
0529 01E0 0066              ADD     TEMP            * add calculated random digit
0530 01E1 6863              SACH    RNUMH           * restore random upper 16-bits
0531 01E2 6064              SACL    RNUML           * restore random lower 16-bits
0532 01E3 2A66              LAC     TEMP,10         * acc with random digit*>400
0533 01E4 F180              BGZ     CONTN           * if digit was zero load >FC00
     01E5 01E7
0534 01E6 2062              LAC     LHF000          * accum now=>FC00
0535 01E7 558B      CONTN   LARP    3               *=>bandpass filt past sample#0
0536 01E8 6080              SACL    *               * store noise gen out in #0
0537 01E9 7E28              ADRK    >28             *=>bandpass filt past sample#40
0539              *
0540              * START OF BANDPASS FILTER
0541              *
0542 01EA CA00              ZAC                     * clear accum before summation
0543 01EB 3C9C              LT      *-,4            * load past sample X40
0544 01EC 389B              MPY     *-,3            * A40*X40
0545 01ED 3F9C              LTD     *-,4
0546 01EE 389B              MPY     *-,3            * A39*X39
0547 01EF 3F9C              LTD     *-,4
0548 01F0 389B              MPY     *-,3            * A38*X38
0549 01F1 3F9C              LTD     *-,4
0550 01F2 389B              MPY     *-,3            * A37*X37
0551 01F3 3F9C              LTD     *-,4
0552 01F4 389B              MPY     *-,3            * A36*X36
0553 01F5 3F9C              LTD     *-,4
0554 01F6 389B              MPY     *-,3            * A35*X35
0555 01F7 3F9C              LTD     *-,4
0556 01F8 389B              MPY     *-,3            * A34*X34
0557 01F9 3F9C              LTD     *-,4
0558 01FA 389B              MPY     *-,3            * A33*X33
0559 01FB 3F9C              LTD     *-,4
0560 01FC 389B              MPY     *-,3            * A32*X32
0561 01FD 3F9C              LTD     *-,4
0562 01FE 389B              MPY     *-,3            * A31*X31
0563 01FF 3F9C              LTD     *-,4
0564 0200 389B              MPY     *-,3            * A30*X30
0565 0201 3F9C              LTD     *-,4
0566 0202 389B              MPY     *-,3            * A29*X29
0567 0203 3F9C              LTD     *-,4
0568 0204 389B              MPY     *-,3            * A28*X28
0569 0205 3F9C              LTD     *-,4
0570 0206 389B              MPY     *-,3            * A27*X27
0571 0207 3F9C              LTD     *-,4
0572 0208 389B              MPY     *-,3            * A26*X26
0573 0209 3F9C              LTD     *-,4
0574 020A 389B              MPY     *-,3            * A25*X25
0575 020B 3F9C              LTD     *-,4
0576 020C 389B              MPY     *-,3            * A24*X24
0577 020D 3F9C              LTD     *-,4
0578 020E 389B              MPY     *-,3            * A23*X23
0579 020F 3F9C              LTD     *-,4
0580 0210 389B              MPY     *-,3            * A22*X22
0581 0211 3F9C              LTD     *-,4
0582 0212 389B              MPY     *-,3            * A21*X21
0583 0213 3F9C              LTD     *-,4
0584 0214 389B              MPY     *-,3            * A20*X20
0585 0215 3F9C              LTD     *-,4
0586 0216 389B              MPY     *-,3            * A19*X19
0587 0217 3F9C              LTD     *-,4
0588 0218 389B              MPY     *-,3            * A18*X18
0589 0219 3F9C              LTD     *-,4
0590 021A 389B              MPY     *-,3            * A17*X17
0591 021B 3F9C              LTD     *-,4
0592 021C 389B              MPY     *-,3            * A16*X16
0593 021D 3F9C              LTD     *-,4
0594 021E 389B              MPY     *-,3            * A15*X15
0595 021F 3F9C              LTD     *-,4
0596 0220 389B              MPY     *-,3            * A14*X14
0597 0221 3F9C              LTD     *-,4
0598 0222 389B              MPY     *-,3            * A13*X13
0599 0223 3F9C              LTD     *-,4
0600 0224 389B              MPY     *-,3            * A12*X12
0601 0225 3F9C              LTD     *-,4
```

FIGURE H.3. (continued)

```
0602 0226 389B        MPY     *-,3          * A11*X11
0603 0227 3F9C        LTD     *-,4
0604 0228 389B        MPY     *-,3          * A10*X10
0605 0229 3F9C        LTD     *-,4
0606 022A 389B        MPY     *-.3          * A9*X9
0607 022B 3F9C        LTD     *-,4
0608 022C 389B        MPY     *-,3          * A8*X8
0609 022D 3F9C        LTD     *-,4
0610 022E 389B        MPY     *-,3          * A7*X7
0611 022F 3F9C        LTD     *-,4
0612 0230 389B        MPY     *-,3          * A6*X6
0613 0231 3F9C        LTD     *-,4
0614 0232 389B        MPY     *-,3          * A5*X5
0615 0233 3F9C        LTD     *-,4
0616 0234 389B        MPY     *-,3          * A4*X4
0617 0235 3F9C        LTD     *-,4
0618 0236 389B        MPY     *-,3          * A3*X3
0619 0237 3F9C        LTD     *-,4
0620 0238 389B        MPY     *-,3          * A2*X2
0621 0239 3F9C        LTD     *-,4
0622 023A 389B        MPY     *-,3          * A1*X1
0623 023B 3F8C        LTD     *,4
0624 023C 388A        MPY     *,2           * A0*X0(no decr,and => IX0)
0625 023D CE15        APAC                  * accumulate final result
0626 023E 698C        SACH    *,1,4         * filter out at interp input #0
0627 023F 7E28        ADRK    >28           * restore coeff pointer to A40
0628 0240 558A        LARP    2             * reg pntr to inter input #0
0629 0241 4068        ZALH    FLAG          * check if this is stage 1
0630 0242 F380        BLZ     CONTF1        * branch if no previous stages
     0243 0248
0631 0244 208E        LAC     *,0,6         * filt out back,=>to work buff
0632 0245 009A        ADD     *-,0,2        * sum past stage out,=> IX0
0633 0246 6080        SACL    *             * store sum at interp input #0
0634 0247 2068        LAC     FLAG          * check if this is stage 9
0635 0248 F280 CONTF1 BLEZ    INTERP        * to interp if not stage 9
     0249 024D
0636 024A 408D        ZALH    *,5           * get summed value back
0637 024B FF80        B       CONTI1        * branch arround interpolators
     024C 027E
0639            *
0640            * START OF INTERPOLATOR FILTER
0641            *
0642 024D 7E0A INTERP ADRK    >A            * => oldest interp input (IX10)
0643 024E CA00        ZAC
0644 024F 3D99        LTA     *-,1          *  IX10, DO NOT MOVE IT DOWN
0645 0250 389A        MPY     *-,2          * I20*IX10
0646 0251 3D99        LTA     *-,1
0647 0252 389A        MPY     *-,2          * I18*IX9
0648 0253 3D99        LTA     *-,1
0649 0254 389A        MPY     *-,2          * I16*IX8
0650 0255 3D99        LTA     *-,1
0651 0256 389A        MPY     *-,2          * I14*IX7
0652 0257 3D99        LTA     *-,1
0653 0258 389A        MPY     *-,2          * I12*IX6
0654 0259 3D99        LTA     *-,1
0655 025A 3D99        LTA     *-,1
0656 025A 389A        MPY     *-,2          * I10*IX5
0656 025B 3D99        LTA     *-,1
0657 025C 389A        MPY     *-,2          * I8*IX4
0658 025D 3D99        LTA     *-,1
0659 025E 389A        MPY     *-,2          * I6*IX3
0660 025F 3D99        LTA     *-,1
0661 0260 389A        MPY     *-,2          * I4*IX2
0662 0261 3D99        LTA     *-,1
0663 0262 389A        MPY     *-,2          * I2*IX1
0664 0263 3D99        LTA     *-,1
0665 0264 389D        MPY     *-,5          * I0*IX0
0666 0265 CE15        APAC                  * accumulate final result
0667 0266 699A        SACH    *-,1,2        * interp out in working buffer
0668 0267 7E0A        ADRK    >A            * point to IX9
0669 0268 CA00        ZAC
0670 0269 3F99        LTD     *-,1          * load IX9, NOW MOVE IT DOWN
0671 026A 389A        MPY     *-,2          * I19*IX9
0672 026B 3F99        LTD     *-,1
0673 026C 389A        MPY     *-,2          * I17*IX8
0674 026D 3F99        LTD     *-,1
0675 026E 389A        MPY     *-,2          * I15*IX7
0676 026F 3F99        LTD     *-,1
0677 0270 389A        MPY     *-,2          * I13*IX6
0678 0271 3F99        LTD     *-,1
0679 0272 389A        MPY     *-,2          * I11*IX5
0680 0273 3F99        LTD     *-,1
0681 0274 389A        MPY     *-,2          * I9*IX4
0682 0275 3F99        LTD     *-,1
0683 0276 389A        MPY     *-,2          * I7*IX3
0684 0277 3F99        LTD     *-,1
```

FIGURE H.3. (continued)

```
0685 0278 389A          MPY     *-,2           * I5*IX2
0686 0279 3F99          LTD     *-,1
0687 027A 389A          MPY     *-,2           * I3*IX1
0688 027B 3F89          LTD     *,1
0689 027C 389D          MPY     *-,5           * I1*IX0
0690 027D CE15          APAC                   * accumulate final result
0691 027E 6998  CONTI1  SACH    *-,1,0         * store in working buffer
0692 027F C17D          LARK    1,I20          * resore interp coeff pointer
0693 0280 FB90          BANZ    FILT           * for more filters in this band
     0281 01D4
0694 0282 CE26          RET                    * return to calling routine
0695 0283
0696 0283
0697 0283
0698 0283
0699 0283
0700 0283
NO ERRORS, NO WARNINGS
```

FIGURE H.3. (concluded)

```
* PARTIAL MULTIRATE FILTER PROGRAM USING THE AIC
          IDT     'AICMUL'
          .
          .
          .
          AORG    >1A             INTERRUPT VECTOR (RECEIVE)
          B       RINT            BRANCH TO RECEIVE ROUTINE
          AORG    >1C             INTERRUPT VECTOR (SEND)
XINT      LAC     OUTPUT          ACC = SECONDARY DATA
          SACL    >01             DXR =      "           "
          RET                     RETURN TO MAIN PROGRAM
RINT      SST     STAT0           SAVE CURRENT STATUS REGISTER
          SACL    ACCL            SAVE CURRENT LOWER ACC
          SACH    ACCH            SAVE CURRENT UPPER ACC
          LARP    7               POINT TO AR7 (OUTPUT BUFFER POINTER)
          LAC     *-              ACC = OUTPUT SAMPLE
          ANDK    >FFFC           MAKE SURE LAST TWO BITS ARE ZERO
          SACL    >01             DXR = OUTPUT SAMPLE
          ZALS    ACCL            RETRIEVE PAST LOWER ACC
          ADDH    ACCH            RETRIEVE PAST UPPER ACC
          LST     STAT0           RETRIEVE PAST STATUS REGISTER
          EINT                    RE-ENABLE INTERRUPTS
          RET                     RETURN TO MAIN PROGRAM
          .
          .
          .
IMASK     DATA    >0010           ENABLE RECEIVE INTERRUPT ONLY
*
*         END OF MEMORY ALLOCATION/DEFINITION
*
*         BEGIN INITIALIZATION
*
********************************************************************
*         ====================================================     *
*         AIC - TMS320C25 COMMUNICATION INITIALIZATION             *
*         ====================================================     *
*    INITIALIZATION USED TO SET UP COMMUNICATIONS  WITH AIC        *
********************************************************************
OUTPUT    EQU     >60             STORAGE FOR OUTPUT SAMPLES
* SECONDARY COMMUNICATION XMIT DATA TABLE FOR AIC REGISTERS
TABLE     DATA    >03,>67,>03,>448A,>03,>1224
* INITIALIZE PROCESSOR
INIT      LALK    >03E0           INIT ST1,FRAME SYNC PULSES AS INPUT
          SACL    >67             TXM=0,XF=0(LOW) CONN TO AIC(RESET)PIN 2
          LST1    >67             FSM=1 FRAME SYNC PULSES ARE NOT IGNORED
          ZAC                     SET ACC=0
          SACL    >01             CLEAR DXR (DATA XMIT REG) IN DMA 01
          SACL    OUTPUT          INIT OUTPUT TO 0
          LACK    >20             LOAD IMR(INTER MASK REG)= 0010 0000 TO
          SACL    INTMSK          MASK ALL INTER EXCEPT 'XINT', IN DMA 4
          LARP    0               SELECT AR0
          LARK    AR0,>62         TRANSFER XMIT DATA
          RPTK    >05             FROM PMA 20-25
          BLKP    TABLE,*+        INTO DMA 62-67
```

FIGURE H.4. Partial multirate filter program using the AIC (AICMUL)

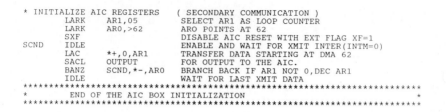

```
* INITIALIZE AIC REGISTERS    ( SECONDARY COMMUNICATION )
        LARK    AR1,05        SELECT AR1 AS LOOP COUNTER
        LARK    AR0,>62       AR0 POINTS AT 62
        SXF                   DISABLE AIC RESET WITH EXT FLAG XF=1
SCND    IDLE                  ENABLE AND WAIT FOR XMIT INTER(INTM=0)
        LAC     *+,0,AR1      TRANSFER DATA STARTING AT DMA 62
        SACL    OUTPUT        FOR OUTPUT TO THE AIC.
        BANZ    SCND,*-,AR0   BRANCH BACK IF AR1 NOT 0,DEC AR1
        IDLE                  WAIT FOR LAST XMIT DATA
*****************************************************************************
*             END OF THE AIC BOX INITIALIZATION                             *
*****************************************************************************
```

FIGURE H.4. (concluded)

the PRN will repeat every $(2^{32} - 1)/(32K) \simeq 36$ hours and the discrete frequencies will be spaced at 7.6 μHz intervals. As a result, the PRN appears as a uniformly distributed random noise. The unfiltered output of the PRN generator, sampled at 16,393 Hz, is shown in Figures H.1 and H.2. Figure H.1 shows that the PRN spectrum is flat over the range of 0–5 kHz which is the range that will subsequently be filtered by the multirate filter. Figure H.2, which shows the PRN generator output in a 50 Hz window, demonstrates that no discrete frequencies are detectable. Figure H.3 shows a complete listing of the multirate filtering program using the AIB. A partial program using the AIC is shown in Figure H.4.

REFERENCES

[1] R. W. Schafer and L. R. Rabiner, "A Digital Signal Processing Approach to Interpolation," *Proceedings of the IEEE*, **61** (6), June 1973.

[2] R. E. Crochiere and L. R. Rabiner, "Optimum FIR Digital Filter Implementations for Decimation, Interpolation and Narrow-Band Filtering," *IEEE Transactions on Acoustics, Speech, and Signal Processing*, **ASSP-23,** October 1975.

[3] R. E. Crochiere and L. R. Rabiner, "Further Considerations in the Design of Decimators and Interpolators," *IEEE Transactions on Acoustics, Speech, and Signal Processing*, **ASSP-24,** August 1976.

[4] R. E. Crochiere and L. R. Rabiner, *Multirate Digital Signal Processing*, Prentice-Hall, Englewood Cliffs, N.J., 1983.

[5] M. G. Bellanger, J. L. Dagnet and G. P. Lepagnol, "Interpolation, Extrapolation, and Reduction of Computation Speed in Digital Filters," *IEEE Transactions on Acoustics, Speech, and Signal Processing*, **ASSP-22,** August 1974.

[6] L. B. Jackson, *Digital Filters and Signal Processing*, Kluwer Academic, Nowell, Mass., 1986.

[7] J. G. Proakis and D. G. Manolakis, *Introduction to Digital Signal Processing*, Macmillan, New York, 1988.

[8] D. J. DeFatta, J. G. Lucas, and W. S. Hodgkiss, *Digital Signal Processing: A System Approach*, Wiley, New York, 1988.

[9] R. Chassaing, "Digital Broadband Noise Synthesis by Multirate Filtering Using the TMS320C25," *1988 ASEE Annual Conference Proceedings*, Vol. 1.

I

Pressure Measurement Using Resonant Wire

A complete listing of the program is included in Figure I.1. A brief description of the program follows. EQU statements are used to reserve various data memory addresses (DMA).

1. DMA >62 to 66, for the wave generation coefficients used in the sweep portion of the algorithm.

2. DMA >67 (label STEP), for the value used to step the frequency to the next desired frequency.

3. DMA >68 (label HIGH), for the input comparison for peak detection.

4. DMA >69 to 6E, for the wave generation coefficients used in the steady frequency oscillator portion of the algorithm.

5. DMA >200 to 250, TABLEC, for the sine value associated with each frequency within the swept range, denoted in the program's comments as either (Y1) or (C). The size of TABLEC is dependent on the operating range of the selected sensor. Although the sensor had an operational range of 1700 to 3000 Hz, the table represents a range of values from 1770 to 1850 Hz.

6. DMA >260 to 2B0, TABLEA, for the cosine value associated with each frequency within the swept range, denoted as (A). Like TABLEC, the size of TABLEA would also be dependent upon the operating range of the sensor.

A further description follows.

1. TABLE contains the initialization data for the digital sine-wave generation algorithms. It also contains the criteria data for the peak amplitude detection used in the frequency detection routine. Tables TABC and TABA contain the scaled (C) value (sine parameter) and the scaled (A) value (cosine parameter), respectively, calculated for each frequency in the swept band. These values are used prior to exiting the detection routine to match a proper . sine parameter for the detected frequency.

2. Starting at program memory address (PMA) D5, label START1, data are transferred from program memory (values in TABC and TABA) to data memory (TABLEC and TABLEA, respectively).

3. The codes starting at PMA E0, label START, initialize the MODE and RATE of the AIB, the coefficients associated with sweeping the sine waveform, the amplitude detection, and the steady frequency waveform.

4. The instructions starting at PMA E6 output the MODE and RATE to the AIB, set the sign extension mode and the frequency stepping delay loop that controls the time at each frequency, and initializes the counter that keeps track of the sweep range.

5. The three instructions starting at PMA F1 complete the frequency stepping delay loop, initialize the delay counter, and determine the actual length of time during which the output signal remains at each frequency. The chosen value of (28) is selected to enable a visual observation of the frequency sweeping through the range.

6. The instructions starting at PMA F6 do the actual stepping to the next frequency by adding a step value (chosen as 1 Hz) to the cosine value (A), then branch to MAIN after decrementing the frequency-range loop counter if the counter value is not zero. If the counter value is zero, the frequency-range loop counter is reset and an unconditional branch to the steady frequency resonator portion of the program at label RES is executed.

7. The instructions starting at PMA >100 (MAIN) generate the sine wave with the frequency controlled by the values of the cosine (A) and the sine (C) functions.

8. The instructions for the amplitude detection are at PMA 10F–11A. The signal emitted by the vibrating wire is compared to a predetermined voltage level used for the peak detection.

9. PMA 11B (TESTA) through PMA >126 are the instructions to find the detected frequency's associated (C) value, comparing the saved (A) parameter contained in data memory location FREQ with the values in table TABA.

10. The instructions starting at PMA >127 (RES) set up a delay loop, then proceed to the actual steady frequency oscillator at PMA >132 (MAIN2).

```
0001          *WIRE28.ASM
0002          *RESONANT WIRE WAVEFORM GENERATION PROGRAM    Fs=22.5 KHZ
0003          *Y(n)=A Y(n-1) + B Y(n-2) + C X(n-1)
0004          * YO =A Y1 + B Y2   FOR N>1    X(n-1)=1 FOR n=1 (= 0 OTHERWISE)
0005          *
0006          * THIS IS THE EQUATE STATEMENTS FOR THE DATA MEMORY LOCATIONS
0007          * ASSOCIATED WITH THE DIGITAL SIGNAL SINE AND COSINE FUNCTIONS
0008          *
0009  0060  MODE      EQU   >60    *TRANSPARENT MODE FOR AIB
0010  0061  RATE      EQU   >61    *RATE=(10 MHZ / Fs) -1=443.4=01BB
0011  0062  YO        EQU   >62    *INITIALLY YO=0, OUTPUT DMA
0012  0063  Y1        EQU   >63    *Y(1)=C=[SIN(2x3.14xFd/Fs)]/2
0013  0064  Y2        EQU   >64    *Y(2)=0
0014  0065  A         EQU   >65    *A=2 COS(WT)
0015  0066  B         EQU   >66    *B=-1
0016  0067  STEP      EQU   >67    *STEP INCREMENT
0017  0068  HIGH      EQU   >68    *COMPARISON SIGNAL LEVEL
0018  0069  AMP       EQU   >69    *INPUT SIGNAL
0019  006A  Y10       EQU   >6A    *TEMPORARY STORAGE OF YO
0020  006B  Y11       EQU   >6B    *TEMPORARY STORAGE OF Y1
0021  006C  Y12       EQU   >6C    *TEMPORARY STORAGE OF Y2
0022  006D  FREQ      EQU   >6D    *HIGHEST FREQUENCY VALUE OF 'A'
0023  006E  B1        EQU   >6E    *B1=-1
0024          *                    *A, B, C DIV BY 2 (MAX VALUE <1)
0025          *                    *A=678D, B=C000, C=3CDE   Fd=1.7KHZ
0026          *
0027          *   TABLEC
0028          *   THIS IS THE EQUATE STATEMENTS FOR THE SINE (C) FUNCTION VALUES
0029          *
0030  0200  C70C      EQU   >200   *THIS IS FREQUENCY 1770
0031  0201  C71C      EQU   >201
0032  0202  C72C      EQU   >202
0033  0203  C73C      EQU   >203
0034  0204  C74C      EQU   >204
0035  0205  C75C      EQU   >205
0036  0206  C76C      EQU   >206
0037  0207  C77C      EQU   >207
0038  0208  C78C      EQU   >208
0039  0209  C79C      EQU   >209
0040  020A  C80C      EQU   >20A   *THIS IS FREQUENCY 1780
0041  020B  C81C      EQU   >20B
0042  020C  C82C      EQU   >20C
0043  020D  C83C      EQU   >20D
0044  020E  C84C      EQU   >20E
0045  020F  C85C      EQU   >20F
0046  0210  C86C      EQU   >210
0047  0211  C87C      EQU   >211
0048  0212  C88C      EQU   >212
0049  0213  C89C      EQU   >213
0050  0214  C90C      EQU   >214   *THIS IS FREQUENCY 1790
0051  0215  C91C      EQU   >215
0052  0216  C92C      EQU   >216
0053  0217  C93C      EQU   >217
0054  0218  C94C      EQU   >218
0055  0219  C95C      EQU   >219
0056  021A  C96C      EQU   >21A
0057  021B  C97C      EQU   >21B
0058  021C  C98C      EQU   >21C
0059  021D  C99C      EQU   >21D
0060  021E  C00C      EQU   >21E   *THIS IS FREQUENCY 1800
0061  021F  C01C      EQU   >21F
0062  0220  C02C      EQU   >220
0063  0221  C03C      EQU   >221
0064  0222  C04C      EQU   >222
0065  0223  C05C      EQU   >223
0066  0224  C06C      EQU   >224
0067  0225  C07C      EQU   >225
0068  0226  C08C      EQU   >226
0069  0227  C09C      EQU   >227
0070  0228  C10C      EQU   >228   *THIS IS FREQUENCY 1810
0071  0229  C11C      EQU   >229
0072  022A  C12C      EQU   >22A
0073  022B  C13C      EQU   >22B
0074  022C  C14C      EQU   >22C
0075  022D  C15C      EQU   >22D
```

FIGURE I.1. Frequency determination with resonant wire (RESWIRE.LST)

```
0076    022E  C16C      EQU    >22E
0077    022F  C17C      EQU    >22F
0078    0230  C18C      EQU    >230
0079    0231  C19C      EQU    >231
0080    0232  C20C      EQU    >232    *THIS IS FREQUENCY 1820
0081    0233  C21C      EQU    >233
0082    0234  C22C      EQU    >234
0083    0235  C23C      EQU    >235
0084    0236  C24C      EQU    >236
0085    0237  C25C      EQU    >237
0086    0238  C26C      EQU    >238
0087    0239  C27C      EQU    >239
0088    023A  C28C      EQU    >23A
0089    023B  C29C      EQU    >23B
0090    023C  C30C      EQU    >23C    *THIS IS FREQUENCY 1830
0091    023D  C31C      EQU    >23D
0092    023E  C32C      EQU    >23E
0093    023F  C33C      EQU    >23F
0094    0240  C34C      EQU    >240
0095    0241  C35C      EQU    >241
0096    0242  C36C      EQU    >242
0097    0243  C37C      EQU    >243
0098    0244  C38C      EQU    >244
0099    0245  C39C      EQU    >245
0100    0246  C40C      EQU    >246    *THIS IS FREQUENCY 1840
0101    0247  C41C      EQU    >247
0102    0248  C42C      EQU    >248
0103    0249  C43C      EQU    >249
0104    024A  C44C      EQU    >24A
0105    024B  C45C      EQU    >24B
0106    024C  C46C      EQU    >24C
0107    024D  C47C      EQU    >24D
0108    024E  C48C      EQU    >24E
0109    024F  C49C      EQU    >24F
0110    0250  C50C      EQU    >250    *THIS IS FREQUENCY 1850
0111          *
0112          *  TABLEA
0113          *  THIS IS THE EQUATE STATEMENTS FOR THE COSINE (A) FUNCTION VALUES
0114          *
0115    0260  A70A      EQU    >260        *THIS IS FOR FREQUENCY 1770
0116    0261  A71A      EQU    >261
0117    0262  A72A      EQU    >262
0118    0263  A73A      EQU    >263
0119    0264  A74A      EQU    >264
0120    0265  A75A      EQU    >265
0121    0266  A76A      EQU    >266
0122    0267  A77A      EQU    >267
0123    0268  A78A      EQU    >268
0124    0269  A79A      EQU    >269
0125    026A  A80A      EQU    >26A        *THIS IS FOR FREQUENCY 1780
0126    026B  A81A      EQU    >26B
0127    026C  A82A      EQU    >26C
0128    026D  A83A      EQU    >26D
0129    026E  A84A      EQU    >26E
0130    026F  A85A      EQU    >26F
0131    0270  A86A      EQU    >270
0132    0271  A87A      EQU    >271
0133    0272  A88A      EQU    >272
0134    0273  A89A      EQU    >273
0135    0274  A90A      EQU    >274        *THIS IS FOR FREQUENCY 1790
0136    0275  A91A      EQU    >275
0137    0276  A92A      EQU    >276
0138    0277  A93A      EQU    >277
0139    0278  A94A      EQU    >278
0140    0279  A95A      EQU    >279
0141    027A  A96A      EQU    >27A
0142    027B  A97A      EQU    >27B
0143    027C  A98A      EQU    >27C
0144    027D  A99A      EQU    >27D
0145    027E  A00A      EQU    >27E        *THIS IS FOR FREQUENCY 1800
0146    027F  A01A      EQU    >27F
0147    0280  A02A      EQU    >280
0148    0281  A03A      EQU    >281
```

FIGURE I.1. (continued)

```
0149    0282 A04A         EQU     >282
0150    0283 A05A         EQU     >283
0151    0284 A06A         EQU     >284
0152    0285 A07A         EQU     >285
0153    0286 A08A         EQU     >286
0154    0287 A09A         EQU     >287
0155    0288 A10A         EQU     >288            *THIS IS FOR FREQUENCY 1810
0156    0289 A11A         EQU     >289
0157    028A A12A         EQU     >28A
0158    028B A13A         EQU     >28B
0159    028C A14A         EQU     >28C
0160    028D A15A         EQU     >28D
0161    028E A16A         EQU     >28E
0162    028F A17A         EQU     >28F
0163    0290 A18A         EQU     >290
0164    0291 A19A         EQU     >291
0165    0292 A20A         EQU     >292            *THIS IS FOR FREQUENCY 1820
0166    0293 A21A         EQU     >293
0167    0294 A22A         EQU     >294
0168    0295 A23A         EQU     >295
0169    0296 A24A         EQU     >296
0170    0297 A25A         EQU     >297
0171    0298 A26A         EQU     >298
0172    0299 A27A         EQU     >299
0173    029A A28A         EQU     >29A
0174    029B A29A         EQU     >29B
0175    029C A30A         EQU     >29C            *THIS IS FOR FREQUENCY 1830
0176    029D A31A         EQU     >29D
0177    029E A32A         EQU     >29E
0178    029F A33A         EQU     >29F
0179    02A0 A34A         EQU     >2A0
0180    02A1 A35A         EQU     >2A1
0181    02A2 A36A         EQU     >2A2
0182    02A3 A37A         EQU     >2A3
0183    02A4 A38A         EQU     >2A4
0184    02A5 A39A         EQU     >2A5
0185    02A6 A40A         EQU     >2A6            *THIS IS FOR FREQUENCY 1840
0186    02A7 A41A         EQU     >2A7
0187    02A8 A42A         EQU     >2A8
0188    02A9 A43A         EQU     >2A9
0189    02AA A44A         EQU     >2AA
0190    02AB A45A         EQU     >2AB
0191    02AC A46A         EQU     >2AC
0192    02AD A47A         EQU     >2AD
0193    02AE A48A         EQU     >2AE
0194    02AF A49A         EQU     >2AF
0195    02B0 A50A         EQU     >2B0
0196         *
0197         *  THIS IS THE START OF THE RESONATING WIRE PROGRAM
0198         *
0199 0000                 AORG    >0      *
0200 0000 FF80            B       START1  *
     0001 00D5
0201         *
0202         *  TABLE
0203         *  THIS TABLE CONTAINS THE INITIALIZATION VALUES FOR THE PROGRAM ALGORITHMS
0204         *
0205 0020    TABLE        AORG    >20     *TABLE STARTS AT PM >20
0206 0020 00FA MDE        DATA    >00FA   *INITIALIZE MODE FOR AIB
0207 0021 01BB RTE        DATA    >01BB   *INITIALIZE RATE
0208 0022 0000 Y00        DATA    >0000   *INITIALIZE YO = 0
0209 0023 3CDE Y01        DATA    >3CDE   *INITIALIZE Y1 = 1
0210 0024 0000 Y02        DATA    >0000   *INITIALIZE Y2 = 0
0211 0025 678D AA         DATA    >678D   *START A AT 1.7 KHZ
0212 0026 C000 BB         DATA    >C000   *INITIALIZE B = -1
0213 0027 0001 ST         DATA    >0001   *INITIALIZE STEP = 1
0214 0028 1999 HI         DATA    >1999   *INITIALIZE HIGH TO 2.0 VOLTS
0215 0029 0000 AMPP       DATA    >0000   *INITIALIZE INPUT SIGNAL
0216 002A 0000 Y1100      DATA    >0000   *INITIALIZE Y10
0217 002B 0000 Y1111      DATA    >0000   *INITIALIZE Y11
0218 002C 0000 Y1122      DATA    >0000   *INITIALIZE Y12
```

FIGURE I.1. (continued)

```
0219 002D 0000 FREQQ    DATA  >0000  *INITIALIZE FREQ
0220 002E C000 B11      DATA  >C000  *INITIALIZE B1
0221              *
0222              *    TABC
0223              *  THIS TABLE ALLOCATES SINE EQUATION (C) VALUES TO FREQUENCIES
0224              *
0225 0034 TABC        AORG  >34
0226 0034 1E5C C770   DATA  >1E5C  *THIS IS FREQUENCY 1770 Hz
0227 0035 1E60 C771   DATA  >1E60
0228 0036 1E64 C772   DATA  >1E64
0229 0037 1E68 C773   DATA  >1E68
0230 0038 1E6C C774   DATA  >1E6C
0231 0039 1E70 C775   DATA  >1E70
0232 003A 1E74 C776   DATA  >1E74
0233 003B 1E78 C777   DATA  >1E78
0234 003C 1E7C C778   DATA  >1E7C
0235 003D 1E80 C779   DATA  >1E80
0236 003E 1E84 C780   DATA  >1E84  *THIS IS FREQUENCY 1780 Hz
0237 003F 1E88 C781   DATA  >1E88
0238 0040 1E8C C782   DATA  >1E8C
0239 0041 1E90 C783   DATA  >1E90
0240 0042 1E94 C784   DATA  >1E94
0241 0043 1E98 C785   DATA  >1E98
0242 0044 1E9C C786   DATA  >1E9C
0243 0045 1EA0 C787   DATA  >1EA0
0244 0046 1EA4 C788   DATA  >1EA4
0245 0047 1EA8 C789   DATA  >1EA8
0246 0048 1EAC C790   DATA  >1EAC  *THIS IS FREQUENCY 1790 Hz
0247 0049 1EB0 C791   DATA  >1EB0
0248 004A 1EB4 C792   DATA  >1EB4
0249 004B 1EB8 C793   DATA  >1EB8
0250 004C 1EBC C794   DATA  >1EBC
0251 004D 1EC0 C795   DATA  >1EC0
0252 004E 1EC5 C796   DATA  >1EC5
0253 004F 1EC9 C797   DATA  >1EC9
0254 0050 1ECD C798   DATA  >1ECD
0255 0051 1ED1 C799   DATA  >1ED1
0256 0052 1ED5 C800   DATA  >1ED5  *THIS IS FREQUENCY 1800 Hz
0257 0053 1ED9 C801   DATA  >1ED9
0258 0054 1EDD C802   DATA  >1EDD
0259 0055 1EE1 C803   DATA  >1EE1
0260 0056 1EE5 C804   DATA  >1EE5
0261 0057 1EE9 C805   DATA  >1EE9
0262 0058 1EED C806   DATA  >1EED
0263 0059 1EF1 C807   DATA  >1EF1
0264 005A 1EF5 C808   DATA  >1EF5
0265 005B 1EF9 C809   DATA  >1EF9
0266 005C 1EFD C810   DATA  >1EFD  *THIS IS FREQUENCY 1810 Hz
0267 005D 1F01 C811   DATA  >1F01
0268 005E 1F05 C812   DATA  >1F05
0269 005F 1F09 C813   DATA  >1F09
0270 0060 1F0D C814   DATA  >1F0D
0271 0061 1F11 C815   DATA  >1F11
0272 0062 1F15 C816   DATA  >1F15
0273 0063 1F19 C817   DATA  >1F19
0274 0064 1F1D C818   DATA  >1F1D
0275 0065 1F21 C819   DATA  >1F21
0276 0066 1F25 C820   DATA  >1F25  *THIS IS FREQUENCY 1820 Hz
0277 0067 1F29 C821   DATA  >1F29
0278 0068 1F2D C822   DATA  >1F2D
0279 0069 1F31 C823   DATA  >1F31
0280 006A 1F35 C824   DATA  >1F35
0281 006B 1F39 C825   DATA  >1F39
0282 006C 1F3D C826   DATA  >1F3D
0283 006D 1F41 C827   DATA  >1F41
0284 006E 1F45 C828   DATA  >1F45
0285 006F 1F49 C829   DATA  >1F49
0286 0070 1F4D C830   DATA  >1F4D  *THIS IS FREQUENCY 1830 HZ
0287 0071 1F51 C831   DATA  >1F51
0288 0072 1F55 C832   DATA  >1F55
0289 0073 1F59 C833   DATA  >1F59
0290 0074 1F5D C834   DATA  >1F5D
0291 0075 1F61 C835   DATA  >1F61
```

FIGURE I.1. (continued)

```
0292 0076 1F64  C836    DATA    >1F64
0293 0077 1F68  C837    DATA    >1F68
0294 0078 1F6C  C838    DATA    >1F6C
0295 0079 1F70  C839    DATA    >1F70
0296 007A 1F74  C840    DATA    >1F74    *THIS IS FREQUENCY 1840 Hz
0297 007B 1F78  C841    DATA    >1F78
0298 007C 1F7C  C842    DATA    >1F7C
0299 007D 1F80  C843    DATA    >1F80
0300 007E 1F84  C844    DATA    >1F84
0301 007F 1F88  C845    DATA    >1F88
0302 0080 1F8C  C846    DATA    >1F8C
0303 0081 1F90  C847    DATA    >1F90
0304 0082 1F94  C848    DATA    >1F94
0305 0083 1F98  C849    DATA    >1F98
0306 0084 1F9C  C850    DATA    >1F9C    *THIS IS FREQUENCY 1850 Hz
0307       *
0308       *  TABA
0309       *  THIS TABLE ALLOCATES COSINE EQUATION (A) VALUES TO FREQUENCIES
0310       *
0311 0084       TABA    AORG    >84
0312 0084 70AE  A770    DATA    >70AE    *THIS IS FREQUENCY 1770 Hz
0313 0085 70A9  A771    DATA    >70A9
0314 0086 70A5  A772    DATA    >70A5
0315 0087 70A1  A773    DATA    >70A1
0316 0088 709C  A774    DATA    >709C
0317 0089 7098  A775    DATA    >7098
0318 008A 7093  A776    DATA    >7093
0319 008B 708F  A777    DATA    >708F
0320 008C 708B  A778    DATA    >708B
0321 008D 7086  A779    DATA    >7086
0322 008E 7082  A780    DATA    >7082    *THIS IS FREQUENCY 1780 Hz
0323 008F 707E  A781    DATA    >707E
0324 0090 7079  A782    DATA    >7079
0325 0091 7075  A783    DATA    >7075
0326 0092 7071  A784    DATA    >7071
0327 0093 706C  A785    DATA    >706C
0328 0094 7068  A786    DATA    >7068
0329 0095 7063  A787    DATA    >7063
0330 0096 705F  A788    DATA    >705F
0331 0097 705B  A789    DATA    >705B
0332 0098 7056  A790    DATA    >7056    *THIS IS FREQUENCY 1790 Hz
0333 0099 7052  A791    DATA    >7052
0334 009A 704E  A792    DATA    >704E
0335 009B 7049  A793    DATA    >7049
0336 009C 7045  A794    DATA    >7045
0337 009D 7040  A795    DATA    >7040
0338 009E 703C  A796    DATA    >703C
0339 009F 7038  A797    DATA    >7038
0340 00A0 7033  A798    DATA    >7033
0341 00A1 702F  A799    DATA    >702F
0342 00A2 702A  A800    DATA    >702A    *THIS IS FREQUENCY 1800 Hz
0343 00A3 7026  A801    DATA    >7026
0344 00A4 7021  A802    DATA    >7021
0345 00A5 701D  A803    DATA    >701D
0346 00A6 7019  A804    DATA    >7019
0347 00A7 7014  A805    DATA    >7014
0348 00A8 7010  A806    DATA    >7010
0349 00A9 700B  A807    DATA    >700B
0350 00AA 7007  A808    DATA    >7007
0351 00AB 7003  A809    DATA    >7003
0352 00AC 6FFE  A810    DATA    >6FFE    *THIS IS FREQUENCY 1810 Hz
0353 00AD 6FFA  A811    DATA    >6FFA
0354 00AE 6FF5  A812    DATA    >6FF5
0355 00AF 6FF1  A813    DATA    >6FF1
0356 00B0 6FEC  A814    DATA    >6FEC
0357 00B1 6FE8  A815    DATA    >6FE8
0358 00B2 6FE3  A816    DATA    >6FE3
0359 00B3 6FDF  A817    DATA    >6FDF
0360 00B4 6FDB  A818    DATA    >6FDB
0361 00B5 6FD6  A819    DATA    >6FD6
0362 00B6 6FD2  A820    DATA    >6FD2    *THIS IS FREQUENCY 1820 Hz
0363 00B7 6FCD  A821    DATA    >6FCD
0364 00B8 6FC9  A822    DATA    >6FC9
```

FIGURE I.1. (continued)

```
0365 0089 6FC4  A823        DATA    >6FC4
0366 008A 6FC0  A824        DATA    >6FC0
0367 008B 6FBB  A825        DATA    >6FBB
0368 008C 6FB7  A826        DATA    >6FB7
0369 008D 6FB2  A827        DATA    >6FB2
0370 008E 6FAE  A828        DATA    >6FAE
0371 008F 6FAA  A829        DATA    >6FAA
0372 00C0 6FA5  A830        DATA    >6FA5      *THIS IS FREQUENCY 1830 Hz
0373 00C1 6FA1  A831        DATA    >6FA1
0374 00C2 6F9C  A832        DATA    >6F9C
0375 00C3 6F98  A833        DATA    >6F98
0376 00C4 6F93  A834        DATA    >6F93
0377 00C5 6F8F  A835        DATA    >6F8F
0378 00C6 6F8A  A836        DATA    >6F8A
0379 00C7 6F86  A837        DATA    >6F86
0380 00C8 6F81  A838        DATA    >6F81
0381 00C9 6F7D  A839        DATA    >6F7D
0382 00CA 6F78  A840        DATA    >6F78      *THIS IS FREQUENCY 1840 Hz
0383 00CB 6F74  A841        DATA    >6F74
0384 00CC 6F6F  A842        DATA    >6F6F
0385 00CD 6F6B  A843        DATA    >6F6B
0386 00CE 6F66  A844        DATA    >6F66
0387 00CF 6F62  A845        DATA    >6F62
0388 00D0 6F5D  A846        DATA    >6F5D
0389 00D1 6F59  A847        DATA    >6F59
0390 00D2 6F54  A848        DATA    >6F54
0391 00D3 6F50  A849        DATA    >6F50
0392 00D4 6F4B  A850        DATA    >6F4B      *THIS IS FREQUENCY 1850 Hz
0393            *
0394            *
0395            *
0396            *     THIS SECTION INITIALIZES DATA MEMORY ASSOCIATED WITH THE SINE AND
0397            *     COSINE FUNCTIONS AND IS ONLY EXECUTED ON INITIAL START
0398            *
0399 00D5 5588  START1      LARP    0        *SET ARP TO AR0 FOR INDIRECT ADDRESSING
0400 00D6 D000              LRLK    0,>0200  *SET AR0 TO POINT AT TOP OF TABLEC IN DMA
     00D7 0200
0401 00D8 CB4F              RPTK    79       *REPEAT THE NEXT INSTRUCTION 80 TIMES
0402 00D9 FCA0              BLKP    TABC,*+ *
     00DA 0034
0403 00DB D000              LRLK    0,>0260  *SET AR0 TO POINT AT TOP OF TABLEA IN DMA
     00DC 0260
0404 00DD CB4F              RPTK    79       *REPEAT THE NEXT INSTRUCTION 80 TIMES
0405 00DE FCA0              BLKP    TABA,*+ *
     00DF 0084
0406            *
0407            *     THIS SECTION INITIALIZES THE ALGORITHM DMA EVERY TIME THE PROGRAM
0408            *     IS REINITIALIZED TO START
0409            *
0410 00E0 5588  START       LARP    0        *SET ARP TO AR0 FOR INDIRECT ADDRESSING
0411 00E1 D000              LRLK    0,>0060  *SET AR0 TO POINT AT TOP OF TABLE IN DATA MEMORY
     00E2 0060
0412 00E3 CB0E              RPTK    14       *REPEAT THE NEXT INSTRUCTION FIFTEEN TIMES
0413 00E4 FCA0              BLKP    TABLE,*+
     00E5 0020
0414            *
0415            *     THIS SECTION OUTPUTS MODE AND RATE TO THE AIB, SETS SIGN EXTENTION MODE,
0416            *     THE DELAY LOOP COUNTER, AND THE RANGE LOOP COUNTER
0417            *
0418 00E6 E060              OUT     MODE,0   *OUTPUT MODE = >FA TO PORT 0
0419 00E7 E161              OUT     RATE,1   *OUTPUT RATE = >01BB TO PORT 1
0420 00E8 CE07              SSXM             *SET SIGN EXTENSION MODE
0421 00E9 D200              LRLK    2,>28    *SET INITIAL VALUE FOR DELAY LOOP COUNTER (AR2-->40)
     00EA 0028
0422 00EB D300              LRLK    3,>0979  *SET INITIAL VALUE FOR SWEEP RANGE COUNTER (AR3-->2425)
     00EC 0979
0423 00ED FA80  WAIT        BIOZ    GO       *WAIT TIL BIO=0 (EOC OF A/D LOW)
     00EE 00F1
0424            *                            *BRANCH TO MAIN AFTER EOC
0425 00EF FF80              B       WAIT     *A/D PROVIDES TIMING
     00F0 00ED
0426            *
0427            *     THIS SECTION SETS THE LENGTH OF TIME AT EACH FREQUENCY
```

FIGURE I.1. (continued)

```
0001                     * ENCODER PROGRAM FOR NARROW BAND VOICE MODULATION
0002                     * NOTE: fs = 9.3 KHZ
0003 0000                     AORG    0
0004 0000 FF80               B       START
     0001 011B
0005 0020                     AORG    >20                    *P.M. START AT >20
0006         *
0007         *
0008 0020
0009 0020 000A  MD          DATA    >000A                  *MODE FOR AIB
0010 0021 0432  RATE        DATA    1074                   *SAMPLING FREQUENCY Fs=9.3 KHZ
0011 0022
0012                     * COEFFICIENTS FOR 1.5-2.4 KHZ BP FILTER START HERE
0013         *
0014         *
0015                     *   FIR COEFFICIENTS
0016         *
0017 0022  TABLE
0018         *
0019 0022 FF63               DATA    -157                   *C094
0020 0023 FF5F               DATA    -161                   *C093
0021 0024 FD31               DATA    -719                   *C092
0022 0025 00BC               DATA    188                    *C091
0023 0026 0626               DATA    1574                   *C090
0024 0027 0263               DATA    611                    *C089
0025 0028 FA6C               DATA    -1428                  *C088
0026 0029 FB81               DATA    -1151                  *C087
0027 002A 01E5               DATA    485                    *C086
0028 002B 0208               DATA    520                    *C085
0029 002C FFF9               DATA    -7                     *C084
0030 002D 031F               DATA    799                    *C083
0031 002E 0311               DATA    785                    *C082
0032 002F FA9C               DATA    -1380                  *C081
0033 0030 F812               DATA    -2030                  *C080
0034 0031 0257               DATA    599                    *C079
0035 0032 0888               DATA    2184                   *C078
0036 0033 01AB               DATA    427                    *C077
0037 0034 FCA1               DATA    -863                   *C076
0038 0035 FFBE               DATA    -66                    *C075
0039 0036 FE2E               DATA    -466                   *C074
0040 0037 F928               DATA    -1752                  *C073
0041 0038 0087               DATA    135                    *C072
0042 0039 0C64               DATA    3172                   *C071
0043 003A 0600               DATA    1536                   *C070
0044 003B F613               DATA    -2541                  *C069
0045 003C F709               DATA    -2295                  *C068
0046 003D 0283               DATA    643                    *C067
0047 003E 023F               DATA    575                    *C066
0048 003F FF76               DATA    -138                   *C065
0049 0040 08E5               DATA    2277                   *C064
0050 0041 096F               DATA    2415                   *C063
0051 0042 F3B7               DATA    -3145                  *C062
0052 0043 EAE3               DATA    -5405                  *C061
0053 0044 0390               DATA    912                    *C060
0054 0045 152A               DATA    5418                   *C059
0055 0046 058C               DATA    1420                   *C058
0056 0047 F9C6               DATA    -1594                  *C057
0057 0048 02C4               DATA    708                    *C056
0058 0049 F867               DATA    -1945                  *C055
0059 004A E020               DATA    -8160                  *C054
0060 004B FD89               DATA    -631                   .*C053
0061 004C 3DAA               DATA    15786                  *C052
0062 004D 2A06               DATA    10758                  *C051
0063 004E C083               DATA    -16253                 *C050
0064 004F A6E9               DATA    -22807                 *C049
0065 0050 1B13               DATA    6931                   *C048
0066 0051 6E1C               DATA    28188                  *C047
0067 0052 1B13               DATA    6931                   *C046
0068 0053 A6E9               DATA    -22807                 *C045
0069 0054 C083               DATA    -16253                 *C044
0070 0055 2A06               DATA    10758                  *C043
0071 0056 3DAA               DATA    15786                  *C042
0072 0057 FD89               DATA    -631                   *C041
0073 0058 E020               DATA    -8160                  *C040
0074 0059 F867               DATA    -1945                  *C039
0075 005A 02C4               DATA    708                    *C038
0076 005B F9C6               DATA    -1594                  *C037
0077 005C 058C               DATA    1420                   *C036
0078 005D 152A               DATA    5418                   *C035
0079 005E 0390               DATA    912                    *C034
0080 005F EAE3               DATA    -5405                  *C033
0081 0060 F3B7               DATA    -3145                  *C032
0082 0061 096F               DATA    2415                   *C031
```

FIGURE J.3. Program for encoder (ENCODER.LST)

```
0083 0062 08E5          DATA    2277        *C030
0084 0063 FF76          DATA    -138        *C029
0085 0064 023F          DATA    575         *C028
0086 0065 0283          DATA    643         *C027
0087 0066 F709          DATA    -2295       *C026
0088 0067 F613          DATA    -2541       *C025
0089 0068 0600          DATA    1536        *C024
0090 0069 0C64          DATA    3172        *C023
0091 006A 0087          DATA    135         *C022
0092 006B F928          DATA    -1752       *C021
0093 006C FE2E          DATA    -466        *C020
0094 006D FFBE          DATA    -66         *C019
0095 006E FCA1          DATA    -863        *C018
0096 006F 01AB          DATA    427         *C017
0097 0070 0888          DATA    2184        *C016
0098 0071 0257          DATA    599         *C015
0099 0072 F812          DATA    -2030       *C014
0100 0073 FA9C          DATA    -1380       *C013
0101 0074 0311          DATA    785         *C012
0102 0075 031F          DATA    799         *C011
0103 0076 FFF9          DATA    -7          *C010
0104 0077 0208          DATA    520         *C009
0105 0078 01E5          DATA    485         *C008
0106 0079 FB81          DATA    -1151       *C007
0107 007A FA6C          DATA    -1428       *C006
0108 007B 0263          DATA    611         *C005
0109 007C 0626          DATA    1574        *C004
0110 007D 00BC          DATA    188         *C003
0111 007E FD31          DATA    -719        *C002
0112 007F FF5F          DATA    -161        *C001
0113 0080 FF63          DATA    -157        *C000
0114            *
0115 0081
0116 0081
0117            *
0118            *
0119 0081      TAI
0120            *
0121            *  COEFFICIENTS FOR .6 KHZ LP START HERE
0122            *
0123            *
0124            *  FIR COEFFICIENTS
0125            *
0126 0081 02E8          DATA    744         *C094
0127 0082 02FD          DATA    765         *C093
0128 0083 027E          DATA    638         *C092
0129 0084 017A          DATA    378         *C091
0130 0085 001D          DATA    29          *C090
0131 0086 FEA5          DATA    -347        *C089
0132 0087 FD5A          DATA    -678        *C088
0133 0088 FC7F          DATA    -897        *C087
0134 0089 FC47          DATA    -953        *C086
0135 008A FCC6          DATA    -826        *C085
0136 008B FDEF          DATA    -529        *C084
0137 008C FF95          DATA    -107        *C083
0138 008D 016C          DATA    364         *C082
0139 008E 031B          DATA    795         *C081
0140 008F 044D          DATA    1101        *C080
0141 0090 04BC          DATA    1212        *C079
0142 0091 0444          DATA    1092        *C078
0143 0092 02EC          DATA    748         *C077
0144 0093 00E5          DATA    229         *C076
0145 0094 FE87          DATA    -377        *C075
0146 0095 FC42          DATA    -958        *C074
0147 0096 FA89          DATA    -1399       *C073
0148 0097 F9C0          DATA    -1600       *C072
0149 0098 FA23          DATA    -1501       *C071
0150 0099 FBB9          DATA    -1095       *C070
0151 009A FE52          DATA    -430        *C069
0152 009B 0184          DATA    388         *C068
0153 009C 04C0          DATA    1216        *C067
0154 009D 0763          DATA    1891        *C066
0155 009E 08D9          DATA    2265        *C065
0156 009F 08B4          DATA    2228        *C064
0157 00A0 06C6          DATA    1734        *C063
0158 00A1 0334          DATA    820         *C062
0159 00A2 FE74          DATA    -396        *C061
0160 00A3 F944          DATA    -1724       *C060
0161 00A4 F496          DATA    -2922       *C059
0162 00A5 F168          DATA    -3736       *C058
0163 00A6 F0A3          DATA    -3933       *C057
0164 00A7 F2F2          DATA    -3342       *C056
0165 00A8 F8A2          DATA    -1886       *C055
0166 00A9 0191          DATA    401         *C054
0167 00AA 0D27          DATA    3367        *C053
```

FIGURE J.3. (continued)

```
0168  00AB  1A64        DATA    6756        *C052
0169  00AC  2800        DATA    10240       *C051
0170  00AD  348F        DATA    13455       *C050
0171  00AE  3EB3        DATA    16051       *C049
0172  00AF  4549        DATA    17737       *C048
0173  00B0  4792        DATA    18322       *C047
0174  00B1  4549        DATA    17737       *C046
0175  00B2  3EB3        DATA    16051       *C045
0176  00B3  348F        DATA    13455       *C044
0177  00B4  2800        DATA    10240       *C043
0178  00B5  1A64        DATA    6756        *C042
0179  00B6  0D27        DATA    3367        *C041
0180  00B7  0191        DATA    401         *C040
0181  00B8  F8A2        DATA    -1886       *C039
0182  00B9  F2F2        DATA    -3342       *C038
0183  00BA  F0A3        DATA    -3933       *C037
0184  00BB  F168        DATA    -3736       *C036
0185  00BC  F496        DATA    -2922       *C035
0186  00BD  F944        DATA    -1724       *C034
0187  00BE  FE74        DATA    -396        *C033
0188  00BF  0334        DATA    820         *C032
0189  00C0  06C6        DATA    1734        *C031
0190  00C1  08B4        DATA    2228        *C030
0191  00C2  08D9        DATA    2265        *C029
0192  00C3  0763        DATA    1891        *C028
0193  00C4  04C0        DATA    1216        *C027
0194  00C5  0184        DATA    388         *C026
0195  00C6  FE52        DATA    -430        *C025
0196  00C7  FBB9        DATA    -1095       *C024
0197  00C8  FA23        DATA    -1501       *C023
0198  00C9  F9C0        DATA    -1600       *C022
0199  00CA  FA89        DATA    -1399       *C021
0200  00CB  FC42        DATA    -958        *C020
0201  00CC  FE87        DATA    -377        *C019
0202  00CD  00E5        DATA    229         *C018
0203  00CE  02EC        DATA    748         *C017
0204  00CF  0444        DATA    1092        *C016
0205  00D0  04BC        DATA    1212        *C015
0206  00D1  044D        DATA    1101        *C014
0207  00D2  031B        DATA    795         *C013
0208  00D3  016C        DATA    364         *C012
0209  00D4  FF95        DATA    -107        *C011
0210  00D5  FDEF        DATA    -529        *C010
0211  00D6  FCC6        DATA    -826        *C009
0212  00D7  FC47        DATA    -953        *C008
0213  00D8  FC7F        DATA    -897        *C007
0214  00D9  FD5A        DATA    -678        *C006
0215  00DA  FEA5        DATA    -347        *C005
0216  00DB  001D        DATA    29          *C004
0217  00DC  017A        DATA    378         *C003
0218  00DD  027E        DATA    638         *C002
0219  00DE  02FD        DATA    765         *C001
0220  00DF  02E8        DATA    744         *C000
0221  00E0
0222  00E0
0223  00E0
0224              *
0225              * COEFFICIENTS FOR 1.6 KHZ LP START HERE
0226              *
0227  00E0        TAB1
0228  00E0
0229              *
0230              *  FIR COEFFICIENTS
0231              *
0232  00E0  011A        DATA    282         *C058
0233  00E1  FE41        DATA    -447        *C057
0234  00E2  FD28        DATA    -728        *C056
0235  00E3  FF31        DATA    -207        *C055
0236  00E4  0252        DATA    594         *C054
0237  00E5  0316        DATA    790         *C053
0238  00E6  0068        DATA    104         *C052
0239  00E7  FD01        DATA    -767        *C051
0240  00E8  FCB0        DATA    -848        *C050
0241  00E9  0022        DATA    34          *C049
0242  00EA  03CF        DATA    975         *C048
0243  00EB  0385        DATA    901         *C047
0244  00EC  FF21        DATA    -223        *C046
0245  00ED  FB2F        DATA    -1233       *C045
0246  00EE  FC4D        DATA    -947        *C044
0247  00EF  01E9        DATA    489         *C043
0248  00F0  0628        DATA    1576        *C042
0249  00F1  03D9        DATA    985         *C041
0250  00F2  FC8A        DATA    -886        *C040
0251  00F3  F7E5        DATA    -2075       *C039
```

FIGURE J.3. (continued)

```
0252 00F4 FC08          DATA    -1016       *C038
0253 00F5 0614          DATA    1556        *C037
0254 00F6 0B72          DATA    2930        *C036
0255 00F7 040F          DATA    1039        *C035
0256 00F8 F465          DATA    -2971       *C034
0257 00F9 ECAF          DATA    -4945       *C033
0258 00FA FBE4          DATA    -1052       *C032
0259 00FB 2109          DATA    8457        *C031
0260 00FC 488A          DATA    18570       *C030
0261 00FD 5976          DATA    22902       *C029
0262 00FE 488A          DATA    18570       *C028
0263 00FF 2109          DATA    8457        *C027
0264 0100 FBE4          DATA    -1052       *C026
0265 0101 ECAF          DATA    -4945       *C025
0266 0102 F465          DATA    -2971       *C024
0267 0103 040F          DATA    1039        *C023
0268 0104 0B72          DATA    2930        *C022
0269 0105 0614          DATA    1556        *C021
0270 0106 FC08          DATA    -1016       *C020
0271 0107 F7E5          DATA    -2075       *C019
0272 0108 FC8A          DATA    -886        *C018
0273 0109 03D9          DATA    985         *C017
0274 010A 0628          DATA    1576        *C016
0275 010B 01E9          DATA    489         *C015
0276 010C FC4D          DATA    -947        *C014
0277 010D FB2F          DATA    -1233       *C013
0278 010E FF21          DATA    -223        *C012
0279 010F 0385          DATA    901         *C011
0280 0110 03CF          DATA    975         *C010
0281 0111 0022          DATA    34          *C009
0282 0112 FCB0          DATA    -848        *C008
0283 0113 FD01          DATA    -767        *C007
0284 0114 0068          DATA    104         *C006
0285 0115 0316          DATA    790         *C005
0286 0116 0252          DATA    594         *C004
0287 0117 FF31          DATA    -207        *C003
0288 0118 FD28          DATA    -728        *C002
0289 0119 FE41          DATA    -447        *C001
0290 011A 011A          DATA    282         *C000
0291 011B
0292 011B
0293 011B
0294            *
0295 011B
0296            *
0297      0000  MODE     EQU     >0          *MODE FOR AIB (NO OFFSET)
0298      0001  CLOCK    EQU     >1          *Fs IN DMA+1 (OFFSET BY 1)
0299 011B
0300      0002  YN       EQU     >2          *OUTPUT (BP) IN DP6 (OFF 2)
0301      0003  XN       EQU     >3          *NEW SAMPLE IN DP6 (OFF 3)
0302 011B
0303      007C  ZN       EQU     >7C         *OUTPUT (LP) IN DMA+4 (OFF 4)
0304      007D  S0       EQU     >7D         *LOAD SINE VALUES INTO
0305      007E  S1       EQU     >7E         *LOCATIONS 037D-037F
0306      007F  S2       EQU     >7F         *DP 6
0307 011B
0308      0005  SN       EQU     >5          *INPUT TO 1600 LP DP7 (OFF 5)
0309 011B
0310 011B C806 START    LDPK    >06         *SELECT PAGE 6, TOP OF B1 300
0311 011C D001          LALK    >0000
     011D 0000
0312 011E 607D          SACL    S0
0313 011F D001          LALK    >1BB6
     0120 1BB6
0314 0121 607E          SACL    S1
0315 0122 D001          LALK    >E44A
     0123 E44A
0316 0124 607F          SACL    S2
0317 0125
0318 0125 D600          LRLK    AR6,>2      *LOAD SINE WAVE POS POINTER
     0126 0002
0319 0127 D700          LRLK    AR7,>037D
     0128 037D
0320 0129
0321 0129 CA20          LACK    MD          *LOAD ACC WITH (MD)
0322 012A 5800          TBLR    MODE        *TRANSFER INTO DMA 300
0323 012B E000          OUT     MODE,0      *OUT VALUE MODE TO PORT 0
0324 012C CA21          LACK    RATE        *LOAD ACC WITH (RATE)
0325 012D 5801          TBLR    CLOCK       *TRANSFER INTO DMA 301
0326 012E E101          OUT     CLOCK,1     *OUT VALUE CLOCK TO PORT 1
0327 012F E301          OUT     CLOCK,3     *DUMMY OUTPUT TO PORT 3
0328 0130
0329           **************************************************
0330           * LOAD FILTER COEFFICIENTS FOR BP 1.5-2.4KHZ *
0331           **************************************************
0332 0130
```

FIGURE J.3. (continued)

```
0333 0130 5588          LARP   AR0            *SELECT AR0 FOR INDIR ADDR
0334 0131 D000          LRLK   AR0,>200       *AR0 POINT TO TOP OF BLOCK B0
     0132 0200
0335 0133 CB5E          RPTK   94             *REPEAT FOR COEFFICIENTS
0336 0134 FCA0          BLKP   TABLE,*+       *P.M. 20-7F=> D.M. 200-25F
     0135 0022
0337 0136
0338                    *********************************************
0339                    * LOAD FILTER COEFFICIENTS FOR LP 600 HZ     *
0340                    *********************************************
0341 0136
0342 0136 5588          LARP   AR0            *SELECT AR0 FOR INDIR ADDR
0343 0137 D000          LRLK   AR0,>260       *AR0 POINT TO >260 OF BLOCK B0
     0138 0260
0344 0139 CB5E          RPTK   94             *REPEAT FOR COEFFICIENTS
0345 013A FCA0          BLKP   TABL,*+        *P.M. 80-DF => D.M. 260-2BF
     013B 0081
0346 013C
0347                    *********************************************
0348                    * LOAD FILTER COEFFICIENTS FOR LP 1600 HZ    *
0349                    *********************************************
0350 013C
0351 013C 5588          LARP   AR0            *SELECT AR0 FOR INDIR ADDR
0352 013D D000          LRLK   AR0,>2C0       *AR0 POINT TO >2C0 OF BLOCK B1
     013E 02C0
0353 013F CB3A          RPTK   58             *REPEAT FOR COEFFICIENTS
0354 0140 FCA0          BLKP   TAB1,*+        *P.M. E0-14F => D.M. 2C0-2FB
     0141 00E0
0355 0142
0356                    *
0357                    *********************************************
0358                    * INPUT XN AND BANDPASS FILTER 1.5-2.4 KHZ   *
0359                    *********************************************
0360                    *
0361 0142
0362 0142
0363 0142 CE05          CNFP                  *CONFIGURE BLOCK B0 AS P.M.
0364 0143 FA80   WAIT   BIOZ   MAIN           *NEW SAMPLE WHEN BIO PIN LOW
     0144 0147
0365 0145 FF80          B      WAIT           *CONTINUE UNTIL EOC IS DONE
     0146 0143
0366                    *
0367 0147 8203   MAIN   IN     XN,2           *INPUT NEW SAMPLE IN DMA 303
0368 0148 2003          LAC    XN
0369 0149 CE19          SFR
0370 014A 6003          SACL   XN
0371 014B D100          LRLK   AR1,>303+94    *AR1 = LAST SAMPLE
     014C 0361
0372 014D 5589          LARP   AR1            *SELECT AR1 FOR INDIR ADDR
0373 014E A000          MPYK   0              *SET PRODUCT REGISTER TO 0
0374 014F CA00          ZAC                   *SET ACC TO 0
0375                    * THE BEEF
0376 0150 CB5E          RPTK   94             *MULTIPLY THE COEFFICIENTS
0377 0151 5D90          MAC    >FF00,*-       *H0*X(n-40) + H1*X(n-39) +...
     0152 FF00
0378                    *                     *FIRST PMA FF00,LAST DMA 32B
0379 0153 CE15          APAC                  *LAST ACCUMULATE
0380 0154 6902          SACH   YN,1           *SHIFT LEFT 1, UPPER 16 BITS
0381 0155
0382                    *
0383                    *********************************************
0384                    * THIS SECTION MULTIPLIES BP OUT BY 3.1KHZ   *
0385                    *********************************************
0386                    *
0387 0155
0388 0155 558F          LARP   AR7            *POINT TO SINE WAVE VALUE
0389 0156 3CA0          LT     *+             *LOAD T REG WITH SINE VALUE
0390 0157 3802          MPY    YN             *MULT 3.1 KHZ VALUE WITH YN
0391 0158 CE14          PAC
0392 0159 6C02          SACH   YN,4           *YN = YN X SINE VALUE
0393 015A 558E          LARP   AR6
0394 015B FB90          BANZ   CON            *RELOCATE POINTER IF AT END
     015C 0161
0395 015D D600          LRLK   AR6,2          *OF SINE VALUES
     015E 0002
0396 015F D700          LRLK   AR7,>037D
     0160 037D
0397 0161
0398                    *********************************************
0399                    * START OF LP 600HZ SECTION                  *
0400                    *********************************************
0401 0161
0402 0161 D100   CON    LRLK   AR1,>303+94    *AR1 = LAST SAMPLE
     0162 0361
0403 0163 5589          LARP   AR1            *SELECT AR1 FOR INDIR ADDR
0404 0164 A000          MPYK   0              *SET PRODUCT REGISTER TO 0
```

FIGURE J.3. (continued)

```
0405 0165 CA00            ZAC                 *SET ACC TO 0
0406                  * THE BEEF
0407 0166 CB5E            RPTK    94          *MULTIPLY THE COEFFICIENTS
0408 0167 5C90            MACD    >FF60,*-    *H0*X(n-40) + H1*X(n-39) +...
     0168 FF60
0409                  *                       *FIRST PMA FF00,LAST DMA 32B
0410 0169 CE15            APAC                *LAST ACCUMULATE
0411 016A 697C            SACH    ZN,1        *SHIFT LEFT 1, UPPER 16 BITS
0412 016B
0413                  *
0414                  ********************************************
0415                  * YN = LP OUT + MULTIPLIED BP OUT          *
0416                  ********************************************
0417                  *
0418 016B
0419 016B 207C            LAC     ZN
0420 016C 0002            ADD     YN
0421 016D C807            LDPK    7           *POINT TO DP 7
0422 016E 6005            SACL    SN
0423 016F
0424                  *
0425                  ********************************************
0426                  * START OF LP 1600 SECTION                 *
0427                  ********************************************
0428                  *
0429 016F
0430 016F D100            LRLK    AR1,>385+58 *AR1 = LAST SAMPLE
     0170 03BF
0431 0171 5589            LARP    AR1         *SELECT AR1 FOR INDIR ADDR
0432 0172 A000            MPYK    0           *SET PRODUCT REGISTER TO 0
0433 0173 CA00            ZAC                 *SET ACC TO 0
0434 0174
0435 0174 CB3A            RPTK    58          *MULTIPLY THE COEFFICIENTS
0436 0175 5C90            MACD    >FFC0,*-    *H0*X(n-40) + H1*X(n-39) +...
     0176 FFC0
0437                  *                       *FIRST PMA FF00,LAST DMA 32B
0438 0177 CE15            APAC                *LAST ACCUMULATE
0439 0178 C806            LDPK    6           *POINT TO DP 6
0440 0179 6902            SACH    YN,1        *SHIFT LEFT 1, UPPER 16 BITS
0441 017A E202            OUT     YN,2        *OUT RESULT Y(n) FROM DMA 302
0442 017B FF80            B       WAIT        *BRANCH FOR NEXT SAMPLE
     017C 0143
0443                      END
NO ERRORS, NO WARNINGS
```

FIGURE J.3. (concluded)

REFERENCE

[1] R. W. Harris and J. F. Cleveland, "A Baseband Communications System," *QST*, American Radio Relay League, Newington, Conn., November 1978.

K

Guitar Tuning

The program listing in Turbo Pascal for communication between an IBM PC and a HP spectrum analyzer (Model 70000) is shown in Figures K.1 and K.2.

```
Program Analyzer (input,output);

{ This program is specifically designed to configure a HP 70000 }
{ spectrum analyzer for the purpose of tuning a guitar.         }


VAR
 satisfied,done : boolean;
 mkA, mkH, Q, X, SA, index, t, p, WResult, YResult : integer;
 StringLen, n, num, w, WParameter, j              : integer;
 YParameter,y,z,DeltaMax,DeltaMin,TUNED           : real;
 answr : string[1];
 readstring,dat : string[8];
 NewW : string[10];
 nstring : string[50];

{$I tpdecl.pas }

function itohex (i:integer):str4;
(* Convert integer to hex string. *)
(* Useful for printing hex value of IBSTA *)
var  k : integer; s : string[4]; nib : integer;
begin
 k := 12;  s := '';
 while (k >= 0) do
  begin
    nib := (i shr k) and $f;
    if ($A <= nib) and (nib <= $F)
    then nib := nib + $37
    else nib := nib + $30;
    s := s + chr(nib);
    k := k - 4;
  end;
 itohex := s;
end;

procedure prvars (ibsta:integer;iberr:integer;ibcnt:integer);
var  stas : string[4];
begin
stas := itohex(ibsta);
writeln ('ibsta=0x',stas,'iberr=0x',iberr,'ibcnt=0x',ibcnt);
end;


procedure error;
begin
 writeln (' Error');
 prvars (ibsta,iberr,ibcnt);
end;


procedure pause;
VAR
  ch : char;
```

FIGURE K.1. Pascal program for communication between PC and H-P spectrum analyzer (SPECT-COM.PAS)

```
BEGIN
 write ('Press any key to continue.');
 while not keypressed do;
    read (KBD,ch);
 clreol;
end;


Procedure send;
begin
 StringLen := length(nstring);
 index := 1;
 while index <= StringLen do begin
  ibbuf[index] := nstring[index];
  index := index + 1;
 end;
 ibwrt (SA,ibbuf,StringLen);
 { if ((ibsta AND ERR) <> 0) then error;}
end;


Procedure receive ;
 begin
  insert ('0000000000',readstring,1);
  ibcnt := 8;
  ibrd (SA,ibbuf,ibcnt);
{  if ((ibsta AND ERR) <> 0) then error;    }
  index := 1;
  while index <= ibcnt  do begin
   readstring[index] := ibbuf[index];
   index := index + 1;
  end;
  t := index - 1;
  p := 80 - t;
  delete (readstring,t,p);            { generate char length }
  val (readstring,YParameter,YResult);            { YP = real value    }
  val (readstring,WParameter,WResult);            { WP = integer value }
 end;

Procedure rcvdbg;  { receiver in debug mode }
 begin
  insert ('0000000000',readstring,1);
  ibcnt := 8;
  ibrd (SA,ibbuf,ibcnt);
{  if ((ibsta AND ERR) <> 0) then error;    }
  index := 1;
  while index <= ibcnt  do begin
   readstring[index] := ibbuf[index];
   index := index + 1;
  end;
  t := index - 1;
  p := 80 - t;
  delete (readstring,t,p);            { generate char length }
  val (readstring,YParameter,YResult);            { YP = real value    }
  val (readstring,WParameter,WResult);            { WP = integer value }
  writeln('readstring = ',readstring);
  writeln('YParameter = ',YParameter);
  writeln('YResult = ',YResult);
  writeln('WParameter = ',WParameter);
  writeln('WResult = ',WResult);
 end;


begin    { start here }
 clrscr;
 textbackground (black);
 textcolor (white);
 for j := 1 to 50 do
 writeln;
 gotoXY (1,1);
 bdname := 'SA';
 SA:= ibfind (bdname);
 ibsre(SA,1);
 if (SA < 0 ) then error;
 ibclr(SA);
 if (SA < 0) then error;
   done := true;
 TUNED := 440;            { Tuning frequency for "A" note }
 nstring := 'ip';         { HP instrument Preset }
 send;
 nstring := 'ss 0hz';     { Center frequency step to MANUAL }
 send;
 nstring := 'vb 300hz';   { Video bandwith to MANUAL }
 send;
 nstring := 'rb 10hz';    { resolution bandwith to MANUAL }
 send;
 nstring := 'sp 1000hz';  { Set frequency span }
 send;
```

FIGURE K.1. (continued)

```
          nstring := 'cf 440hz';   ( Set center frequency )
          send;
          nstring := 'fa 300hz';   ( Set Start frequency )
          send;
          nstring := 'fb 500hz';   ( Set Stop  frequency )
          send;
          nstring := 'st 1s';      ( Set Sweep time )
          send;
          nstring := 'rl 10dbm';   ( Set the reference level to 10 DBM )
          send;
          nstring := 'at 0db';     ( Set the input attenuator to 0DB )
          send;
          delay (10000);
          writeln('Directions for tuning a guitar in the "A" note. (440 HZ)');
          writeln;
          writeln('The strings are numbered 1-6; 1 being the thickest string.');
          writeln('To tune string 1 "E", depress 5th fret and strike.');
          writeln('To tune string 2 "A", strike.');
          writeln('To tune string 3 "D", depress 7th fret and strike.');
          writeln('To tune string 4 "G", depress 2nd fret and strike.');
          writeln('To tune string 5 "B", depress 10th fret and strike.');
          writeln('To tune string 6 "E", depress 5th fret and strike.');

          while done do begin

            nstring := 'mkpk hi';   ( Set marker HIGH )
            send;
            nstring := 'mkf?';      ( READ the marker HIGH )
            send;
            receive;
            mkH := WParameter;
            nstring := 'mka?';      ( read marker Amplitude  )
            send ;
            receive;
            mkA := WParameter;
            if ( mkA < -70 ) then      ( If the amplitude is less than -70 DO NOTHING )
            else begin
              writeln('Current FREQ = ',mkH, ' HZ.  ');
              if ( mkH = 440 ) then
                writeln('TUNED                            ');
              if ( mkH < TUNED ) then
                writeln ('TOO LOW! Tighten the string.    ');
              if ( mkH > TUNED ) then
                writeln('TOO HIGH! Loosen the string.     ');
            end;
          gotoXY(1,15);
          delay (1500);
          end;
        end.
```

FIGURE K.1. (concluded)

```
(* TURBO Pascal Declarations                                     *)


(*$V-*)           (* relax string length restrictions       *)


Const
      MAXIBBUF = $100;  (* maximum buffer size for I/O functions        *)
      MAXINTBUF = $100; (* maximum buffer size for integer I/O functions *)

Type
      iobuf  =  array[1..MAXIBBUF] of char;
      iolbuf = array[1..MAXINTBUF] of integer;
      ibstring =  string[50];
      str4 = string[4];
Var
    ibsta : integer;  (* status word            *)
    iberr : integer;  (* GPIB error code        *)
    ibcnt : integer;  (* number of bytes sent or, in the event of *)
                      (* DOS error, the DOS error code           *)
    ibbuf : iobuf;    (* I/O buffer for commands/data   *)
    intbuf : iolbuf;  (* integer I/O buffer for data    *)
    bdname : ibstring;(* board or device name           *)
    bd,dvm : integer; (* Board descriptor               *)
    vcnt: integer;    (* v or byte count                *)
    flname : ibstring;(* file name                      *)
    mask : integer;   (* wait mask for IBWAIT ftn.      *)
    ppr,spr:integer;  (* parallel,serial poll responses *)

(* Ibfn is the common entry point into the language interface, tpib.com.   *)
(* Its arguments are generalized to meet the needs of each individual GPIP *)
(* function, and are decoded as follows:                                   *)
```

FIGURE K.2. Turbo Pascal declarations

```
(*              ibfnasm (name,iberr,ibcnt,buf,ibuf,vcnt,bd,fcode,iberr,ibcnt)  *)
(* where:      name  = string for bdname, flname, bname                        *)
(*             iberr = GPIB-PC error code                                      *)
(*             ibcnt = GPIB-PC count                                           *)
(*             ibuf  = integer array for (var) rdi, wrti buffers               *)
(*             buf   = integer array for (var) rd, wrt, cmd buffers            *)
(*             vcnt  = integer for v, cnt, (var) spr, (var) ppr                *)
(*             bd    = integer for bd                                          *)
(*             fcode = integer for function code                              *)

function ibfn (name:ibstring;var iberr,ibcnt:integer;var buf:iobuf;var buf1:io1b
    var vcnt:integer;bd,fncode:integer):integer; external 'tpib.com';

(* You MUST include the appropriate declaration, as          *)
(* given below, for each procedure or function you call.     *)
(* You may omit declarations for functions you do not call.  *)

procedure    ibbna (bd:integer;bname:ibstring);
   var
   name : ibstring;
   begin
   name := bname + chr(0);
   ibsta := ibfn(name,iberr,ibcnt,ibbuf,intbuf,vcnt,bd,26);
   end;
procedure    ibcac (bd:integer;v:integer);
   begin
   ibsta := ibfn(bdname,iberr,ibcnt,ibbuf,intbuf,v,bd,16);
   end;
procedure    ibclr (bd:integer);
   begin
   ibsta := ibfn(bdname,iberr,ibcnt,ibbuf,intbuf,vcnt,bd,22);
   end;
procedure    ibcmd (bd:integer;ibbuf:iobuf;cnt:integer);
   begin
   ibsta := ibfn(bdname,iberr,ibcnt,ibbuf,intbuf,cnt,bd,32);
   end;
procedure    ibcmda (bd:integer;ibbuf:iobuf;cnt:integer);
   begin
   ibsta := ibfn(bdname,iberr,ibcnt,ibbuf,intbuf,cnt,bd,33);
   end;
procedure    ibdiag (bd:integer;ibbuf:iobuf;cnt:integer);
   begin
   ibsta := ibfn(bdname,iberr,ibcnt,ibbuf,intbuf,cnt,bd,34);
   end;
procedure    ibdma (bd:integer;v:integer);
   begin
   ibsta := ibfn(bdname,iberr,ibcnt,ibbuf,intbuf,v,bd,11);
   end;
procedure    ibeos (bd:integer;v:integer);
   begin
ibsta := ibfn(bdname,iberr,ibcnt,ibbuf,intbuf,v,bd,12);
   end;
procedure    ibeot (bd:integer;v:integer);
   begin
   ibsta := ibfn(bdname,iberr,ibcnt,ibbuf,intbuf,v,bd,14);
   end;
function     ibfind (bdname:ibstring):integer;
   var
   name : ibstring;
   begin
   name := bdname + chr(0);
   ibfind := ibfn(name,iberr,ibcnt,ibbuf,intbuf,vcnt,bd,27);
   end;
procedure    ibgts (bd:integer;v:integer);
   begin
   ibsta := ibfn(bdname,iberr,ibcnt,ibbuf,intbuf,v,bd,15);
   end;
procedure    ibist (bd:integer;v:integer);
   begin
   ibsta := ibfn(bdname,iberr,ibcnt,ibbuf,intbuf,v,bd,10);
   end;
procedure    ibloc (bd:integer);
   begin
   ibsta := ibfn(bdname,iberr,ibcnt,ibbuf,intbuf,vcnt,bd,5);
   end;
procedure    ibonl (bd:integer;v:integer);
   begin
   ibsta := ibfn(bdname,iberr,ibcnt,ibbuf,intbuf,v,bd,1);
   end;
procedure    ibpad (bd:integer;v:integer);
   begin
   ibsta := ibfn(bdname,iberr,ibcnt,ibbuf,intbuf,v,bd,8);
   end;
procedure    ibpct (bd:integer);
   begin
```

FIGURE K.2. (continued)

```
          ibsta := ibfn(bdname,iberr,ibcnt,ibbuf,intbuf,vcnt,bd,24);
          end;
procedure     ibppc (bd:integer;v:integer);
          begin
          ibsta := ibfn(bdname,iberr,ibcnt,ibbuf,intbuf,v,bd,7);
          end;
procedure     ibrd (bd:integer;var ibbuf:iobuf;cnt:integer);
          begin
          ibsta := ibfn(bdname,iberr,ibcnt,ibbuf,intbuf,cnt,bd,28);
          end;
procedure     ibrdi (bd:integer;var intbuf:iolbuf;cnt:integer);
          begin
          ibsta := ibfn(bdname,iberr,ibcnt,ibbuf,intbuf,cnt,bd,36);
          end;
procedure     ibrda (bd:integer;var ibbuf:iobuf;cnt:integer);
          begin
          ibsta := ibfn(bdname,iberr,ibcnt,ibbuf,intbuf,cnt,bd,29);
          end;
procedure     ibrdf (bd:integer;flname:ibstring);
          var
          name : ibstring;
          begin
          name := flname + chr(0);
          ibsta := ibfn(name,iberr,ibcnt,ibbuf,intbuf,vcnt,bd,17);
          end;
procedure     ibrpp (bd:integer;var ppr:integer);
          begin
          ibsta := ibfn(bdname,iberr,ibcnt,ibbuf,intbuf,ppr,bd,19);
          end;
procedure     ibrsc (bd:integer;v:integer);
          begin
          ibsta := ibfn(bdname,iberr,ibcnt,ibbuf,intbuf,v,bd,2);
          end;
procedure     ibrsp (bd:integer;var spr:integer);
          begin
          ibsta := ibfn(bdname,iberr,ibcnt,ibbuf,intbuf,spr,bd,25);
          end;
procedure     ibrsv (bd:integer;v:integer);
          begin
          ibsta := ibfn(bdname,iberr,ibcnt,ibbuf,intbuf,v,bd,6);
          end;
procedure     ibsad (bd:integer;v:integer);
          begin
          ibsta := ibfn(bdname,iberr,ibcnt,ibbuf,intbuf,v,bd,9);
          end;
procedure     ibsic (bd:integer);
          begin
          ibsta := ibfn(bdname,iberr,ibcnt,ibbuf,intbuf,vcnt,bd,3);
          end;
procedure     ibsre (bd:integer;v:integer);
          begin
          ibsta := ibfn(bdname,iberr,ibcnt,ibbuf,intbuf,v,bd,4);
          end;
procedure     ibstop (bd:integer);
          begin
          ibsta := ibfn(bdname,iberr,ibcnt,ibbuf,intbuf,vcnt,bd,21);
          end;
procedure     ibtmo (bd:integer;v:integer);
          begin
          ibsta := ibfn(bdname,iberr,ibcnt,ibbuf,intbuf,v,bd,13);
          end;
procedure     ibtrg (bd:integer);
          begin
          ibsta := ibfn(bdname,iberr,ibcnt,ibbuf,intbuf,vcnt,bd,23);
          end;
procedure     ibwait (bd:integer;mask:integer);
          begin
          ibsta := ibfn(bdname,iberr,ibcnt,ibbuf,intbuf,mask,bd,0);
          end;
procedure     ibwrt (bd:integer;ibbuf:iobuf;cnt:integer);
          begin
          ibsta := ibfn(bdname,iberr,ibcnt,ibbuf,intbuf,cnt,bd,30);
          end;
procedure     ibwrti (bd:integer;intbuf:iolbuf;cnt:integer);
          begin
          ibsta := ibfn(bdname,iberr,ibcnt,ibbuf,intbuf,cnt,bd,37);
          end;
procedure     ibwrta (bd:integer;ibbuf:iobuf;cnt:integer);
          begin
          ibsta := ibfn(bdname,iberr,ibcnt,ibbuf,intbuf,cnt,bd,31);
          end;
procedure     ibwrtf (bd:integer;flname:ibstring);
          var
          name : ibstring;
          begin
          name := flname + chr(0);
          ibsta := ibfn(name,iberr,ibcnt,ibbuf,intbuf,vcnt,bd,18);
          end;
```

FIGURE K.2. (concluded)

L

Musical Tone Octave Generator

Table L.1 shows the frequencies and the associated coefficients values for the oscillation method. The musical tone octave generator programs, for both the oscillator and the sawtooth methods, are shown in Figures L.1 and L.2, respectively.

TABLE L.1 Frequencies and Coefficient Values for Different Tones within an Octave

Tone Harmonic	Required Frequency	Measured Frequency	A	C
C4-1st	261.63	262.22	7F91	02A0
C4-2nd	523.26	523.51	7E46	029E
C4-3rd	784.89	785.16	7C20	02A7
D4-1	293.66	294.12	7F74	02F2
D4-2	587.32	587.69	7DD3	02EF
D4-3	880.98	881.37	7B20	02F9
E4-1	329.63	330.05	7F50	034E
E4-2	659.26	659.61	7D43	034A
E4-3	988.89	989.02	79DF	0353
F4-1	349.23	350.05	7F3A	0380
F4-2	698.46	698.53	7CEE	0376
F4-3	1047.69	1047.80	7921	0384
G4-1	392.00	392.64	7F07	03EE
G4-2	784.00	784.02	7C23	03E6
G4-3	1176.00	1176.23	775C	03ED
A4-1	440.00	440.20	7EC7	0468
A4-2	880.00	880.29	7B23	045E
A4-3	1320.00	1320.20	7526	0462
B4-1	493.88	494.09	7E76	04F1
B4-2	987.76	988.08	79E2	04E2
B4-3	1481.64	1481.64	7261	04E2
C5-1	523.25	523.28	7E46	053C
C5-2	1046.50	1046.57	7924	052A
C5-3	1569.75	1570.00	70BF	0526

B = C000

```
0001                    * DIGITAL MUSIC OCTAVE TONE OSCILLATOR PROGRAM              *
0002                    * WITH ENVELOPE AND DELAY - REPEAT TONE                      *
0003                    * Y(n)=Ay(n-1)+By(n-2)+Cy(n-1), fs= 20khz                    *
0004                    * Y(0)=Ay1+by2 for N>1, x(n-1)=1 for n=1(=0 otherwise)       *
0005                    * No external input or Data Initialization Required          *
0006                    ************************************************************
0007      0060  MODE    EQU    >60      * TRANSPARENT MODE FOR AIB (>00FA)
0008      0061  RATE    EQU    >61      * RATE1=(10MHZ/FS)-1=449=>01F3
0009            *                       * FUNDAMENTAL OSC VARIABLES
0010      0062  A       EQU    >62      * A=2COS(WT)
0011      0063  B       EQU    >63      * b=-1=>C000
0012      0064  Y0      EQU    >64      *      Y(n)
0013      0065  Y1      EQU    >65      *      Y(n-1)
0014      0066  Y2      EQU    >66      *      Y(n-2)
0015            *                       * 1st HARMONIC VARIABLES
0016      0067  AH      EQU    >67      *
0017      0068  BH      EQU    >68      *
0018      0069  Y0H     EQU    >69      *      Y(n)
0019      006A  Y1H     EQU    >6A      *      Y(n-1)
0020      006B  Y2H     EQU    >6B      *      Y(n-2)
0021            *                       * 2nd HARMONIC VARIABLES
0022      006C  AH2     EQU    >6C      *
0023      006D  BH2     EQU    >6D      *
0024      006E  Y0H2    EQU    >6E      *      Y(n)
0025      006F  Y1H2    EQU    >6F      *      Y(n-1)
0026      0070  Y2H2    EQU    >70      *      Y(n-2)
0027      0079  SEL     EQU    >79      * TONE SELECT ADDR, MEM LOC FOR AR4
0028      007A  TONLEN  EQU    >7A      * TONE LENGTH COUNT STORAGE
0029      007B  CNTLEN  EQU    >7B      * LENGTH COUNTER
0030      007C  CNT     EQU    >7C      * COUNTER MEM LOC FOR AR3
0031      007D  YV      EQU    >7D      * variable Y0 for loop routine
0032            *              PAGE 6   * COMBINED OUTPUT SIGNAL(Y0+Y0H+Y0H2)
0033      0000  Y0T     EQU    >00      * realtime output signal BLK B1 >300
0034      000A  Y10T    EQU    >0A      *              0.5ms delay  PG 6
0035      0032  Y50T    EQU    >32      *              2.5ms
0036      0064  Y100T   EQU    >64      *              5.0ms
0037            *              PAGE 7   *
0038      0017  Y150T   EQU    >17      *              7.5ms BLK B1 PG 7
0039      0049  Y200T   EQU    >49      *              10.0ms
0040      0076  Y250T   EQU    >76      *              12.5ms
0041      007F  YS      EQU    >7F      * Y0+Y() delayed reverb signal
0042 0000               AORG   >0
0043 0000 FF80          B      START
     0001 00AA
0044                    ************************************************************
0045 0020       TABLE   AORG   >20
0046 0020 00FA  MD      DATA   >00FA    * MODE(00FA OR 000A)
0047 0021 01F3  CLK     DATA   >01F3    * RATE=20KHZ
0048 0024       TABLEC  AORG   >24      *********** TONE C4    ***************
0049            *                       ******** fundamental freq **********
0050 0024 7F91          DATA   >7F91    * A        261.63HZ Coffs
0051 0025 C000          DATA   >C000    * B
0052 0026 0000          DATA   >0       * Y(0)
0053 0027 02A0          DATA   >02A0    * Y(n-1) (Scaled for 5v)
0054 0028 0000          DATA   >0       * Y(n-2)
0055            *                       ******* 1st harmonic freq **********
0056 0029 7E46          DATA   >7E46    * A        523.26HZ
0057 002A C000          DATA   >C000    * B
0058 002B 0000          DATA   >0       * Y(0)
0059 002C 029E          DATA   >029E    * Y(n-1) (Scaled for 2.5v)
0060 002D 0000          DATA   >0       * Y(n-2)
0061            *                       ******* 2nd harmonic freq **********
0062 002E 7C20          DATA   >7C20    * A        784.89HZ
0063 002F C000          DATA   >C000    * B
0064 0030 0000          DATA   >0       * Y(0)
0065 0031 02A7          DATA   >02A7    * Y(n-1) (Scaled for 1.7v)
0066 0032 0000          DATA   >0       * Y(n-2)
0067 0035       TABLED  AORG   >35      *********** TONE D4    ***************
0068            *                       ******** fundamental freq **********
0069 0035 7F74          DATA   >7F74    *        293.66HZ
0070 0036 C000          DATA   >C000
0071 0037 0000          DATA   >0
0072 0038 02F2          DATA   >02F2
0073 0039 0000          DATA   >0
0074            *                       ******* 1st harmonic freq **********
0075 003A 7DD3          DATA   >7DD3    *        587.32HZ
0076 003B C000          DATA   >C000
0077 003C 0000          DATA   >0
0078 003D 02EF          DATA   >02EF
0079 003E 0000          DATA   >0
0080            *                       ******* 2nd harmonic freq **********
0081 003F 7B20          DATA   >7B20    *        880.98HZ
0082 0040 C000          DATA   >C000
0083 0041 0000          DATA   >0
0084 0042 02F9          DATA   >02F9
```

FIGURE L.1. Tone generator program for oscillation method (TONEGENR.LST)

```
0085 0043 0000           DATA    >0
0086 0046       TABLEE    AORG    >46          ************ TONE E4   **************
0087            *                              ******** fundamental freq ***********
0088 0046 7F50           DATA    >7F50         *           329.63HZ
0089 0047 C000           DATA    >C000
0090 0048 0000           DATA    >0
0091 0049 034E           DATA    >034E
0092 004A 0000           DATA    >0
0093            *                              ******* 1st harmonic freq ***********
0094 004B 7D43           DATA    >7D43         *           659.26HZ
0095 004C C000           DATA    >C000
0096 004D 0000           DATA    >0
0097 004E 034A           DATA    >034A
0098 004F 0000           DATA    >0
0099            *                              ******* 2nd harmonic freq ***********
0100 0050 79DF           DATA    >79DF         *           988.89HZ
0101 0051 C000           DATA    >C000
0102 0052 0000           DATA    >0
0103 0053 0353           DATA    >0353
0104 0054 0000           DATA    >0
0105 0057       TABLEF    AORG    >57          ************ TONE F4   **************
0106            *                              ******** fundamental freq ***********
0107 0057 7F3A           DATA    >7F3A         *           349.23HZ
0108 0058 C000           DATA    >C000
0109 0059 0000           DATA    >0
0110 005A 0380           DATA    >0380
0111 005B 0000           DATA    >0
0112            *                              ******* 1st harmonic freq ***********
0113 005C 7CEE           DATA    >7CEE         *           698.46HZ
0114 005D C000           DATA    >C000
0115 005E 0000           DATA    >0
0116 005F 0376           DATA    >0376
0117 0060 0000           DATA    >0
0118            *                              ******* 2nd harmonic freq ***********
0119 0061 7921           DATA    >7921         *           1047.69HZ
0120 0062 C000           DATA    >C000
0121 0063 0000           DATA    >0
0122 0064 0384           DATA    >0384
0123 0065 0000           DATA    >0
0124 0068       TABLEG    AORG    >68          ************ TONE G4   **************
0125            *                              ******** fundamental freq ***********
0126 0068 7F07           DATA    >7F07         *           392HZ
0127 0069 C000           DATA    >C000
0128 006A 0000           DATA    >0
0129 006B 03EE           DATA    >03EE
0130 006C 0000           DATA    >0
0131            *                              ******* 1st harmonic freq ***********
0132 006D 7C23           DATA    >7C23         *           784HZ
0133 006E C000           DATA    >C000
0134 006F 0000           DATA    >0
0135 0070 03E6           DATA    >03E6
0136 0071 0000           DATA    >0
0137            *                              ******* 2nd harmonic freq ***********
0138 0072 775C           DATA    >775C         *           1176HZ
0139 0073 C000           DATA    >C000
0140 0074 0000           DATA    >0
0141 0075 03ED           DATA    >03ED
0142 0076 0000           DATA    >0
0143 0079       TABLEA    AORG    >79          ************ TONE A4   **************
0144            *                              ******** fundamental freq ***********
0145 0079 7EC7           DATA    >7EC7         *           440HZ
0146 007A C000           DATA    >C000
0147 007B 0000           DATA    >0
0148 007C 0468           DATA    >0468
0149 007D 0000           DATA    >0
0150            *                              ******** 1st harmonic freq **********
0151 007E 7B23           DATA    >7B23         *           880HZ
0152 007F C000           DATA    >C000
0153 0080 0000           DATA    0
0154 0081 045E           DATA    >045E
0155 0082 0000           DATA    0
0156            *                              ******** 2nd harmonic freq **********
0157 0083 7526           DATA    >7526         *           1320HZ
0158 0084 C000           DATA    >C000
0159 0085 0000           DATA    0
0160 0086 0462           DATA    >0462
0161 0087 0000           DATA    0
0162 008A       TABLEB    AORG    >8A          ************ TONE B4 **************
0163            *                              ******** fundamental freq ***********
0164 008A 7E76           DATA    >7E76         *           493.88HZ
0165 008B C000           DATA    >C000
0166 008C 0000           DATA    >0
0167 008D 04F1           DATA    >04F1
0168 008E 0000           DATA    >0
0169            *                              ******* 1st harmonic freq ***********
0170 008F 79E2           DATA    >79E2         *           987.76HZ
0171 0090 C000           DATA    >C000
0172 0091 0000           DATA    >0
```

FIGURE L.1. (continued)

448

```
0173 0092 04E2          DATA    >04E2
0174 0093 0000          DATA    >0
0175                *                       ****** 2nd harmonic freq ************
0176 0094 7261          DATA    >7261       *        1481.64HZ
0177 0095 C000          DATA    >C000
0178 0096 0000          DATA    >0
0179 0097 04E2          DATA    >04E2
0180 0098 0000          DATA    >0
0181 009B     TABLC5    AORG    >9B         *********** TONE C5 ****************
0182                *                       ******** fundamental freq ***********
0183 009B 7E46          DATA    >7E46       *        523.25HZ
0184 009C C000          DATA    >C000
0185 009D 0000          DATA    >0
0186 009E 053C          DATA    >053C
0187 009F 0000          DATA    >0
0188                *                       ****** 1st harmonic freq ************
0189 00A0 7924          DATA    >7924       *        1046.5HZ
0190 00A1 C000          DATA    >C000
0191 00A2 0000          DATA    >0
0192 00A3 052A          DATA    >052A
0193 00A4 0000          DATA    >0
0194                *                       ****** 2nd harmonic freq ************
0195 00A5 70BF          DATA    >70BF       *        1569.75HZ
0196 00A6 C000          DATA    >C000
0197 00A7 0000          DATA    >0
0198 00A8 0526          DATA    >0526
0199 00A9 0000          DATA    >0
0200                ****************************************************************
0201 00AA 5588 START    LARP    AR0         * select AR0 for indir addr
0202 00AB D000          LRLK    AR0,>60     * AR0 point to DMA60
     00AC 0060
0203 00AD CB01          RPTK    >1          * repeat for MODE + RATE values
0204 00AE FCA0          BLKP    TABLE,*+    * transfer PM 20,21 to DM 60,61
     00AF 0020
0205 00B0 E060          OUT     MODE,0      * output MODE (>FA) to port 0
0206 00B1 E161          OUT     RATE,1      * output RATE (>01f3) to port 1
0207 00B2 E361          OUT     RATE,3      * dummy output to port 3
0208 00B3 C303          LARK    AR3,3       * set AR3 to 3 as osci.count
0209 00B4 D001          LALK    >5000       * load acc w length value
     00B5 5000
0210 00B6 607A          SACL    TONLEN      * store in memory
0211 00B7 D400          LRLK    4,>C4       * load AR4 w/C4 addr to SEL tone
     00B8 00C4
0212                ************* TONE REPEAT GEN ********************************
0213 00BA 207A NWTON    LAC     TONLEN      * load acc w/tone length value
0214 00BA 607B          SACL    CNTLEN      * store in counter
0215 00BB 558C          LARP    4           * select tone addr in AR4
0216 00BC 7479          SAR     AR4,SEL     * store AR4 to SEL
0217 00BD 2079          LAC     SEL         * load acc w/SEL
0218 00BE D003          SBLK    >104        * sub last tone addr + 8,test end
     00BF 0104
0219 00C0 F680          BZ      START       * branch if acc=0
     00C1 00AA
0220 00C2 2079          LAC     SEL         * load acc w/SEL addr
0221 00C3 CE25          BACC                * branch to addr at acc
0222                ************* TONE TABLE LOAD ROUTINES ********************
0223 00C4 5588 C4       LARP    AR0         * select AR0 ****** LOAD C4 COFFS
0224 00C5 D000          LRLK    AR0,>62     * AR0 point to DMA 62
     00C6 0062
0225 00C7 CB0E          RPTK    >E          * repeat for 15 values
0226 00C8 FCA0          BLKP    TABLEC,*+   * transfer PM TO DM 62-70
     00C9 0024
0227 00CA FF80          B       WAIT
     00CB 0102
0228 00CC 5588 D4       LARP    AR0         * LOAD D4 COFFS
0229 00CD D000          LRLK    AR0,>62
     00CE 0062
0230 00CF CB0E          RPTK    >E
0231 00D0 FCA0          BLKP    TABLED,*+
     00D1 0035
0232 00D2 FF80          B       WAIT
     00D3 0102
0233 00D4 5588 E4       LARP    AR0         * LOAD E4 COFFS
0234 00D5 D000          LRLK    AR0,>62
     00D6 0062
0235 00D7 CB0E          RPTK    >E
0236 00D8 FCA0          BLKP    TABLEE,*+
     00D9 0046
0237 00DA FF80          B       WAIT
     00DB 0102
0238 00DC 5588 F4       LARP    AR0         * LOAD F4 COFFS
0239 00DD D000          LRLK    AR0,>62
     00DE 0062
0240 00DF CB0E          RPTK    >E
0241 00E0 FCA0          BLKP    TABLEF,*+
     00E1 0057
0242 00E2 FF80          B       WAIT
     00E3 0102
```

FIGURE L.1. (continued)

449

```
0243 00E4 5588  G4     LARP    AR0         * LOAD G4 COFFS
0244 00E5 D000         LRLK    AR0,>62
     00E6 0062
0245 00E7 CB0E         RPTK    >E
0246 00E8 FCA0         BLKP    TABLEG,*+
     00E9 0068
0247 00EA FF80         B       WAIT
     00EB 0102
0248 00EC 5588  A4     LARP    AR0         * LOAD A4 COFFS
0249 00ED D000         LRLK    AR0,>62
     00EE 0062
0250 00EF CB0E         RPTK    >E
0251 00F0 FCA0         BLKP    TABLEA,*+
     00F1 0079
0252 00F2 FF80         B       WAIT
     00F3 0102
0253 00F4 5588  B4     LARP    AR0         * LOAD B4 COFFS
0254 00F5 D000         LRLK    AR0,>62
     00F6 0062
0255 00F7 CB0E         RPTK    >E
0256 00F8 FCA0         BLKP    TABLEB,*+
     00F9 008A
0257 00FA FF80         B       WAIT
     00FB 0102
0258 00FC 5588  C5     LARP    AR0         * LOAD C5 COFFS
0259 00FD D000         LRLK    AR0,>62
     00FE 0062
0260 00FF CB0E         RPTK    >E
0261 0100 FCA0         BLKP    TABLC5,*+
     0101 009B
0262           ************* MAIN LOOP *************************************
0263 0102 FA80  WAIT   BIOZ    MAIN        * wait till BIO=0 (EOC of A/D low)
     0103 0106
0264           *                           * branch to MAIN after EOC
0265 0104 FF80         B       WAIT        * A/D provides timing
     0105 0102
0266 0106 C066  MAIN   LARK    0,Y2        * AR0 point to DMA>66
0267 0107 C163         LARK    1,B         * AR1 point to DMA>63
0268 0108 2064         LAC     Y0
0269 0109 607D         SACL    YV          * set YV to Y0
0270 010A CA00  LOOP   ZAC                 * zero the acc
0271 010B 5588         LARP    0           * select AR0 for indir addr
0272 010C 3C99         LT      *-,1        * TR=Y2, dec AR0, select AR1
0273 010D 3898         MPY     *-,0        * PR=B*Y2, dec AR1, select AR0
0274 010E 3F89         LTD     *,1         * TR=Y1, ACC=B*Y2, move down Y1
0275 010F 3888         MPY     *,0         * PR=A*Y1, select AR0
0276 0110 CE15         APAC                * ACC=B*Y2+A*Y1
0277 0111 697D         SACH    YV,1        * shift 1 due to extra bit
0278           *                           * store upper 16 bits to >7D
0279 0112 207D         LAC     YV          * since A,B,C was divided by 2
0280 0113 007D         ADD     YV          * double results
0281 0114 607D         SACL    YV          * store acc
0282 0115 558B         LARP    3           * select AR3
0283 0116 7F01         SBRK    1           * subt 1 from AR3
0284           *********** TEST FOR OSC CYCLE COUNT ************************
0285 0117 CA00  TEST   ZAC                 * zero the acc  test for fund freq
0286 0118 737C         SAR     AR3,CNT     * store AR3 to CNT >7C
0287 0119 207C         LAC     CNT         * load acc w/ >7C
0288 011A CD02         SUBK    2           * subt 2 from acc
0289 011B F680         BZ      SET01       * branch if acc =0
     011C 0126
0290 011D CA00         ZAC                 * zero the acc  test for 1st harm
0291 011E 207C         LAC     CNT         * load acc w/ >7C
0292 011F CD01         SUBK    1           * subt 1 from acc
0293 0120 F680         BZ      SET02       * branch if acc=0
     0121 012F
0294 0122 CA00         ZAC                 * zero the acc  test for 2nd harm
0295 0123 207C         LAC     CNT         * load acc w/ >7C
0296 0124 F680         BZ      SET03       * branch if acc=0
     0125 0138
0297 0126 207D  SET01  LAC     YV
0298 0127 6064         SACL    Y0          * store acc(YV) at Y0
0299 0128 5664         DMOV    Y0          * data move osci 1 coff to Y1
0300 0129 C06B         LARK    0,Y2H       * set AR0 to Y2H >6B (for OSC2)
0301 012A C168         LARK    1,BH        * set AR1 to BH >68 (for OSC2)
0302 012B 2069         LAC     Y0H
0303 012C 607D         SACL    YV          * set YV to Y0H (for OSC2)
0304 012D FF80         B       LOOP
     012E 010A
0305 012F 207D  SET02  LAC     YV
0306 0130 6069         SACL    Y0H         * store acc(YV) at Y0H
0307 0131 5669         DMOV    Y0H         * data move osci 2 coff to Y1H
0308 0132 C070         LARK    0,Y2H2      * set AR0 to Y2H2 >70 (for OSC3)
0309 0133 C16D         LARK    1,BH2       * set AR1 to BH2 >6D (for OSC3)
0310 0134 206E         LAC     Y0H2
0311 0135 607D         SACL    YV          * set YV to Y0H2 (for OSC3)
0312 0136 FF80         B       LOOP        * return to determine next osc pt
     0137 010A
```

FIGURE L.1. (continued)

```
0313 0138 207D  SET03   LAC   YV
0314 0139 606E          SACL  YOH2        * store acc(YV) at YOH2
0315 013A 566E          DMOV  YOH2        * data move osci 3 coff to Y1H2
0316 013B C303          LARK  AR3,3       * reset counter for next 3 OSC points
0317                    ************ SUM 3 OSC POINTS FOR TOTAL YOT ******************
0318 013C CA00  SUM     ZAC               * zero acc
0319 013D 2064          LAC   YO          * load fund osc point
0320 013E 0069          ADD   YOH         * add 1st harmonic osc point
0321 013F 006E          ADD   YOH2        * add 2nd harmonic osc point
0322 0140 C806          LDPK  6           * load data mem page 6 for >300
0323 0141 6000          SACL  YOT         * store sum of YO's to total YOT
0324                    ************* ADD TIME DELAY SIGNAL TO REAL SIGNAL ************
0325                    * change data page and Y()T delay for diff delay *************
0326 0142 CA00          ZAC               * zero acc
0327 0143 C806          LDPK  6           * load d-mem pg 7 for Y250(6 for<150)
0328 0144 200A          LAC   Y10T        * change Y()T for diff delay (VARIABL
0329 0145 CE19          SFR               * divide delay signal by 2
0330 0146 CE19          SFR               * divide delay signal by 2 (.25 total
0331 0147 C806          LDPK  6           * load data mem page 6
0332 0148 0000          ADD   YOT         * sum YO with delay for reverb signal
0333 0149 C807          LDPK  7           * load data mem page 7
0334 014A 607F          SACL  YS          * store sum YOT+Y()T in YS
0335 014B D600          LRLK  6,>3FA      * set AR6 to point to DMA >3FA
     014C 03FA
0336 014D 558E          LARP  6           * select AR6 for indir addr
0337 014E CBFA          RPTK  250         * data move entire block B0+B1
0338 014F 5690          DMOV  *-          * shift block down for next N
0339                    * change data page+select YS for summed real+delayed sig ******
0340 0150 C806          LDPK  6           * sel page 7 for YS(pg6 for YOT)
0341 0151 E200          OUT   YOT,2       * change out sig(YOT-REAL)(YS-delay)
0342 0152 C800          LDPK  0           * set data page pointer to 0
0343                    ************* TONE ENVELOPE CONTROL ********************
0344 0153 2065          LAC   Y1          * decrement OSC1 present C value
0345 0154 CD1F          SUBK  >1F         * change for diff decay envelope OSC1
0346 0155 6065          SACL  Y1          * restore at Y1
0347 0156 206A          LAC   Y1H         * decrement OSC2 present C value
0348 0157 CD1F          SUBK  >1F         * change for diff decay envelope OSC2
0349 0158 606A          SACL  Y1H         * restore at Y1H
0350 0159 206F          LAC   Y1H2        * decrement OSC2 present C value
0351 015A CD1F          SUBK  >1F         * change for diff decay envelope OSC3
0352 015B 606F          SACL  Y1H2        * restore at Y1H2
0353 015C 207B          LAC   CNTLEN      * decrement tone length counter
0354 015D CD01          SUBK  1
0355 015E 607B          SACL  CNTLEN      * restore decremented value
0356 015F F180          BGZ   WAIT        * branch if acc>0 wait for next Fs
     0160 0102
0357                    ************* TEMPORARY STALL AT END ********************
0358 0161 207A          LAC   TONLEN      * TO SEPARATE TONE SIGNALS
0359 0162 607B          SACL  CNTLEN
0360 0163 207B  STALL   LAC   CNTLEN
0361 0164 CD01          SUBK  1
0362 0165 607B          SACL  CNTLEN
0363 0166 F180          BGZ   STALL
     0167 0163
0364                    ************* RESET SELECT FOR NEXT TONE ********************
0365 0168 558C          LARP  4           * sel AR4 to change SEL addr
0366 0169 7E08          ADRK  8           * add 8 to AR4 for next tone sel
0367 016A FF80          B     NWTON       * branch if acc(CNTLEN)=0
     016B 00B9
0368                    END
NO ERRORS, NO WARNINGS
```

FIGURE L.1. (concluded)

```
0001          ***************************************************************
0002          * DIGITAL MUSIC SAWTOOTH GENERATOR PROGRAM - SAWTONER        *
0003          *    NO EXTERNAL INPUT OR DATA INITIALIZATION REQUIRED       *
0004          ***************************************************************
0005    0060  MODE    EQU   >60         * MODE FOR AIB >A
0006    0061  RATE    EQU   >61         * RATE = (10 MHZ/fS) -1 = >01F3(20K)
0007    0062  TONE    EQU   >62         * STORAGE FOR WAVE GENERATION
0008    0063  STEP    EQU   >63         * STORAGE OF STEP INCREMENT VALUE
0009    0064  CNT     EQU   >64         * STORAGE OF COUNT VALUE
0010    0065  SEL     EQU   >65         * STORAGE OF NEW TONE ADDR
0011    0066  CNTLEN  EQU   >66         * STORAGE OF ENVELOPE TIME LENGTH
0012          *             PAGE 6
0013    0000  YOT     EQU   >00         * REALTIME OUTPUT SIGNAL BLK B1 >300
0014    000A  Y10T    EQU   >0A         *            0.5MS DELAY PG6
0015    0032  Y50T    EQU   >32         *            2.5MS
0016    0064  Y100T   EQU   >64         *            5.0MS
0017          *             PAGE 7
0018    0017  Y150T   EQU   >17         *            7.5MS DELAY PG7
0019    0049  Y200T   EQU   >49         *            10.0MS
0020    0076  Y250T   EQU   >76         *            12.5MS
0021    007F  YS      EQU   >7F         * TONE+Y() DELAYED REVERB SIGNAL
0022 0000             AORG  0           *
```

FIGURE L.2. Tone generator program for sawtooth method (SAWTONER.LST)

```
0023 0000 FF80              B       START       *
     0001 0033
0024 0020          TABLE    AORG    >20         * PROG MEM START AT PMA 20
0025 0020 00FA              DATA    >00FA       * MODE VALUE FOR AIB
0026 0021 01F3              DATA    >01F3       * SAMPLE RATE = 20 KHZ
0027 0022 4000    INIT      DATA    16384       * INITIAL POINT OF WAVEFORM
0028                        *************** STEP VALUES FOR TONES ***********************
0029 0023 FE56    S1        DATA    -426        * TONE C4
0030 0024 FE1E    S2        DATA    -482        * TONE D4
0031 0025 FDE7    S3        DATA    -537        * TONE E4
0032 0026 FDC1    S4        DATA    -575        * TONE F4
0033 0027 FD7D    S5        DATA    -643        * TONE G4
0034 0028 FD28    S6        DATA    -728        * TONE A4
0035 0029 FCE1    S7        DATA    -799        * TONE B4
0036 002A FCA2    S8        DATA    -862        * TONE C5
0037                        *************** COUNT VALUES FOR TONES **********************
0038 002B 004C    C4        DATA    76          * TONE C4
0039 002C 0043    D4        DATA    67          * TONE D4
0040 002D 003C    E4        DATA    60          * TONE E4
0041 002E 0038    F4        DATA    56          * TONE F4
0042 002F 0032    G4        DATA    50          * TONE G4
0043 0030 002C    A4        DATA    44          * TONE A4
0044 0031 0028    B4        DATA    40          * TONE B4
0045 0032 0026    C5        DATA    38          * TONE C5
0046                        *************** INITIALIZING AIB ****************************
0047 0033 C800    START     LDPK    0           * SELECT DM PAGE 0
0048 0034 5588              LARP    AR0         * SELECT AR0 FOR INDIR ADDR
0049 0035 C060              LARK    AR0,>60     * AR0 POINT TO DMA60
0050 0036 CB02              RPTK    >2          * REPEAT FOR MODE,RATE,INIT
0051 0037 FCA0              BLKP    TABLE,*+    * TRANSFER PM 20-22 TO DM 60-62
     0038 0020
0052 0039 E060              OUT     MODE,0      * OUTPUT MODE = >FA TO PORT 0
0053 003A E161              OUT     RATE,1      * OUTPUT RATE = >1F3 TO PORT 1
0054 003B E361              OUT     RATE,3      * DUMMY OUTPUT TO PORT 3
0055 003C C37D    RSTART    LARK    3,>7D       * LOAD AR3 WITH >7D TO RERUN TONES
0056                        *************** SELECTION OF TONE  **************************
0057 003D D100    NWTON     LRLK    1,>5000     * AR1= COUNT VALUE FOR ENVELOPE
     003E 5000
0058 003F 558B              LARP    3           * SELECT AR3 FOR TONE ADDR
0059 0040 7365              SAR     3,SEL       * STORE TONE ADDR TO SEL >65
0060 0041 2065              LAC     SEL         * LOAD ACC WITH TONE ADDR
0061 0042 CDAD              SUBK    >AD         * SUBTRACT >AD FROM ADDR
0062 0043 F680              BZ      RSTART      * IF ACC=0 BRANCH TO RSTART
     0044 003C
0063                        *************** START OF NEW TONE ***************************
0064 0045 2065              LAC     SEL         * RELOAD ACC WITH TONE ADDR
0065 0046 CE25              BACC                * BRANCH TO TONE ADDR
0066 0047 5588    LOOP      LARP    0           * SELECT AR0
0067 0048 3064              LAR     AR0,CNT     * LOAD AR0 WITH COUNT VALUE
0068 0049 FA80    WAIT      BIOZ    MAIN        * BRANCH IF BIO LOW
     004A 004D
0069 004B FF80              B       WAIT        * IF NOT JUMP BACK AND WAIT
     004C 0049
0070                        ********* ADD TIME DELAY SIGNAL TO REAL SIGNAL ***************
0071                        ***** Change data page and Y()T delay for diff delay **********
0072 004D 2062    MAIN      LAC     TONE        * LOAD ACC WITH TONE
0073 004E C806              LDPK    6           * SELECT DM PG 6 FOR >300
0074 004F 6000              SACL    Y0T         * STORE TONE TO Y0T
0075 0050 CA00              ZAC                 * ZERO ACC
0076 0051 C807              LDPK    7           * SELECT DM PG 7 FOR Y250(6 FOR <150)
0077 0052 200A              LAC     Y10T        * LOAD ACC W/Y250OUT DELAY (VARIBLE)
0078 0053 CE19              SFR                 * DEVIDE DELAY BY 2
0079 0054 CE19              SFR                 * DEVIDE DELAY BY 2 (.25 TOTAL)
0080 0055 C806              LDPK    6           * SELECT DM PG 6
0081 0056 0000              ADD     Y0T         * SUM TONE WITH DELAY FOR REVERB
0082 0057 C807              LDPK    7           * SELECT DM PG 7
0083 0058 607F              SACL    YS          * STORE SUM IN YS
0084 0059 D600              LRLK    6,>3FA      * SET AR6 TO POINT TO DMA >3FA
     005A 03FA
0085 005B 558E              LARP    6           * SELECT AR6 FOR INDIR ADDR
0086 005C CBFA              RPTK    250         * DATA MOVE BLOCK B0+B1
0087 005D 5690              DMOV    *-          * SHIFT DOWN FOR NEXT N
0088                        *** Change data page+select YS for summed real+delayed sig ****
0089 005E C806              LDPK    6           * SEL DM PG 6 FOR Y0T(PG 7 FOR YS)
0090 005F E200              OUT     Y0T,2       * OUTPUT CURRENT VALUE TO PORT 2
0091                        *************** TONE ENVELOPE *******************************
0092 0060 C800              LDPK    0           * SELECT DM PG 0
0093 0061 CA00              ZAC                 * ZERO ACC
0094 0062 5589              LARP    1           * SELECT AR1
0095 0063 7F01              SBRK    1           * SUBTRACT 1 FROM VALUE IN AR1
0096 0064 7166              SAR     1,CNTLEN    * STORE AR1 TO >66
0097 0065 2066              LAC     CNTLEN      * LOAD ACC WITH >66
0098 0066 F680              BZ      STCNT       * IF ACC=0 GOTO STALL COUNT
     0067 0074
0099 0068 5588              LARP    0           * SELECT AR0
0100                        *************** NEW STEP VALUE FOR TONE *********************
0101 0069 CA00              ZAC                 * ZERO ACC
0102 006A 2062              LAC     TONE        * CURRENT WAVE VALUE IN ACC
```

FIGURE L.2 (continued)

```
0103 006B 0063          ADD     STEP      * ADD STEP VALUE IN DMA >63
0104 006C 6062          SACL    TONE      * STORE NEXT VALUE OF WAVE TO >62
0105 006D FB90          BANZ    WAIT      * IF AR0 NOT=0 GO TO WAIT FOR BIO LOW
     006E 0049
0106           *                          * -1 FROM AR0 (COUNT VALUE)
0107 006F D001          LALK    INIT      * RELOAD INITIAL POINT TO ACC
     0070 0022
0108 0071 5862          TBLR    TONE      * LOAD STARTING POINT TO >62
0109 0072 FF80          B       LOOP      * BRANCH TO LOOP TO REPEAT CYCLE
     0073 0047
0110           ************* DELAY BEFORE NEW TONE ***************************
0111 0074 D200  STCNT   LRLK    2,>5000   * LOAD AR2=>5000 FOR STALL
     0075 5000
0112 0076 558A  STALL   LARP    2         * SELECT AR2
0113 0077 FB90          BANZ    STALL     * IF AR2 NOT=0 GOTO STALL, AR2-1
     0078 0076
0114 0079 558B          LARP    3         * SELECT AR3
0115 007A 7E06          ADRK    6         * ADD 6 TO AR3
0116 007B FF80          B       NWTON     * BRANCH BACK FOR NEW TONE
     007C 003D
0117           ***** SUBROUTINES FOR LOADING COUNT AND STEP VALUES ***********
0118 007D FC64  TONEC4  BLKP    C4,CNT    * LOAD COUNT FOR TONE C4 (261.63HZ)
     007E 002B
0119 007F FC63          BLKP    S1,STEP   * LOAD STEP VALUE FOR C4
     0080 0023
0120 0081 FF80          B       LOOP      * BRANCH TO LOOP FOR TONE
     0082 0047
0121 0083 FC64  TONED4  BLKP    D4,CNT    * LOAD COUNT FOR TONE D4 (293.66HZ)
     0084 002C
0122 0085 FC63          BLKP    S2,STEP   * LOAD STEP VALUE FOR D4
     0086 0024
0123 0087 FF80          B       LOOP      * BRANCH TO LOOP FOR TONE
     0088 0047
0124 0089 FC64  TONEE4  BLKP    E4,CNT    * LOAD COUNT FOR TONE E4 (329.63HZ)
     008A 002D
0125 008B FC63          BLKP    S3,STEP   * LOAD STEP VALUE FOR E4
     008C 0025
0126 008D FF80          B       LOOP      * BRANCH TO LOOP FOR TONE
     008E 0047
0127 008F FC64  TONEF4  BLKP    F4,CNT    * LOAD COUNT FOR TONE F4 (349.23HZ)
     0090 002E
0128 0091 FC63          BLKP    S4,STEP   * LOAD STEP VALUE FOR F4
     0092 0026
0129 0093 FF80          B       LOOP      * BRANCH TO LOOP FOR TONE
     0094 0047
0130 0095 FC64  TONEG4  BLKP    G4,CNT    * LOAD COUNT FOR TONE G4 (392.00HZ)
     0096 002F
0131 0097 FC63          BLKP    S5,STEP   * LOAD STEP VALUE FOR G4
     0098 0027
0132 0099 FF80          B       LOOP      * BRANCH TO LOOP FOR TONE
     009A 0047
0133 009B FC64  TONEA4  BLKP    A4,CNT    * LOAD COUNT FOR TONE A4 (440.00HZ)
     009C 0030
0134 009D FC63          BLKP    S6,STEP   * LOAD STEP VALUE FOR A4
     009E 0028
0135 009F FF80          B       LOOP      * BRANCH TO LOOP FOR TONE
     00A0 0047
0136 00A1 FC64  TONEB4  BLKP    B4,CNT    * LOAD COUNT FOR TONE B4 (493.88HZ)
     00A2 0031
0137 00A3 FC63          BLKP    S7,STEP   * LOAD STEP VALUE FOR TONE
     00A4 0029
0138 00A5 FF80          B       LOOP      * BRANCH TO LOOP FOR TONE
     00A6 0047
0139 00A7 FC64  TONEC5  BLKP    C5,CNT    * LOAD COUNT FOR TONE C5 (523.25HZ)
     00A8 0032
0140 00A9 FC63          BLKP    S8,STEP   * LOAD STEP VALUE FOR TONE
     00AA 002A
0141 00AB FF80          B       LOOP      * BRANCH TO LOOP FOR TONE
     00AC 0047
0142           END
NO ERRORS, NO WARNINGS
```

FIGURE L.2. (concluded)

REFERENCES

[1] C. A. Taylor, *The Physics of Musical Sounds*, English Universities Press, London, 1965.

[2] K. C. Pohlmann, *Principles of Digital Audio*, H. W. Sams, Indianapolis, IN,. 1989.

Index